Intersecting Environmental Governance With Technological Advancements

Imran Hossain
Varendra University, Bangladesh

A.K.M. Mahmudul Haque
University of Rajshahi, Bangladesh

S.M. Akram Ullah
University of Rajshahi, Bangladesh

Vice President of Editorial	Melissa Wagner
Managing Editor of Acquisitions	Mikaela Felty
Managing Editor of Book Development	Jocelynn Hessler
Production Manager	Mike Brehm
Cover Design	Phillip Shickler

Published in the United States of America by
IGI Global Scientific Publishing
701 East Chocolate Avenue
Hershey, PA, 17033, USA
Tel: 717-533-8845
Fax: 717-533-8661
E-mail: cust@igi-global.com
Website: https://www.igi-global.com

Copyright © 2025 by IGI Global Scientific Publishing. All rights reserved. No part of this publication may be reproduced, stored or distributed in any form or by any means, electronic or mechanical, including photocopying, without written permission from the publisher.
Product or company names used in this set are for identification purposes only. Inclusion of the names of the products or companies does not indicate a claim of ownership by IGI Global Scientific Publishing of the trademark or registered trademark.

Library of Congress Cataloging-in-Publication Data

CIP Pending
ISBN: 979-8-3693-7001-8
EISBN: 979-8-3693-7003-2

Vice President of Editorial: Melissa Wagner
Managing Editor of Acquisitions: Mikaela Felty
Managing Editor of Book Development: Jocelynn Hessler
Production Manager: Mike Brehm
Cover Design: Phillip Shickler

British Cataloguing in Publication Data
A Cataloguing in Publication record for this book is available from the British Library.

All work contributed to this book is new, previously-unpublished material.
The views expressed in this book are those of the authors, but not necessarily of the publisher.
This book contains information sourced from authentic and highly regarded references, with reasonable efforts made to ensure the reliability of the data and information presented. The authors, editors, and publisher believe the information in this book to be accurate and true as of the date of publication. Every effort has been made to trace and credit the copyright holders of all materials included. However, the authors, editors, and publisher cannot assume responsibility for the validity of all materials or the consequences of their use. Should any copyright material be found unacknowledged, please inform the publisher so that corrections may be made in future reprints.

Table of Contents

Foreword xviii

Preface xx

Chapter 1
From Policy to Practice: Environmental Governance Framework for
Sustainable Future in India 1
    Ramesh Chandra Sethi, KIIT University, India
    Shashwata Sahu, KIIT University, India

Chapter 2
Global Environmental Agreements: Shaping International Policy Frameworks. 31
    Moriom Akter, Daffodil International University, Bangladesh

Chapter 3
Contingency Framework of Structural Factors for Public Participation Spaces
and Consultation for Improved Governance 63
    José G. Vargas-Hernandez, Tecnològico Nacional de Mèxico, ITS
        Fresnillo, Mexico
    Francisco J. González-Avila, Tecnològico Nacional de Mèxico, ITSF,
        Mexico
    Selene Castañeda-Burciaga, Universidad Politécnica de Zacatecas,
        Mexico
    Omar Alejandro Guirette, Universidad Politècnica de Zacatecas,
        Mexico
    Omar C. Vargas-Gonzàlez, Tecnològico Nacional de Mèxico, Cd.
        Guzmàn, Mexico

Chapter 4
A Global Policy Design for Sustainable Human Development Vis-à-vis
Environmental Security 91
    Jipson Joseph, Christ University, India
    Ananya Pandey, Christ University, India

Chapter 5
Environmental Governance in Bangladesh: Challenges and Opportunities of Solid Waste Management in Semi-Urban Areas... 119
Shafiul Islam, Rajshahi University, Bangladesh
A.N. Bushra, Lalon University of Science and Arts, Kushtia, Bangladesh
Abrar Bin Shafi, Sir Salimullah Medical College, Dhaka, Bangladesh

Chapter 6
Encouraging Public Private Collaborations for Low Carbon Green Infrastructure in India: The Impact of Government Policies 149
Divya Bansal, Amity University, Noida, India
Naboshree Bhattacharya, Amity University, Ranchi, India
Indrajit Ghosal, Brainware University, Kolkata, India

Chapter 7
Sustainable Water and Sanitation Management: An Analysis of City Corporations' Challenges.. 181
Imran Hossain, Varendra University, Bangladesh
Asfaq Salehin, Varendra University, Bangladesh
A.K.M. Mahmudul Haque, University of Rajshahi, Bangladesh
Abdur Rahman Mia, University of Rajshahi, Bangladesh

Chapter 8
Innovative Strategic Integration of IoT and Digital Marketing for Sustainable Environmental Governance 211
Madhusudan Narayan, Amity Business School, Amity University Jharkhand, Ranchi, India
Ashok Kumar Srivastava, Amity University Jharkhand, Ranchi, India
Ashutosh Sharma, Amity Institute of Applied Sciences, Amity University Jharkhand, Ranchi, India

Chapter 9
Challenges and Opportunities in Implementing Tech-Enabled Environmental Governance: A Systematic Literature Review 237
Durdana Ovais, BSSS Institute of Advanced Studies, India
Richa Jain, Prestige Institute of Management and Research, Bhopal, India

Chapter 10
Data Analytics, Machine Learning, and IoT for Environmental Governance ... 265
Indranil Mutsuddi, JIS University, India

Chapter 11
The Role of Technology in Environmental Governance 287
    Manas Kumar Jha, Hexagon Geosystem, India
    Dilip Kumar Markandey, Central Pollution Control Board, Delhi, India
    Ruchi Gupta, GIZ India, India
    Pranavi Mishra, School of Biotechnology and Bioinformatics, Navi Mumbai, India

Chapter 12
Integrating Climate Resilience Into Educational Curricula: Frameworks, Challenges, and Opportunities 323
    Md Ikhtiar Uddin Bhuiyan, Jahangirnagar University, Bangladesh
    Sawmeem Sajia, Jahangirnagar University, Bangladesh

Chapter 13
Gamification in Climate Action: Understanding the Role of Game Technologies and Participatory Engagement 359
    Yigang Liu, Nanjing Forestry University, China

Chapter 14
Future Trends and Innovations in Environmental Governance for Sustainable Development 381
    K. Anitha, Meenakshi Academy of Higher Education and Research, India
    Indrajit Ghosal, Brainware University, India
    J. Amala, DDGD Vaishnav College, India
    Imran Hossain, Varendra University, Bangladesh

Chapter 15
Environmental Governance for Promoting and Application of Integrated Water Resources Management 413
    Shahriar Shams, Universiti Teknologi Brunei, Brunei & INTI International University, Malaysia

Chapter 16
Green Socio-Ecological-Technological Innovation Policymaking and
Strategies 435
    *José G. Vargas-Hernandez, Tecnològico Nacional de Mèxico, ITS
        Fresnillo, Mexico*
    *Francisco Javier J. González, Tecnológico Nacional de México, ITSF,
        Mexico*
    *Selene Castañeda-Burciaga, Universidad Politécnica de Zacatecas,
        Mexico*
    Omar Guirette, Universidad Politécnica de Zacatecas, Mexico
    *Omar C. Vargas-González, Tecnológico Nacional de México, Ciudad
        Guzmán, Mexico*

Chapter 17
Local Government Challenges in Implementing the 3R Strategy for
Sustainable Waste Management 465
    Imran Hossain, Varendra University, Bangladesh
    A.K.M. Mahmudul Haque, University of Rajshahi, Bangladesh
    S.M. Akram Ullah, University of Rajshahi, Bangladesh
    *Abdul Kadir, Dhaka University of Engineering and Technology,
        Bangladesh*
    Kabir Hossain, Varendra University, Bangladesh

Chapter 18
Environmental Governance for Promoting Dental Public Health 491
    Sadia Chowdhury, Independent Researcher, Brunei

Compilation of References 513

About the Contributors 607

Index 615

Detailed Table of Contents

Foreword xviii

Preface xx

Chapter 1
From Policy to Practice: Environmental Governance Framework for
Sustainable Future in India 1
    Ramesh Chandra Sethi, KIIT University, India
    Shashwata Sahu, KIIT University, India

India faces significant environmental challenges i.e. air and water pollution, biodiversity loss, and impacts of climate change. A robust environmental governance framework that translates policy goals into tangible actions is needed to address these issues. It explores the current state of environmental governance in India. The study focuses on three objectives:(i) to understand the environmental problems and challenges faced by India, (ii) to evaluate environmental policy and its implementation process of SDGs in India, and (iii) to propose an environmental governance framework with practical strategies to tackle future environmental challenges. It adopts a doctrinal methodology, relying on secondary data sources. It covers various aspects of India's environmental governance structure, focusing on environmental concerns, policy analysis, and implementation gaps. It emphasizes bridging the gaps between policy and practices in environmental governance to achieve sustainable development.

Chapter 2
Global Environmental Agreements: Shaping International Policy Frameworks. 31
Moriom Akter, Daffodil International University, Bangladesh

The substantial contribution that international treaties provide to the development of successful global environmental policies is examined in the chapter "Global Environmental Agreements: Shaping International Policy Frameworks". These accords, which were made possible by international cooperation, provide the framework for coordinated efforts to address pressing environmental problems including pollution, climate change, and biodiversity loss. These accords allow the interchange of information, resources, and technology that are essential for sustainable development by promoting international collaboration. Policies are current and beneficial because they are flexible enough to be altered in response to new scientific discoveries and technological advances. To provide just and complete solutions, the chapter highlights the value of inclusion and promotes the engagement of different stakeholders, particularly underprivileged populations.

Chapter 3
Contingency Framework of Structural Factors for Public Participation Spaces and Consultation for Improved Governance 63
    José G. Vargas-Hernandez, Tecnològico Nacional de Mèxico, ITS Fresnillo, Mexico
    Francisco J. González-Avila, Tecnològico Nacional de Mèxico, ITSF, Mexico
    Selene Castañeda-Burciaga, Universidad Politécnica de Zacatecas, Mexico
    Omar Alejandro Guirette, Universidad Politècnica de Zacatecas, Mexico
    Omar C. Vargas-Gonzàlez, Tecnològico Nacional de Mèxico, Cd. Guzmàn, Mexico

This study has the aim to analyse the structural factors for public participation spaces in a contingency framework for consultation and improved governance. It departs from the assumption that the structural factors framed by a contingency construct can support the public participation spaces for consultation and improved governance. The method employed is the meta-analysis supported by the reflective and descriptive analysis based on the conceptual, theoretical and empirical research literature. The analysis concludes that the creation and development of some structural factors such as the infrastructure, structure, power, resources, methods, and tools framed in a contingency model contribute to enhance the public and political participation spaces and consultation for governance in any organization, community, and society.

Chapter 4
A Global Policy Design for Sustainable Human Development Vis-à-vis
Environmental Security 91
    Jipson Joseph, Christ University, India
    Ananya Pandey, Christ University, India

Development and environmental security are mostly considered contradictory principles. The post-Second World War period profoundly needed development but as its by-product the environmental stability was lost. Environmentalists and Ecological movements called for environmental justice. They emphasized the need for a quality environment not only for the present but also for the coming generations. Sustainable development thus evolved as a suitable developmental model. Towards the end of the 20th century there originated the human security paradigm, which advanced the idea of sustainable human development. This neologism proposed a people-centered developmental model without disregarding ecological concerns. Since the globe is in industry 4.0 and almost started to actively engage in industry 5.0, the sustainable human development model can function as a perfect mechanism to link both development and environmental security. In this perspective, this chapter focuses on a global policy design that promotes sustainable human development without endangering the environment.

Chapter 5
Environmental Governance in Bangladesh: Challenges and Opportunities of
Solid Waste Management in Semi-Urban Areas 119
    Shafiul Islam, Rajshahi University, Bangladesh
    A.N. Bushra, Lalon University of Science and Arts, Kushtia, Bangladesh
    Abrar Bin Shafi, Sir Salimullah Medical College, Dhaka, Bangladesh

This study explores the challenges and opportunities of solid waste management in semi-urban areas in Bangladesh, focusing on understanding the roles and experiences of various stakeholder groups using the Power Interest Matrix as a framework. Field investigations reveal significant disparities in waste management practices, particularly in semi-urban municipalities where institutional incapacity, resource constraints, and governance failures exacerbate waste accumulation and environmental degradation. The findings highlight a lack of comprehensive solid waste management policies for semi-urban areas, inadequate financial and human resources, and limited technological innovation, all of which contribute to the inefficiency of current solid waste management systems. Local elites often resist waste management reforms due to personal interests. In contrast, informal waste workers remain marginalized despite their significant role in waste collection, lacking formal recognition and proper safety measures. The absence of data systems within municipal authorities impedes effective planning.

Chapter 6
Encouraging Public Private Collaborations for Low Carbon Green
Infrastructure in India: The Impact of Government Policies 149
    Divya Bansal, Amity University, Noida, India
    Naboshree Bhattacharya, Amity University, Ranchi, India
    Indrajit Ghosal, Brainware University, Kolkata, India

The purpose of this study is to look into how government policies affect the development of public-private partnerships (PPPs) for low-carbon, green infrastructure in India. The research will provide a thorough examination of current regulations pertaining to waste management, green buildings, renewable energy, and sustainable transportation in order to assess how well they work to encourage private sector cooperation and investment. The research aims to identify opportunities, problems, and best practices for strengthening the role of government in promoting PPPs for sustainable infrastructure development through the analysis of case studies and interviews with important stakeholders. This will furnish policymakers, practitioners, and scholars with significant perspectives to propel India's shift towards a low-carbon economy. This research intends to inform the creation of more effective strategies for promoting sustainable infrastructure expansion and expediting India's transition to a greener future by examining the interplay between government policies and PPPs.

Chapter 7
Sustainable Water and Sanitation Management: An Analysis of City
Corporations' Challenges... 181
    Imran Hossain, Varendra University, Bangladesh
    Asfaq Salehin, Varendra University, Bangladesh
    A.K.M. Mahmudul Haque, University of Rajshahi, Bangladesh
    Abdur Rahman Mia, University of Rajshahi, Bangladesh

This chapter investigates the challenges faced by Rajshahi and Gazipur City Corporations, Bangladesh, in ensuring sustainable water and sanitation practices. A qualitative method was used to collect data from various sources, including in-depth interviews, key informant interviews, focus group discussions, and existing literature. The study has identified several major challenges faced by City Corporations (CCs) in Bangladesh. These challenges include inadequate infrastructure, population growth, unplanned urbanization, insufficient financial resources, and institutional incapacity, issues related to water reuse and conservation, and improper waste management. These challenges have a significant impact on sustainable water and sanitation management. The study identified that the wastage of water is caused by inadequate infrastructure, population growth, unplanned urbanization, institutional incapacity, and poor coordination. Furthermore, the scarcity of safe water is also caused by population growth, unplanned urbanization, and the absence of water reuse and conservation systems.

Chapter 8
Innovative Strategic Integration of IoT and Digital Marketing for Sustainable Environmental Governance 211
    Madhusudan Narayan, Amity Business School, Amity University Jharkhand, Ranchi, India
    Ashok Kumar Srivastava, Amity University Jharkhand, Ranchi, India
    Ashutosh Sharma, Amity Institute of Applied Sciences, Amity University Jharkhand, Ranchi, India

Purpose: This study explores how integrating IoT and digital marketing can enhance sustainable environmental governance, addressing environmental challenges and supporting the achievement of sustainable development goals (SDGs). It aims to show how this combination can tackle environmental challenges and support (SDGs). Methodology: A literature review and case studies evaluate how real-time IoT data can enhance targeted digital marketing strategies, promoting sustainable behaviour. Findings: Integration improves public engagement and campaign effectiveness, using IoT data for timely, impactful messaging. Challenges like data privacy and technical interoperability must be addressed. Practical Implications: Insights for stakeholders help refine sustainability initiatives and advance global environmental goals. Originality/Value: This research offers a framework that combines IoT and digital marketing, empowering innovative, data-driven solutions for sustainability. Keywords: IoT, Digital Marketing, Environmental Governance, Sustainability, Data Analytics, SDG

Chapter 9
Challenges and Opportunities in Implementing Tech-Enabled Environmental Governance: A Systematic Literature Review 237
    Durdana Ovais, BSSS Institute of Advanced Studies, India
    Richa Jain, Prestige Institute of Management and Research, Bhopal, India

As technological advancements rapidly evolve, they present both opportunities and challenges for environmental governance. Despite the significant promise of technology in enhancing environmental governance, gaps remain in understanding its effective implementation. This study addresses these gaps through a systematic literature review to elucidate how technology can be effectively utilized to advance environmental governance, overcome current challenges, and promote transformative change. This study employs a systematic literature review methodology, analyzing a broad range of academic sources including journals, conference proceedings, and reports. The review will follow rigorous search strategies and transparent inclusion/exclusion criteria to ensure a representative sample. Thematic analysis is applied to identify key trends, knowledge gaps, and emerging frameworks, providing a comprehensive understanding of tech-enabled environmental governance.

Chapter 10
Data Analytics, Machine Learning, and IoT for Environmental Governance ... 265
Indranil Mutsuddi, JIS University, India

Effective environmental governance is crucial for addressing complex & interconnected environmental challenges of the 21st century. It requires collaboration of multiple sectors & integration of diverse perspectives to create resilient & sustainable societies. Organizations, governments, and NGOs need to adopt a collaborative approach to implement sustainable solutions for environmental governance. While governments are responsible for policymaking, NGOs excel in implementing these policies due to their strong connections with grassroots communities. Corporations play a critical role by leveraging their capital, technological resources, and research and development capabilities to address existing environmental challenges with sustainable solutions. The book chapter presents theoretical & practical insights regarding the use of Data Analytics, ML and IoT for environmental governance.

Chapter 11
The Role of Technology in Environmental Governance 287
Manas Kumar Jha, Hexagon Geosystem, India
Dilip Kumar Markandey, Central Pollution Control Board, Delhi, India
Ruchi Gupta, GIZ India, India
Pranavi Mishra, School of Biotechnology and Bioinformatics, Navi Mumbai, India

Technology's integration into environmental governance signifies a fundamental change in how societies safeguard and manage their natural resources. This chapter will present the significance of technology in advancing environmental governance frameworks in a variety of ways, with a focus on data collection, policy enforcement, and monitoring. Modern technologies allow more accurate tracking of environmental changes and more efficient management techniques. Examples of these technologies include big data analytics, geographic information systems (GIS), and remote sensing. Furthermore, blockchain technology and digital platforms increase environmental policy accountability and transparency, which encourages increased public participation and compliance. The convergence of these technologies addresses complicated ecological concerns in a world that is changing rapidly, while also improving decision-making processes and enabling more robust and adaptive environmental governance.

Chapter 12
Integrating Climate Resilience Into Educational Curricula: Frameworks,
Challenges, and Opportunities 323
    Md Ikhtiar Uddin Bhuiyan, Jahangirnagar University, Bangladesh
    Sawmeem Sajia, Jahangirnagar University, Bangladesh

Climate change is one of the most vital challenges of the 21st century which pertains, directly or indirectly, to virtually every sphere of human life and natural systems. As education determines the values, behavior, and responses of society and individuals towards ecological issues, it is important to address the existing gap between the traditional curriculum and incorporation of climate change-focused knowledge in the existing framework. There are several defining reasons for this study in the area of incorporation of climate resilience into the visionary curricula. First, a growing necessity to adopt and implement educational frameworks which could assist in the process of coping with or adapting to climate change effects is driven by a specific social need. The key focus of this research is to generate a framework for the integration of climate resilience into educational curricula at multiple levels. The study follows a mixed-methods pathway in order to create an inclusive structure for climate resilience integration in the curricula of education.

Chapter 13
Gamification in Climate Action: Understanding the Role of Game
Technologies and Participatory Engagement... 359
    Yigang Liu, Nanjing Forestry University, China

The imperative to address climate change has led to innovative approaches, one of which is the application of gamification. This paper explores the potential of gamification to mitigate climate change by enhancing eco-education, simulating scientific scenarios, and providing policy feedback. It argues that gamification, through its ability to engage users in a "magic circle" of play, can foster a participatory culture that encourages behavioral changes aligned with climate mitigation strategies. The objective is to shift the focus from merely improving gamification's effectiveness to leveraging the gaming industry's potential to directly contribute to reducing personal carbon footprints and shaping power consumption habits through gaming. It concludes that gamification and the gaming industry can play a significant role in climate change mitigation by transforming players' behaviors and by establishing a virtual carbon credit market that incentivizes sustainable practices.

Chapter 14
Future Trends and Innovations in Environmental Governance for Sustainable
Development 381
    *K. Anitha, Meenakshi Academy of Higher Education and Research,
        India*
    Indrajit Ghosal, Brainware University, India
    J. Amala, DDGD Vaishnav College, India
    Imran Hossain, Varendra University, Bangladesh

In response to escalating environmental challenges, this research explores emerging trends and innovations poised to revolutionize environmental governance for sustainable development. Key areas of focus include digital technology integration, green finance, circular economy models, community-driven governance, and international cooperation. Digital innovations, particularly artificial intelligence (AI) and blockchain, enhance environmental monitoring, ensure transparency, and foster trust in governance processes. Green finance channels investments towards sustainable projects, with tools like green bonds facilitating a transition to low-carbon economies. The circular economy, emphasizing resource efficiency and waste reduction, is vital for sustainable growth, while community-driven governance empowers local involvement in decision-making for more responsive and innovative solutions. The chapter emphasizes a multifaceted approach in creating resilient and adaptive governance frameworks capable of addressing today's environmental crises.

Chapter 15
Environmental Governance for Promoting and Application of Integrated
Water Resources Management 413
    *Shahriar Shams, Universiti Teknologi Brunei, Brunei & INTI
        International University, Malaysia*

Water crisis has been ranked amongst the top 10 global challenges with estimates predicting up to a 40 percent fall in freshwater resources by 2030. On top of that, water demand has far outpaced current population growth and experiencing water scarcity due to rising average global temperatures exacerbated by climate change. Hence, the environmental and sustainable management of water resources is a global imperative, and Integrated Water Resource Management (IWRM) has emerged as a key paradigm to address the multifaceted challenges associated with water management. This chapter discusses the dynamic interplay between IWRM and environmental governance through capacity developments emphasising their pivotal roles in promoting sustainable water resource management. In terms of the governance process, it is only possible to promote and apply IWRM through a well-defined policy for capacity development, ensuring transparency and accountability in water management and addressing inequality, gender biasness supported by available tools under Global Water Partnership (GWP).

Chapter 16
Green Socio-Ecological-Technological Innovation Policymaking and
Strategies 435
    José G. Vargas-Hernandez, Tecnològico Nacional de Mèxico, ITS
        Fresnillo, Mexico
    Francisco Javier J. González, Tecnológico Nacional de México, ITSF,
        Mexico
    Selene Castañeda-Burciaga, Universidad Politécnica de Zacatecas,
        Mexico
    Omar Guirette, Universidad Politécnica de Zacatecas, Mexico
    Omar C. Vargas-González, Tecnológico Nacional de México, Ciudad
        Guzmán, Mexico

This study aims to analyze the relationships between policymaking and strategies for socio-ecological, environmental, and green socio-technological innovation. The analysis departs from the assumption that the environment of a socio-ecological and socio-technological innovation can be fostered by policymaking and strategies aimed to provide the transformation of pure research on scientific knowledge into innovative commercial products and services. The method employed is the analytic-reflective based on the theoretical and empirical literature review. It is concluded that policymaking and strategies accelerates the creation and development processes of socio-ecological, environmental, and green socio-technological innovation.

Chapter 17
Local Government Challenges in Implementing the 3R Strategy for
Sustainable Waste Management 465
    Imran Hossain, Varendra University, Bangladesh
    A.K.M. Mahmudul Haque, University of Rajshahi, Bangladesh
    S.M. Akram Ullah, University of Rajshahi, Bangladesh
    Abdul Kadir, Dhaka University of Engineering and Technology,
        Bangladesh
    Kabir Hossain, Varendra University, Bangladesh

This research delves into the multifaceted landscape of sustainable waste management in the context of Barisal City Corporation, emphasizing the challenges of implementing the 3R strategy. The city's waste management practices and perceptions among its residents are examined through a comprehensive primary data collection process, offering critical insights into the current state of waste management. The study reveals formidable challenges such as insufficient local government initiatives and supervision, irregular waste transportation and segregation, insufficient manpower, unconventional waste storage and disposal practices, unaddressed environmental impact, and the lack of a secondary transfer station. At the same time, the research uncovers promising prospects, including source segregation, energy-efficient waste transportation, and an emerging recycling culture. By addressing these challenges, the city can significantly advance its waste management initiatives, aligning them with the 3R strategy and striving for a cleaner, more sustainable urban environment.

Chapter 18
Environmental Governance for Promoting Dental Public Health 491
Sadia Chowdhury, Independent Researcher, Brunei

Environmental change has a profound effect on the physical environment as well as all elements of natural and human systems. This includes social and economic conditions, as well as the functioning of health systems. Climate change poses an additional risk to the environment, since it leads to more frequent and severe storms, floods, high temperatures, droughts, and wildfires. This chapter examines the importance and role of environmental governance in promoting oral public health. The discussion will centre around the multiple determinants that affect dental public health, encompassing the importance of fluoridation, the consequences of air and water pollution, the accessibility of uncontaminated water, sustainable approaches in dentistry, minimising sugar intake and advocating for nutritious diets, the availability of dental care (including healthcare facilities and workforce training), and the role of education and awareness. It is essential to include sustainability education into the curriculum of both undergraduate and postgraduate students.

Compilation of References 513

About the Contributors 607

Index 615

Foreword

It is with great pleasure that I introduce *Intersecting Environmental Governance with Technological Advancements*. This timely book addresses a critical and dynamic area in environmental studies, recognizing the growing need for innovative, interdisciplinary approaches to tackle the challenges posed by environmental degradation, climate change, and resource management. As traditional governance frameworks struggle to keep pace with complex global challenges, technology emerges as a powerful enabler. This book skillfully explores the convergence of governance and technology, demonstrating how this intersection offers new possibilities for sustainable development.

The contributions within this volume reflect a diverse range of perspectives, with case studies spanning regions such as India, Bangladesh, China, Brunei, and Mexico. These chapters not only identify the challenges inherent in implementing effective environmental governance but also offer practical solutions that integrate cutting-edge technologies like artificial intelligence, the Internet of Things (IoT), blockchain, and data analytics. By focusing on real-world applications, such as public-private partnerships, smart cities, and water management frameworks, this book bridges the gap between theory and practice, providing readers with actionable insights.

The chapters of the book present a forward-looking vision, exploring trends such as green finance, smart cities, and integrated water resource management. These chapters provide readers with insights into the future of environmental governance, emphasizing the need for adaptive frameworks that can respond to emerging challenges.

Intersecting Environmental Governance with Technological Advancements is an essential read for anyone interested in the future of sustainability, whether they are policymakers, researchers, students, or industry professionals. The editors have curated a remarkable collection that not only offers new perspectives but also challenges readers to rethink how governance frameworks and technologies can work together to address the environmental challenges of our time.

This book arrives at a crucial moment when environmental issues demand urgent attention and collaborative solutions. Its contributions inspire action by showing that the intersection of governance and technology is not merely an academic concept but a practical pathway to a sustainable future. I am confident that readers will find within these pages both the knowledge and the inspiration needed to contribute meaningfully to this global endeavor.

Mourade Azrour
Moulay Ismail University, Morocco

Preface

The intersection of environmental governance and technological advancements offers a profound opportunity to address some of the most pressing challenges of our time. Our world is at a critical juncture- confronting issues such as climate change, resource depletion, water scarcity, and environmental degradation, while also experiencing unprecedented technological growth. This book, *Intersecting Environmental Governance with Technological Advancements*, brings together diverse perspectives from researchers, policymakers, and practitioners, illustrating how technology can empower governance systems to meet these challenges more effectively. Through detailed case studies, innovative frameworks, and forward-thinking strategies, the chapters collectively provide a comprehensive view of the synergies between governance and technology, highlighting how these intersections are essential to achieving sustainable development.

The first section of the book focuses on the foundational frameworks of environmental governance. **Chapter 1,** *From Policy to Practice: Environmental Governance Framework for a Sustainable Future in India* by Ramesh Chandra Sethi and Shashwata Sahu, opens the discussion by delving into the complexities of translating environmental policies into actionable strategies. India, with its vast and diverse environmental challenges, offers a compelling case study for examining the gap between policy and practice. This chapter provides practical solutions for closing these gaps, making it a valuable resource for policymakers and practitioners alike. In **Chapter 2,** *Global Environmental Agreements: Shaping International Policy Frameworks*, Moriom Akter explores the critical role of international treaties in fostering collaboration across nations. This chapter emphasizes how global cooperation is essential for addressing transboundary environmental issues, reinforcing the importance of coordinated governance.

Building on these global perspectives, **Chapter 3,** *Contingency Framework of Structural Factors for Public Participation and Governance* by José G. Vargas-Hernandez and his colleagues, offers a thoughtful analysis of public participation in governance. This chapter highlights the importance of creating inclusive frameworks that empower citizens to actively engage in environmental decision-making, ensuring that governance is not only top-down but also participatory. The focus shifts in **Chapter 4,** *A Global Policy Design for Sustainable Human Development Vis-à-vis Environmental Security* by Jipson Joseph and Ananya Pandey, to the intersection of human development and environmental security. This chapter presents innovative policy solutions for integrating environmental considerations into human development frameworks, reflecting the interconnected nature of social and environmental challenges.

The next set of chapters presents regional case studies, offering insights into how environmental governance frameworks are implemented at the local level. **Chapter 5,** *Environmental Governance in Bangladesh: Challenges and Opportunities of Solid Waste Management in Semi-Urban Areas*, by Shafiul Islam, AN Bushra, and Abrar Bin Shafi, examines the complexities of managing waste in semi-urban settings. This chapter highlights the governance challenges posed by rapid urbanization and resource constraints, while also identifying opportunities for innovation and collaboration among stakeholders. **Chapter 6,** *Encouraging Public-Private Collaborations for Low-Carbon Green Infrastructure in India*, by Divya Bansal, Naboshree Bhattacharya, and Indrajit Ghosal, builds on this theme by showcasing how public-private partnerships can drive sustainable infrastructure projects. This chapter emphasizes the role of government policies in facilitating collaboration between sectors, making it a valuable read for those involved in policy development and urban planning.

The book continues with an exploration of the challenges faced by urban centers in managing essential resources. In **Chapter 7,** *Sustainable Water and Sanitation Management: An Analysis of City Corporations' Challenges in Bangladesh*, Imran Hossain and his team provide a detailed analysis of the difficulties faced by city corporations in maintaining water and sanitation systems. This chapter underscores the importance of governance frameworks that are adaptive and responsive to the needs of urban populations, particularly in the face of population growth and climate-induced water scarcity.

The focus then shifts to technology-enabled innovations for sustainability. **Chapter 8,** *Innovative Strategic Integration of IoT and Digital Marketing for Sustainable Environmental Governance* by Madhusudan Narayan, Ashok Kumar Srivastava, and Ashutosh Sharma, introduces readers to the power of real-time data and targeted digital campaigns in promoting sustainable behaviors. This chapter provides a roadmap for leveraging the Internet of Things (IoT) to enhance environmental governance, making it a must-read for technology enthusiasts and environmental advocates alike. **Chapter 9,** *Challenges and Opportunities in Implementing Tech-Enabled Environmental Governance: A Systematic Literature Review*, by Durdana Ovais and Richa Jain, concludes this section with a comprehensive review of the emerging trends in tech-enabled governance. This chapter identifies the barriers to adoption and offers practical recommendations for overcoming these challenges, ensuring that technology serves as an enabler rather than an obstacle.

The journey into the intersection of governance and technology continues with an exploration of advanced technologies that hold the potential to revolutionize environmental management. **Chapter 10,** *Data Analytics, Machine Learning, and IoT for Environmental Governance* by Indranil Mutsuddi, delves into the transformative role of predictive analytics, smart monitoring, and interconnected sensors. This chapter demonstrates how these technologies can facilitate real-time environmental monitoring, improving decision-making processes. Mutsuddi illustrates the power of data-driven governance frameworks, helping cities and institutions respond proactively to environmental threats, making it a crucial chapter for those exploring practical technological applications.

Building on these themes, **Chapter 11,** *The Role of Technology in Environmental Governance: Data Analytics, AI, and Blockchain Solutions* by Manas Kumar Jha, Dilip Kumar Markandey, Ruchi Gupta, and Pranavi Mishra, emphasizes the potential of blockchain in fostering transparency and accountability within governance systems. The chapter provides insightful discussions on how artificial intelligence (AI) can predict environmental risks, while blockchain ensures traceability and public trust in environmental policies. This chapter bridges the technological divide between smart tools and responsible governance, encouraging stakeholders to leverage these technologies to enhance policy implementation and compliance.

The book next turns to education and behavioral change as essential components of sustainable governance. In **Chapter 12,** *Integrating Climate Resilience into Educational Curricula: Frameworks, Challenges, and Opportunities* by Md Ikhtiar Uddin Bhuiyan and Sawmeem Sajia, the authors advocate for incorporating climate resilience into educational systems. This chapter highlights the importance of preparing future generations to tackle environmental challenges with knowledge and practical tools, promoting a shift toward sustainability through early engagement in schools and universities. Education, as this chapter argues, is the foundation upon which long-term environmental change can be built.

Chapter 13, *Gamification in Climate Action: Understanding the Role of Game Technologies and Participatory Engagement* by Yigang Liu, offers an innovative approach to engaging the public in environmental efforts. Liu explores how gamification- the use of game elements in non-game contexts- can motivate behavioral changes and foster participatory governance. This chapter highlights case studies where gamified platforms have been used to influence personal consumption behaviors and support climate change initiatives, making it a fascinating read for those interested in new ways to engage communities in environmental action.

The final section of the book focuses on future trends and forward-looking strategies in environmental governance. **Chapter 14,** *Future Trends and Innovations in Environmental Governance for Sustainable Development* by Anitha K, Indrajit Ghosal, Amala J, and Imran Hossain, provides an insightful analysis of the emerging trends shaping environmental governance, including green finance, circular economy models, and international cooperation. This chapter paints an optimistic picture of the future, identifying opportunities for governments and organizations to adopt more sustainable practices by embracing innovation.

Water management, a critical area in environmental governance, takes center stage in **Chapter 15,** *Environmental Governance for Promoting and Applying Integrated Water Resources Management* by Shahriar Shams. This chapter emphasizes the importance of integrated water resource management (IWRM) in addressing water scarcity and climate change-induced water stress. Shams offers practical recommendations for policymakers and water managers, underscoring the need for cross-sector collaboration and adaptive governance to secure water resources for future generations.

Chapter 16, *Green Socio-Ecological-Technological Innovation: Policymaking and Strategies* by José G. Vargas-Hernandez, Francisco J. González-Avila, Selene Castañeda-Burciaga, Omar Alejandro Guirette, and Omar C. Vargas-González, presents strategies for accelerating socio-technological innovation through effective policymaking. The authors explore how governance structures can foster innovation by creating ecosystems that support research, development, and commercialization of green technologies, encouraging a shift toward more sustainable economies.

In **Chapter 17,** *Local Government Challenges in Implementing the 3R Strategy for Sustainable Waste Management* by Imran Hossain, A.K.M. Mahmudul Haque, S.M. Akram Ullah, Abdul Kadir, and Kabir Hossain, the authors focus on the challenges faced by local governments in implementing the Reduce, Reuse, Recycle (3R) strategy. This chapter examines the specific case of Barisal City Corporation in Bangladesh, highlighting both obstacles and success stories. It provides a practical guide for municipalities looking to adopt the 3R approach, emphasizing the need for supportive governance frameworks and public participation.

The book concludes with **Chapter 18,** *Environmental Governance for Promoting Dental Public Health* by Sadia Chowdhury, which offers an unexpected but compelling case study on the intersection of environmental governance and public health. Chowdhury explores how environmental policies can influence dental health outcomes, making the case for more integrated approaches to public health and environmental management. This chapter exemplifies the broad applicability of environmental governance frameworks across different sectors, encouraging readers to think beyond traditional boundaries.

Through these chapters, *Intersecting Environmental Governance with Technological Advancements* demonstrates the need for holistic, forward-looking governance frameworks that can adapt to new challenges and opportunities. From smart technologies and innovative financing models to participatory governance and cross-sector collaboration, this book provides a comprehensive toolkit for building sustainable futures. It encourages readers to embrace new ways of thinking about environmental management, urging collaboration across disciplines and sectors.

Ultimately, this book serves as a call to action for policymakers, practitioners, researchers, and students to rethink how governance and technology can intersect to address the environmental challenges of our time. It shows that by harnessing the power of innovation and collective effort, we can create governance systems that are not only more efficient and effective but also more inclusive and responsive to the needs of people and the planet.

Imran Hossain

Varendra University, Bangladesh

A.K.M. Mahmudul Haque

University of Rajshahi, Bangladesh

S.M. Akram Ullah

University of Rajshahi, Bangladesh

Chapter 1
From Policy to Practice:
Environmental Governance Framework for Sustainable Future in India

Ramesh Chandra Sethi
https://orcid.org/0009-0005-8302-4896
KIIT University, India

Shashwata Sahu
https://orcid.org/0009-0003-6843-9554
KIIT University, India

ABSTRACT

India faces significant environmental challenges i.e. air and water pollution, biodiversity loss, and impacts of climate change. A robust environmental governance framework that translates policy goals into tangible actions is needed to address these issues. It explores the current state of environmental governance in India. The study focuses on three objectives:(i) to understand the environmental problems and challenges faced by India, (ii) to evaluate environmental policy and its implementation process of SDGs in India, and (iii) to propose an environmental governance framework with practical strategies to tackle future environmental challenges. It adopts a doctrinal methodology, relying on secondary data sources. It covers various aspects of India's environmental governance structure, focusing on environmental concerns, policy analysis, and implementation gaps. It emphasizes bridging the gaps between policy and practices in environmental governance to achieve sustainable development.

DOI: 10.4018/979-8-3693-7001-8.ch001

I. INTRODUCTION

"We won't have a society if we destroy the environment" – Margaret Mead

India, a booming economy of over a billion people, faces significant environmental challenges that could destroy many parts of its natural ecosystems and negatively impact healthcare in India. Many of these include pollution, which has become a global problem affecting the quality of air, water, and soil from rural to urban landscapes (Dandona, 2020). Hence, the World Health Organization (WHO) identified several Indian cities, including India's capital city (Delhi), as having the most polluted air. It has become a challenge for India because of the increased burden on chronic disease management, influencing urgency requirements beyond any regulations that need regulation without allowing collateral health damage. Climate change is recognized as a rapidly growing threat, coupled with ever-changing weather patterns, rising temperatures, and increased occurrences of extreme events such as cyclones and floods that have engulfed the country. These phenomena undermine agricultural productivity and food security and heighten socioeconomic risks, especially for resource-based livelihoods of poor communities (Woodhill et al., 2022).

This is further compounded by the immense threat to biodiversity that India's diverse habitats, from glaciers of the Himalayas to the Western Ghats, which harbor a third of species listed on the International Union for Conservation of Nature (IUCN) list and half of the worldwide threatened species, are facing due to habitat conversions for housing or economic growth. Increasing large-scale infrastructure projects, industrial activities, and agricultural intensification threaten other organisms endemic to the region already on their way to the endangered list (Perrings & Halkos, 2015). Fragmented regulatory frameworks and poor enforcement of environmental laws are two obstacles to these conservation efforts, leading to biodiversity loss and ecosystem instability. Establishing a resilient environmental governance framework is essential, given the layered environmental challenges. Indeed, good governance is not about passing laws and policies but mainly about effective law enforcement (Mostafavi et al., 2021). Good governance is crucial to ensuring that environmental policies lead to action on the ground and help transitional economies realize their potential for sustainable development in a manner compatible with future resource conservation. It requires a whole-of-government response that combines regulators, civil society organizations, and private sector stakeholders to tackle the complex problems holistically (Bansard & Schroder, 2021). In addition, a robust environmental governance system in India is needed within the borders of this country. India has the global significance of its population, which is second only to China in size and economic development level; therefore, Indian environmental issues relate directly or indirectly to biodiversity conservation and sustainability measures of climate change mitigation programs worldwide. By the global direction of aligning itself to international mandates such as "Sustainable Development Goals (SDGs)" emanating from the United Nations, India aims at improving environmental governance on the one hand by its domestic imperatives and also contributing towards a larger canvas role-play vis-a-vis Global Environment (Jasmine, 2022).

II. OBJECTIVES OF THE STUDY

This paper aims to achieve three objectives: First, the paper attempts to understand the complicated environmental problems in contemporary India. Second, the paper critically evaluates the environmental policy landscape in India and its relation to and acceptance of the international norms of environment and sustainability, such as the notion of the UN's SDGs. Third, the paper aims to develop the best-suited environmental governance approach to India's socio-economic and natural factors. By synthesizing overall discussion and the best environmental practices or understanding more and less applicable solutions or propositions for modern India, the paper aims to suggest several credible, appropriate, measurable, and relevant solutions to enhancing policy integration and its implementation means, gaining better and stronger regulatory implementation, enhancing stakeholder participation, and managing an environment with the help of technologies. The research paper aims to suggest recommendations that would work by considering factors associated with an existing gap between policy development and law implementation. Thus, through these objectives, the present paper aims to suggest particular solutions or a perspective for making India more environmentally sustainable.

III. RESEARCH METHODOLOGY

The present research is conducted on issues of environmental governance in India employing the doctrinal method. The study's significant traits use diverse secondary data and systematic analysis to single out patterns, the effectiveness of current statutory and administrative apparatus, and the gaps in their implementation. From this angle, the research unites laws and regulations concerning the environment and ways of implementation with policy papers and their critical analysis. As a result, by uniting legal texts, management policy papers, and academic discussions concerning this issue, the approach proves to be a way of achieving a comprehensive understanding of environmental governance in India. The choice of a doctrinal method can be reasoned by its focus on legal and regulatory aspects of any governmental operations to solve environmental issues. Such an approach allows for systematizing statutory provisions, judicial judgments, and management practices. It enables us to examine mechanisms to protect the natural environment and ensure environmentally sustainable development in India.

IV. ENVIRONMENTAL ISSUES AND CHALLENGES IN INDIA

One crucial issue India faces on the way to sustainable development is undoubtedly environmental challenges. These include air and water pollution, ineffective waste management, and a critical decrease in biodiversity. Each of these concerns has the potential to cause multiple threats to people's health, socioeconomic welfare, and natural balance, many of which are exceptionally hazardous. Here (Figure-1) are some key environmental issues and challenges India faces at the country level.

Figure 1. Key environmental issues and challenges facing India

Air Pollution
Water Pollution
Ineffective Waste Management
Biodiversity Decline
Soil Degradation
Deforestation
Desertification
Climate Change
Ocean Pollution and Marine Ecosystem Degradation
Groundwater Depletion
Plastic Pollution
E-waste Management
Urban Heat Islands
Agricultural Pollution
Mining and Industrial Pollution
Noise Pollution
Lack of Environmental Awareness

(A) **Air Pollution:** Air pollution is one of India's most crucial environmental problems. The WHO reports that about twenty Indian cities, including Delhi, Mumbai, and Kolkata, suffer from having the highest pollution levels in the world. The most common sources of pollutants in these megalopolises are exhaust fumes, discharges from large factories, and biomass that most of the local population uses for cooking or heating their homes (Sarkar, 2024). The Central Pollution Control Board estimates that the levels of particulate matter (PM2.5 and PM10) regularly exceed established safe norms and cause multiple respiratory and cardiovascular diseases (Singh et al., 2021). Given that a study in the Lancet Planetary Health states that air pollution is responsible for 1.67 million deaths annually in India, this problem is a significant threat to public health.

(B) **Water Pollution:** Water pollution is one of India's most critical environmental issues, as its surface and ground sources are heavily contaminated. Industrial wastewater, agricultural chemicals, and untreated city sewage pollute the country's water bodies, posing a tremendous threat to human health and living organisms (Rakhecha, 2020). According to the parameter measurements of the National Water Quality Monitoring Program, 70% of the country's surface water has been severely polluted and is unsafe for irrigation or drinking (Mahbubul et al., 2023). Among the Indian rivers, the Ganges, Yamuna, and Brahmaputra are particularly chemically polluted with heavy metals, pesticides, and pathogens.

(C) **Waste Management:** In India, the problem of ineffective waste management is crucial because of rapid urbanization and population growth. The country produces approximately 62 million tons of municipal solid waste annually, but that part must be collected or adequately removed from the contaminated areas. The Swachh Bharat Mission is developing and has already achieved some success, but it is essential to state that the problem still needs to be more serious. Thus, open dumping and unregulated landfills in India lead to soil and groundwater pollution. In addition to this, the inadequate disposal and dumping of electronic, medical, and plastic waste also have significant adverse effects (Behera et al., 2021). The gap described between the existing and the required level of waste management results in severe environmental and health risks, including diseases and certain hazardous and harmful chemicals released into the environment.

(D) **Biodiversity Decline:** India is famous for its rich biodiversity, stretching through various terrains from the Himalayas to mangrove swamps. But that is under threat: as India's wealth is pulled apart for revenue and sliced into smaller fragments of prosperity by development projects - habitat loss follows. At the same time, the air becomes constantly dirtier from all manner of factory working, including coal-fired boilers that are most abundant in certain parts of India; its waterways are savagely contaminating, and the unchecked growth in human population will generate more wastewater, which translates onto land perceptibly as filthy rivers covered with foam. Such pollution slowly suffocates rivers and their inhabitants. The destruction of natural habitats, deforestation, urban expansion, and encroachment upon agricultural land are some of the significant factors in the decline of biodiversity in India.

V. EVALUATION OF ENVIRONMENTAL POLICY AND IMPLEMENTATION

(A) Current Environmental Policies in India

India's environmental policy domain is marked by complex laws, acts, rules, and notifications created to address environmental issues and promote sustainable development. This framework is mainly guided and supported by India's key legislations: the National Environmental Policy, 2006; the Environment (Protection) Act, 1986; the Air (Prevention and Control of Pollution) Act, 1981; the Water (Prevention and Control of Pollution) Act, 1974; and the Forest (Conservation) Act, 1980. These acts provide a legal base for regulating and controlling pollution issues, preserving forests, and protecting the environment. The Environment (Protection) Act of 1986, often perceived as an umbrella legislation, provides a comprehensive definition of the main measures to coordinate environmental protection. This Act empowers the Central government to give rules to set standards related to emissions and discharges, prescribe methods for data collection, set procedures to control hazardous substances, and, hands down, create authorities that will manage the environment.

Furthermore, the Act specifies that the rules and notifications are to be established to address specific environmental issues and provide two examples of the E-Waste (Management) Rules, 2016, and the Plastic Waste Management Rules, 2016 (Sahdev et al., 2024). The main set of acts was enlarged and developed by the Air and Water Acts, which dealt mainly with the problem of pollution related to industrial objects and vehicles. Thus, the Air Act hands down the establishment of State Pollution Control Boards and Central Pollution Control Boards to ensure that air quality is being systematically observed and permits to these sources are given according to fixed standards or withdrawn along with keeping in check these standards' perseverance (Gulia et al., 2022). The Water Act essentially mirrors these Boards' Authority in the issue of water pollution, providing that the Boards are given the power to control water pollution prevention and control the sources of pollution, where industrial effluents and sewage consistently maintain the standards and elaborate measures (Kapp et al., 2014). Even though this legal base is relatively comprehensive, its observance is highly unenforceable, and many industries that source pollution do not follow control norms because of insufficient control and regulation.

The Forest Act aims to regulate the diversion of forest land to non-forestry uses to conserve forest resources; indeed, the Act has slowed deforestation and contributed to the acceleration of afforestation. However, the Act has also been criticized as interfering with development projects and not taking adequate consideration of the rights of Indigenous people who may depend on forest resources. India has also introduced several policy initiatives to integrate the environment into the planning and development process (Jocelyn et al., 2018). Thus, the National Action Plan on Climate Change (NAPCC) was launched in 2008 and includes eight National Missions: the National Solar Mission and the National Mission of Enhanced Energy Efficiency, the National Mission on Sustainable Habitat, including energy efficiency in buildings, the National Water Mission, the National Mission for a Green India, the National Mission for Sustainable Agriculture and the National Mission for Sustaining the Himalayan Ecosystem (Pandve, 2009). The NAPCC exemplifies India's commitment to balancing environmental goals with economic development. One of the compelling reasons for pursuing sustainable development is the government's commitment to achieving the SDGs defined by the UN. Table 1 shows different key environmental legislations in India.

Table 1. Key environmental legislation in India

Legislation	Objective	Key Provisions
Water (Prevention and Control of Pollution) Act, 1974	Control and abatement of water pollution	SPCBs and CPCB, water quality standards, penalties
Forest (Conservation) Act, 1980	Conservation of forests	Regulation of deforestation, conversion of forest land
Air (Prevention and Control of Pollution) Act, 1981	Control and abatement of air pollution	SPCBs and CPCB, air quality standards, penalties
Environment (Protection) Act, 1986	Protection and improvement of the environment	Central Authority, standards for emissions, penalties
Forest Rights Act, 2006	Rights of forest-dwelling communities	Recognition of community rights, conservation responsibilities
National Environmental Policy, 2006	Commitment to create a clean environment and positive contribution to international efforts	Covers conservation, use of biological resources and knowledge for bio-survey and bi-utilization
National Green Tribunal Act, 2010	Expeditious disposal of environmental cases	Establishment of NGT, jurisdiction over environmental issues

India has announced its national development goals into the framework of the SDGs and has made specific commitments under SDG 13, Climate Action; SDG 15, Life on Land; and SDG 6, Clean Water and Sanitation. Setting biodiversity conservation targets in the National Biodiversity Action Plan and formulating the National Water Policy are examples of how India works towards implementing the SDGs. Despite the various integrated legislative and policy frameworks, India's environmental governance framework has regularly been hampered by fragmented institutional arrangements, overlapping jurisdictions, and ineffective enforcement means. Generally, the number and diversity of public and private agencies, Departments, and organizations involved in enforcing and regulating environmental issues differ nationally, where the Ministry of Environment, Forest and Climate Change (MoEFCC), The Central Pollution Control Board (CPCB), State Pollution Control Boards (SPCBs), and vital Central Ministries are active. At the State level, where Administrative Bodies and other, more specialized Departments are involved in the process. This integrated approach narrows the scope of policy-related tasks but fails to eliminate surplus agencies, resulting in severe coordination-related and administrative problems in India.

(B) Implementation Process Concerning Sustainable Development Goals (SDGs)

Environmental governance is a complex challenge in India, involving policy formulation, administrative actions, and stakeholder participation. Adopted by all UN Member States in 2015, the SDGs provide a shared blueprint for peace and prosperity from people to planet. The commitment of India to these global goals requires some governance mechanisms that have clear rules and are robust and well-coordinated to transform policy directions into actionable outcomes (Dhanapal et al., 2023). India has been proactive in action towards the SDGs, especially with the Indian Think Tank - NITI Aayog, playing a significant role in coordinating initiatives at the national and State levels. Mapping every target against current government schemes and policies on the facility, forming a comprehensive mapping of the SDG targets that align with existing national development strategies (Perinchery, 2024). This strategic mapping will also mean that national initiatives like the Swachh Bharat Mission, NAPCC, and Pradhan Mantri Ujjwala Yojana can feed directly into accomplishing specific SDG targets - specifically those relevant to clean water and sanitation (SDG 6), climate action (SDG 13) and affordable and clean energy (SGD7).

However, the most significant challenges are faced during these implementations. A key one is the resource allocation and funding restrictions that hamper efforts to scale up SDG projects. There are, of course, government drives to alleviate poverty, but the funding is usually not enough, and when it falls short, it is difficult to say how many received help, which affects the scale on which these programs can be carried out. Also, India's governance issues make deploying resources ineffective due to bureaucracy and administrative inefficiency, which invariably delay projects needed for economic growth. Such fragmentation delays implementation and creates inconsistencies that weaken the overall impact of SDGs (Verma, 2020). Another critical dimension of implementing SDGs is ensuring that State and local governments are part of the process. Central and State Authorities in India must collaborate and coordinate actively for the seamless execution of a policy as they function within a federal structure (Saigal et al., 2024). Disagreement of priorities, capacities, and resources at the State level decides progress accordingly. The alleys of development become obvious—some States with excellent governance frameworks and available resources make great strides in implementing SDGs.

In contrast, others must catch up on sharpening regional imbalances to reach the SDG targets (Kundu et al., 2010). Public participation and community engagement are the most critical factors for the successful implementation of SDGs. Nonetheless, the local communities and civil society organizations still need to be more represented in decision-making. This void offers a lack of ownership and accountability at the regional level, which is very much needed to sustain long-term environmental triggers (Mallk, 2023). Active participation of local stakeholders leads to more openness, relevancy in policy, and a collective sense of responsibility toward sustainable development-monitoring and evaluation rating of progress, as well as the identification where needed. India has set up an extensive SDG index to monitor progress across States in India, offering essential insights into the impact of policies and programs (NITI Aayog, 2024). Data can still be unreliable since different methods of gathering and reporting this type of information create inconsistencies. More robust implementation of these mechanisms, better data collection, regular audits, and transparent reporting will significantly improve the assessment process, which can drive policy changes.

(C) Strengths, Weaknesses, Opportunities, and Threats (SWOT Analysis) of the Existing Framework

A thorough SWOT analysis regarding present-day Indian environmental governance arrangements offers a perspective on the key strengths, weaknesses, and recent opportunities and challenges influencing their functioning (Benzaghta et al., 2021). India has an impressive network of academicians, particularly economists focused on natural resource and growth-related issues (Pandey, 2018). However, strategies that ultimately span the gulf between policy and practice create a path toward holistic transformation to identify societal innovations to ensure a sustainable future (Silvestre, 2019). Indian legislative instruments form a solid legal basis for addressing different environmental issues. India's diverse vegetation and wildlife even bear similarities to the country's biodiversity, but despite such rich ecosystems in both countries, decisions recently taken by their respective governments indicate they need to catch up on this matter (Broughton, 2005). In contrast, at least the judiciary has played a proactive role in interpreting and enforcing environmental laws to be more environmentally friendly, as landmark judgments attest to a further extent how its judicial institutions can benefit the environment due to the proper implementation of the rule of law. Specialized institutions such as the National Green Tribunal (NGT) also strengthen their ability to deal with environmental conflicts efficiently and quickly (Gil, 2015).

This is true, but the framework suffers from many weaknesses, making environmental governance in India challenging to master (Aswathy, 2022). Implementation remains a weak link, as bureaucratic sluggishness and red tape prevent the effective enforcement of laws vital to preserving India's environmental heritage (Singh, 2024). Often, a big challenge for the regulatory agencies at the Central and State levels is monitoring compliance with environmental regulations (Bedoya et al., 2020). Additionally, the framework is suffering from fragmentation and no harmonious linkage between many governmental Bodies, which causes duplication of efforts and dissimilar execution of policies. Public participation in environmental decision-making is also minimal, undermining community involvement and potential buy-in of the sustainability projects. Among the existing framework, however, there are enough opportunities and entry points that can be utilized to strengthen environmental governance in India. The international focus on sustainability and climate change mitigation provides an excellent context for India to mold its policies. The SDGs, particularly SDG 13, Climate Action, and SDG 15, Life on Land, were developed as an international blueprint for integrating environmental objectives into national development strategies. Technological advancements, environmental monitoring, and management innovations will enhance regulatory compliance and enforcement opportunities (Yan et al., 2024). Moreover, rises in public awareness and environmental activism can be used to mobilize better policy reform and increase stakeholder accountability.

On the other hand, several threats can limit its effectiveness as a framework. One of the biggest threats is rapid industrialization and urbanization, which has led to tremendous demands on natural resources (industrial processes) and ecosystems (urban expansion) (Xu, 2021). For policymakers, the conflict between economic growth for poverty reduction and environmentally sustainable development poses essential challenges (Ames et al., 2001). Climate change impacts, such as extreme weather events and sea-level rise, will only compound existing environmental issues, while new emerging vulnerabilities can be expected (Bolan et al., 2024). Conversely, political and economic interests can often determine environmental policies created to enforce stricter standards. Still, due to such compromises, laws are laid back and empty of satisfactory regulatory enforcement (Zúniga et al., 2021). Also, corruption and secrecy in governance processes deplete the credibility and influence of these regulations. This SWOT contemplates the condition of environmental governance in India while highlighting what should be a multipartite response to its strengths, weaknesses, opportunities, and threats (Bull et al., 2016). Improving implementation mechanisms, coordinating between the Regulatory Bodies, and promoting public institutionalization are necessary steps to overcome this policy-practice divide. New doors for environmental governance can be explored by adopting the latest technologies and adhering to global sustainability objectives (Glass, 2019). The threats of industrialization, urbanization, and climate change require resilient and adaptive policy frameworks that reconcile economic development with environmental conservation (Satterthwaite et al., 2020). Through a big inclusive lens, India can strengthen its notion of environmental governance to guarantee a sustainable and resilient future for both the people and natural sustainability (Hariram et al., 2023).

VI. RELEVANT CASE STUDIES CHALLENGING ASPECTS OF ENVIRONMENTAL GOVERNANCE IN INDIA

It is wise to understand case studies, examples of best practices, environmental movements, and the strategies the Government of India adopted for successful environmental governance.

(A) **The Chipko Movement:** This movement in the Himalayas of the Uttarakhand region is one of India's environmental governance success stories from the 1970s. It was a tree-hugger grassroots movement led mainly by local women who physically embraced the trees loggers were attempting to cut down. Cachuma created a compelling movement that exposed the environmental and social dangers of destroying the local forests, ultimately resulting in this susceptible region having no business logging. The Chipko movement is an example that highlights the importance of community involvement and people's participation in protection, conservation, and environmental governance. The movement also echoed the importance of integrating gender-inclusive perspectives in protecting the environment and emphasized women's role as stewards for conserving our environment. A famous case is the **"Joint Forest Management (JFM),"** which started in the 1980s. JFM means that local people would help in managing the forest resources. De-notified forests are protected by JFM, which formally assigns vulnerable forest areas to communities and Departments, ensuring they share responsibilities and benefits. JFM has, in turn, been praised for its role in increasing forest cover and securing the livelihoods of localized communities. JFM's success illustrates the value of incorporating local stakeholders into decentralized governance and promoting community-based natural resource management.

(B) Forest Rights Act (FRA) 2006: The FRA was passed to acknowledge and secure the forest rights of Indigenous Forest-dwelling communities over ancestral lands where they have been traditionally residing and safeguarding forest resources for generations. Although meant to be progressive legislation, the FRA has been plagued by implementation problems. Most of the claims came from individuals already eligible under international law, with those rights being hampered by bureaucracy and long delays or outright rejections to have them recognized. Conflicts between conservation priorities and community rights have further compounded this. The rejection of many community claims across the State in Chhattisgarh because wildlife habitats have to be protected has led to discord and mutual suspicion between local communities and conservation officials. This example underscores the dilemma of protecting biodiversity while pursuing social justice and suggests a need for mechanisms to resolve these types of conflicts more effectively.

(C) National Green Tribunal (NGT): The NGT began operation in 2010 and NGT has resorted to pathways such as fining industries that have flouted pollution norms or ordering a ban on sand mining in certain areas to protect the river ecosystems. However, the tribunal has come under criticism for its limited jurisdiction and resources, which hinder it from enforcing rulings effectively. For example, the NGT continues to be opposed by a powerful section of industrial and political circles that is challenging its decisions, thereby questioning its authority over implementing its orders. The air pollution case **in Delhi** offers a compelling illustration of ramified and complex aspects that environmental governance must tackle. The government has taken several policy measures to reduce pollution, such as introducing the Graded Response Action Plan (GRAP) and the Odd-Even Rule in Delhi. The annual increase in pollution levels during winter months, further catalyzed by stubble burning by farmers of Punjab and Haryana, correctly underscores the transboundary nature of environmental problems that need solutions through cooperative governance routes. Curbing Delhi's air pollution crisis requires strong inter-State cooperation, deep policy integration, and extensive public awareness and engagement.

(D) The Clean Ganga Mission: It was started in 2014 and is an excellent initiative to work towards the only national river and deceptively one of the most polluted water bodies on earth. The program, called **Namami Gange Programme,** includes utilizing the available phases of the 'Ganga Action Plan' and integrating ongoing efforts comprehensively to address two significant issues - solid municipal pollution means meeting treatment facilities at least 50% capacity generation reduction from a point source that is generated in cities located along riverfront flow average control sewage minimum within targeted limits action contain biodegradable matter present stream-locally/town other body priority land use change. The program has succeeded in a few areas, such as expanding and improving sewage treatment capacity and riverfront development. Still, several fundamental challenges remain on the road to holistic, sustainable results. The mission highlights that large-scale environmental challenges require sustained political will, adequate funding, and integrated planning and execution.

(E) Supreme Court of and Environmental Jurisprudence Scenario in India: In several path-breaking judgments, the Courts have underlined the import of solid environmental governance systems and highlighted that sustainable development cannot be kept hostage to rampant exploitation. The "**Taj Trapezium Case - MC Mehta v. Union of India (1997)**" is often cited since it was the first time any Court had taken such a far-reaching decision concerning air pollution. This case further accentuated the concept of "polluter pays" and laid down stringent anti-pollutant norms to prevent environmental degradation, demonstrating how the judiciary protects the environment. One of the landmark judgments is "**Vellore Citizens Welfare Forum v. Union of India (1996),**" wherein the Supreme Court propounded principles such as sustainable development and the precautionary principle. The Court held that the industries that were discharging untreated effluents into water bodies had violated Article 21 of the Constitution, i.e., the Right to Life, and it emphasized adopting preventive measures for maintaining environmental quality. The Court struck this balance in the "**Narmada Bachao Andolan v. Union of India (2000)**" decision, which allowed the construction of such a dam while setting stringent conditions to mitigate environmental harm and provide for the resettlement of displaced persons. The judgment, however, highlighted those environmental aspects had often been overlooked in developmental projects, and compliance with environmental laws could not be compromised. The application of the polluter pays principle gained further impetus through the judgment in "**Indian Council for Enviro-Legal Action v. Union of India (1996)**" where, amongst other things, industries that cause pollution have been held to be liable towards villagers affected by it and thus directed to pay compensation. This kind of ruling strengthens the responsibility of industries to adhere to and respect environmental norms and raises questions on stringent measures enforcing environmental laws. Combined, these judgments highlight the judiciary's apparent role as a bridge between policy and practice in environmental governance. The judiciary is serious about protecting Constitutional rights and promoting sustainable development.

VII. DISCUSSION AND ANALYSIS

(A) Complexity of Environmental Governance in India, Considering Legal, Administrative, and Socio-Economic Factors

The complexity of governance over the environment in India is multi-dimensional, as various legal, administrative, and socio-economic factors lie at the center stage. India has a sound legislative framework for the environment through various environmental laws and the rules that come under their preview, like The Environmental Protection Act of 1986, the Forest Conservation Act (FCA) of 1980, the Air Acts and Water Acts, etc. However, this wide range of laws often results in overlapping jurisdictions and regulatory uncertainties. The latter, however, is compounded by the fragmented enforcement across multiple Ministries/Departments at both Central and State levels. What emerges from all three, with their overlapping mandates and even diffuse powers over different areas of economic life necessary for real change, is a multitude of administrative bottlenecks and policy incoherence that seriously hinders effective law implementation or compliance enforcement. In addition, limited institutional capacity and the resource constraints of already overstretched aid-borderline failed humanitarian systems. These agencies are also often understaffed and do not have the technical expertise to enforce their laws properly. Regulatory efficacy deteriorates when institutions are affected by corruption and political interference, giving the government space for selective enforcement or widespread non-compliance.

Additionally, with India´s federal structure and States possessing ample autonomy in fragmented environmental management, there is enormous variability within regions regarding implementation and enforcement. Environmental governance in India is further complicated due to the involvement of socio-economic factors. This environmental governance challenge is linked to (a) the fragmented legal frameworks governing different components of urban planning and (b) issues related to highly overlapping administrative responsibilities between municipalities, State governments, and nationally constituted commissions. Overcoming these challenges involves a comprehensive approach that builds on legal certainty, strengthens the institutions, and pursues development policies held up by environmental sustainability.

(B) Congruence Between India's Environmental Policies and International Frameworks

In this context, the alignment between India's environmental policies and international frameworks, such as the SDGs, becomes crucial in understanding the nature of its environmental governance. The SDGs, which the United Nations adopted in 2015, are a universal call for action to end poverty, protect our environment, and achieve prosperity for all. India's commitment to these goals is manifested in its national policies and programs about environmental sustainability. The alignment (or lack thereof) between those policies and the SDGs highlights unity and gaps. Table 3 shows environmental policy implementation challenges that can be addressed by building capacities at institutional levels, encouraging multi-stakeholder collaborations, and improving policy coherence to achieve sustainable development targets. National Biodiversity Action Plan (NBAP) would align with SDG 15, which focuses on protecting, restoring, and promoting sustainable use of terrestrial ecosystems and combating biodiversity loss. Initiatives like Swachh Bharat Abhiyan (Clean India Mission) and Namami Gange (Clean Gangs Mission) aim to achieve SDGs 6 and 11 to ensure clean water, sanitation, and safe cities, respectively.

Table 2. Analysis of policy implementation challenges

Challenge	Description	Examples
Bureaucratic Inefficiencies	Delays and inefficiencies in administrative processes	Slow recognition of rights under FRA, delayed project approvals
Inadequate Enforcement	Weak implementation of environmental laws	Low compliance with pollution standards, ineffective penalties
Resource Constraints	Limited financial and technical resources	Underfunded regulatory bodies, lack of technical expertise
Fragmented Institutional Frameworks	Overlapping jurisdictions and lack of coordination	Multiple agencies with unclear mandates, inconsistent policy execution
Socio-economic Disparities	Inequitable access to resources and decision-making	Marginalized communities excluded from consultations, disproportionate impacts

VIII. FINDINGS

India has exhaustive laws for environmental protection, but enforcing these legal frameworks still needs to be improved in its porous and fragmented implementation. These are primarily bureaucratic inefficiencies, a need for coordination between Central and State agencies, and a weak enforcement mechanism. All policies recommend mandatory public participation and community involvement, but they have only symbolic value in decision-making. This undermines the value of local knowledge and the need for more engaged communities in environmental conservation. There is a lack of policy and practice. Well-intended policies often cannot be reached because of the lack of resources, limited administrative capacities, or socioeconomic constraints. In addition, more extensive integration of SDGs is required in national and local environmental policy to provide a coherent approach to environmental governance. This includes better-coordinated stakeholder engagement, notably that of private enterprises, CSOs, and NGOs, with a practical framework. This confirms the need for reforms that strengthen implementation mechanisms, enhance public participation in a real sense of transparency and accountability, and are conducive to stakeholder cooperation and inclusive environmental governance.

(A) Critical Issues Hindering Effective Environmental Governance in India

There is a need for adequate enforcement mechanisms to enforce environmental laws. In contrast, fragmented institutional frameworks are available at different levels. Bureaucratic delays and lack of coordination among government agencies at various levels typically lead to delayed implementation and enforcement of environmental regulations. More financial and technical resources in place for monitoring environmental standards have impeded effective regulation. Corruption and unenforced laws are weakening environmental enforcement, putting compliance at stake, and causing general violations of rules. Overlapping jurisdictions and unclear mandates among multiple regulatory bodies also contribute to weaknesses in these systems and create gaps and redundancies that ultimately lead to institutional fragmentation across many aspects of environmental governance.

Further, social and economic inequalities and low public awareness lead to communities with poor skills application in decision-making processes, contributing to unsustainability and the weakness of governance mechanisms in being inclusive and effective. The central-local gap has led to the implementation of ill-fitted policies for the local environment and poorly satisfying needs on the ground. Resolving these pressing problems demands a series of unparalleled remedies to rationalize administrative procedures, better resource planning, stronger legal foundations, and improved inclusive participation with relevant stakeholders to ensure a practical framework for environmental governance in India.

(B) Policy Practice Gaps and Areas of Improvement

Environmental governance analysis examines the gaps between policies and practices that delay successful environmental management in India. Although they have a robust legal backbone, the real cause of failure in enforcing environmental policies lies in bureaucratic hassles, red-tapism, lack of resources, and the need for coordination among government mechanisms. More compliance and enforcement and solid public access to information about implementation may be just one piece of the puzzle regarding these gaps. In addition, socio-economic disparities and poor accessibility to information compromise equitable stakeholder participation in making crucial environmental governance decisions. Weaknesses include enforcement mechanisms, inter-agency coordination, and agencies for implementation that need strengthening of their own funding and capacity building. There is also a need to interlink traditional knowledge with modern scientific approaches to deliver better policy outcomes closer to the grassroots level. Solving these challenges is a prerequisite for India to meet sustainable development goals and environmental sustainability.

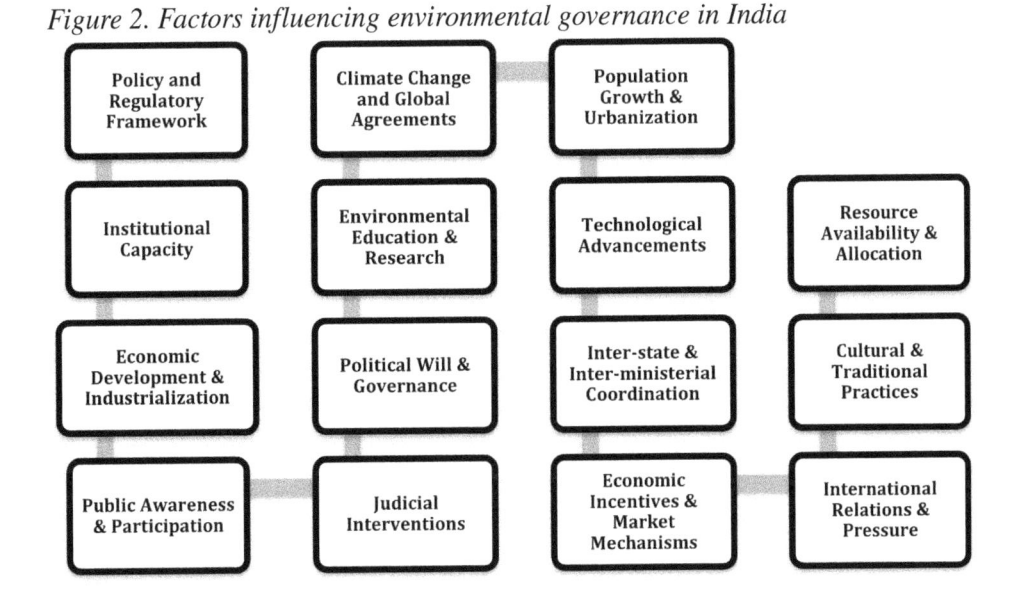

Figure 2. Factors influencing environmental governance in India

IX. ENVIRONMENTAL GOVERNANCE FRAMEWORK AND STRATEGIES

(A) Proposed Enhanced Environmental Governance Framework for India

An enhanced environmental governance framework is critical to mitigate India's diverse environmental challenges and guarantee sustainable development (Wu et al., 2023). This paper underlines a systematic, integrated planning process to improve the policy-to-practice interface and create an environment where individual players can be accountable, encouraging participation from all concerned agencies/sectors. The first one is the framework required to ensure a convergence between environmental laws and policies at various levels of government: Central, State, and Local. The current fractured, uneven regulatory authority system results in inefficient ways of environmental management. Establishing an independent and responsible Central-level Regulator with clear powers and functions will help eliminate procedural delays caused by the multi-layered decision-making process among different Ministries (Somanathan, 2024). This Institutional Body shall have powers to oversee the implementation of policy, monitoring compliance, and enforcement of environmental regulation (Gunningham, 2011).

Secondly, the framework needs to have a robust monitoring and evaluation system in place. The system must be developed using Geographic Information Systems (GIS), remote sensing, and data analytics technologies, ensuring constant real-time observation of environmental indicators. These technologies can quickly and precisely gather data, enabling an evidence-based decision-making process and rapid response against people involved in environmental offenses. Routine audits and impact assessments are required to measure the efficacy of environmental policies and programs, leading to best practices for real-time improvement and adaptability (Pei et al., 2021). Environmental decision-making processes should be transparent and inclusive, allowing local communities, Civil Society Organizations (CSOs), and other stakeholders to participate actively (Sharma, 2023). Establishing mechanisms for public consultations, feedback, and grievances can build a sense of ownership and accountability among citizens towards environmental policies, rendering them more accepted. At the same time, enabling civil society and local governments to have a proactive environmental management approach could translate into more site-specific solutions with prospects of continuity (Zarsky et al., 2000). Implementation reinforces the required governance system, execution, and asset distribution and builds technically and operationally savvy environmental regulators, policymakers, and enforcement authorities through training. Enough financial resources should be set aside to support environmental actions, such as research and development, infrastructural developments, and community-based conservation projects (Clark et al., 2018). Land trusts can also form public-private partnerships to attract new resources and encourage more creative environmental management practices (Thackway et al., 1999).

The governance framework must build on international cooperation and alignment with global environmental frameworks, including the SDGs and the Paris Agreement. India must also actively participate in global environmental governance platforms and disseminate best practices to inform its learning from international experiences (Jong et al., 2021). Transnational collaboration on issues like air quality, climate change, and river pollution brings standard solutions with quick fixes that promote regional environmental security and sustainability (Afifa et al., 2024). Finally, fostering a culture of sustainability via education and awareness programs is essential. The environment is integrated with the national curriculum and awareness programs, which leads to a sense of respect and responsibility from childhood (Ardoin et al., 2020). The relevant media/digital platform should disseminate information on environmental issues and promote sustainable behavior changes in society (Shabalala, 2023).

(B) Strategies Recommendations to Bridge the Gap Between Policy Formulation and Implementation

To reduce the sizable divide between policymaking and policy implementation within Indian environmental governance, a mix of legal, administrative, and participatory measures should be adopted (Ganguly, 2016). Strengthening the enforcement of environmental regulations, mainly through powerful institutional mechanisms, is an excellent place to begin. This includes maintaining the capabilities of regulatory agencies, providing adequate resources, and developing mechanisms for stringent monitoring and compliance. Audits and evaluations of environmental policies are regularly needed to identify gaps in implementation, lessons being learned, or where improvements should be made. High Court, enforcement of environmental laws by the judiciary, and prompt adjudication in environmental matters would also act as preventive measures/incentives to comply with regulations.

Further, it will help restore the broken-up nature of environmental governance in India, and these days, more focus is on promoting inter-agency coordination in Nepal (Ojha et al., 2019). Integrating environmental management requires the development of a framework and platforms to help build coherence among different government Departments, Regulatory Bodies, and local authorities. This will enable coordination between Central and State agencies to synchronize policies, prevent duplication, etc. It also allows for a more coherent application of environmental protection measures and helps ensure national goals align with regional priorities and capabilities. As crucial as the policy-implementation gap is, bridging this requires equally rigorous input regarding public participation and community involvement. Involving local communities, CSOs, and other relevant organizations in environmental decision-making can create a stronger sense of ownership and accountability (Gemmill et al., 2002). This participatory process helps to make policies better reflect reality on the ground and empower communities in their implementation. Public consultations, participatory planning, and community monitoring are crucial in promoting transparency, inclusivity, and the legitimacy of environmental governance. There is also potential for enhancing sustainability and social acceptability of environmental management by producing formal governance systems that incorporate Indigenous communities' traditional knowledge and practices (Boiral et al., 2020).

Environmental policy implementation benefits from technological innovations and digital tools. Automation with regular monitoring and data collection using technology improves the accuracy of information, which is crucial for effective environmental management (Javaid, 2022). For example, real-time data on deforestation and pollution levels, among other environmental impact parameters, can be obtained using satellite imagery, remote sensing, and GIS. Online portals and mobile apps should be developed where environmental violations can be reported and progressed with the public to improve immediate response levels to communication. In addition, implementing environmental policies can be significantly assisted through financial mechanisms and incentives. This involves ensuring that the requisite budgetary resources are allocated to these programs and likewise for efficiency in their implementation (Qadir et al., 2021). Economic incentives such as providing grants for renewable energy projects, the favorable tax treatment of environmentally friendly practices, and penalties in the event of non-compliance can give impetus to change behavior among businesses, calming down towards sustainable usage amongst business entities or individuals.

(C) Role of Public Participation, Community Involvement, and Stakeholder Engagement

Public participation, community engagement, and stakeholder involvement are central to environmental governance architecture, especially for a country like India, which is fraught with complexities and diversities. These components of good governance are essential to make sustainable development meaningful, as they guarantee that all diverse voices and interests get their due weight in environmental deliberation. Public engagement in environmental management is critical to upholding democratic principles, ensuring affected people have a say in managing these natural resources (Agrawal et al., 1999). Public hearings, consultations, or other participatory decision-making can perform this participation. The Environmental

Impact Assessment (EIA) process in India requires public consultations. The center for Clean Air Policy 20 large project can go forward to seek approval and give communities a voice in concerns or suggestions. However, for democracy to work, it has to be more than tokenistic (Dilay et al., 2019).

Table 3. Stakeholder roles in environmental governance

Stakeholder	Role	Key Contributions
Government Agencies	Policy formulation, enforcement, and regulation	Development, enforcement of environmental laws, and inter-agency coordination
Local Communities	Resource management and conservation initiatives	Community-based conservation efforts and traditional knowledge integration
Non-Governmental Organizations (NGOs)	Advocacy, awareness, and capacity building	Raising public awareness, monitoring, and reporting environmental issues
Private Sector	Sustainable practices and innovation	Adoption of green technologies and corporate social responsibility initiatives
International Bodies	Technical, financial support, and policy alignment	Support for sustainable development projects alignment with global best practices

Nonetheless, the success of community involvement depends on policies and an institutional structure that enables it. The Forest Rights Act of 2006 illustrates this, granting legal recognition to forest-dwelling communities over in-situ management rights to minor and non-timber producers (Sarkar, 2011). But implementation burdens persist, too delayed rights recognition, and conflicts between conservation and community aims remain. Addressing this requires legal and institutional reforms, an enhanced capacity-building program, and equitable benefit-sharing mechanisms. Engaging multi-stakeholder platforms can foster dialogue and coordination among diverse actors, creating more complete and durable solutions. NGOs and CSOs play an important role in environmental advocacy, awareness creation, and demanding government accountability.

X. FUTURE RESEARCH DIRECTIONS

Future research might explore real-time data capture and policy application in ways that increase transparency and accountability. This might involve urban planning, pollution tracking, and disaster response. Studying local, State, and Central government policy coherence can also alleviate questions on fragmented environmental governance. For starters, an in-depth examination of modeling locally grounded community-based governance might help. Investigate the role of traditional and Indigenous knowledge systems in promoting sustainability, particularly in forest-related and water resource management. A potential solution may lie in how these practices can complement formal governance systems, balancing respect for cultural heritage with improved environmental protection. Research could also examine the role of public-private partnerships in enhancing environmental sustainability. Future research on environmental governance in India has several promising directions, including addressing some significant gaps and ongoing challenges. There is an urgent need for more international cooperation through knowledge-sharing and participation on global fronts, along with money investment on technical grounds, which can fortify and strengthen environmental governance in India. Finally, the complex role of climate change in biodiversity and ecosystem services in India needs to be studied. Future research may focus on adaptive management practices, potential ecosystem responses to climate change, and vulnerable or at-risk species/ habitats. The following may help inform policy design to minimize biodiversity loss while addressing climate adaptation requirements and lead to a broader, more flexible framework for sustainable environmental governance in India.

XI. CONCLUSION

This research paper reveals the need for robust environmental governance in India. It emphasizes that there is a long way to go if it aims to achieve the SDGs by bridging the policy-practice gap. Among India's most pressing environmental challenges are pollution (severe to catastrophic levels), climate change effects, and biodiversity loss. Despite reasonably elaborate baseline policy and legal framework, many of these measures face severe challenges during implementation. The study points out the fact that though India has an array of environmental laws and regulations in place, these achieve only moderate levels of success because their implementation is hamstrung by bureaucratic inefficiencies, scarce resources, and lackadaisical due diligence from a host of government authorities as well non-government actors. The findings suggest that environmental governance requires public participation and community involvement. Local communities play a crucial role in conserving biodiversity, and the study goes on to underscore that conservation policies ought to be designed so as not only to acknowledge but also to include traditional knowledge and practice alongside modern scientific approaches. Efforts to check the policy-practice gap must be made by strengthening existing environmental laws on the enforcement side. At the same time, the Private-Public-Partnership model is crucial in fostering innovation and efficiency in environmental management, notably regarding waste management or renewable energy development. Collaboration is, therefore, pivotal, especially amongst different nations. To strengthen the institutional structure of environmental governance in India, Indian leaders must work closely with stakeholders globally and be aligned with international best practices. These initiatives will establish natural boundaries that protect and preserve the earth while leaving a legacy for the sustainable development of future generations.

REFERENCES

Aayog, N. I. T. I. (2024), Release of SDG India Index 2023-24, India Accelerates Progress towards the SDGs Despite Global Headwinds, https://pib.gov.in/PressReleasePage.aspx?PRID=2032857

Afifa, , Arshad, K., Hussain, N., Ashraf, M. H., & Saleem, M. Z. (2024). Air pollution and climate change as grand challenges to sustainability. *The Science of the Total Environment*, 928, 172370. DOI: 10.1016/j.scitotenv.2024.172370 PMID: 38604367

Agrawal, A., & Gibson, C. C. (1999). Enchantment and Disenchantment: The Role of Community in Natural Resource Conservation. *World Development*, 27(4), 629–649. DOI: 10.1016/S0305-750X(98)00161-2

Ames, B., (2001), Macroeconomic Policy and Poverty Reduction, prepared by the International Monetary Fund and the World Bank, https://www.imf.org/external/pubs/ft/exrp/macropol/eng/

Ardoin, N. M., Bowers, A. W., & Gaillard, E. (2020). Environmental education outcomes for conservation: A systematic review. *Biological Conservation*, 24, 108224. DOI: 10.1016/j.biocon.2019.108224

Aswathy, Y. (2022), Environmental Governance in India: Problems and Prospects, *Rajasthali Journal*, Vol. 1, Issue. 2, 22-26, https://rajasthali.marudharacollege.ac.in/papers/Volume-1/Issue-2/02-04.pdf

Bansard, Jennifer. & Schröder, Mika. (2021). Deep Dive - The Sustainable Use of Natural Resources: The Governance Challenge, https://www.iisd.org/articles/deep-dive/sustainable-use-natural-resources-governance-challenge

Bedoya, Franco, Mani, Sebastian & Muthukumara. (2020-2011) World Bank Group, The Drivers of Firms' Compliance to Environmental Regulations: The Case of India, https://openknowledge.worldbank.org/entities/publication/3aa0e778-1cfe-5329-a208-59d6a1c2ee12

Behera, M. R., Pradhan, H. S., Behera, D., Jena, D., & Satpathy, S. K. (2021). Achievements and challenges of India's sanitation campaign under clean India mission: A commentary. *Journal of Education and Health Promotion*, 10(1), 350. DOI: 10.4103/jehp.jehp_1658_20 PMID: 34761036

Benzaghta, M. A., Elwalda, A., & Mous, M. M. (2021). SWOT analysis applications: An integrative literature review. *Journal of Global Business Insights*, 6(1), 1–21. https://digitalcommons.usf.edu/cgi/viewcontent.cgi?article=1148&context=globe. DOI: 10.5038/2640-6489.6.1.1148

Boiral, O., Heras-Saizarbitoria, I., & Brotherton, M.-C. (2020). Improving environmental management through Indigenous peoples' involvement. *Environmental Science & Policy*, 103, 10–20. DOI: 10.1016/j.envsci.2019.10.006

Bolan, S., Padhye, L. P., Jasemizad, T., Govarthanan, M., Karmegam, N., Wijesekara, H., Amarasiri, D., Hou, D., Zhou, P., Biswal, B. K., Balasubramanian, R., Wang, H., Siddique, K. H. M., Rinklebe, J., Kirkham, M. B., & Bolan, N. (2024). Impacts of climate change on the fate of contaminants through extreme weather events. *The Science of the Total Environment*, 909, 168388. DOI: 10.1016/j.scitotenv.2023.168388 PMID: 37956854

Broughton, E. (2005). The Bhopal disaster and its aftermath: A review. *Environment & Health*, 4(1), 6. https://www.ncbi.nlm.nih.gov/pmc/articles/PMC1142333/. DOI: 10.1186/1476-069X-4-6 PMID: 15882472

Bull, J., Jobstvogt, N., Böhnke-Henrichs, A., Mascarenhas, A., Sitas, N., Baulcomb, C., Lambini, C. K., Rawlins, M., Baral, H., Zähringer, J., Carter-Silk, E., Balzan, M. V., Kenter, J. O., Häyhä, T., Petz, K., & Koss, R. (2016). Strengths, Weaknesses, Opportunities and Threats: A SWOT analysis of the ecosystem services. *Ecosystem Services*, 17, 99–111. DOI: 10.1016/j.ecoser.2015.11.012

Byravan, S., & Rajan, S. C. (2013), An Evaluation of India's National Action Plan on Climate Change, DOI:DOI: 10.2139/ssrn.2195819

Clark, R., Reed, J., & Sunderland, T. (2018). Bridging funding gaps for climate and sustainable development: Pitfalls, progress and potential of private finance. *Land Use Policy*, 71, 335–346. DOI: 10.1016/j.landusepol.2017.12.013

Dandona, L. (2020). Health and economic impact of air pollution in the States of India: The Global Burden of Disease Study 2019. *The Lancet. Planetary Health*, 5(1), e25–e-38. DOI: 10.1016/S2542-5196(20)30298-9 PMID: 33357500

de Jong, E., & Vijge, M. J. (2021). From Millennium to Sustainable Development Goals: Evolving discourses and their reflection in policy coherence for development. *Earth System Governance*, 7, 100087. DOI: 10.1016/j.esg.2020.100087

Dhanapal, G. (2023). Barriers and opportunities in achieving climate and sustainable development goals in India: A multilevel analysis. *Journal of Integrative Environmental Sciences*, 20(1), 1–16. Advance online publication. DOI: 10.1080/1943815X.2022.2163665

Ganguly, Sunayana. (2016), Deliberating Environmental Policy in India: Participation and the Role of Advocacy, DOI: 10.4324/9781315744476

Gemmill-Herren, B., & Bamidele-Izu, A. (2002), The role of NGOs and Civil Society in Global Environmental Governance, https://www.researchgate.net/publication/228786506_The_role_of_NGOs_and_Civil_Society_in_Global_Environmental_Governance

Gil, G. N. (2015). Environmental Justice in India: The National Green Tribunal and Expert Members, Translational. *Environmental Law (Northwestern School of Law)*, 5(1). https://www.cambridge.org/core/journals/transnational-environmental-law/article/environmental-justice-in-india-the-national-green-tribunal-and-expert-members/2E26B50742FFB8BB743557132DC7DD66

Glass, L.-M., & Newig, J. (2019). Governance for achieving the Sustainable Development Goals: How important are participation, policy coherence, reflexivity, adaptation, and democratic institutions? *Earth System Governance*, 2, 100031. DOI: 10.1016/j.esg.2019.100031

Gulia, S., Shukla, N., Padhi, L., Bosu, P., Goyal, S. K., & Kumar, R. (2022). Evolution of air pollution management policies and related research in India. *Environmental Challenges*, 6, 100431. Advance online publication. DOI: 10.1016/j.envc.2021.100431

Gunningham, N. (2011, July). Enforcing Environmental Regulation. *Journal of Environmental Law*, 23(2), 169–201. DOI: 10.1093/jel/eqr006

Hariram, N. P., Mekha, K. B., Suganthan, V., & Sudhakar, K. (2023). Sustainalism: An Integrated Socio-Economic-Environmental Model to Address Sustainable Development and Sustainability. *Sustainability (Basel)*, 15(13), 10681. DOI: 10.3390/su151310682

Jasmine, B. (2022), Environmental Governance in India: Issues, Concerns and Opportunities, https://www.teriin.org/article/environmental-governance-india-issues-concerns-and-opportunities

Javaid, M., Haleem, A., Singh, R. P., Suman, R., & Gonzalez, E. S. (2022). Understanding the adoption of Industry 4.0 technologies in improving environmental sustainability. *Sustainable Operations and Computers*, 3, 203–217. DOI: 10.1016/j.susoc.2022.01.008

Kapp, R. W., Jr. (2014) Safe Drinking Water Act, Encyclopedia of Toxicology, https://www.sciencedirect.com/topics/agricultural-and-biological-sciences/clean-water-act

Kundu, Amitabh & Varghese, K. (2010), Regional Inequality and 'Inclusive Growth' in India under Globalization: Identification of Lagging States for Strategic Intervention, Oxfam India, Oxfam India working papers series, OIWPS – VI.

Lee, J. I., & Wolf, S. A. (2018). Critical assessment of implementation of the Forest Rights Act of India. *Land Use Policy*, 79, 834–844. https://www.sciencedirect.com/science/article/abs/pii/S0264837717311705. DOI: 10.1016/j.landusepol.2018.08.024

Mallk, K. (2023). Global Development by Public Participation: An Approach to Achieve SDGs. *Journal of Public Administration and Governance*, 13(1). Advance online publication. https://www.researchgate.net/publication/369110515_Global_Development_by_Public_Participation_An_Approach_to_Achieve_SDGs. DOI: 10.5296/jpag.v13i1.20590

Mostafavi, Nariman. et al., (2021). Resilience of environmental policy amidst the rise of conservative populism, *J Environ Stud Sci*. 2022; 12(2): 311–326, DOI: 10.1007/s13412-021-00721-1

Ojha, H. R., (2019), Governance: Key for Environmental Sustainability in the Hindu Kush Himalaya, *The Hindu Kush Himalaya Assessment,* https://link.springer.com/chapter/10.1007/978-3-319-92288-1_16

Pandey, K. (2018), Urbanisation: India loses natural resources to economic growth: report, *Down To Earth*, Retrieved from: https://www.downtoearth.org.in/urbanisation/india-loses-natural-resources-to-economic-growth-report-61836

Pandve, H. T. (2009). India's National Action Plan on Climate Change. *Indian Journal of Occupational and Environmental Medicine*, 13(1), 17–19. https://www.ncbi.nlm.nih.gov/pmc/articles/PMC2822162/. DOI: 10.4103/0019-5278.50718 PMID: 20165607

Pei, T., Xu, J., Liu, Y., Huang, X., Zhang, L., Dong, W., Qin, C., Song, C., Gong, J., & Zhou, C. (2021). GI Science and remote sensing in natural resource and environmental research: Status quo and future. *Geography and Sustainability*, 2(3), 207–215. DOI: 10.1016/j.geosus.2021.08.004

Perinchery, A. (2024), Climate Action, and Also Contradictions: NITI Aayog's SDG India Index 2024 Report is a Mixed Bag, *WIRE*, Retrieved from: https://thewire.in/environment/climate-action-and-also-contradictions-niti-aayogs-sdg-india-index-2024-report-is-a-mixed-bag

Perrings, Charles. & Halkos, George. (2015). *Environmental Research. Letters*. Vol. 10, No.9,1-10, https://iopscience.iop.org/article/10.1088/1748-9326/10/9/095015

Qadir, S. A., Al-Motairi, H., Tahir, F., & Al-Fagih, L. (2021). Incentives and strategies for financing the renewable energy transition: A review. *Energy Reports*, 7, 3590–3606. DOI: 10.1016/j.egyr.2021.06.041

Rakhecha, P. R. (2020). Water environment pollution with its impact on human diseases in India. *Int J Hydro*, 4(4), 152–158. DOI: 10.15406/ijh.2020.04.00240

Sahdev, I., Kumar, S., & Sahu, T. (2024), E-waste management in India, Sustainability, *Agri, Food and Environmental Research*, Vol. 12., http://dx.doi.org/

Saigal, A., & Bawa, D. (2024), Role of Local Leadership in attaining Sustainable Development Goals, *Terra Green (TERI)*, Vol. 16, https://shaktifoundation.in/role-of-local-leadership-in-attaining-sustainable-development-goals/

Sarkar, Debnarayan. (2011), The implementation of the forest rights act in India: Critical issues Economic Affairs 31(2):25-29, 1111/j.1468-0270.2011.02097.x

Sarkar, N. (2024), Delhi gets the attention - but Kolkata's air pollution is just as dangerous, https://news.mongabay.com/2024/04/delhi-gets-the-attention-but-kolkatas-air-pollution-is-just-as-dangerous/

Satterthwaite, D., Archer, D., Colenbrander, S., Dodman, D., Hardoy, J., Mitlin, D., & Patel, S. (2020). Building Resilience to Climate Change in Informal Settlements. *One Earth*, 2(2), 143–156. DOI: 10.1016/j.oneear.2020.02.002

Shabalala, Nonkanyiso Pamella. (2023), Environmental Education as a Catalyst to Teach Students About Their Economy and Politics, *Journal Pendidikan Indonesia Gemilang* 3(2): 306-322, . v3i2.229DOI: 10.53889/jpig

Sharma, R. (2023). Civil society organizations' institutional climate capacity for community-based conservation projects: Characteristics, factors, and issues. *Current Research in Environmental Sustainability*, 5, 100218. DOI: 10.1016/j.crsust.2023.100218

Silvestre, B. S., & Țîrcă, D. M. (2019). Innovations for sustainable development: Moving toward a sustainable future. *Journal of Cleaner Production*, 208, 325–332. https://www.sciencedirect.com/science/article/abs/pii/S0959652618329834. DOI: 10.1016/j.jclepro.2018.09.244

Singh, R. (2024), Environmental Policies in India, Environmental Problems, Protection and Policies, 299-317, https://www.researchgate.net/publication/378590048_ENVIRONMENTAL_POLICIES_IN_INDIA

Singh, V., Singh, S., & Biswal, A. (2021). Exceedances and trends of particulate matter (PM2.5) in five Indian megacities. *The Science of the Total Environment*, 750, 141461. DOI: 10.1016/j.scitotenv.2020.141461 PMID: 32882489

Somanathan, T. V. The Administrative and Regulatory State, Retrieved from: https://icpp.ashoka.edu.in/wp-content/uploads/2024/04/Somanathan-The-Administrative-and-Regulatory-State-1.pdf

Syeed, M. M. M., Hossain, M. S., Karim, M. R., Uddin, M. F., Hasan, M., & Khan, R. H. (2023). Surface water quality profiling using the water quality index, pollution index and statistical methods. *Critical Review*, 18(June), 100247. DOI: 10.1016/j.indic.2023.100247

Thackway, R., & Olsson, K. (1999). Public/private partnerships and protected areas: Selected Australian case studies. *Landscape and Urban Planning*, 44(2–3), 87–97. DOI: 10.1016/S0169-2046(99)00003-1

Verma, A., & Saurabh, M. (2020), An Analysis of Poverty Alleviation Programmes in India with Special Reference to Sustainable Development Goals, https://papers.ssrn.com/sol3/papers.cfm?abstract_id=3637927

Woodhill, J., Kishore, A., Njuki, J., Jones, K., & Hasnain, S. (2022). Food systems and rural wellbeing: Challenges and opportunities. *Food Security*, 14(5), 1099–1121. DOI: 10.1007/s12571-021-01217-0 PMID: 35154517

Wu, Y., & Tham, J. (2023). The impact of environmental regulation, Environment, Social and Government Performance, and technological innovation on enterprise resilience under a green recovery. *Heliyon*, 9(10), e20278. DOI: 10.1016/j.heliyon.2023.e20278 PMID: 37767495

Xu, J. Y. (2021). Impact of urbanization on ecological efficiency in China: An empirical analysis based on provincial panel data. *Ecological Indicators*, 129, 107827. DOI: 10.1016/j.ecolind.2021.107827

Yan, Z., Yu, Y., Du, K., & Zhang, N. (2024). How does environmental regulation promote green technology innovation? Evidence from China's total emission control policy. *Ecological Economics*, 219, 108137. DOI: 10.1016/j.ecolecon.2024.108137

Zarsky, Lyuba, & Tay, Simon SC. (2000), Civil Society and the Future of Environmental Governance in Asia, https://nautilus.org/eassnet/civil-society-and-the-future-of-environmental-governance-in-asia/

Zúniga-González, C. A. (2021). 2021, The impact of economic and political reforms on environmental performance in developing countries. *PLoS One*, 16(10), e0257631. DOI: 10.1371/journal.pone.0257631 PMID: 34610016

Chapter 2
Global Environmental Agreements:
Shaping International Policy Frameworks

Moriom Akter
Daffodil International University, Bangladesh

ABSTRACT

The substantial contribution that international treaties provide to the development of successful global environmental policies is examined in the chapter "Global Environmental Agreements: Shaping International Policy Frameworks". These accords, which were made possible by international cooperation, provide the framework for coordinated efforts to address pressing environmental problems including pollution, climate change, and biodiversity loss. These accords allow the interchange of information, resources, and technology that are essential for sustainable development by promoting international collaboration. Policies are current and beneficial because they are flexible enough to be altered in response to new scientific discoveries and technological advances. To provide just and complete solutions, the chapter highlights the value of inclusion and promotes the engagement of different stakeholders, particularly underprivileged populations.

DOI: 10.4018/979-8-3693-7001-8.ch002

1. INTRODUCTION

a) Importance of Environmental Governance

Ensuring sustainable management of natural resources and ecosystems depends on environmental governance. It includes all of the laws, rules, and organizations that control how people interact with the environment, trying to strike a balance between ecological preservation and economic progress. Good environmental governance promotes a better Earth for coming generations by reducing pollution, biodiversity loss, and the consequences of climate change. Decision-making processes are made more open, accountable, and transparent when a variety of stakeholders are involved, such as governments, non-governmental organizations, the commercial sector, and the general public. Together, we can reinforce and modify governance structures while also enhancing the application and enforcement of environmental regulations. Effective environmental governance ultimately refers to the systems and organizations that facilitate the decision-making and execution of environmental management and policy, as well as the achievement of global sustainability goals and the preservation of a constructive relationship between human societies and the environment. By striking a balance between the need to preserve natural resources and ecosystems and the advancement of the economy and society, effective governance guarantees environmental preservation. It includes the intricate interactions between international treaties, domestic laws, and non-state players' proactive involvement, including the public, business community, and non-governmental organizations.

b) Objectives of the Chapter

1. **To follow the progression and historical development of environmental laws and regulations.**

Examining the path from early, sometimes disjointed environmental efforts to the complete frameworks in existence today is necessary to trace the historical evolution of environmental laws and regulations. Environmental rules were first reactive, tackling particular problems like water and air pollution after noticeable harm had been done. The first significant regulatory attempts were prompted by the tremendous environmental deterioration caused by rapid industrialization, which made the Industrial Revolution a watershed moment.

Modern environmental governance was greatly influenced by major worldwide turning points like the Stockholm Conference in 1972 and the Rio Earth Summit in 1992. These incidents brought attention to the necessity of international collaboration and led to the creation of important accords such as Agenda 21 and the Stockholm

Declaration. National policies changed as well; nations created laws to safeguard their resources—land, water, and air. It is essential to comprehend this historical background to fully appreciate the importance and complexity of the environmental governance institutions that exist today.

2. **To assess the functioning of modern policy frameworks and the way they affect sustainable development.**

Assessing the effectiveness of current international agreements, national laws, and regional initiatives in supporting sustainable development is a necessary step in evaluating modern policy frameworks. Notable agreements such as the Paris Agreement, which aims to limit global warming to far below 2°C over pre-industrial levels, and the Convention on Biological Diversity, which deals with the preservation of biological diversity, are two examples of this. This objective aims to determine the strengths, limitations, and gaps in the effectiveness of policies by examining the implementation and results of these frameworks. Reductions in greenhouse gas emissions, advancements in biodiversity conservation, and the encouragement of sustainable economic practices are important evaluation metrics. This goal also takes political will, financing, and execution of policies into account, providing a thorough understanding of how modern policies are influencing environmental results and promoting sustainable development.

3. **To examine the way various stakeholders influence and carry out environmental policy.**

A wide range of players, each with a distinct function in the formulation and execution of policies, are involved in the multifaceted process of environmental governance. Regulations are made and implemented by governments and multilateral organizations like the United Nations Environment Programme (UNEP). Civil society organizations and non-governmental organizations (NGOs) push for more robust environmental regulations, increase public awareness, and hold authorities responsible. The private sector drives innovation and investment in sustainable technologies, while public-private partnerships leverage resources and expertise for environmental projects. Public participation and grassroots movements are also critical, as they ensure that policies are inclusive and reflective of societal needs and values. Through an analysis of these stakeholders' roles, this purpose strives to bring attention to the intricate and collaborative nature of environmental governance, highlighting the ways in which cross-sector collaboration enhances the efficacy of policies.

4. **To provide thorough case studies that highlight effective tactics and point out typical problems.**

   Case studies offer specific instances of environmental policy implementation in various settings, illustrating both achievements and obstacles. Germany's Energiewende (energy transition) strategy, for example, demonstrates how public support and government incentives may propel a major move toward renewable energy. On the other hand, conservation initiatives in Costa Rica show how high levels of biodiversity conservation can result from an emphasis on ecotourism and protected areas. It is possible to uncover best practices, creative tactics, and contextual elements that support effective environmental governance by analyzing these and other case studies. Examining less successful initiatives, such as the continuous deforestation in the Amazon, also sheds light on common problems like economic pressures, corruption, and the implementation of policies. The purpose of this paper is to present a fair and impartial viewpoint on environmental governance by highlighting replicable tactics and issues that need to be resolved in order to build stronger frameworks.

5. **Investigate new trends and potential paths for environmental policy while offering suggestions for efficient governance.**

   The changing opportunities and difficulties of global environmental governance are reflected in emerging themes in environmental policy. Innovative approaches to sustainability can be found in ideas like the circular economy, which emphasizes avoiding waste and maximizing resource efficiency. Large-scale environmental projects are increasingly being funded through the use of green finance, which includes investments in ecologically sustainable projects. While technological innovations like blockchain and artificial intelligence provide new capabilities for environmental enforcement and monitoring, they also pose regulatory issues. The subject matter looks into these changes and how they might affect how policies are made, taking into account how more adaptable and flexible governance structures might better handle today's problems. Emerging global challenges, such as climate-induced migration, climate justice, and the use of indigenous knowledge in conservation techniques, represent substantial additions to the governance agenda. The chapter also looks at how these problems necessitate innovations in policy that represent equity, flexibility, and resilience, pointing future frameworks in the direction of a more inclusive and thorough approach to sustainability.

2. HISTORICAL EVOLUTION OF ENVIRONMENTAL POLICIES

a) Early Environmental Regulations

As civilizations became aware of the negative effects of industrialization on public health and natural resources, the first environmental rules came into being. These laws were initially reactive, addressing particular problems like water and air pollution only after serious harm had become apparent. For example, the British Alkali Act of the 19th century sought to limit acid emissions from manufacturers. One of the first federal initiatives to stop river pollution in the United States was the Rivers and Harbors Act of 1899. Rather than encouraging general ecological sustainability, these early regulations frequently focused on alleviating urgent environmental damage in a fragmented and limited manner. However, they established a foundation for later, more comprehensive environmental protection initiatives by laying the foundations for more organized and integrated approaches to environmental governance. Gaining knowledge of these early regulations can help one better understand how environmental policy has changed over time and how strong regulatory frameworks are becoming increasingly important to protect the environment. Ad hoc approaches to particular environmental challenges typified the early era of environmental management. Early laws were frequently reactionary, addressing resource depletion and pollution only after noticeable harm had been done. Examples include the regulation of air pollution in response to the Industrial Revolution and the establishment of national parks to protect natural landscapes from industrial exploitation. Regional bodies are needed that enable greater buy-in to global mechanisms, increase compliance and have greater emphasis on the South. The UN Economic Commission for Europe (UN-ECE) was a relatively successful example of such a mechanism, particularly at the time of transition of many European countries from centrally-planned communist regimes to western economies. (Mee, 2005)

This early legislation mostly concentrated on immediate environmental damage in a fragmented fashion, even if they laid the groundwork for later environmental protection programs.

Today's environmental governance must deal with serious issues like pollution, biodiversity loss, and climate change. Adaptive governance frameworks that can handle the intricacies of a globalized environment are necessary to address these concerns. In order to improve compliance and efficacy, future legislation should prioritize international collaboration and integrate diverse stakeholder perspectives, especially those from the Global South. Developing innovative solutions to manage climate-induced migration, safeguard indigenous knowledge, and advance climate justice is crucial to creating comprehensive plans to save our world for future generations.

b) Key International Milestones

- **Stockholm Conference (1972):** It was the first significant global meeting to discuss environmental challenges, and it took place in Stockholm, Sweden. This international conference was devoted to environmental issues, producing the Stockholm Declaration that established the framework for contemporary environmental policy. Global environmental governance underwent a sea change with the 1972 Stockholm Conference, formerly the United Nations Conference on the Human Environment. Representatives from 113 nations, as well as numerous non-governmental organizations and UN agencies, attended the summit. The United Nations Environment Programme (UNEP) was established as a result of its emphasis on the relationship between environmental stewardship and development. The Stockholm Declaration, one of the primary products of the conference, stressed the need for international cooperation and the inclusion of environmental considerations in development planning while enumerating 26 principles for sustainable environmental management. In addition to sparking a global environmental movement, the conference laid the foundation for future international agreements and policies aimed at environmental conservation.
- **Rio Earth Summit (1992):** By putting in place significant agreements that emphasize global cooperation and sustainable development, such as the Convention on Biological Diversity, Agenda 21, and the Rio Declaration. The Rio Earth Summit of 1992, also known as the United Nations Conference on Environment and Development (UNCED), was a historic event in the history of environmental governance. It was held in Rio de Janeiro, Brazil, and brought together thousands of participants from government, non-governmental organizations, and the private sector, in addition to leaders from 178 other nations. The goal of the conference was to balance environmental preservation with global economic development. Two of the main milestones were Agenda 21, a comprehensive action plan for sustainability on a global, national, and local level, and the Rio Declaration on the Environment and Development, which listed 27 principles for sustainable development. Two other significant treaties that were up for signature were the Convention on Biological Diversity and the Framework Convention on Climate Change. The Rio Earth Summit significantly advanced global environmental policy by promoting the merger of environmental and socioeconomic goals and international cooperation to address pressing ecological challenges.

- **Kyoto Protocol (1997):** A worldwide accord requiring developed countries to reduce their greenhouse gas emissions. Kyoto, Japan, hosted the 1997 ratification of the historic international treaty known as the Kyoto Protocol, which is a component of the United Nations Framework Convention on Climate Change (UNFCCC). Its primary objective was to avert global warming by reducing greenhouse gas emissions. The convention established legally binding goals for 37 developed nations and the European Community to reduce their emissions by an average of 5% below 1990 levels between 2008 and 2012. It offered market-based instruments, such as emissions trading, the Clean Development Mechanism (CDM), and Joint Implementation (JI), that allow countries to achieve their objectives with flexible, affordable solutions. Despite its noble goals—such as the departure of some parties and the non-participation of major emitters like the United States—the Kyoto Protocol encountered challenges. Nonetheless, it was a turning point in the history of climate policy, paving the way for further international climate agreements, such as the Paris Agreement of 2015

c) Evolution of National and Regional Policies

National and regional environmental policies have developed along a spectrum, from stopgap measures to all-encompassing, integrated frameworks targeted at sustainable development. Environmental policies were initially largely reactionary measures meant to address pressing environmental issues like industrial pollution and habitat degradation. Early legislation was primarily concerned with reducing particular damages, often without taking long-term sustainability or wider ecological effects into account.

The late 20th century marked a pivotal shift towards more proactive environmental governance. Key milestones included the establishment of dedicated environmental agencies and the enactment of foundational legislation. For instance, the United States' formation of the Environmental Protection Agency (EPA) in 1970 and the adoption of the Clean Air Act and Clean Water Act exemplify this era of regulatory expansion aimed at addressing pollution at a national level.

Internationally, the 1992 Rio Earth Summit underscored the importance of integrating environmental considerations into development policies through agreements like Agenda 21 and the Rio Declaration on Environment and Development. These frameworks emphasized sustainable development as a global imperative, influencing subsequent national policies worldwide.

Regionally, programs like the Environmental Action Programmes of the European Union have standardized environmental standards among member states, encouraging cooperation on matters like biodiversity preservation, air and water quality, and climate change mitigation. These initiatives show a tendency toward cooperative government, in which regional blocs work together to successfully handle common environmental issues.

Many environmental negotiators express a concern about "negotiation fatigue." This refers to the sense that the proliferation of environmental negotiations has taken on a life of its own and created a nearly "permanent" global environmental negotiation enterprise where negotiation itself seems to have become the goal, sometimes to the detriment of actual treaty implementation. (Munoz Cabre, 2009)

In recent decades, national and regional policies have evolved to encompass broader environmental goals, including biodiversity conservation, ecosystem restoration, and climate resilience. Increasingly, policies integrate economic incentives, technological innovation, and public participation to foster sustainable practices across sectors. The evolution continues with ongoing efforts to align environmental policies with global sustainability goals, ensuring that future generations inherit a healthy and resilient planet.

3. CONTEMPORARY POLICY FRAMEWORKS

a) Overview of Major International Agreements

Paris Agreement (2015): A historic agreement made under the UNFCCC with the goal of keeping global warming far below 2°Cover pre-industrial levels. The Paris Agreement, approved in December 2015 under the auspices of the United Nations Framework Convention on Climate Change (UNFCCC), is a major multinational effort to tackle climate change. It builds on decades of worldwide talks and agreements to reduce greenhouse gas emissions and moderate global warming. The deal aims to keep warming far below 2 degrees Celsius, with attempts to limit it to 1.5 degrees Celsius over pre-industrial levels, which experts consider vital for averting catastrophic climate impacts.

The Paris Agreement's salient aspects are:

NDCs, or nationally determined contributions: Each participating country sets its own emissions reduction targets and outlines its strategy for achieving them. Every five years, the NDCs are modified to reflect each nation's evolving climate actions.

1. **Transparency and Accountability:** To foster confidence and guarantee that nations keep their promises, the agreement places a strong emphasis on transparency in reporting and assessing progress toward NDCs.

2. **Global Inventory:** The five-year-cycle global stock take assesses the implementation of the agreement, identifies areas in need of enhancement, and directs future climate action.
3. **Climate Finance:** The accord acknowledges the significance of wealthy nations providing financial help to developing nations in order to aid in activities related to capacity building, adaptation, and mitigation.
4. **Adaptation:** It encourages adaptation initiatives to increase resistance to the effects of climate change, especially in communities and areas that are vulnerable.

With 196 parties having signed, the Paris Agreement became operative in November 2016 and stands as one of the most widely endorsed international agreements ever. Its cooperative framework encourages nations to gradually strengthen their climate pledges over time, even if they are not legally obligatory.

The adoption of the agreement signaled a dramatic turn in the direction of multilateral cooperation on climate change and demonstrated the urgency with which it is imperative to confront climate risks and make the transition to a low-carbon, sustainable future. Notwithstanding obstacles and disparities in national implementation, the Paris Agreement continues to be a crucial tool for global climate governance, directing initiatives to lessen the effects of climate change and promote climate-resilient development across the board.

Convention on Biological Diversity (CBD)

Adopted in 1992 during the Earth Summit in Rio de Janeiro, the Convention on Biological Diversity (CBD) is a significant international agreement designed to support sustainable development while preserving Earth's biodiversity. It is one of the most extensively ratified treaties in the world, with 196 parties as of 2022. The CBD highlights the essential role of ecosystem services for human well-being and acknowledges the inherent value of biodiversity for sustainable development.

The principal goals of the CBD consist of:
1. **Conservation of Biological Diversity:** The goals of CBD are to ensure fair and just distribution of the benefits derived from genetic resources, promote sustainable use of all the components that make up life, and preserve biological diversity.
2. **Sustainable Use of Biological Resources:** It encourages sustainable practices that maintain ecosystems, species, and genetic diversity for future generations.

3. **Fair and Equitable Benefit-Sharing:** The CBD promotes the fair and equitable sharing of benefits derived from genetic resources, particularly with indigenous and local communities that possess traditional knowledge related to biodiversity.
4. **Access to Genetic Resources:** It addresses access to genetic resources and the need for prior informed consent and mutually agreed terms between countries and stakeholders.
5. **Integration of Biodiversity into National Policies:** Incorporating biodiversity considerations into national development strategies, regulations, and sectors including farming, fishing, and forestry is a commitment made by parties to the CBD.

The Oslo Principles on Climate Change Obligations (Oslo Principles) articulate a set of principles comprising the essential obligations of states and enterprises to avert the critical level of global warming.170 An international group of jurists, academics, and experts in international law, human rights law, and environmental law adopted the Oslo Principles in March 2015. Principle 1 of the Oslo Principles references the precautionary principle. In the context of climate change, the precautionary principle requires that GHG emissions be reduced to the extent, and at a pace, necessary to protect against the threats of climate change that can still be avoided. (Sicangco, 2020)

Through several measures, such as the creation of protected areas, programs for the conservation of species, and sustainable development projects, the CBD has encouraged international collaboration on biodiversity conservation since its ratification. Additionally, it has aided with the ratification of associated accords and guidelines, like the Nagoya Protocol on Access and Exchange of Benefits.

Despite persistent obstacles including habitat loss, climate change, and invasive species, the CBD continues to be an essential framework for international efforts to conserve biodiversity. In order to preserve Earth's rich biological legacy for future generations, it highlights the connections between sustainable development, biodiversity conservation, and human well-being.

Timeline of Key International Environmental Agreements

Table 1. Outlines a timeline of key international environmental agreements, from the Stockholm Declaration (1972) to the Paris Agreement (2015), emphasizing the importance of global collaboration in addressing environmental challenges (UNEP, 2020; Greenpeace, 2021).

Year	Agreement	Key Provisions
1972	Stockholm Declaration	Foundation of international environmental governance
1992	Rio Declaration, Agenda 21, CBD	Sustainable development, biodiversity conservation.
1997	Kyoto Protocol	Binding targets for reducing emissions of greenhouse gases.
2015	Paris Agreement	Restrict global warming to less than 2°C.

b) Regional Initiatives and National Laws

Numerous national legislation and regional partnerships have resulted from countries and regions customizing international agreements to fit their unique settings. China's strong renewable energy policy and the European Union's Green Deal, which aims for climate neutrality by 2050, are two examples. In order to solve local environmental concerns and coordinate efforts across jurisdictions, national legislation and regional initiatives play crucial roles in developing environmental governance.

National Regulations: To protect their surroundings, nations all around the world have passed laws. These laws frequently represent the goals and difficulties of the country, which can include everything from biodiversity preservation and climate change mitigation to pollution control and natural resource management. For example, the United States Clean Air Act regulates emissions from automobiles and industry and establishes standards for air quality in order to safeguard the environment and human health. Comparably, the Renewable Energy Sources Act of Germany encourages the growth of renewable energy by offering financial incentives to producers of renewable energy, supporting both national energy security and climate goals.

Regional Initiatives: To more successfully address common environmental challenges, adjacent countries can work together through regional initiatives. The European Union's air and water quality directives are among the examples; they set uniform criteria for all member states and encourage cooperative research and technical innovation. The Great Green Wall Initiative of the African Union intends to use tree planting and the promotion of sustainable land management techniques to counteract desertification and land degradation throughout the Sahel region.

Collaboration and Integration: To obtain synergistic results, effective environmental governance frequently entails coordinating regional and national actions. Addressing transboundary environmental concerns including air and water pollution, biodiversity loss, and climate change requires the facilitation of information sharing, capacity building, and cooperative problem-solving, all of which are made possible by this integration. Further illustrating the value of multilateral cooperation in preserving natural resources and advancing sustainable development on a larger scale, regional initiatives offer forums for nations to negotiate agreements, pool resources, and coordinate responses to environmental emergencies.

Regional Agreements and Indigenous Contributions: In areas with distinct biological and cultural contexts, regional accords are especially important for tackling particular environmental issues. These agreements facilitate the development of frameworks that take into account regional ecological, social, and economic elements, allowing for more specialized and frequently more successful responses to environmental problems. Treaties such as the Amazon Cooperation Treaty Organization (ACTO) aim to coordinate actions among member states in order to conserve the Amazon rainforest. The African Union's Agenda 2063, which addresses challenges that are particularly urgent in the African context of desertification, deforestation, and biodiversity loss, also places a strong emphasis on the sustainable use of natural resources.

Because of their enduring relationship with the natural world and their commitment to sustainable resource management, indigenous groups are being more and more acknowledged as vital contributions to environmental protection. Indigenous communities' involvement in the conservation of the Amazon rainforest in Latin America has generated accords that incorporate protections for the forest and the people who live there. Numerous United Nations projects that support indigenous voices in climate change demonstrate how the generations-long ecological knowledge that indigenous populations possess can improve conservation efforts. These communities support laws that protect biodiversity and the wise use of natural resources, frequently in line with regional sustainability objectives.

Regional accords and indigenous groups' contributions combine to shape global policy in the context of wider environmental frameworks. Respect for indigenous rights and knowledge is mentioned in a number of agreements, including the Convention on Biological Diversity (CBD) and the Paris Agreement. Priorities for climate resilience, reforestation, and ecosystem management have been established in many areas by indigenous participation, which has prompted a shift in policy toward more inclusive, culturally appropriate measures.

c) Evaluation of Policy Efficacy and Deficiencies

There are still big gaps in the efficacy and application of policies, notwithstanding advancements. Lack of political will, inadequate resources, and the difficulty of coordinating efforts across several levels of governance are some of the difficulties.

When evaluating the efficacy of environmental policies, it is important to consider how they will affect the attainment of specific goals, such as lowering pollution, protecting biodiversity, and slowing down climate change, as well as any weaknesses or potential areas for development.

Evaluation of Effectiveness: Measurably beneficial effects on environmental quality and sustainability indicators are exhibited by effective environmental policies. Policies aimed at improving air quality, for instance, may result in lower emissions of pollutants such as sulfur dioxide and particulate matter, improving public health outcomes and giving urban sky a better appearance. Similar to this, conservation laws that create protected areas or control fishing methods can aid in the restoration and preservation of important habitats and hotspots for biodiversity.

Despite advancements, environmental regulations frequently encounter obstacles and weaknesses that reduce their efficacy. Typical problems consist of:

1. **Implementation and Enforcement**: The intended impact of a policy may be undermined by limited policy compliance and enforcement due to inadequate resources or weak enforcement procedures.
2. **Lack of Integration:** Policies that are not well-integrated across sectors may have competing goals or pass up chances to work together. Energy policy, for example, may favor economic expansion over environmental sustainability, which would raise greenhouse gas emissions.
3. **Complexity and Adaptation:** Because environmental issues like climate change are dynamic and multifaceted, policies must be flexible enough to change as scientific understanding and socioeconomic circumstances do.
4. **Measuring and Assessment:** Inadequate procedures for monitoring and evaluating policies can make it more difficult for decision-makers to effectively assess policy performance and make necessary adjustments.

Increasing Policy Effectiveness: To improve policy effectiveness, many approaches must be taken to solve these issues. These include:

1. **Integrated Approaches:** Using integrated frameworks for policy adoption that take into account social, economic, and environmental factors can help to promote comprehensive solutions and prevent unforeseen repercussions.
2. **Stakeholder Engagement:** Including communities, businesses, and non-governmental organizations (NGOs) in policy implementation promotes ownership, teamwork, and accountability while also enhancing transparency.
3. **Innovation and Technology:** Using innovation and technology can promote environmentally friendly behaviors and make it easier to monitor, comply with, and enforce laws pertaining to the environment.
4. **Building Ability:** Putting money into helping legislators, regulators, and the general public develop their ability helps to fortify institutional structures and improve the way environmental policies are implemented.

Governments and stakeholders can advance environmental governance and achieve more sustainable outcomes for current and future generations by regularly evaluating the effectiveness of policies, identifying gaps, and implementing targeted improvements.

4. PARTICIPATION IN ENVIRONMENTAL GOVERNANCE

a) The Function of Intergovernmental Bodies and Governments

One of the most important things that governments do is pass and administer environmental laws. Intergovernmental institutions like the United Nations Environment Programme (UNEP) and the Intergovernmental Panel on Climate Change (IPCC) provide scientific assessments and policy recommendations aid in international collaboration. Governments and intergovernmental organizations (IGOs) play a crucial role in environmental governance since it is their responsibility to create, carry out, and uphold laws that address the world's environmental issues. Within their respective jurisdictions, national and subnational governments hold substantial power to establish regulatory frameworks and standards that impact environmental results. To effectively implement environmental rules, they pass legislation, set up regulatory bodies, and allot funding for these purposes.

Intergovernmental organizations are essential for promoting international coordination and collaboration between states. Global environmental accords are negotiated and developed on platforms provided by organizations like the European Environment Agency (EEA), the Intergovernmental Panel on Climate Change (IPCC), and the United Nations Environment Programme (UNEP). These agreements promote cooperative action on problems including pollution management, biodiversity conservation, and climate change by establishing standards, goals, and guidelines that member states are required to follow.

One of the most important things that governments do is pass and administer environmental laws. Intergovernmental institutions like the United Nations Environment Programme (UNEP) and the Intergovernmental Panel on Climate Change (IPCC) provide scientific assessments and policy recommendations aid in international collaboration. Assistance is necessary to ensure that all countries participate equally in global environmental projects and to overcome the differences in resources and capacities between them.

In addition, governments work in tandem with the private sector, civic society, and non-governmental organizations (NGOs) to secure wider backing and involvement for environmental projects. The implementation of a collaborative approach enhances the legitimacy and efficacy of environmental governance initiatives by fortifying transparency, accountability, and public engagement in decision-making processes.

The regulatory roles that countries and IGOs play in environmental governance are essentially secondary to their leadership roles in international policymaking, international effort coordination, and global advocacy of sustainable development principles. Their deeds are crucial in guaranteeing the protection of natural resources for both the current and future generations, as well as in influencing the direction of environmental policy in the future.

b) Impact of Civil Society and Non-Governmental Organizations (NGOs)

In environmental governance, non-governmental organizations (NGOs) and civil society are essential because they supplement, and occasionally contradict, the acts of governments and intergovernmental organizations (IGOs). Their power comes from their capacity to sway public opinion, push for legislative changes, and carry out neighborhood-based campaigns that tackle global, regional, and local environmental problems.

NGOs can concentrate on certain environmental issues like pollution reduction, biodiversity conservation, or climate change. Generally, they function independently of the government. They frequently close the gaps in government action by carrying out studies, overseeing the application of laws, and lending their knowledge to improve environmental decision-making. NGOs make sure that a variety of viewpoints are taken into account during policy talks by amplifying voices that could otherwise be ignored by using their networks and experience.

A crucial role is also played by civil society, which includes a wider spectrum of community organizations, activists, and individuals. Through direct action, public education campaigns, and community-based conservation efforts, they support environmental governance. Through grassroots activism, civil society projects frequently encourage creativity and locally specific solutions to particular environmental problems, supporting sustainable behaviors and influencing policy.

In addition, NGOs and civil society groups improve environmental governance's accountability and openness. They perform the role of watchdogs, keeping an eye on the activities of IGOs and governments, promoting openness in the decision-making process, and holding interested parties responsible for their environmental pledges. Their involvement encourages democratic participation and gives communities the power to actively participate in the formulation and application of environmental policies.

In summary, civil society and NGOs play a critical role in environmental governance. Their knowledge, experience, and grassroots initiatives support national and international initiatives for environmental policies and practices that are more inclusive, efficient, and sustainable. In order to achieve long-term environmental sustainability, NGOs and civil society will continue to play a crucial role in formulating policy and fostering environmental stewardship as environmental issues change.

c) Public-Private Partnerships and the Private Sector's Contributions

The private industry and public-private partnerships (PPPs) play a crucial role in environmental governance by virtue of their ability to innovate, mobilize resources, and implement sustainable practices. Private businesses provide financial resources and technological innovations to environmental projects, increasing productivity and stimulating the economy while reducing their negative effects on the environment. PPPs combine the resources and knowledge of the private sector with the goals of public policy, thereby utilizing the advantages of both sectors. This partnership produces a variety of projects that improve environmental sustainability, including waste management strategies, conservation campaigns, and renewable energy projects. Governments can leverage private sector resources and creativity through Public-Private Partnerships (PPPs), which can expedite the attainment of environmental objectives and encourage conscientious corporate conduct. These collaborations are essential for tackling difficult environmental issues and accomplishing long-term sustainability goals through shared accountability and teamwork.

d) Public Engagement and Initiated Movements

For environmental policy to be legitimate and effective, public engagement is necessary. As seen by the international climate strikes headed by young activists, grassroots movements have been crucial in bringing attention to issues and influencing policy reforms. The amplification of community voices in policy processes and the promotion of democratic decision-making are two important functions of public engagement and grassroots movements in environmental governance. These movements, which encourage people to support sustainable practices and hold businesses and governments responsible, frequently start as a reaction to regional environmental issues. Grassroots movements bring attention to environmental problems and advocate for legislative changes that put social justice and ecological health first through demonstrations, petitions, and community projects. By ensuring that policies take into consideration the many needs and beliefs of society, public engagement fosters accountability and openness in governance.

Furthermore, community-based initiatives enable localities to carry out grassroots projects like waste reduction initiatives or community gardens, which support larger environmental objectives. These movements build environmental governance frameworks and promote a shared commitment to sustainable development and conservation initiatives by bridging gaps between policymakers and the public.

e) Technology's Influence on Environmental Policy Frameworks

Technology is changing environmental governance and policy frameworks, especially blockchain and artificial intelligence (AI). Through data analysis, predictive modeling, and astute resource management, artificial intelligence improves environmental monitoring and management. AI systems, for instance, use satellite data to estimate environmental repercussions and track deforestation, allowing policymakers to make well-informed decisions. Blockchain technology enhances environmental governance's accountability and transparency. It offers safe provenance tracing, encouraging moral consumption, and makes trading carbon credit easier by guaranteeing safe, substantiated transactions. Additionally, because it is decentralized, inclusive decision-making is supported, giving communities a say in environmental policy.

As these technologies develop, they encourage stakeholder collaboration, stimulate innovation in sustainable practices, and tackle difficult issues like equitable governance and migration brought on by climate change. In the end, blockchain and AI are necessary to develop egalitarian, flexible, and successful environmental policies in response to new global issues.

5. CASE STUDIES

a) Implementing Policies Successfully

German Initiatives for Renewable Energy

The goal of Germany's Energiewende (energy transition) strategy is to drastically cut greenhouse gas emissions by switching from fossil fuels to renewable energy sources. With its ambitious objectives to shift to sustainable energy sources, Germany has been at the forefront of renewable energy programs. The Energiewende, or energy transition, which was started in the early 2000s, is crucial to its plan. This program aims to replace nuclear power and fossil fuels with renewable energy sources like wind, solar, and biomass power. Germany's Renewable Energy Act (EEG) introduced feed-in tariffs to incentivize the production of renewable energy, leading to a significant rise in renewable capacity and a reduction in greenhouse gas emissions.

The nation has made significant investments in research and development, supporting the development of innovative renewable technologies and building a strong infrastructure for grid integration. Additionally, contributing to decentralized energy generation and increasing local participation in energy policy are community-owned energy cooperatives. Germany's renewable energy programs serve as a model for worldwide sustainability efforts, illustrating the viability and benefits of shifting to a low-carbon economy, despite obstacles like cost management and grid stability.

Costa Rica's Efforts to Conserve the Environment

Costa Rica's creative legislation and environmentally friendly tourism strategies have made it a global leader in the conservation of biodiversity. Costa Rica has gained international acclaim for its dedication to environmental sustainability and is well-known for its aggressive conservation efforts. The nation is home to an abundance of species, varied ecosystems, and lush rainforests, all contributing to its great biodiversity. Policies in Costa Rica that place a high priority on biodiversity conservation and sustainable development are what motivate conservation efforts there. Proposals include paying landowners for maintaining natural habitats through payment for ecosystem services (PES) programs, developing ecotourism, and growing national parks and protected areas.

Strong government regulations, public awareness initiatives, and cooperation with non-governmental organizations and international organizations are all credited with Costa Rica's success in conservation. The nation has set high standards for itself, such as becoming carbon neutral by 2050, which calls for creative ways to strike a balance between environmental stewardship and economic growth. Costa

Rica is a global leader in environmental governance and sustainable development, utilizing its natural resources and involving communities in conservation initiatives

Case Study Comparison

Table 2. Compares various case studies of environmental initiatives across different regions, highlighting the effectiveness and challenges faced in each context

Case Study	Key Strategies	Outcomes	Challenges
Germany's Renewable Energy	Policy incentives, subsidies	Increased renewable energy usage	High initial costs, grid integration
Costa Rica's Conservation	Protected areas, eco-tourism	High biodiversity conservation	Balancing development and conservation
Amazon Deforestation	Enforcement of laws, monitoring	Mixed success, continued deforestation	

b) Advancing Knowledge and Difficulties

The intricate and interrelated nature of global environmental crises presents challenges for environmental governance. Achieving fair participation across national boundaries, managing political disagreements in international discussions, and balancing economic development with sustainability goals are among the major obstacles. The significance of integrated strategies that take into account social, economic, and environmental factors is highlighted by lessons learned. Stakeholder participation to promote accountability and transparency, flexible policy frameworks, and strong monitoring and evaluation systems are all necessary for effective governance. Innovative approaches are needed to address these issues, such as utilizing technology for compliance and monitoring, forming partnerships to strengthen international cooperation, and including local people in decision-making processes. Policymakers can better manage future obstacles, promote sustainable development goals, and protect the environment for future generations, by drawing lessons from the past and modifying plans accordingly.

There is a table to summarize the challenges in environmental governance, lessons learned, and strategies for addressing these challenges (Table 3):

Table 3. Summarizing the challenges in environmental governance, lessons learned, and strategies for addressing these challenges

Challenges	Lessons Learned	Strategies
Reconciling economic growth with sustainability goals	Integrated approaches are crucial for balancing economic and environmental priorities.	Implement adaptive policy frameworks that promote sustainability.
Ensuring equitable participation across nations	Equitable participation enhances global cooperation and legitimacy of policies.	Foster stakeholder engagement and inclusive decision-making processes.
Navigating political differences in international negotiations	Addressing political differences through diplomacy and consensus-building.	Enhance global cooperation and partnerships for collective action.
Strategies for Effective Governance	Monitoring and Evaluation	Adaptive Policy Frameworks

c) A Comparative Examination of Various Governance Models

To comprehend how different approaches affect the efficacy of policies and environmental outcomes in diverse nations and areas, a comparative analysis of governance models in environmental management is vital. With distinct advantages and disadvantages, centralized, decentralized, and hybrid governance models can be broadly divided into three categories.

Centralized governance regimes, which are frequently distinguished by robust regulatory frameworks and enforcement mechanisms, concentrate decision-making authority at the national level. This strategy is demonstrated by nations such as China, where environmental rules can be quickly put into effect by top-down directives, but local implementation and compliance may be difficult owing to regional differences in competence and resources.

Decentralized governance models, prevalent in nations such as Germany and Sweden, place a strong emphasis on community involvement and municipal sovereignty.

These strategies encourage creativity and grassroots participation by promoting customized responses to environmental problems depending on regional requirements and circumstances. Coordination between several administrative levels and maintain standardization in norms, however, can be difficult.

In an effort to capitalize on the advantages of both, hybrid governance models incorporate components of both decentralized and centralized systems. Federal systems such as the United States and Australia are prime examples of this, as they combine state-level policy implementation flexibility with national norms. These methods aim to strike a balance between local responsiveness and adaptability and national goal consistency.

A comparative examination reveals that no governance paradigm is ideal for every situation. A number of variables, including institutional strength, stakeholder participation, political stability, and financial resources, are frequently necessary for success.

Diverse governance models offer valuable insights that highlight the significance of adaptability, adaptive management, and ongoing education in effectively tackling changing environmental issues.

Through an examination of these models, policymakers may pinpoint optimal methodologies, foresee possible obstacles, and customize tactics for governance to augment ecological robustness and accomplish worldwide objectives for sustainable development.

Here's a structured table summarizing the comparative analysis of centralized, decentralized, and hybrid governance models in environmental management (table 4):

Table 4. Summarizing the comparative analysis of centralized, decentralized, and hybrid governance models in environmental management

Governance Model	Characteristics	Examples	Strengths	Weaknesses
Centralized	- Decision-making authority at national level	China	Strong regulatory frameworks and enforcement mechanisms	Challenges in local implementation due to regional variations in capacity and resources
	- Top-down directives for swift policy implementation			
Decentralized	- Emphasizes local autonomy and community involvement	Sweden, Germany	Tailored solutions based on local needs and conditions	Coordination challenges across different administrative levels
	- Promotes grassroots support and innovation			
Hybrid	- Combines elements of centralized and decentralized	USA, Australia	Balances national standards with local flexibility	Ensuring uniformity in national goals
	- Aims for flexibility and responsiveness			Addressing conflicting priorities

6. FUTURE TRENDS AND CHALLENGES

a) Recent Developments in the Field of Environmental Law

Green technology, climate change mitigation, and sustainability are the three main focuses of emerging trends in environmental policy. More and more governments are embracing the circular economy, tightening rules on carbon emissions, and utilizing renewable energy sources. Furthermore, the preservation of natural habitats and biodiversity is becoming more and more important. Global environmental concerns are largely addressed through international agreements and alliances, while local regulations are changing to include community-based conservation efforts.

Future environmental policy will also be shaped by important trends, including increased public knowledge and involvement in environmental decision-making.

Directive Economic Situation: An economic framework that maximizes resource utilization and minimizes waste while advancing sustainability via recycling and reuse. A circular economy redefines growth by emphasizing advantages for the entire community. Extending the life of materials entails developing goods with durability, recyclability, and reuse in mind. This model emphasizes continuous resource utilization and waste elimination, which sets it apart from the typical linear economy. Rethinking product design, encouraging sharing and mending, and putting in place extensive recycling networks are important tactics.

Sustainable Financial Services: Financial instruments that prioritize environmental sustainability include green bonds and sustainable investing. "Green finance" is the term used to describe financial investments that support environmentally sustainable projects and initiatives. This includes funding for renewable energy, pollution control, energy savings, and environmentally friendly agriculture. By integrating environmental, social, and governance (ESG) considerations into financial decision-making, green finance aims to reduce environmental risks and promote sustainable economic growth. Financial mechanisms, such as green bonds and loans tied to sustainability, are important tools in this field. Green finance fosters innovation and opens up new avenues for sustainable growth in addition to assisting in the mitigation of environmental effects.

b) Technological Advancements and Their Regulatory Implications

Technological innovations like blockchain and artificial intelligence could improve environmental enforcement and monitoring, but they also present new regulatory concerns. Artificial intelligence, blockchain, and the Internet of Things (IoT) are examples of technological innovations that have revolutionized a number of industries by fostering connectedness, efficiency, and innovation. These developments do, however, present serious regulatory issues.

Recently, UNEP has undertaken a major overhaul of its media outreach (UNEP, 2019a). Presently, UNEP has twenty-one social media accounts representing all six UN languages and more than two million followers (UNEP, 2019a). UNEP is running three public awareness campaigns: the Clean Seas campaign against plastic pollution, the Wild for Life campaign against illegal wildlife trafficking, and the Breathe Life campaign against air pollution (UNEP 2018) (Urho, 2019)

Concerns including data privacy, cybersecurity, the moral application of AI, and the legality of online transactions must be addressed by governments and regulatory organizations. It is critical to strike a balance between protecting consumers and promoting innovation. Creating flexible and progressive regulatory frameworks can aid in risk reduction and encourage ethical usage of technology. Sufficient monitoring, constant stakeholder communication, and international collaboration are necessary to properly handle the regulatory fallout from quick technical advancements.

c) Global Issues and the Requirement for Coordinated Actions

Global concerns including pollution, biodiversity loss, and climate change require international coordination. Global issues like economic injustice, pandemics, climate change, and cybersecurity threats require coordinated responses due to their complexity and transnational nature. These are problems that no one country can successfully solve on its own. To provide complete solutions, cooperation between governments, businesses, public society, and international organizations is crucial. For example, whereas pandemic responses benefit from coordinated public health initiatives and shared resources, climate change mitigation necessitates global accords such as the Paris Agreement. Coordinated policy to support fair trade, sustainable development, and equitable growth is necessary to address economic inequality. Developing resilient, adaptable systems and encouraging international cooperation are ultimately essential to addressing these global issues.

d) Policy Recommendations for Future Governance Frameworks

Future governance frameworks should emphasize sustainability, inclusivity, and adaptability to effectively address emerging global challenges. Key policy recommendations include:

1. **Enhancing International Partnership:** By pooling resources and coordinating efforts, strengthen international alliances to address cross-border problems like pandemics, cybercrime, and climate change.
2. **Encouraging Inclusive Governance**: To develop more representative and equitable policies, make sure a variety of stakeholders, including underrepresented communities, are involved in the decision-making process.
3. **Promoting Innovation and Adaptability:** Establish adaptable legislative frameworks that promote technological advancement while protecting against possible threats, including security and data privacy issues.
4. **Strengthening Transparency and Accountability:** To foster public confidence and uphold moral behavior, establish procedures for increased transparency in governance procedures, and hold institutions accountable.
5. **Setting Environmental Sustainability as a Top Priority:** Integrate sustainable practices into all aspects of governance to prevent climate change, protect natural resources, and promote a circular economy.
6. **Consolidating Social Safety Networks:** Establish strong social safety nets to assist disadvantaged groups, guaranteeing their economic stability and mitigating disparities.

By implementing these suggestions, governance frameworks in the future will be better equipped to handle the complexity of a world that is changing quickly, promoting stability and sustainable development on a worldwide scale.

7. CONCLUSION

a) Overview of the Main Findings

The chapter emphasizes how important it is to have strong environmental governance frameworks in order to handle today's urgent issues, like pollution, biodiversity loss, and climate change. The primary conclusions highlight how crucial sound environmental governance is to both global stability and sustainable development.

The significance of international cooperation is a major subject. Nations must cooperate and share resources in order to address transnational environmental concerns. Strengthening international collaborations can result in better-coordinated tactics, utilizing pooled expertise and technology to more successfully tackle environmental challenges like climate change.

One important element that is emphasized is **inclusive governance**. More representative and equitable policies can be achieved by ensuring that a variety of stakeholders, including underrepresented communities, are involved in the decision-making process. By guaranteeing that the advantages and disadvantages of environmental regulations are equitably spread, this inclusion contributes to the advancement of social and environmental justice.

The necessity of **creativity and flexibility** within regulatory systems is also emphasized in this chapter. As technology advances, governance frameworks need to be adaptable enough to foster creativity while guarding against threats like security and data privacy issues. This flexibility is essential for responding to environmental concerns that are dynamic in nature.

It is stated that **accountability and transparency** are the cornerstones of good governance. Establishing procedures that encourage openness in decision-making and holding organizations responsible helps foster ethical behavior and increase public confidence. The effectiveness and legitimacy of environmental policy depend heavily on these kinds of actions.

Prioritizing the sustainability of the environment is another important result. It is imperative to include sustainable practices in all facets of governance in order to effectively tackle climate change, preserve natural resources, and advance the circular economy. Sustainable governance techniques promote long-term ecological health and lessen their negative effects on the environment.

Finally, it is believed that to **protect vulnerable populations**, social safety nets must be strengthened. Ensuring that all communities can endure and adapt to environmental changes requires strong social protection systems, which can also improve economic resilience and reduce inequality.

The chapter concludes by showing how important it is to have inclusive and comprehensive environmental governance frameworks in order to promote sustainable development, achieve global stability, and navigate the complexity of today's global

b) The Significance of Inclusive and Adaptive Policy Frameworks

For policy frameworks to effectively address the complex issues of our quickly changing world, they must be inclusive and adaptable. In order to ensure sustainable development and fair progress, these frameworks offer the adaptability and inclusivity required to adjust to changing environmental, social, and economic situations.

Adaptability in policy frameworks needs to be flexible since global challenges are ever-changing. Technological developments, economic ups and downs, and climate change all necessitate governing structures that are flexible enough to react rapidly to fresh data and changing patterns. Governments can implement pertinent and timely remedies thanks to adaptive policies, which promote resilience in the face of unpredictability. For instance, adaptable legal frameworks can foster technological innovation while controlling security and privacy threats, guaranteeing that advances advance society without undermining moral principles.

Inclusivity in policymaking ensures that varied viewpoints are taken into account during the policy-making process, resulting in more thorough and equitable solutions. Policies that include disadvantaged and vulnerable communities in the process of making decisions are more likely to take into account the requirements and difficulties faced by all social segments. Public trust in institutions is increased, social justice is promoted, and inequality is decreased through inclusive governance. It makes sure that the advantages of development are distributed fairly, and that marginalized people do not bear an unfair share of the costs.

Including inclusive and flexible components in policy frameworks can greatly improve **accountability and transparency**. Adaptive policies offer transparent and accountable governance procedures by being periodically evaluated and altered in response to new information and results. By ensuring transparent stakeholder participation and the inclusion of varied viewpoints, inclusivity enhances the legitimacy of policy decisions.

An additional crucial component of these frameworks is the **inclusion of environmental sustainability.** Adaptive policies have the potential to successfully address climate change and resource depletion by quickly integrating scientific discoveries and emerging best practices in environmental management. Policies that are inclusive make sure that sustainability initiatives take into account the social aspects of environmental problems, encouraging a well-rounded strategy that promotes the health of the environment and the welfare of people.

Furthermore, **inclusive and flexible frameworks** might improve social safety nets by offering dependable assistance to populations that are more susceptible. By doing this, communities are better prepared to deal with social, economic, and environmental changes, which promotes resilience and lowers inequality.

To sum up, inclusive and flexible policy frameworks are essential for negotiating the complexity of today's international issues. By fostering adaptability, fairness, openness, and sustainability, they help society attain long-term stability and sustainable growth.

c) Concluding Remarks Regarding the Future of Sustainable Environmental Governance

Concluding remarks regarding the direction of **sustainable environmental governance** stress the need for inclusive, flexible, and integrated strategies to successfully address today's environmental issues. The pressing need for coordinated and creative governance solutions that cross traditional boundaries and involve a wide range of stakeholders stems from concerns like pollution, biodiversity loss, and climate change.

Encouraging international cooperation is an essential component of this progressive route. Global environmental concerns necessitate shared resources and collaborative response. Enhancing international collaborations can improve the efficiency of programs meant to lessen the effects of climate change, protect ecosystems, and cut down on pollution. Working together makes it possible to combine financial, technological, and intellectual resources, producing results that are more reliable and long-lasting.

Another essential element is **adaptive governance**. Policies need to be adaptable and sensitive to new information and changing conditions because of how quickly the environmental environment is changing. This flexibility enables prompt modifications and inventions that can more effectively handle unforeseen difficulties and seize new opportunities. By ensuring that policies continue to be applicable and efficient throughout time, adaptive governance promotes resilience and ongoing development.

The principle **of inclusivity in environmental governance** guarantees that the viewpoints of all relevant parties, particularly those belonging to marginalized and vulnerable communities, are acknowledged and considered. Inclusive decision-making processes promote social justice and equity while ensuring a fair allocation of the costs and rewards related to the environment. This approach promotes more public trust and engagement, both of which are essential for the successful implementation of environmental policy.

It is essential to **prioritize sustainability** in all facets of governance. Including sustainable practices in legislative frameworks encourages the health of ecosystems and addresses environmental degradation. Sustainable governance techniques lessen waste, protect natural resources for future generations, and facilitate the shift to a circular economy.

In conclusion, **adopting global collaboration**, flexible tactics, inclusivity, and a resolute dedication to sustainability are the keys to the future of sustainable environmental governance. By using these guidelines, governance frameworks can successfully handle the intricacies of environmental issues and promote a more just, resilient, and healthy global community.

REFERENCES

African Climate Summit. (2023, September). *Africa Climate Week promotes regional initiatives for sustainable growth ahead of COP28*. Retrieved from https://www.africaclimatesummit.org

Biermann, F., (Eds.). (2009). *Global Environmental Governance Reconsidered*. MIT Press.

Biermann, F., & Pattberg, P. (2008). Global environmental governance: Taking stock, moving forward. *Annual Review of Environment and Resources*, 33(1), 277–294. DOI: 10.1146/annurev.environ.33.050707.085733

Bodansky, D. (2010). The history of the global climate change regime. *Review of European, Comparative & International Environmental Law*, 19(1), 63–77.

Bodansky, D. (2016). *The Art and Craft of International Environmental Law*. Harvard University Press.

Bodansky, D. (2016). The Paris Climate Agreement: A New Hope? *The American Journal of International Law*, 110(2), 288–319. DOI: 10.5305/amerjintelaw.110.2.0288

Busch, J., Engelmann, J., Fisher, B., & Sanchez-Azofeifa, G. A. (2023). The rise of global environmental treaties in the Anthropocene: How to understand the new frameworks. *Environmental Science & Policy*, 145, 345–357.

Chasek, P. S., (2017). *Global Environmental Politics* (7th ed.). Westview Press.

Chasek, P. S., Downie, D. L., & Brown, J. W. (2013). *Global environmental politics* (6th ed.). Routledge.

Conca, K., & Dabelko, G. D. (Eds.). (2017). *Green Planet Blues: Environmental Politics from Stockholm to Johannesburg* (4th ed.). Westview Press.

Gupta, J., Arts, K. Achieving the 1.5 °C objective: just implementation through a right to (sustainable) development approach. *International Environmental Agreements: Politics, Law and Economics*, 18(1), 11-28. DOI: 10.1007/s10784-017-9376-7

IPCC. (2014). [*Synthesis Report. Contribution of Working Groups I, II and III to the Fifth Assessment Report of the Intergovernmental Panel on Climate Change. Geneva, Switzerland: IPCC.*]. *Climatic Change*, 2014.

Ivanova, M. (2021). *The Untold Story of the World's Leading Environmental Institution: UNEP at Fifty*. MIT Press. DOI: 10.7551/mitpress/12373.001.0001

Kim, R. E. (2015). Transnational networks and global environmental governance: The cities for climate protection program. *Global Environmental Politics*, 15(1), 89–107.

Mayer, B., & Randazzo, M. (2023). Environmental justice in the global south: Policy frameworks and indigenous contributions. *Journal of Environmental Law*, 35(3), 467–495.

Mee, L. D. (2005). The Role of UNEP and UNDP in Multilateral Environmental Agreements. *International Environmental Agreement: Politics, Law and Economics*, 5(1), 227–263. DOI: 10.1007/s10784-005-3805-8

Mitchell, R. B. (2003). International environmental agreements: A survey of their features, formation, and effects. *Annual Review of Environment and Resources*, 28(1), 429–461. DOI: 10.1146/annurev.energy.28.050302.105603

Mitchell, R. B., (2015). *Global Environmental Politics: Dilemmas in World Politics* (7th ed.). CQ Press.

Morgera, E., & Kulovesi, K. (Eds.). (2016). *Research Handbook on International Law and Climate Change*. Edward Elgar Publishing.

Munoz Cabre, M., Thrasher, R., & Najam, A. (2009). Measuring the negotiation burden of multilateral environmental agreements. *Global Environmental Politics*, 9(4), 1–13. DOI: 10.1162/glep.2009.9.4.1

Oberthür, S., & Gehring, T. (2006). Reforming international environmental governance: An institutional perspective. *International Environmental Agreement: Politics, Law and Economics*, 6(4), 359–376. DOI: 10.1007/s10784-004-3095-6

Oberthür, S., & Groen, L. (2017). Explaining goal achievement in international negotiations: The EU and the Paris Agreement. *Journal of European Public Policy*, 24(10), 1481–1500.

Pattberg, P., & Widerberg, O. (2015). Theorizing global environmental governance: Key findings and future questions. *Millennium*, 40(2), 421–448. DOI: 10.1177/0305829814561773

Rajamani, L. (2006). The changing nature of international environmental law. *International Environmental Agreement: Politics, Law and Economics*, 6(2), 115–139.

Sands, P. (2017). *Principles of International Environmental Law* (4th ed.). Cambridge University Press.

Sicangco, M. C. T. (2020). *Climate change, coming soon to a court near you: International climate change legal frameworks*. Asian Development Bank., DOI: 10.22617/TCS200365-2

Steffen, W., (2015). *Global Change and the Earth System: A Planet Under Pressure.* Springer.

United Nations Environment Programme. (2023, June). *Global pact on plastic pollution: steps toward an inclusive, legally binding agreement. World Environment Day.* Retrieved from https://www.unep.org/plastics-instrument

Urho, N., Ivanova, M., Dubrova, A., & Escobar-Pemberthy, N. (2019). *International environmental governance: Accomplishments and way forward.* Nordic Council of Ministers.

Voigt, C. (2016). The Paris Agreement: What is the standard of conduct for parties? *Questions of International Law*, 24, 17–28.

Young, O. R., (2014). *The Institutional Dimensions of Global Environmental Change: Science, Economics, Politics, and Ethics.* MIT Press.

KEY TERMS AND DEFINITIONS

Global Environmental Agreements: Global environmental agreements are treaties and agreements between several nations that address global environmental problems like pollution control, biodiversity conservation, and climate change.

Sustainable Development: Sustainable development is progress that meets current needs without compromising the ability of future generations to meet their own. It incorporates economic, social, and environmental issues.

Climate Change: The warming of fossil fuels and devastation of forests, along with variations in surface temperatures, precipitation, wind patterns, and other factors, are the main causes of the gradual changes in the Earth's climate system brought about by human activity.

Biodiversity Loss: A decrease in the variety and variability of living forms within a species, ecosystem, or genetic level is known as biodiversity loss. This is frequently brought on by human actions including habitat destruction, pollution, and climate change.

Pollution Control: Pollution control is the process of controlling and minimizing the discharge of toxic materials into the environment in order to safeguard ecosystems and public health.

International Cooperation: International cooperation is the cooperative efforts of nations to solve problems together and accomplish common objectives; this is frequently accomplished through the drafting and implementation of treaties and accords.

Technological Innovation: Technological innovation is the creation and use of new technologies to address issues, boost productivity, and improve people's quality of life—especially when it comes to sustainability and environmental preservation.

Circular Economy: A closed-loop economy known as the "circular economy" is predicated on the concepts of reusing, mending, recycling, and converting in order to eliminate waste and make constant use of resources.

Inclusivity in Governance: To achieve fair and representative policies, inclusive governance refers to the process of incorporating a variety of stakeholders, especially disadvantaged and marginalized communities.

Adaptive Policy Frameworks: Adaptive policy frameworks are frameworks for governance that are dynamic and sensitive to fresh data, scientific advancements, and shifting circumstances. This promotes ongoing development and pertinence in the face of environmental issues.

Chapter 3
Contingency Framework of Structural Factors for Public Participation Spaces and Consultation for Improved Governance

José G. Vargas-Hernandez
https://orcid.org/0000-0003-0938-4197
Tecnològico Nacional de Mèxico, ITS Fresnillo, Mexico

Francisco J. González-Avila
Tecnològico Nacional de Mèxico, ITSF, Mexico

Selene Castañeda-Burciaga
https://orcid.org/0000-0002-2436-308X
Universidad Politécnica de Zacatecas, Mexico

Omar Alejandro Guirette
https://orcid.org/0000-0003-1336-9475
Universidad Politècnica de Zacatecas, Mexico

Omar C. Vargas-Gonzàlez
https://orcid.org/0000-0002-6089-956X
Tecnològico Nacional de Mèxico, Cd. Guzmàn, Mexico

ABSTRACT

This study has the aim to analyse the structural factors for public participation spaces in a contingency framework for consultation and improved governance. It departs

DOI: 10.4018/979-8-3693-7001-8.ch003

from the assumption that the structural factors framed by a contingency construct can support the public participation spaces for consultation and improved governance. The method employed is the meta-analysis supported by the reflective and descriptive analysis based on the conceptual, theoretical and empirical research literature. The analysis concludes that the creation and development of some structural factors such as the infrastructure, structure, power, resources, methods, and tools framed in a contingency model contribute to enhance the public and political participation spaces and consultation for governance in any organization, community, and society.

INTRODUCTION

Participatory governance emerges as a descriptive notion more than a methodology (Mosse, 2000:32). The efficiency of public participation is being explored in some research studies (Callahan, 2007) from the standpoint of how to achieve a more effective public participation. The emerging form of public participatory governance creates new possibilities of power forms suggesting productive anticipation and implying a state both reflective and public participatory (Rogoff and Schneider, 2008: 347). The transaction costs of rational actors determine their form of governance. As transaction costs decrease, agencies tend to spend more on advisory committee. The relationship between members of advisory committees and the agency is very close.

The theoretical perspectives of the institutional participatory governance arrangements as the object of research are based on the traditional politics and government framework (Peters & Pierre 2001). Configuration of the participatory framework and the legal entity results in the founding and operating format of the organizational and institutional framework. Tendencies of change in institutional participatory governance are related to the creation of institutional frameworks attempting to meet the political interests and the needs of citizens to incorporate new forms. The regional participatory governance is supported by an analytical framework. Some sources to develop a framework of participatory governance arrangements and practices for local implementation can be used sources such as document analysis from governments, interviewing, participatory observation, knowledge generated in research institutes and the organizational sectors.

Notions of state and public participation as dominant issues in political science are on the shift from government to network and participatory governance arrangements under a relativist perspective and the conditions of unstable status of universal truths and permanence. In the perspective of the last century the state was the manifestation of the politics as the dominant entity intrinsically manifested in the will and reason, assuming contingency between nation-state, governance, and people. These actors are the central government or the state that experiences a centralist trend and instrumentally rationalistic at ordering the new change towards the implementation of participatory modes of governance.

Participatory institutions are subject to the level of public resources and authority granted to citizens leading to public debates in local political arenas based on specific cases. The informal participation is not always incited by government institutions not being selected by the organizers but self-selected or self-appointed and entering the controversy or debate from the partisan participation. Participatory learning and action implemented by networks of practitioners and researchers has contributed to shape the debates on participatory governance. A hybrid regulatory approach leads to hybrid governance responses to political debates and controversies over the different concerns and issues. The post-national debate on broader participatory governance in functional territories is in favor of institutional building through processes of reordering (Newman, 2006), rescaling the territory (Perkmann, 2007) and consolidation of institutional governance (Nelles and Durand, 2014, p. 106).

The implementation of some initiatives requires mainly to be governed by properly professional self-regulation and giving rise to public debate that may become more intensive and controversial depending on the specific situation. Dialogue must be instructive regarding the chances, ambivalences, and limits of participatory governance arrangements, such as public debates. At any public meeting supporting the public debates, and the participants who could be self-selected, the feedback forms should be available to participants to be able to express and manifest their concerns and views.

Research has identified diverse forms of public participation spanning from traditional and formal participatory arrangements to institutional participatory practices of governance. Research has a relevant role in the creation of regulation where science is at the center of regulation and political debate, although arguments are limited to political participation. Research on institutional and public participatory governance requires more focus on the identification of associations of these institutions with the well-being. Description of politics and public participation can be described in terms of shapes and cannot be used in empirical research intellectually empty-handed (Geertz 1973, p.27).

However, participatory governance focuses on involving citizens and other stakeholders in the decision-making process, and in this regard, traditional and modern models are recognized. Traditional participatory governance frameworks include local council systems, popular assemblies, and local consultation mechanisms; while current or modern frameworks can be based on digital governance and online participation platforms, citizen consultation forums, participatory budgets, among others. Regarding the above, it is possible to visualize how modern approaches allow greater inclusion, accessibility and facilitate transparency, although traditional frameworks have greater adaptability to local cultural norms; as shown in Figure 1, there are aspects that differentiate both models. Currently, there is a trend towards more modern models.

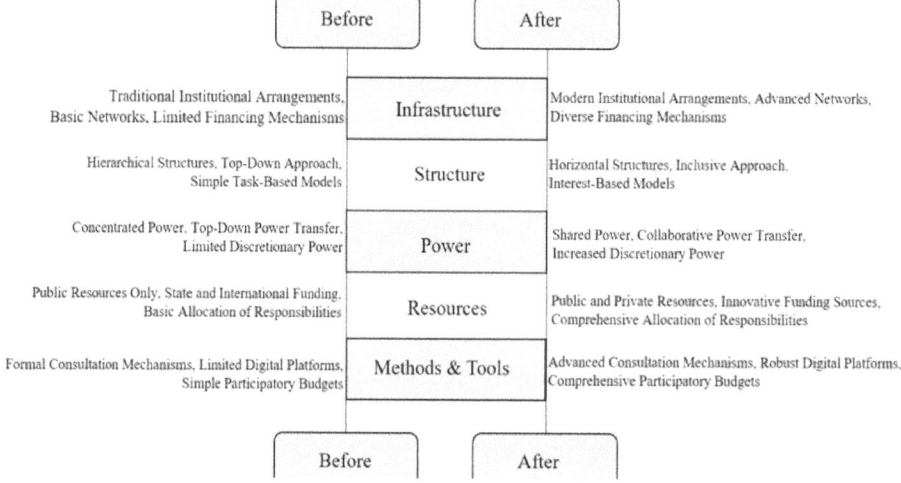

Figure 1. Description of participatory governance frameworks

Before		After
Traditional Institutional Arrangements, Basic Networks, Limited Financing Mechanisms	Infrastructure	Modern Institutional Arrangements, Advanced Networks, Diverse Financing Mechanisms
Hierarchical Structures, Top-Down Approach, Simple Task-Based Models	Structure	Horizontal Structures, Inclusive Approach, Interest-Based Models
Concentrated Power, Top-Down Power Transfer, Limited Discretionary Power	Power	Shared Power, Collaborative Power Transfer, Increased Discretionary Power
Public Resources Only, State and International Funding, Basic Allocation of Responsibilities	Resources	Public and Private Resources, Innovative Funding Sources, Comprehensive Allocation of Responsibilities
Formal Consultation Mechanisms, Limited Digital Platforms, Simple Participatory Budgets	Methods & Tools	Advanced Consultation Mechanisms, Robust Digital Platforms, Comprehensive Participatory Budgets
Before		After

This study is planned to start with the analysis of public participation followed by the political participation spaces framed by the contingency factors of infrastructure, structure and power that give support to public consultation and resources leading to concentrating on the methods and tools of public participation. Finally, some conclusive remarks are offered.

PUBLIC PARTICIPATION

The practical theory explains the components of public participation as access, standing and influence (Cronen 1995). A comprehensive decision-making process analysis must be completed to explore the necessity of public participation in the decision-making process. The institutional participatory governance extends public access to services and infrastructure to benefit the state and local communities that results in political inclusion and facing difficulties from opposition of powerful groups and bureaucracies resisting change. Citizens may have access to local government officials seeking support through clientelist exchanges of public services. Participatory governance of public services delivery raises consciousness of human rights and accessibility to the best quality of public services provided and the enhancement of accountability in the governance of public services providers, from the states and markets in their complexity of the contextual relations.

Participatory governance practices are subject to the different interests of the different actors and stakeholders. Actors and stakeholders must discuss and negotiate their different interests in participative governance that may influence their orientation and engagement in their activities as their incentives to be included in these processes. Institutional rules and regulations govern the behavior of humans which influence any choice they make individually or in organizations. A legitimate public participation governance issue of civil society, non-governmental and social organizations relate to monitoring, regulating and empowerment initiatives. When the actors have an inability to fulfill the mandates, they are perceived by the citizens that have limited power to influence government in concerned issues such as the monitoring role of the environment.

The different uses of public participation have an influence on the construction of citizenship, membership in the communities, beneficiaries, users, customers, etc., in dissimilar and somewhat contradictory postulations framed by a wide range of theoretical backgrounds of participation. The territorial governance and self-governance participation is achieved thorough the inclusion of people as active citizenship. Public participation in the production of knowledge can influence research by the distribution of funding giving priority to some specific issues, which may be new such as participation in research of environmental organizations following formal and informal rules and regulations established by government organizations and funded by interested stakeholders. The securing of funding means provide motivation for the involved actors and stakeholders in the processes of participatory governance to resolve the problems and maintain the status quo by increasing. More independent organizations tend to be trendsetters in applying international funding for their activities considering that institutions have applied when available.

Public participation and participatory governance are dependent on mutual trust and horizontal social and political structure among all the actors and participants (Gustafson and Hertting, 2017). Citizens may participate depending on the levels of trust and sustain mobilization of new processes. Governance refers to any regime of practices and participatory governance denotes an emergent, fluid, and iterative governance performance among various actors and stakeholders that may take several shapes identified in relation to the state and to align frames to practices aimed at delivering problem definitions and solutions, generate trust, and negotiate the legitimation and effectiveness of rules. Participatory governance is meaningful to elicit public legitimacy of processes and the involved inclusive, trustworthy, egalitarian, accountable and transparent parties.

The participatory governance orientations and processes are relevant through the established relationships and interactions among the different actors and stakeholders who learn from each other and build mutual trust. Social and political horizontal trust among the different participant actors is linked to public participation and participatory governance (Gustafson and Hertting, 2017). Horizontal trust is referred to the trust among the citizens in society leading to building commitment, responsibility, and public participation. Institutional participatory governance enhances mutual trust, which limits the negative effects of transaction cost. Institutional embeddedness and interactive relationships between the state agencies and nonstate stakeholders improve their mutual trust and limit their negative effects of transaction costs (Warner, 2006; Willem &Lucidarme, 2014). The new forms of participatory governance arrangements attempt to regenerate trust.

The stakeholders involved in the public sector may misunderstand the institutional participatory governance and distrust the civil sector that only find problems and neither discern the solutions nor attempt to perceive them. Political actors and bureaucrats cannot trust any mechanism of participatory governance to predict what is the best form of participation to take. International and state agencies may hand over authority despite the mistrust, conflict and social divisions creating difficulties for the participatory institutions. The expansion of state legitimacy and authority is a positive sum gain because citizens and government officials must learn to trust each other through their interactions and in relation to other state activities. Local authorities support initiatives secure by funds of the state for participatory governance corroborating the need for trust between the public, private and social sectors which can act on their own best interests and shape the implementation of the institutional participatory governance to the use of public resources and respond to the needs of long-term cooperation.

The public participation patterns are related to distinct publics and their visions affected and oriented by state-driven initiatives. Actors are the public as they are politically active, although in different settings and moments may not be (Verhoeven, 2004). The public participation process meets the needs of all participants involved and communicates their interests. The spontaneous public participation is engaged and concerned publics. The public participation process facilitates the involvement of affected actors and stakeholders.

Individual participants with no particular interest in specific governance issues or are not spoiled by partisan ideologies and views and open rational education of youths. These participants are pure publics as an alternative of engaged publics and issue publics. The under-termination problem of voluntary and involuntary participants of public in participative governance is more practical based on empirical solutions rather than ideological of actor groups which must be observed to become manifest in practices focused on people to be identified themselves in specific cases related to the issues at stake.

The public participation of citizens in decision making processes might be more rational to improve participatory governance (Hertting and Klijn, 2018). Civil society organizations have a significant impact on participatory governance publics before the adoption of participatory institutions. Concerned publics may be excluded from social movements. Participation in knowledge production involves more than the stakeholders for broader social changes in egalitarian societies and the opportunities for participatory governance are increasing. Some of the participatory governance issues are transnational in character and require the involvement of various publics and participating actors with different cultural backgrounds and codes which are relevant in the ordering and governing of issues.

The emergence of other actors, contingent multifaceted publics and transient are facing issues at stake in a dicentric modes of ordering. Participatory governance institutions contribute and are built upon the growth of publics and the emergence of new publics. The emergence of these public spaces for public involvement and participation are sites of issue and engaged publics for formal public participatory governance arrangements for enhancing the political discourses on civic participation and participatory governance. Public participatory governance may emerge in different types and forms involving citizens, government officials, experts, and other new actors and stakeholders that do not belong to the traditional state, civil society, etc. However, traditional institutions of local government and community, councils, civil associations, and organizations tend to be biased towards institutional participatory governance in the local context.

Political Participation Spaces

Another level of participation is political participation mixed with scientific research initiative in knowledge production. Political participation in knowledge production is a form of participation that may be state initiated but rather may emerge from civil society and participants have not been invited and selected by government institutions or by any other organizations. Actors in this form of participation tend to be self-selected and self-appointed entering the debates with their respective personal views. Public involvement and political participation take place at unexpected, politicized sites and more civil society-led rather than state-led characterized by antagonistic and adversarial arguments. Civil society-led participation set up by state institutions in response to informal political participation. It is a formal participatory governance arrangement to mediate adversarial public involvement. Coherent and proactive political participation yields more success than half-hearted, contradictory, and delayed participation approaches.

Another form of participation is the creation of political spaces designated as public arenas for public involvement on direct civic and political participation leading to participatory governance. Public participation spaces broaden the base of the involved participants to understand excluded people from local governments in their concerns and issues that require advocacy pressing governments to enact commitments to their constituents. Some hearings can be a hybrid space for participatory knowledge production and a powerful act of political participation. However, there are some other cases where the boundaries between political participation and knowledge production are blurred. Other forms of political participation in funded research and where citizens may participate are often crucial to influence other areas and engage in new forms of participatory governance. Participation in knowledge production is a powerful form of political participation. Other types of participatory governance practices based on the orthodox notion of political participation require rethinking the political space.

Events causing situations of institutional ambiguity impelling re-ordering trigger the emergence of new initiatives for political spaces to frame the rules of the game and the roles of the involved actors in public participatory governance arrangements. The polity is disconnected from the territorial space under the assumptions that the actors from a political community governance arrangement operating in unstable and erratic network configurations. Formal governmental institutions may be the wrong place for politics and sub politics knowledge, as the elaborate notion of displacement of politics in the network society, which according to Castells, (1997) the functions are organized pertaining to a space of flows linking them up around the world, fragmenting people in multiple space of places with subordinate functions and locals disconnected and segregated from each other.

Public participation is questioned to be a right of citizens and shared experiences stemming from being of active social movements and groups. However, public participatory governance arrangements are a hegemonic tool for active citizenship to get consensus on urban spaces restricted by some economic actors but preferred to be a wider urban society. Different groups have different expectations from public political participation in formal spaces. Civil society and social organizations as mediators between citizens, communities and government officials are not always doing well and become dysfunctional and questioning who may participate in invited and formal spaces of participatory governance. It is required strong actors of civil society organizations working in social and public spaces achieving their own interests but willing to work with other local government officials, civil society organizations, other community actors and activists that have a potential impact on participatory institutions.

Local communities are the political participatory space where social relations among a mass of people take place and where the empowerment of transformation struggles for institutional change of rules governing the relations and determining allocation of resources, entitlements, and property rights, etc. Stakeholders, and broader-based interested parties and other affected groupings, such as the environmental movements, are invited and included to participate but partially represented in the structures of governance spaces.

Citizens engage in different forms with government to open new spaces for public participation and demand the right of information. New political spaces are emerging as an attempt of governance related to political institutions and formal participatory governance arrangements. The new political spaces are sites for political participation and judgment and the formation in public issues and concerns. The creation of political spaces may unexpectedly turn into new politicized public spaces to debate about the facts mingled with other political arguments. Civic interaction as the basis of participatory governance separates politics and science, matter of concern and facts (Latour and Woolgar, 1979) leading into informal public participation in forums, while some other relevant debates suddenly can become highly politicized and open more formal political spaces for informal public participation. These situations of civic dislocation, institutional ambiguity and the subsequent effects and impact on new political spaces for public participatory governance may occur as the result of sudden disruptive events and pervasive perceptions of alienation and skepticism (Jasanoff 2004a, 2004b).

Emerging forms of participatory governance practices may develop the application forms of organizational and individual cooperation for sharing the governance responsibilities, the establishment of public-private-civil partnerships and the common use of resources. Public participation as a principle of governance is an approach for the establishment of an innovative hybrid model of institutional participatory governance based on public-private-civil partnerships. The public participatory process articulates the challenges of the actors and stakeholders to institutionalize the governance practices through the establishment of public-private-civil partnerships including the civil society organizations.

Public-civil partnership forms are the practice of common governance as the prominent characteristic of the common good (Ostrom, 2006) with strong local authority and community participation. Local authorities are characterized by highly politicized and cronyism that is adverse and rejects the implementation of more inclusive, equitable, open, and sustainable forms of governance. Community participation is the crucial element of institutional governance to enabling the needs and interests through the implementation of concrete programs. Local administration may consider the public-civil partnership in natural resources participatory governance a cooperation mechanism with stakeholders and actors in achieving goals

Institutional participatory institutions are engaged with poor citizens who are constrained in public participation due to structural and motivational problems. Public participatory governance is the progressive realignment of institutions protecting and favoring the excluded and poor people. From the demand side of public participatory governance processes the weaknesses of the accountability approach become clearer in cases of social and gender exclusions that challenge the institutional norms of service delivery. However, if the civil social organizations are weak, the participatory governance is limited due to uncapable partners with local government officials and international funding agencies. Government at all levels and international agencies create or change structures and their relationships for public participatory governance to allow that happens with poor groups (Mittling, 2004). Public participatory institutions have at their core to determine the use of scarce public resources and goods to find solutions to large percentage of population living in underdeveloped and poor communities.

Organizational structure is the causal condition of regional governance and other organizational factors that may foster participative governance (Perkmann, 2007; Medeiros, 2011; Knippschild, 2011). Participative governance assumes that apply to institutions with determined organizational structure and the degree of participation in governance is higher if the organizational set-up of institutions is elaborated.

The emerging coalitions supporting disadvantaged groupings do not always achieve effective outcomes out of participation processes, a reason why the inclusion of an independent third party as an impartial, objective, and integral part of public participation processes. Dynamics of active institutional participatory governance focuses on a society that is increasingly sliding towards higher inequalities and sharp divisions. Most of the minority and disadvantaged groups enter the institutional participative governance without any kind of ideological and political experience, training skills, etc.

Moreover, the inclusion of an independent party may provide the countervailing power needed to enable and empower marginalized groups. Fragmentation of a governing political coalition facilitates and privileges relationships with other different interests of urban society and the use of institutional resources. Human, financial, technical, and material resources are relevant to establish the coordination and administration of the participatory governance model assuming the perception of all the interests in compliance with the rules of the game. The participation of civil society favors the organization inhering financial and legal resources.

The discussion of impacts produced by participatory governance institutions regarding the potential and possibilities to distribute more equitably the public resources, create and inculcate values and foster accountability, which generate high expectations for outcomes. The outcomes of engagements in participatory governance include reductions in malpractices and corruption, improvement of political behavior, changes in spending, redistribution, and reallocation of resources (Navarro 1998:68). Individuals and groups make choices related to institutions that results in the combination of resources and rules (Alsop et al, 2006; Giddens, 1993).

Citizens invest resources, energy and time in participatory institutions seeking to collaborate to improve the quality of life. Institutional participatory governance is a relevant topic to support the conditions for advanced collaborations based on values of freedom, parity, and equality. Participatory governance may ensure equal participation and all involved actors and stakeholders who may share equitable responsibilities. An imbalance between investments in non-institutional and institutional participatory governance may have dysfunctional and negative effects on the provision and allocations of public resources delimited by the tension of a political grip. Conflict is always a component present in institutional participatory governance programs because the different interests of stakeholders and citizens need to determine the allocation of authority levels and scarce resources to be exercised.

INFRASTRUCTURE, STRUCTURE, AND POWER

Institutional participatory governance infrastructure is the ground to address the strengths and deficits through developing trust and cooperation and creativity encouraging public participation (Obuljen et al., 2014). Local participatory governance infrastructure consists of institutional arrangements established and supported by local government and communities, civil organizations and associations, networks of local and international organizations and citizens. This approach requires the establishment of new means of financing public infrastructure for participatory governance.

The historical institutional design is divided into arrangements, they are classified and structured in schemes that modern institutions use to manage and administer the different areas that deal with the lives of humans. The contextures are the instantiated structures that have an influence on the governance practices (Lynch 1991). Design of the organizational structure model for participatory governance is subject to the available resources and dependent on the institutional arrangements considering the best practices applicable to the situations. Usually, institutional participatory governance models are based on tasks where all the stakeholders participate. The interest-based participatory governance model reduces the hierarchical and hegemonic structural tendencies of institutions leading to more horizontal participative approach. This type of models of local participatory governance and decision-making bring both benefits and challenges in structure and processes to improve collaboration and efficiency of public services.

The implementation of this model of participatory decision making and governance has both benefits and challenges in structures and processes of collaboration and efficiency in public service. The efficiency of governance is determined by transaction costs and the form of participatory governance is of low transaction costs and hierarchical structures of governance have higher transaction costs (Williamson, 1985; Williamson, 1991, 1996). Practices of participatory governance may be identified in situations of institutional ambiguity with a dominant discursive structure to reorder disruptions of subject matters at issues dealing with legitimation and effectiveness. Dislocation is the emergence of a series of events that cannot be represented, symbolized, or domesticated by the disruption of discursive structure (Laclau 1990, p. 41). Communication structures from classical-modernist institutions to the citizens are quite dysfunctional and deficient. Emotions can be structured and canalized to develop participatory governance practices.

An emergent innovative model of institutional participatory governance can be useful an assessment of structure and process.

The broader contextual structure of institutional participatory governance programs together with the structural factors interact with the different individual participatory governance forms. The synergistic conjoint maximize the strengths, commitment, and energy of organic induced participation with the structure and authority. Institutions are persistent because they are constrained by social and political structures of power. Concentrated administrative power in one hand could follow manipulation and therapy, but not the tools of public participation.

Power sharing is essential to enable participatory governance. Transfer of power from the top to the bottom, and from the governor to the governed requires sharing of the power. Macro governance is linked to micro-governance into a more localized governance arrangement characterized by discretionary power. The participatory engagement logic increases the involvement and knowledge on political interests and the empowerment of citizens (Rogers and Weber, 2010; Pateman 2012). Knowledge may be constructed in practice in laboratories using participant-observer methods (Latour & Woolgar 1979, and Latour 1987, 1988) and contextualized with the social phenomena formed by elements mutually constituted and transformed, such as power, institutions, agency, knowledge, etc.

Participatory processes engage and change relationships of power. Participation is linked to a larger process of changing the power relations for social and political transformations. Transformation of institutional participatory governance functions is remedied by operational functioning and established by the different actors and stakeholders involved in the non-institutional and institutional existing positions. A range of approaches are used to strengthen participatory governance and policy, to analyze relationships, to strengthen the voices and power of citizens. The institutional reforms have centered on issues of local governance, citizen engagement, community mobilization and decentralization of delivery services which have been able to transform local decision-making of participatory initiatives aimed to empower the marginal. In times of crisis, participatory institutions tend to adopt constitutional reforms embedded in political agendas based on the public and private interests of designers. The impact of institutional participatory governance reforms must be in all areas of governmental effectiveness.

Organizational reforms lead to innovation forms supported by participation and an improvement of transparency in some specific fields such as food safety control. Institutional reforms focus on local governance, transparency, social accountability, citizen engagement, decentralization of services delivery and community mobilization. Concerns about the issues of nature and environmental sustainability are relevant causes that have triggered the creation of new political constituencies, political parties and non-governmental organizations leading to a political reform.

The empowerment of new actors may create and adopt a diversity of institutional participatory governance. Participation moves beyond the empowerment of citizens to small groups and towards the circles of elite non-governmental organizations, parties, etc. Empowerment of social organizations and inclusion of consumers in expert and advisory committees, are specific cases of involvement of actors in participatory knowledge production that may have a positive political significance and performance impact. Citizen's participation in knowledge production may have powerful effects related to political environment and public forums (Nowotny, Scott and Gibbons, 2001). Popular participation can produce a process that weakens rival regional government powers and strengthens local governments. An institutional void around specific governance of issues is problematic and always leads to conflicts between the dislocations of power and contradictory imperatives with other events.

Decentralization reforms improve governance and citizen engagement is most effective in removing the barriers for community meaningful participation. Institutional arrangements of local public participation and participatory governance require decentralized political structures with authority supported with autonomy and dependence to control political capacities, financial sources, and resources closer to citizens (Wampler, 2012). A comprehensive decentralization reform transfers powers to more regional and local government levels to empower citizens, devolve political power and formalize the participation of civil society. Local government control must be strong in relation to local government autonomy (Sellers & Litström, 2007).

Decentralization is a dynamic process to transfer public resources and services from central authorities and responsibilities to lower levels implying access and connectivity to governance models. Participatory governance defines the sharing of governance responsibilities among the stakeholders involved (Wilcox, 1994: 5) who can be the public and local administrations, institutions, local communities, civil initiatives, non-governmental organizations, etc. However, instable processes of decentralization and hybrid forms of participatory governance may lead to uncertain outcomes. Shifting towards more comprehensive conservation as a response to uncertainty posing some unexpected participatory governance challenges for public administration at all government levels. Among other benefits, public administration improves the quality of decisions and building of consensus, maintain legitimacy and credibility minimizes costs, anticipate public attitudes and concerns, avoid confrontations and conflicts, contribute to develop civil society, etc. (Creighton 2005).

PUBLIC CONSULTATION AND RESOURCES

Induced organic participation of citizens is a requisite for participatory governance arrangements filling the gap in effective institutional public participation practices emphasizing the governance reforms. Concerned citizens use different public spaces than the ones designed for public participation to attract attention to their cause, such as supermarkets, halls, etc. In the long run, among some forms of public participation, public consultation and specialized mechanisms of participatory governance may exist a trade-off limited to single issues, may provide sustainable results. Scientific advisory committees are forums open to public participation.

Creating and advocating an institutional participative governance model supported by an umbrella of participating associations and organizations must be different. A model of participatory governance has a relevant element in the allocation of responsibility. An institutional model of participative governance may be instigated to perform activities, and additional benefits must be proportionated to the additional benefits. This model of governance based on consultations brings decision making on project implementation close to citizens besides levering influence and power, including the networks to have access to information and knowledge. The incorporation of public opinion combined with public consultation and public incorporation inviting to participate and increase the access to the processes.

The notions of consultation and participation must be based on a double-sided approach. Formal consultation is a new technique of governance initiated by governments, in contrast to spontaneous and informal participation of citizens and groups proactively engaged with political spaces and local government. In the design of the framework in which the content model of institutional participatory governance through methods of public consultation, direct engagement, and collaboration in the community and civil, society organizations, citizens, and other stakeholders. Formal public consultation initiatives and controlled exercises in the form of contracted citizen workshops and consensus conferences with assignments and work attached to be done could be used as an element of manipulated engineering.

Through consultations citizens clearly express their needs and desires to be implemented in a project. Institutional participatory governance arrangements and initiatives, such as the discourse intensification used in consensus conferences and consultation processes. Through consultations, citizens may clearly express their desires to support project proposals ready to be implemented and being able to deliver structured, coherent, and integrated projects.

Formal public consultations can replace spontaneous participation to obtain better control of outcome. Economic actors should be included in consultative boards of the city coexisting with political actors in diverse participatory practices. The advantage of this institutional approach to participatory governance enables us to establish consultation relationships with the region, city, community, neighborhood, etc. All these institutions within the participatory governance framework toward the creation of stable institutional forms building the relations and considering they are the relations themselves. For example, city regulations should provide a framework for the inclusive participation and consultation processes of citizens in governance and urban transformation embodying compatibility with the principles of participatory governance.

To modernize urban renewal typologies to include a wide array of formal and informal institutions, the narrow scope of stakeholders and organizations involved in governance schemes and initiatives that resonate with participatory governance model designs. Lower-level institutions of the city administration can provide consultation, regulations, and some incentives as actions to establish and enhance participatory practices.

The institutional void of state is the public participatory governance practice (Hajer, Underhill 2003) reflects on new forms of socio-political association (Fischer, 2006: 20). A state may adapt and implement public participation more successfully than others emphasizing the necessity to balance interest between administrators and the citizens. The silent contract between citizens and political institutions of public participative governance, in cases of ambiguity, the scientific expertise broken and the state license to operate is withdrawn. The emergence of state initiatives supports some forms of public participation (Akkerman et al 2004). However, public involvement and public participation cannot be confined to state-initiated arrangements. Traditional participatory governance asserts the existence of a conflict between state agencies and non-state stakeholders which may result in collaboration relationships between different interests (Ansell, 2011).

Service providers have difficulties in changing state structures to be more accountable to social groups and citizens and empower poor and excluded groups. The inter-institutional dimension focuses on state and government organizations. Competing authorities in the structures may impede the different forms of public participation and organizational and institutional autonomy. Local participatory governance requires a high level of autonomy from local authorities and civil associations and organizations in a community performing the functions supporting the notion. The actions have different effects on the organizations in the transfer of competences and resources in the participatory governance processes.

The inclusion of citizens in the institutional and public participatory governance involves them in constant contact with politicians and government officials in a state-sanctioned institution. Construction of citizen engagement strengthens public participation, responsiveness and accountability in the states and develops a more cohesive and inclusive society (Gaventa & Barrett 2010). Citizen participation leads to positive results and outcomes. Public participatory governance institutions are different from other types of public demand, making government officials and citizens have allocated authority and different responsibilities within the state. Clarity of the rules of the game and norms for the involved parties to determine who has the authority, responsible and accountable of problematic situations related to institutional public participatory governance. When this situation occurs, it is called a civic dislocation expressed as a mismatch of the services to the public offered by institutions of government and the real services delivered (Jasanoff 1997) creating conflicts and causing citizens to disengage from the state.

The non-state-initiated participation may emerge from civil society. Public participatory governance practices reiterate the shift from the state role and activities to the civil society organizations. Between the actions and practices of the state and civil society engagement with public participatory institutions, the process produces the expected and desired outcomes. Participation and public involvement are of an antagonistic nature, more civil-society led than state-led and takes place at non-expected politicized spaces, characterized by adversarial arguments, controversies, and struggles. Organizations of civil society struggling against the institutions of the authoritarian state have the challenge to broaden and deepen public participatory spaces and engage with the state and institutionalize local governance.

Institutional and public participatory governance outcomes are related to the degree of state fragmentation at the levels of regional, horizontal, vertical, and longitudinal. Any shift across vertical levels over time, from agency to agency and within regions. Public participatory institutions may work very well in one state while they may not work in other states. The institutional regional cooperation among state agencies, local governments and communities must be constituted by the regions involved on cross regional territorial governance based on symmetric multilevel governance where the state organization affirms its institutional set up (Ferreira, 2015; Lago, 2015). Resources available to local governments are a component public participatory institution leading to what these governments can deliver goods and services (Boulding & Wampler 2010) that states, and communities need these resources to implement public infrastructure and works.

The formal public participatory governance state-initiated arrangements are more sophisticated in practice and far from uniform between the nation-states as an acceptable model of governance (Fox & Miller 1996:3). Formal state institutions are responding to the new governance practices and new forms of ordering and administering. Formal and arranged public participation and participatory governance is not limited to state-led cases subject to political reality. Formal organization of public participation of both state and non-state initiatives of actors, such as forums initiated by the corporate sector, non-governmental organizations and governmental institutions focusing on participatory governance, may be the objects subject to empirical research. Participatory governance supports the public participation achieved by government officials and citizens who are urged to become more responsible. Publics and other actors get involved in incidents of state-initiated and non-state initiated in moments that may be different to official public participation practices in different but related types of arrangements.

METHODS AND TOOLS OF PUBLIC PARTICIPATION

The methods and tools of an active public participation facilitates the participatory governance. Local governments involved in participatory governance implementation have obligations on the improvement of public participation meaning, and evolving methods, tools, processes, and mechanisms (SKL, 2009). Local authorities may not have different features in governance and participatory practices, forms, methods, regulations, and participative inclusion. New participative methods of governance are being invested and experimented to connect the stakeholders involved in the new institutional participatory governance models.

Participatory governance in institutional arrangements, according to Engl (2013) has the features of public participation at different levels, vertical and diagonal governance, different political institutions in governmental and administrative levels, other public actors, non-public and civil society actors, and ethnic and linguistic diversity. Regulations used in governance provide impartial answers to government officials and public administrators in settling disputes and provide arguments for the public debates beyond the interests of the stakeholders and conceals practice of participatory governance.

Participants to have meaningful participation in participation forums require comprehensive information and unbiased reports, information, and data easily to collect, maintain and understand. The administration uses information in unidirectional way and communication is used bidirectional are not enough to enabling participation as a two-way method, even though it is a double-sided notion. Large scale institutions are applying top-down participatory approaches in mechanistic and rigid, unimaginative ways. If a passive and single-sided process occurs such as the case of information, this is not considered public participation. Public information cannot be accepted as public participation because it is a single-sided communication which makes up a step towards the participation process (Creighton, 2005). Public participation is more than information, adopting a wide perspective on the guidance role regarding involvement in continuous participation. The public participation process can provide all the participants with the required information to become involved in participation in a meaningful way.

The existing mechanisms of participatory governance used for strengthening engagement between citizens and governments were reviewed by Goetz & Gaventa (2001). Government institutions and agencies may use formal participatory mechanisms of governance to complement other mechanisms of participation. Formal participation is a mechanism of participation that complement other mechanisms. Formal participation is only one mechanism that may not be able to substitute spontaneous and informal participation. Formal mechanisms of participatory arrangements of governance have their own limitations, uncertainty, and dynamics. Government agencies need to develop novel mechanisms and tools to deal with uncertainty in responsible commitment. Uncertainty does not always reduce the calculated risks. Local governments and agencies must deal with uncertainty and calculable risk.

The critical governance practices identified by the actors as the public concerns and issues of uncertain, uncontrollable, and complex nature that is reflected by elusive and multi-faced public that require new modes of governance built on the knowledge to deal with unruly issues. The new forms of participatory governance arrangements are far-reaching, technical, widespread, and value-laden nature for knowledgeable citizens about the issues and concerns and informed about complexities and uncertainties leading to multiple solutions beyond the dichotomy of acceptance or rejection.

Participatory governance mechanisms are valued as part of solutions to institutional failures. Citizen engagement is a necessary factor for monitoring and evaluation framework supported by the purpose, process, people, context, and outcomes, according to the Motsi's evaluation framework. The model of good governance must be fair for people directly involved to govern innovatively. The outcomes of any procedural participatory governance institutions must be determined by the designers of the program and measured by the changes of behavior of the actors involved. Responses to regulatory issues in pragmatic terms include some measures that may defy the calculation of risks. Participatory governance technological assessment may be undertaken to support political opinion forming and judgement on any governance issues in the public domains.

Institutional participatory governance analyzed with a framework has some advantages and challenges in indicators such as efficiency, participation, accountability, and transparency. Institutionalizing the participatory governance model has some advantages and challenges when analyzed within a governance framework using the indicators of transparency, efficiency, and participation to assess the quality of projects submitted to the public service. The timeline may be difficult for the assessment of efficiency but facilitates the assessment of the performance of transparency with the involved citizens. Changes in dimensions of government must offer indicators in efficiency, efficacy, and responsiveness (Putnam 1993; Grindle 2007). However, transparency, accountability and participation can be improved.

Institutional participatory governance has limited benefits from government responsiveness, transparency and accountability remaining as a challenge for many localities and communities. Donors and non-governmental organizations tend to invest heavily in the demand side of civil society and citizen's participatory governance considering the complexity and scale of building local accountability and holding local government to account (Gaventa 2006). An integrated participatory governance accountability model links institutional change across the dimensions of society and governance. The demand of public services delivery is more critical in terms of accountability for improving the governance and transforming social and gender relationships. The operationalization of a participatory governance of public services model is more related to co-governance of the interrelationships between the public and private spheres involved in the supply and demand processes (Gaventa 2006: 23). In this operationalization of participatory governance of public services delivery in the content, context, and accountability, is mediated by social and gender issues.

Citizens are motivated to participate for the benefit of having a tangible impact on quality of their lives, which leads to evaluate performance of participatory institutions (Figure 2). Participatory institutions can have measured the impact of public goods, services and works produced by local governments and international funding agencies. However, measuring the impact of participatory governance institutions on social well-being is difficult due to the methodological approach to establish causality and time frame. Despite the difficulties in measuring social wellbeing, it is conceived as part of the impact on the outcomes.

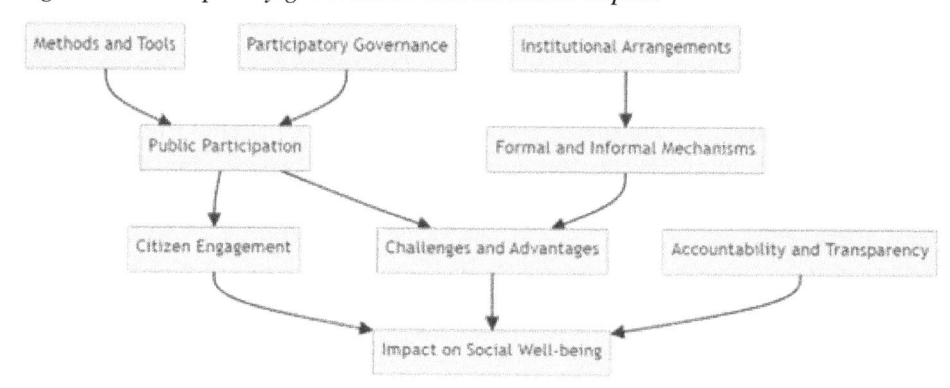

Figure 2. Participatory governance and its social impact

PARTICIPATORY GOVERNANCE: SUCCESS CASES

The analysis of case studies that have used governance includes research in different contexts, such as the study by Pabón (2024), which describes the role of participatory governance in a critical scenario, in which a group of citizens took action to mitigate the environmental impact of the waters of the Bogotá River, getting the government to issue a ruling to delegate responsibilities to those involved and adopt measures to recover the river.

Likewise, another study that takes up the role of participatory governance during a period of crisis is Rofman et al. (2024), who conducted a study on the COVID-19 pandemic in Buenos Aires, since this episode caused the management procedures of urban social policies to be altered. The authors argue that the participatory governance mechanisms of urban social policies were maintained and strengthened during the local management of the pandemic crisis.

Participatory governance also has implications for public management as a tool for sustainable development, as described by Alcivar-Falcones and Miranda-Pichucho (2024), who conducted a study that analyzes this issue. Their findings reflect the incorporation of the concept of networked citizens, agreements from transparency and institutional governance. The authors propose an institutional governance that facilitates citizen participation focused on the local interests of society and allows sustainable development.

Another study that has addressed the issue is that of Palacios et al. (2021), who analyzed public policies and local participatory governance from the point of view of the collaborators in a municipality in Lima. The main results emphasize the need for governance based on citizen participation, involving both public and private participants, as well as civil society, who are aware of the needs of citizens, from which public policies with a positive impact on the population are constructed.

Another research that has addressed participatory governance is the one developed by Mérida and Telleria (2021), who analyzed the transformations in the modes of governance from the implementation of new mechanisms of participation in Spanish city councils. Their results showed how participatory policies modified the pre-existing governance model, with which the new processes allowed the opening of new networks of actors.

On the other hand, participatory governance can also be analyzed from the perspective of environmental governance. In this regard, the study conducted by Munévar-Quintero et al. (2023) stands out; they analyzed the initiatives of the administrative acts that declared the management categories of the areas belonging to the National System of Protected Areas in the Eje Cafetero ecoregion in Colombia. Their findings highlight that participation is an essential criterion in the exercise of environmental governance for the conformation of protected areas.

Within this same approach to participatory governance related to environmental issues is the research carried out by Iñiguez-Gallardo and López-Rodríguez (2024), who analyzed participatory management models in mangroves in Ecuador. The authors point out that multilevel and climate governance has an impact on the protection and management of mangroves, highlighting the agreements that were developed as an important legal instrument for the conservation and sustainable use of ecosystems.

Another area of study of participatory governance has been within the education sector. (2023), they analyzed participatory governance in the implementation of Flexible Educational Models in a Colombian educational institution. Their results focused on generating specific strategies to improve educational quality and offer recommendations to strengthen participatory governance, with the purpose of making timely decisions in the educational field.

According to the different studies, the scope of participatory governance in different contexts is recognized as an area of opportunity to integrate a comprehensive approach to governance.

CONCLUSION

The analysis concludes that the creation and development of some structural factors such as the infrastructure, structure, power, resources, methods, and tools framed in a contingency model contribute to enhance the public and political participation spaces and consultation for governance in any organization, community, and society. The analysis of the field of governance is a new challenge for citizen participation and generation of trust in the changing role of civic public participation. Construction of pure publics for participatory governance arrangements requires more reflexivity in relation to the links with engaged publics recruited, constituted, and selected as a precondition and result shaped by processes of public participatory governance arrangements.

Public participatory institutions would not have an impact in situations of ineffective states categorized as absent or failing. The states with more successful public participatory institutional experiences are effective and robust states in middle income developing countries. The state has been mirrored as a unified political space that is gradually losing its power. Weak local governments tend to lower their expectations regarding the outcomes. The urgency of modes of public participatory governance may result in institutional ambiguity in a decentralized state and the reordering the issues at stake that entail the state and non-state actors' participation, such as the emerging frames of environmental concerns in the formation of geopolitical constellations.

Institutional experimentation is conducted on traditional and classical mechanisms of governance related to the politics are not viable concerning issues in both environmental interpretations and somatic of human identity, individual and collective responsibility. The sense of community for collective action enhances the institutional participative governance arrangements. The experiments involving citizens with participatory governance have triggered several and varied forms subject to the issues and concerns, such as the case of global warming where the costs of regulation in future risks defy the traditional regulation approaches.

Communities for capable governments and active civil societies can support institutional governance institutions than communities with weak states and less active civil societies. The administrative and governing bodies must prevent the monopolistic control of power in the hands of the few guarantees to promote and maintain plural participation. Government agencies and institutions may have formal mechanisms of participatory governance arrangements and look at informal mechanisms as complementary forms of participation. Government agencies tend to use more participatory governance under low transaction costs.

The participatory governance framework deserves further research to ascertain if formal rules can concretize in real experience and practice. Cross national and cross regional studies on institutional participatory governance are difficult and complicated and requires more expansive research projects and agenda beyond reliance and variation of types of programs.

REFERENCES

Akkerman, T., Hajer, M., & Grin, J. (2004). The Interactive State: Democratisation from Above? *Political Studies*, 52(1), 82–95. DOI: 10.1111/j.1467-9248.2004.00465.x

Alcivar-Falcones, J., & Miranda-Pichucho, F. (2024). Contributions of Participatory Governance in Public Management. An Opportunity for Sustainable Local Development. *Digital Publisher CEIT*, 9(1), 746–761. DOI: 10.33386/593dp.2024.1.2237

Alsop, R., Bertelsen, M., & Holland, J. (2006). *Empowerment in Practice: From Analysis to Implementation*. World Bank.

Ansell, C. (2011). *Pragmatist Governance: Re-Imagining Institutions and Democracy*. Oup Usa.

Boulding, C., & Wampler, B. (2010). Voice, Votes, and Resources: Evaluating the Effect of Participatory Democracy on Well-being. *World Development*, 38(1), 125–135. DOI: 10.1016/j.worlddev.2009.05.002

Callahan, K. (2007). *Elements of Effective Governance*. Taylor&Francis Group CRC Press.

Castells, M. (1997). The Information Age. Economy, Society and Culture.: Vol. II. *Power and Identity*. Blackwell.

Creighton, J. L. (2005). *The Public Participation Handbook: Making Better Decisions Through Citizen Involvement*. Jossey-Bass.

Cronen, V. E. (1995). Practical Theory and The Tasks Ahead for Social Approaches to Communication. In Leeds-Hurwitz, L. (Ed.), *Social Approaches to Communication* (pp. 217–242). Guilford Press.

Engl, (2013). *Partizipative Governance, 130ff .85. Committee of the Regions, EGTC Monitoring Report 2013*, 99.EEF379.indb174EEF 379.indb 17408/11/201601:0908/11/2016

Ferreira, G. (2015) Interview by Peter Ulrich, Porto (Portugal), 26th February 2015.

Fischer, F. (2006). Participatory Governance as Deliberative Empowerment. The Cultural politics of Discursive Space. *American Review of Public Administration*, 36(1), 19–40. DOI: 10.1177/0275074005282582

Fox, C. J., & Miller, H. T. (1996). *Postmodern public administration: Towards discourse*. Thousand Oaks (etc.). *Sage (Atlanta, Ga.)*.

Gaventa, J. (2006). Triumph, Deficit or Contestation? Deepening the" Deepening Democracy" Debate, *Working Paper 264*, Brighton, Institute of Development Studies.

Gaventa J and G Barrett, (2010) *So What Difference Does It Make? Mapping the Outcomes of Citizen Engagement*, Development Research Centre: Citizenship, Participation and Accountability, London, DFID.

Geertz, C. (1973). *The Interpretation of Culture*. Basic Books.

Giddens, A. (1993). Profiles and Critiques in Social Theory, in Cassel P. (ed), 1993. *The Giddens Reader*, London, Macmillan, pp 218-225.

Goetz, A. M., & Gaventa, J. (2001). FromConsultation to Influence: Bringing citizenvoice and client focus into service delivery. *IDS Working Paper*, •••, 138.

Grindle, M. (2007). *Going Local: Decentralization, Participation, and the Promise of Good Governance*. Princeton University Press.

Gustafson, P., & Hertting, N. (2017). Understanding Participatory Governance: An Analysis of Participants' Motives for Participation. *American Review of Public Administration*, 47(5), 538–549. DOI: 10.1177/0275074015626298

Hajer, M., & Underhill, G. (2003) Rethinking Politics: Transnational Society, Network Interaction, and Democratic Governance. Working document, University of Amsterdam.

Hertting, N., & Klijn, E. H. (2018). Institutionalization of Local Participatory Governance in France, the Netherlands, and Sweden: Three Arguments Reconsidered. N. Hertting and C. Kugelberg (Eds.) *Local Participatory Governance and Representative Democracy*. New York and London, Routledge.

Iñiguez-Gallardo, V., & López-Rodríguez, F. (2024). Gobernanza participativa para manglares en Ecuador. *Madera y Bosques*, 30(4), e3042612. DOI: 10.21829/myb.2024.3042612

Jasanoff, S. (1997). Civilization and Madness: The Great BSE Scare of 1996. *Public Understanding of Science (Bristol, England)*, 6(3), 221–232. DOI: 10.1088/0963-6625/6/3/002

Jasanoff, S. (2004a) post-sovereign science and global nature. Harvard University, Environmental Politics/ Colloquium Papers.

Jasanoff, S. (2004b). The idiom of co-production. In Jasanoff, S. (Ed.), *States of Knowledge. The coproduction of science and social order* (pp. 1–13). Routledge. DOI: 10.4324/9780203413845-6

Knippschild, R. (2011). Cross-Border Spatial Planning: Understanding. Designing and Managing Cooperation processes in the German-Polish-Czech Borderland. *European Planning Studies*, 19(4), 629–645. DOI: 10.1080/09654313.2011.548464

Laclau, E. (1990). *New reflections on the revolution of our time.* Verso.

Lago, X. (2015) Interview by Peter Ulrich, Vigo (Spain), 26th february 2015

Latour, B. (1987). *Science in Action.* Harvard University Press.

Latour, B. (1988). *The Pasteurization of France.* Harvard University Press.

Latour, B., & Woolgar, S. (1979). *Laboratory Life: The Social Construction of Scientific Facts.* Sage.

Lynch, M. (1991). Laboratory space and the technological complex: An investigation of topical contextures. *Science in Context*, 4(1), 51–78. DOI: 10.1017/S0269889700000156

Medeiros, E. (2011). (Re)defining the Euroregion Concept. *European Planning Studies*, 19(1), 141–158. DOI: 10.1080/09654313.2011.531920

Mérida, J., & Telleria, I. (2021). ¿Una nueva forma de hacer política? Modos de gobernanza participativa y «Ayuntamientos del Cambio» en España (2015-2019). *Gestión y Análisis de Políticas Públicas*, (26), 92–110. DOI: 10.24965/gapp.i26.10841

Mittling, D. (2004). Reshaping local democracy through participatory governance. Environment & Urbanization.16:1 Participatory Governance, April 2004, IIED: London, UK. Also available online at:www.iied.org/urban/pubs/eu_briefs.htmlBox

Mosse, D. (2000). Irrigation and Statecraft in Zamindari India. In Fuller, C. J., & Benei, V. (Eds.), *The Everyday State and Society in Modern India* (pp. 163–193). Social Science Press.

Munévar-Quintero, C., Valencia-Hernández, J. G., Hernández-Gómez, N., Aguirre-Fajardo, A. M., & Ramírez-Ríos, M. (2023). Gobernanza participativa en la conformación del Sistema Nacional de Áreas Protegidas de la ecorregión Eje Cafetero, Colombia. *Jurídicas*, 20(1), 139–157. DOI: 10.17151/jurid.2023.20.1.7

Navarro, Z. (1998). Participation, democratis-ing practices and the formation of a modernpolity – the case of 'participatory budgeting in Porto Alegre, Brazil (1989-1998)'. *Development*, 41(3), 68–71.

Nelles, J., & Durand, F. (2014). Political rescaling and metropolitan governance in cross-border regions: Comparing the cross-border metropolitan areas of Lille and Luxembourg. *European Urban and Regional Studies*, 21(1), 104–122. DOI: 10.1177/0969776411431103

Newman, D. (2006). Borders and bordering: Towards an interdisciplinary dialogue. *European Journal of Social Theory*, 9(2), 171–186. DOI: 10.1177/1368431006063331

Nowotny, H., Scott, P., & Gibbons, M. (2001). *Re-thinking science: knowledge and the public in an age of uncertainty*. Polity Press.

Obuljen Koržinek, N., Žuvela, A., Jelinčić, D. A., & Polić, M. (2014) *Strategija razvoja kulture Grada Dubrovnika 2015.–2025*. Dubrovnik: Grad Dubrovnik. Available online at: https://dura.hr/user_files/admin/strateski%20dokumenti/Kulturna%20strategija%20grada%20Dubrovnika.pdf (20/9/2017).

Ochoa, B., López, M., & Hernández, M. (2023). Gobernanza participativa en la implementación de modelos educativos flexibles (MEF). *Estudio de caso: Institución Educativa San José del Trigal del municipio de San José de Cúcuta*. Universidad Simón Bolivar. https://hdl.handle.net/20.500.12442/14384

Ostrom, E. (2006). *Upravljanje zajedničkim dobrima. Evolucija institucija za kolektivno djelovanje*. Naklada Jesenski i Turk.

Pabón, I. (2024). *Parameters for the Formulation of a Public Policy for Water Sanitation of the Bogotá River with a Participatory Governance Approach: Involving Key Actors in Decision Making* [Master's Thesis]. http://hdl.handle.net/11634/56126

Palacios, J. P., Toledo-Córdova, M. F., Miranda-Aburto, E., & Flores Farro, A. (2021). Políticas públicas y gobernanza participativa local. *Revista Venezolana de Gerencia*, 26(95), 564–577. DOI: 10.52080/rvgluz.27.95.8

Pateman, C. (2012). Participatory Democracy Revised. *Perspectives on Politics*, 10(1), 7–19. DOI: 10.1017/S1537592711004877

Perkmann, M. (2007). Policy entrepreneurship and multi-level governance: A comparative study of European cross-border regions. *Environment and Planning. C, Government & Policy*, 25(6), 861–879. DOI: 10.1068/c60m

Peters, B. Guy & Pierre, J. (eds.) (2001) *Politicians, bureaucrats, and administrative reform*. London: Routledge.

Putnam, R. D. (1993). *Making Democracy Work: Civic Traditions in Modern Italy*. Princeton University Press.

Rofman A., Toscani M. d. l. P. y Ferrari Mango C. (2024). Transformaciones en la gobernanza participativa local de la política social en municipios del Gran Buenos Aires en tiempos pandémicos. Geopolítica(s). *Revista de estudios sobre espacio y poder, 15*(1), 145-166. DOI: 10.5209/geop.93330

Rogers, E., & Weber, E. P. (2010). Thinking Harder About Outcomes for Collaborative Governance Arrangements. *American Review of Public Administration*, 40(5), 546–567. DOI: 10.1177/0275074009359024

Rogoff, I., & Schneider, F. (2008). Productive Anticipation. In Held, D., & Moore, H. L. (Eds.), *Cultural Politics in a Global Age. Uncertainty, Solidarity, and Innovation* (pp. 346–358). Oneworld Publications.

Sellers, J. M., & Litström, A. (2007). Decentralization, Local Government, and the Welfare State. *Governance: An International Journal of Policy, Administration and Institutions*, 20(4), 609–632. DOI: 10.1111/j.1468-0491.2007.00374.x

Verhoeven, I. (2004). Veranderend politiek burgerschap en democratie. In *E.R. Engelen & M. Sie Dhian Ho (red.) De staat van de democratie. Democratie voorbij de staat. WRR verkenningen.* Amsterdam University Press.

Wampler, B. (2012). Entering the State: Civil Society Activisim and Participatory Governance in Brazil. *Political Studies*, 60(2), 341–362. DOI: 10.1111/j.1467-9248.2011.00912.x

Warner, J. F. (2006). More sustainable participation? Multi-stakeholder platforms forintegrated catchment management. *International Journal of Water Resources Development*, 22(1), 15–35. DOI: 10.1080/07900620500404992

Willem, A., & Lucidarme, S. (2014). Pitfalls and challenges for trust and effectiveness Designed to Succeed 21 The Korean Journal of Policy Studies on collaborative networks. *Public Management Review*, 16(5), 733–760. DOI: 10.1080/14719037.2012.744426

Williamson, O. E. (1985). *The Economic institutions of capitalism.* Oxford University Press.

Williamson, O. E. (1991). Comparative economic organization: The analysis of discretestructural alternatives. *Administrative Science Quarterly*, 36(2), 269–296. DOI: 10.2307/2393356

Williamson, O. E. (1996). *The mechanisms of governance.* Oxford UniversityPress. DOI: 10.1093/oso/9780195078244.001.0001

Chapter 4
A Global Policy Design for Sustainable Human Development Vis-à-vis Environmental Security

Jipson Joseph
https://orcid.org/0009-0005-1795-5051
Christ University, India

Ananya Pandey
https://orcid.org/0000-0002-2419-6314
Christ University, India

ABSTRACT

Development and environmental security are mostly considered contradictory principles. The post-Second World War period profoundly needed development but as its by-product the environmental stability was lost. Environmentalists and Ecological movements called for environmental justice. They emphasized the need for a quality environment not only for the present but also for the coming generations. Sustainable development thus evolved as a suitable developmental model. Towards the end of the 20th century there originated the human security paradigm, which advanced the idea of sustainable human development. This neologism proposed a people-centered developmental model without disregarding ecological concerns. Since the globe is in industry 4.0 and almost started to actively engage in industry 5.0, the sustainable human development model can function as a perfect mechanism to link both development and environmental security. In this perspective, this chapter focuses on a global policy design that promotes sustainable human development without endangering the environment.

DOI: 10.4018/979-8-3693-7001-8.ch004

1 INTRODUCTION

Development and environmental security are complementary demanding the attention of global conscience (WCED, 1987). The globe overcame the tragedies and complexities of the Second World War through an unparalleled economic growth, energized by scientific and technological advancements. The post-1950 industrial and economic developments of the United States of America (USA/US) and the United Kingdom (UK) forced other countries of Global North to focus on developmental paradigm without being significantly concerned about its environmental impacts. While witnessing the economic advancements of industrialized countries, the developing countries shifted from agricultural to industrial economy to address poverty and other human insecurities (Roy, 1996). But this paradigm shift resulted in severe environmental issues that the globe never witnessed. Unsystematic and profit-oriented developmental activities are the main causes for various environmental problems like deforestation, soil erosion, non-biodegradable waste, contamination of water and air, climate change, and global warming.

Environmental insecurities became worse with the shift from Keynesian philosophy to neo-liberal worldview. The Keynesians argued for the State's key role in economy, whereas the neo-liberals called for its minimal role and propagated larger freedom to market forces (Roy,1996). When the market powers like multinational and transnational companies (MNCs & TNCs) substituted the State's role, it also shifted the focus from people's welfare to the industrialists' economic benefits. Such profit-oriented activities further deteriorated the so-far existing environmental issues and put the whole globe in danger. As a response, the international community shifted its approach from development to sustainable development and then to sustainable human development (SHD) intending to ensure environmental security without compromising the developmental needs of people. Sustainable human development (SHD) appeared in global thinking with the initiative of the United Nations Developmental Program (UNDP). Its 1994 Human Development Report (HDR) introduced the human security paradigm that stands for universality of life claims and SHD. This neologism affirmed environmental security as a vital component of human security framework.

In this perspective, this chapter revolves around the correlation between security (UN Charter, Art. 1), development (added to national and international discussions in the 1950s), and environment (post-1970s concern of the global community). The chapter tries to explore and critically analyze the need for development in ensuring human security, the impact of unsystematic development on environmental security, the emergence of both sustainable development and SHD for securing both human lives and ecosystem, and the interaction between human security and environmental security. In this line, the chapter proposes the possibilities for a global policy design of SHD that also respects environmental security concerns.

2 TECHNOLOGICAL ADVANCEMENTS VIS-À-VIS ENVIRONMENTAL CONCERNS

The second half of the 20th century witnessed immense scientific and technological progress. Technology refers not only to machines and other related artifacts but also to various procedures and processes by which they are made and used for production and other services (Samad & Manzoor, 2011). Technology has slowly become all-pervading, and its impact is seen in every corner. It is the power-engine of the current economic growth and societal wellbeing (Jørgensen, 2001). Benefits of technological progress are visible in every field ranging from agriculture (usage of drones, intelligent sprayers, robotics) to modern means of communication (livestreaming, podcasts, video conferencing). It produces more goods and services with less investment of time, energy and other resources, advances faster movement of goods and information, and supports human attempts for a deeper understanding of the ecosystem (WCED, 1987). There are also technologies that cleanup pollutants and support the stability of ecosystem (Singh & Prasad, 2015). However, unsystematic and irresponsible technological advancement is the immediate cause of environmental degradation (Jørgensen, 2001). It causes rapid consumption of non-renewable resources, creates new types of pollutants, and sometimes even introduces new variants of life forms that may change the evolutionary pathways as is seen in the case of Corona virus (WCED, 1987). Technological advancements can thus be termed as a double-edged sword (Austin & Macauley, 2001). In its absence, economic growth and the overall development of society will be a convoluted eroteme. But in its uncontrollable use and expansion, it causes severe damage to the ecosystem. Failure in systematic developmental policies causes failures in ecosystem management (WCED, 1987). Both these failures together cause human insecurities.

While the 20th century's third quarter focused on irrepressible developmental patterns for accelerating the State's economic growth, the final quarter faced its repercussions in the form of ecological imbalances even when the States were able to enjoy the fruits of financial stability. It is a well-established fact that during economic and political transitions, social safety nets undergo severe suffering and crisis (Jones, 2004). Economic growth alone cannot mitigate ecological damages. It failed to compensate for environmental adversities that are being produced by its profit-oriented character. The last few decades have testified to this failure and helplessness. The entire globe is adversely affected by the environmental issues. But the poor and marginalized people are more vulnerable to these insecurities.

Environmental issues pose an irreparable threat to human security as they endanger mostly the vulnerable sections of society (UNDP, 1995). However, it does not mean that technological innovation is antithetic to human security and societal wellbeing. Environmentally responsible innovations and technologies are plenty, which can contribute to sustainable economic growth. International community should focus on such ecofriendly technologies that advance economic growth without pressurizing the environment (Kaur & Diwakar, 2022). In this line various positive measures and policies have been suggested and incorporated by almost all the countries and international bodies for ensuring sustainable technological advancements (Austin & Macauley, 2001; Francis, 2015). Human security and societal wellbeing require both developmental model and environmental security design for a sustainable future.

2.1 Underdevelopment Poses Threat to Human Security

Development is the standard of human security. In the absence of development, human life is insecure, and it causes socio-economic marginalization of people (Rentschler & Leonova, 2023). Underdeveloped countries fail to provide standard living conditions for a large portion of their people. They have less access to material goods not because of shortage of goods but because of their inability to obtain these goods (Sen, 1981). Sen's argument is in opposition to the neo-Malthusian claims that poverty in the developing countries is due to the scarcity of natural resources (de Soysa, 2008). These inabilities of people result from unequal and unsystematic developmental policies. The huge gap between the rich and poor testifies that only a small minority of society owns the benefits of development (WCED, 1987; WIR, 2022). This gap is not peculiar to some countries but applicable to the whole globe. Lack of properly designed developmental models force millions of people into various insecurities like unemployment, poverty, lack of education and healthcare, and lack of basic resources and infrastructure. They are marginalized by society's mainstream. These people are more vulnerable to pollution and environmental insecurities as they will be forced to live and work in unhealthy and contaminant conditions (Mikati et al., 2018). Besides, these insecurities force developing countries to make policy choices that may damage the ecosystem not because of less concern for its stability but just because of the imperative of immediate survival of their people (UNDP, 1990). For instance, both Namibia and Zimbabwe for addressing poverty decided to kill many wild animals including elephants to distribute its meat to the poor. But while concentrating on immediate survival they finally endanger their sustainable future. In that sense, it can be said that underdevelopment is a cause for environmental insecurity (Homer-Dixon, 1994). Developing countries should prioritize both immediate survival and sustainable future. Sacrificing one for the other will be against the demands of justice.

Global population requires basic facilities like education, healthcare, communication, and transportation for its welfare. The world has so far witnessed many famines in which the main antagonists were underdevelopment and failures in the supply of basic resources (Mallik, 2023). Researchers criticize the policy failures of respective governments for the great famines in human history like Bengal famine and Chinese famine (Gráda, 2007; Mishra et al., 2019; Smil, 1999). It shows that underdevelopment, poverty/famine, and policy failure are all interconnected and they cause human insecurities. Such a scenario demands developing countries to initiate developmental policies for human security (Sanger, 1972/73).

2.2 Development Poses Threat to Environmental Security

Development became a global priority at the end of Second World War (Kingsbury et al., 2004). Development was mostly equated with growth in Gross National Product (GNP), Gross Domestic Product (GDP), or economy (Bandyopadhyay & Shiva, 1988). No proper attention was given to individual welfare. It was thought that individual security will automatically follow economic growth (UNDP, 1990). Until the 1970s, the developing countries' development model was specifically focused on the State's economic growth. Thereafter, various policy changes like pro-poor development, green revolution (Mexico and India), gender sensitive development, and microfinance developmental model (Bangladesh) evolved. Both neo-liberalism and globalization insisted development as a suitable policy framework for people's security and welfare. However, it was a mistake to define development merely in tune with economic progress. It needs to be better understood as "a process of growth towards self-reliance and contentment" by which individuals, communities, and groups become responsible for their lives (Kingsbury et al., 2004, p. 25). Development should aim at making all fields like economy, industry, agriculture, and environment self-reliant. But the greedy developmental designs that do not respect environmental stability lead to irreparable issues like deforestation, desertification, and pollution of world's waterways (Kingsbury et al., 2004). These insecurities cause irreversible habitat modification that adversely affects millions of species (Harrington & Fisher, 1982). This is like a circular model as it begins with human beings and ends with jeopardizing human security itself. But in between it endangers the security of other species, the quality of water, air and soil, and the stability of environment.

The purpose of development is not income-generation but the creation of an enabling environment that helps individuals to have long, healthy, and creative lives (UNDP, 1990). In the absence of a pollution-free and secure environment, no person would be able to live a secure life. Degradation of environmental health is a human-made reality (UNDP, 2011). Industrialization and large-scale extraction of natural resources are the main causes of environmental insecurity (Singh & Prasad, 2015). Until the last few decades, the world economies relied more on polluting industries. Though this has come down in the developed world, it is still very much dominant in low-and-middle income countries (Rentschler & Leonova, 2023). Since 83% of world population lives in the developing countries, they are forced to follow unsustainable developmental policies for its immediate survival (UNCTAD, 2022). They follow such insecure practices not by choice but by sheer necessity (Kingsbury et al., 2004). It is not development but developmental patterns that drive many human insecurities (UNDP, 2022). Current developmental models of many developing countries increase their economic benefits, but they also cause unsustainable depletion of natural resources and damage its life-supporting capabilities. Environmental deterioration for temporary economic gains is nothing but the present generation's 'theft' of the coming generations' birthright for a healthy environment (Korten, 1989). Such practices are therefore unethical and contradictory to justice.

Developing countries still fight for survival and they sacrifice many of their ecological concerns due to poverty and unemployment (WCED, 1987). Industrial development severely pollutes the air, leading to many environmental insecurities. The World Health Organization (WHO) notes that almost 6.7 million people died of the side-effects of air pollution in 2019 and most of them were from the developing countries (2018; 2024). While making significant economic growth through industrialization, South Korea and Taiwan experienced severe pollution that affected their water and air quality during 1960-1970. When they experienced economic growth through development, they became prone to environmental insecurities. But they tried to confront such heavy pollution through stricter environmental laws and spending millions of dollars for its revival (Kingsbury et al., 2004). This environment revival model should be adopted by the whole developing world to ensure a safe environment. It suggests an environmentally friendly developmental model that guarantees the intrinsic value of both humans and environment (UNDP, 1998).

2.3 Green Technology Substantiates Environmental Security

Ecological insecurities shall be weathered by proper developmental policies. Both poverty and environmental insecurities can be subjugated only through a sustainable developmental paradigm. Technology is mostly seen as an adversary of environmental security. Many discussions revolved around correlating technological advancements with environmental security. The latest greening of technology provides a platform for environmentally responsible technological advancements. 'Clean technology,' 'Greentech,' 'Cleantech' etc., are synonymously used for green technology. It stands for production methods or developmental models that use little or zero non-renewable resources and/or creates very less waste compared to the conventional settings of development (Kim, 2011; Samad & Manzoor, 2011). It consists of reduction of carbon emissions and pollution, enhancement of resources and energy efficiency, prevention of loss of biodiversity, and equitable and improved human welfare without increasing environmental risks (UNEP, 2011).

Technological growth cannot be withheld for environmental concerns and ecosystem's stability cannot be sidelined for technological progress. Both are vital components of human welfare and societal growth. Greening of technology or eco-technology evolved in respect to this vision. It stands for "environmentally benign technology, including systems that operate with a minimum impact on species and habitat" (Jørgensen, 2001, 6393). It insists that technological advancements should be done without damaging the ecosystem. The main objective of evolving this model was not to compromise developmental for environmental concerns and vice versa. Besides, the call for environmental justice is properly addressed by this 'green' movement (Ong, 2012). Compared to the other three industrial revolutions that had deleterious effects on eco-security, the fourth industrial revolution is environment friendly using mostly the green technology (Bradu et.al., 2023). The industry 5.0 also seems to follow this ecofriendly model of development.

The Greentech guided developmental paradigm concentrates on low-carbon emissions and efficient use of resources. The circular economy model is an offshoot of green technology. It leads to minimal use of resources and reduction of waste (Cheberyako et al., 2021). Both the depletion of natural resources and the uncontrollable increase of non-biodegradable waste are the sources of environmental insecurity. Greentech tries to resolve both these insecurities. The goal of a sustainable future is achievable only when the whole globe efficiently implements the green technology model for development (Bradu et.al., 2023). Denmark is the greenest country of the world on account of its extensive environmental policies and commitment to the protection of non-renewable resources. Such prioritization is not an easy task as it requires huge investment of money and other infrastructure. It would be a difficult task for the developing countries. Developed world needs to provide assistance to the greening attempts of the developing world as the environmental issues are not limited to any specific boundaries (Cheberyako et al., 2021). Since it is a global issue, a global policy initiative of Greentech is needed to ensure both environmental and developmental security.

3 TOWARDS SUSTAINABLE HUMAN DEVELOPMENT MODEL

Developmental model in the post-Second World War period has gone through many stages. In the beginning, it solely focused on the State's economic development, then it shifted to socioeconomic development for addressing poverty and other insecurities, and it shifted further to sustainable development for linking development with environmental concerns. But in the last decade of the 20th century, it emphasized the centrality of people and their choices in the developmental model (UNDP, 1990). This human-centered model later on clubbed with the sustainable development model and became the sustainable human development paradigm (UNDP, 1994).

Economic development, environmental protection, and responsible social system (a system that echoes the principles of equality, non-discrimination, and social justice) are the basic elements of sustainable human development. From the last few decades of the 20th century, there is a huge population growth due to many factors like effective and efficient healthcare system, availability of nutritious food, decrease in child mortality rate, good transportation and communication facilities, increase in education level, and increase in the lifespan. Economic growth is crucial to ensure a standard of living for this growing population. It also demands society to function beyond the parameters of discrimination, inequality, and injustice. It is equally applicable to both developed and developing countries. However, together with economic growth, the whole globe had to witness many environmental threats like global warming, climate change, and deterioration of natural resources (Brown, 2008). The Global North is responsible for the 92% of excess emissions (Pardikar, 2020). History also testifies that the developed world is the prime culprit for environmental degradation. Their development and present economic security are the outcome of chaotic use and exploitation of natural resources. In this process, they also exploited and polluted the ecosystem of Global South. But the international protocols and agreements force developing countries to reduce their carbon emissions in the same manner and extent of the developed world (Mott et al., 2021). However, the developing countries need to continue with developmental projects as most of their people are below the poverty line. They do not have enough funds and resources for implementing the Greentech. Besides, the Global North does not make effective steps to provide carbon debt compensations to the developing countries (Fanning & Hickel, 2023). The Paris Agreement also did not give any specific guidelines for such steps. In such a scenario, the developing countries are caught up between focusing on economic development and reducing their environmental impact.

Amidst the Covid-19 pandemic, in a meeting of the UN Department of Economic and Social Affairs (UNDESA), the Deputy Secretary-General Amina Mohammed stated that "no one is safe until everyone is safe" (2020). It is equally applicable to both development and environmental security. From the Stockholm Conference of 1972 there were attempts to club both developmental models and environmental security. The Brundtland Commission Report of 1987 and the Rio de Janeiro Earth Summit of 1992 accelerated this initiative with the sustainable development model. This model proposed an eco-centric perspective of development. Then, the UNDP came up with the sustainable human development model as a people-centered paradigm focusing on equitable distribution of the benefits of development and environmental stability within and among the generations (1994). This paradigm shift was a progressive step towards human security including environmental security. The next four subpoints will explain the gradual growth of the global community towards the sustainable human development paradigm.

3.1 Development for Survival: A Post-Second World War Scenario

The 'Black Tuesday' of October 1929 marked the beginning of an economic crisis. It started in the USA and spread out to the whole world (Gay, 1932; Crafts & Fearson, 2010). This period of great depression challenged the conventional model of development and thereby necessitated a systematic model. The New Deal Policy of Franklin Roosevelt, which focused on relief to unemployment and poverty, recovery of economy, and reform of the economic system, provided a "measure of security" to the Americans (Kennedy, 2009). However, soon before this policy framework's benefits could be achieved, the Second World War had erupted, leading the whole globe to another economic catastrophe. The Post-Second World War period faced severe economic crisis due to unplanned expenditure on war, lack of productivity and interstate trade, and collapse of infrastructures. Except for the USA, almost all the then prominent world economies collapsed. During the time of war itself President Roosevelt asked the various US department heads to prepare postwar projects (Rasmussen & Porter, 1981). Till then, the UK was the leading creditor of the world and after the war it became the greatest debtor. The entire globe faced financial crisis. It necessitated them to concentrate on development for boosting up their economy. They were ready to sacrifice anything for development and economic growth. The USA provided funds to many countries for empowering their economies with more production (Kingsbury et al., 2004). The 1944 founded Bretton Woods institutions initiated and coordinated various plans for post-war reconstruction. Many developmental policies were thus adopted to ensure "faster growth through increased integration of the world economy" (Bayoumi, 1995, p. 48). During this same period, many countries became independent from the clutches of colonial powers and the new indigenous governments insisted on developmental policies for their survival and growth (Chai, 1998).

The post-1945 global scenario linked peace and security with economic development. During the first few decades after the War, the capitalist countries supported global economic growth and development. This was mainly in view of two motives: *firstly*, to support their allies to withstand the growth of communist and totalitarian governments; *secondly*, to contribute to the long-term economic growth of developing countries, which will be ultimately beneficial to the donor countries (Busumtwi-Sam, 2002). This initiative positively and progressively supported the survival needs of developing countries, albeit having some selfish motives for the developed world. The second half of the 20th century thus prioritized developmental paradigm for economic growth, human welfare, and social justice.

3.2 Environmental Justice in the Context of Development

Together with economic growth, there also took place inexorable environmental pollution that led many environmentalists to warn the governments and industrialists about the imminent danger that the whole world is about to confront. Various scholarly writings like *Silent Spring* (Carson, 1962), *Our Synthetic Environment* (Bookchin, 1962), and *Population Bomb* (Ehrlich, 1968) demanded for environmental justice by exposing a realistic picture of environmental degradation. Environmental justice initiatives shall be grouped under three philosophical perspectives. *Firstly*, the anthropocentric attitude that treated environmental issues in view of human interests and benefits. Nature has meaning only when it is beneficial to humans and environmental problems should be understood and solved from an anthropocentric perspective. *Secondly*, the eco-centric approach that considers humans merely as a part of the larger system. Ecosystem has its own value regardless of whether it is beneficial to humanity. Andre Næss with his deep ecology was a strong proponent of ecocentrism. While anthropocentrism emphasizes instrumental value, ecocentrism establishes the intrinsic value of the environment (Xinzhong, 2017). *Thirdly*, the human security approach that advances people-centered vision, which includes environmental security as well. It has preference towards human welfare, but it also respects nature's intrinsic value. It is an all-inclusive approach that respects the intrinsic value of both nature and humans (Brauch, 2009). The UNDP conceptualized human security paradigm to provide safety from chronic threats like hunger, disease, repression, and pollution (1998). If the theoretical foundations of environmental approaches are connected to development, then it can be deemed that anthropocentrism was the philosophical foundation for development, ecocentrism for sustainable development, and human security for sustainable human development.

Ecojustice movements until the mid-1960s focused on preserving the wilderness and purity of nature against industrialization and its impact on world population. However, sensible ecologists of the later decades embraced new technologies and developmental models that were not adversarial to the ecosystem (Kirk, 2001). They stood for ecofriendly developments that support both environmental and economic securities. Undisputably, developmental patterns should guarantee ecological sustainability and social equality. However, history testifies that developments mostly caused environmental insecurity and favored the rich causing social inequality. The developmental paradigm largely endangered the dynamism between growth and survival (Bandyopadhyay & Shiva, 1988). Both should complement each other for a sustainable future. Hence, there was the need to positively correlate growth and survival for affirming human security and societal wellbeing. Developmental models cannot sideline environmental justice as human security will be unthinkable without environmental security.

3.3 Sustainable Development: Anthropocentric or Eco-centric?

Sustainable development originated in confrontation with the hazards of industrial pollution. It is based on a thinking "how industry can both produce products to meet needs and generate wealth in ways which do not degrade the environment" (Roy, 1996, 36). However, the Brundtland commission report defined it in a wider ambit as a "development that meets the needs of the present without compromising the ability of future generations to meet their own needs" (WCED, 1987, no.27). This definition, in reality, underlines an anthropocentric perspective (Sauvé et al., 2016) that revolves around human survival and sustainable human progress (WCED, 1987). Economic growth is the primary drive of sustainable development (Brown, 2008). Nevertheless, sustainable development in its contemporary perspective implies environmental sustainability. It tries to state that there is no future for human being sans a sustainable natural environment (Kingsbury et al., 2004). Though the definition favored anthropocentrism, its usage and implementation prioritized ecocentrism. Sustainable development intersects both sustainability and development. The already existing developmental policy design advanced pure economic growth regardless of other vital concerns. Sustainable development does not criticize or sideline developmental policies but insists that the present generation's developmental needs should be met without jeopardizing the future generation's environmental needs (WCED, 1987). In the beginning this new terminology was considered as a constraint to growth, but later on, it was accepted as a suitable policy framework.

Sustainable development is often seen as a substitute for the profit-oriented business concerns of the industrialists as it invites them to consider the developmental strategies, both at the national and international levels, without undermining the environmental concerns. This paradigm tried to harmonize environmental, economic, and societal dimensions affirming that they are indispensable elements of our common sustainable future. It balanced environmental concerns, economic development, and social wellbeing, which was a sincere demand of time and the protagonists of the environmental justice (Kingsbury et al., 2004). The signs of the time revealed that emphasizing one over the other will not ensure a welfare society. Sustainable development stands for "a process of change in which the exploitation of resources, the direction of investments, the orientation of technological development, and institutional change are made consistent with future as well as present needs" (WCED, 1987, no.30). The inextricable link between development and environment was further accelerated with other international summits (UNCED, 1992). Since sustainable development favored eco-centric perspective in practice, there was the need for a more people-centered developmental paradigm that promotes both economic and environmental securities under the same umbrella without any prioritization and that objective was realized by the 1994 proposal of UNDP.

3.4 Sustainable Human Development: An Eco-Anthropocentric Development

Sustainable human development (SHD) is a people-centered sustainable development model. SHD evolved with the rediscovery that "people should be at the center of development" (UNDP, 1994). SHD evolved not because of any drawbacks in the existing sustainable development model but for guaranteeing a more people-centric approach to both sustainability and development. Its origin can be considered as a progressive evolution from the aspirations of sustainable development. SHD combines both sustainable and human development. If sustainable development stands for the developmental policies of the present generation without compromising the demands of the future generation, human development is the expansion of human capabilities for expanding their choices (UNDP, 1990). Human development has its philosophical foundation on the contributions of Amartya Sen, Mahbub ul-Haq, and Martha Nussbaum. From a human welfare perspective, SHD clubs both sustainability and development with the objective of empowering people rather than the economy or the environment. However, both of them are vital elements of this people empowerment process.

The origin of SHD can be traced back to the UNDP's 1994 HDR. It defined SHD as a developmental model "that not only generates economic growth but distributes its benefits equitably; that regenerates the environment rather than destroying it; that empowers people rather than marginalizing them. It is a development that gives priority to the poor, enlarging their choices and opportunities and providing for their participation that affects their lives" (UNDP 1994, iii). Based on this definition, it can be said that the SHD consists of three basic elements: equitable distribution of the benefits of economic growth, regeneration of environment, and empowerment of people by enlarging their choices. In that sense, it tries to address the aspirations of social justice, environmental security, and economic security respectively. It is a pro-people, pro-nature, and pro-poor developmental policy. Sustainable development was a response towards the demand of the environmentalists for an eco-centric developmental policy, whereas SHD is a gradual evolution of the sustainable development model towards a more people-centered paradigm without ignoring both environmental and economic securities. SHD functions on the claim that "human interests take precedence over everything else" (Sultanovich, 2023, 164). Without environmental and economic securities human interests will not be satisfied. Accordingly, this paradigm shift is a holistic approach that addresses all the essential concerns of human wellbeing.

The 1995 World Summit for Social Development held in Copenhagen also emphasized SHD as a suitable policy framework. However, the achievement of SHD depends on proper policies, national and international cooperation, and active participation of the NGOs and civic societies. This paradigm shift is concerned with the two poles of development: ecocentrism and anthropocentrism (Thompson & Barton, 1994). Ecocentrism calls for sustainability whereas anthropocentrism demands economic growth. SHD connects sustainability and development (economic progress) with human-centeredness. The prime beneficiary of both sustainability and development is people. Accordingly, the suitable term for this paradigm is human security as envisaged by the UNDP (1994). In this perspective, SHD is an eco-anthropocentric development.

4 HUMAN SECURITY PARADIGM: EVOLUTION OF AN INCLUSIVE NEOLOGISM

The UNDP's 1994 HDR proposed a human security paradigm to accelerate the shift towards people-centered developmental policies. Until the end of Cold War, security concerns were focused on the protection of national borders. The Palme

Commission Report of 1982 called for global security as the security concerns extended beyond borders. But it is also confined to the ambit of national security as its main concern was to protect States from nuclear conflicts. Global security in its extended vision focused on the interests of whole planet without being enclosed to specific State boundaries and it tried to resolve nonmilitary threats through international cooperation (Intriligator, 1994). Thereafter, the last decade of the 20th century formulated a people-centered security paradigm affirming that the State, international community, and market should exist and function for the people. The Helsinki Accords of 1975 had already given a comprehensive vision of human security in its final report, but it took almost two decades to make its significant impact on the whole globe (Muguruza, 2017).

Human security is defined as "security of people, not just territory. Security of individuals, not just nations. Security through development, not through arms. Security of all the people everywhere – in their homes, in their jobs, in their streets, in their communities, in their environment" (ul Haq, 1995a, p.115). Human security is an inclusive concept consisting of security, human rights, and development (Muguruza, 2017). It shifts security interests from abstract entities to people (Roznai, 2014). Non-discrimination of individuals on the basis of parameters like gender, race, color, income, and religion is the essence of this neologism. In this perspective, human security is both universal and indivisible (ul Haq, 1995a). This paradigm shift emerged as the best way to guarantee security. The UN thus took the initiative to place individual's emancipation and development at the heart of security agenda (UN, 2004). Human security is designed to ensure protection from both chronic threats like hunger, disease, and suppression and hurtful sudden disruptions in the patterns of one's daily life (UNDP, 1994). Besides, it also has the objective to protect people from severe and widespread threats and situations (CHS, 2003). Its ambit is wider as it covers sudden and enduring, severe and extensive threats and insecure situations. Accordingly, everything that can improve the quality of human life forms part of this paradigm (Roznai, 2014).

Human security's interconnected design states that all people have the right "to live in freedom and dignity, free from poverty and despair…all individuals, in particular vulnerable people, are entitled to freedom from fear and freedom from want, with an equal opportunity to enjoy all their rights and fully develop their human potential" (UN General Assembly, 2005, no. 143). It affirms the right-based approach of security and development. Besides, it also underlines that "the State has the responsibility to protect its populations from genocide, war crimes, ethnic cleansing, and crimes against humanity" (UN General Assembly, 2005, no. 143). Human security lens is a rights-responsibility policy model. Even when human security narrative bases itself on human rights it demands higher and wider responsibility from the State. However, while human rights places legal responsibility on the State, human security remains only in the zone of directive principles of State policy as an integrated policy framework (Muguruza, 2017).

Human security paradigm deals with both objective and subjective insecurities. Objective insecurities include environmental threats, violence, terrorism, displacement, health threats, etc., whereas subjective insecurities are in the areas of income, longevity, education, etc. (Gasper & Gómez, 2023). In this perspective, human security echoes global justice. It addresses people's insecurities (bottom-up thinking) compared to national security (top-down thinking). Human security paradigm does not stand for later interventions to confront insecurities, but it tries to prevent them before taking into place. *Ex-ante* prevention efforts rather than *ex-post facto* intervention methods are the essence of this framework (Roznai, 2014). In contemporary times, both State and non-State actors mostly confront environmental issues and other insecurities when they have already taken place. Later interventions will be helpful to reduce or negotiate the damages. But these preventive methods will be more beneficial to human life, societal well-being, and the entire ecosystem.

Human security stands for the whole security of humankind intending to secure its survival and quality of life, whereas environmental security stands for the security of ecosystem securing its sustainability demands. The former tries to protect people from various agents of insecurity like the State, globalization, and terrorism, whereas the latter endeavors to protect the environment from the greed and unsystematic developmental plans of humankind (Brauch, 2008). However, the ecosystem includes people as well. In that sense, environmental security is a crucial aspect of human security. As human security is an inclusive concept, it includes the essential dimensions of people's choices and claims including their environmental concerns.

4.1 Holistic Vision and Approach of Human Security

Human security and environmental security are not diverse concepts. Environmental security is a vital component of human security. This neologism consists of seven categories like economic security, food security, health security, environmental security, personal security, community security, and political security (UNDP, 1994). It is an inclusive paradigm that includes almost all the main security concerns of people (Blatz, 1966). Economic security addresses poverty, food security tries to ensure people's access to healthy and nutritious food, health security stands for access to basic healthcare facilities and protection from diseases, environmental security attempts to protect people from the dangers of environmental pollution, personal security calls for protection against various physical insecurities like torture, war, domestic violence, and terrorism, community security safeguards cultural, ethnic and community identities and traditions, and political security ensures civil and political rights of people and their freedom from political oppression (Muguruza, 2017). All these categories prioritize the security of people from one of the core areas of concern. This wide range of security measures affirms that the main agenda of human security is to reduce the vulnerabilities of less fortunate people (Busumtwi-Sam, 2002). In order to confront the plethora of security problems that were prevalent both within and across the States in the post-Cold War scenario, a holistic approach (Tadjbakhsh & Chenoy, 2007) with an organizing concept (King & Murray, 2001) was needed. Human security not only draws diverse areas together for effectively understanding the interconnection between diverse insecurities but also strengthens cooperation in the fields of human rights, security, and development (Ewan, 2007). On the one hand, it provides a holistic vision of human insecurities, and on the other hand, it advances a holistic approach of human security specifically focused on SHD.

Categories of human security seem to prioritize every area of human life. According to the critics such kind of prioritizing will not do justice to security narrative (Paris, 2004; Newman, 2010; Roznai, 2014; Khong, 2001). But in reality, all these are indispensable for a meaningful human existence. Polarization and marginalization of one area from others will be detrimental to the demands of holistic human security. Since human security paradigm advances a holistic vision and approach, it can easily suit the demands of environmental security in confrontation with the developmental model. It designs a framework of security in which both developmental and environmental concerns are balanced for a people-centered sustainable future objective.

4.2 Environmental Security Epitomizes Human Security

Instead of national security's focus on the State, its organs, borders, and policies, human security focuses on the survival and welfare of people and their environment (Dalby, 2008). Humans are the decisive elements of ecosystems. They are the primary beneficiaries of the whole ecosystem as they largely depend on the environment for natural resources. In its absence, their security is threatened (CHS, 2003). Various environmental insecurities thus become causes of human insecurity. Human security whether it is freedom from fear, or needs, or indignities cannot be realized sans a sustainable environment. "We cannot protect our environment unless we also protect the needs of the humans that rely on it" (UNDP, 2020, 122). In the absence of a systematic policy framework that aims at the empowerment of people, the overuse of natural resources and environmental pollution will become a normal thing. Hence, environmental security is a visible sign that human life is secure.

Human security and environmental security are complementary. Development is one of the core areas of human security paradigm and with technological advancements it has almost reached its maximum level. Technological progress is often seen contradictory to environmental security. If it causes environmental insecurities, then it will affect human security as well. But human security paradigm, which also includes environmental security, promotes a human-centered development without neglecting environmental concerns (UNDP, 1994). Present society also witnesses environmentally sustainable technologies and pro-environmental behaviors that do not cause GHG emissions or pollution. It thus advances environmental security that leads to human security (Hossain et al., 2022).

4.3 Universality of Life Claims

Previously, the world was compartmentalized within nations, sectors (energy, agriculture, trade, industry), and broader areas of concern (environmental, economic, and social) (WCED, 1987). Besides, there was also compartmentalization among people on the basis of sex, color, caste, religion, and so forth. This kind of segmentation led to the denial of basic rights and security to millions of people. Such a divisive attitude is changed in the present globalized world where not only the whole world is interconnected but also all the sectors, areas, and people are intimately linked. Accordingly, an issue in a specific sector/area/nation will have its repercussions in other fields (UN Secretary General, 2005). This unification of the globe and people also gave rise to the awareness of rights and claims. Regardless of any differentiation, people became aware of their claims. A person in a remote village of Bangladesh, Peru, Sudan, Ukraine, Turkey, or Tanzania believes that he or she has an equal claim for development, dignity, security, and the satisfaction of basic needs just as any other person of Global North. Universality of life claims stands for ensuring developmental opportunities to all without differentiating a child on the basis of "wrong sex", "wrong class" or "wrong country" (UNDP, 1994). It is based on people's equal access to developmental opportunities. Irrespective of economic status, people should have equal access to developmental opportunities, only then will they be able to enlarge their choices. They should have the freedom to make choices for increasing options, equalizing opportunities, enabling competitive spirit in market on an equal footing, and sustaining these same things to future generations (ul Haq, 1995b). This is how human security and environmental security are balanced to give effect to the demands of global justice.

5 SUSTAINABLE HUMAN DEVELOPMENT: PREREQUISITE FOR PEOPLE-CENTERED SECURITY

The UNDP proposed SHD for expanding the frontiers of human security. SHD revolves around three central concerns. *Firstly*, it is a developmental model that focuses not solely on economic growth but also on equitable distribution of its benefits. It calls for intergenerational equity. SHD's vital concern is to promote people's equal access to developmental opportunities regardless of the generation criteria. It functions on the shoulders of distributive justice. *Secondly*, it stands for regeneration of the environment rather than its destruction. It is a moral obligation to safeguard the environment. It thus promotes the demands of environmental justice. *Thirdly*, it underlines people's empowerment and not their marginalization. In this line, it is a pro-poor developmental approach that tries to enhance the choices and decision-making capacities of the poor. Accordingly, it mirrors the social justice perspective (UNDP, 1994). Compared to sustainable development that called for the present generation's developmental needs and the future generation's environmental needs, SHD revolves around the equitable sharing of developmental opportunities and environmental quality with the present and future generations. Protection of every species, every resource, or every form of natural capital is not its motto, but the protection of human life. Both economic growth and environmental protection are only means, human security is its end. SHD is a wider concept as it allows us to sustain at least the same level of well-being for the future generation (ul Haq, 1995b). SHD tries to integrate the protection of environment with the future developmental strategies (UNDP, 1990). In 1999, Elkington argued for a 'triple bottom line' for business and industrial management consisting of three vital elements such as economic prosperity, quality environment, and social justice (385). SHD enlarges this 'triple bottom line' into the coming generations also by proposing an equitable sharing of the benefits of development and environmental sustainability with them. In such a progressive vision, it transforms human security paradigm into intergenerational approach.

Human security shifted security concerns from the borders to the various needs of people in their daily life (UNDP, 1995). Human needs basically revolve around survival and welfare. These two are the basic objectives of sustainability and development. Human security's environmental and economic securities address these two objectives. Both environmental and economic insecurities are correlated as "continued environmental degradation is a source of continued impoverishment" (UNDP, 1997, 32). The legitimate concerns of ordinary people with regard to their daily necessities, education, healthcare, transportation, and communication can be answered only through people-centered development. Human security, Sen argues, "is concerned with reducing and—when possible—removing the insecurities that plague human lives" (1999, 8). Human security is not a threat-based but a people-centered policy and hence "the identification of survival, livelihood, and dignity" should be its main focus (Muguruza, 2017, 26). The Darwinian 'survival of the fittest' is to be redefined as the 'survival of the secured' in today's context. The States and international bodies have the responsibility to secure the livelihood and dignity of people through human security's SHD model that focuses on both environmental and economic securities.

SHD is a new model of development that does not marginalize but enhances human life, promotes grassroot participation of people, and focuses on equitable distribution instead of concentration of income without destroying but replenishing natural resources for future generations for which GNP growth is not an end but only a means (ul Haq, 1995b). It is to be noted that our development patterns are the driving forces of human security (UNDP, 2022) as "there will be no development without security and no security without development. And both development and security also depend on respect for human rights and the rule of law" (Muguruza, 2017, 20). This perspective affirms that human security is the center of SHD, which implies that it is also the center of both sustainability and development models.

6 A GLOBAL POLICY FRAMEWORK

Both development and environmental sustainability are essential for the globe. To overcome the endangering issues like illiteracy, unemployment, hunger, poverty, and gender inequality, the world requires various developmental policies. However, such developmental strategies cannot negate the concerns of environmental stability. No development can substitute humans' rights for a good environment. Accordingly, developmental policies should consider the rights of the present and future generations for a safe environment. This requires collective and responsible actions from the international community and its organizations, State governments, public-private partnerships, civic societies, NGOs, and individuals (Kneese, 1979). The Palme Commission of 1982 had already suggested international cooperation for preventing conflicts (Roznai, 2014) and that should be activated to confront all other threats including environmental insecurity.

The global human security policy framework should advance international and intra-national solidarity for the realization of economic development, environmental protection and equitable society (UNDP, 2022). Solidarity demands the collaborative commitment of various actors, regardless of their national or international identity (Gasper & Gómez, 2023). It is a collective call to assess the global insecurities through "the eyes of humankind" (UNDP, 2022, 141). It is an invitation that emerges from our common human nature and an obligation of interconnectedness that goes beyond the specific boundaries of State and other divisive parameters (Gasper & Gómez, 2023). However, this interconnectedness should not be in the form of a "charitable aid relationship" but a "respectable development relationship" (ul Haq, 1995b). The purpose of development is to empower people with more choices and options (UNDP, 1990). People's claim for choices are met only when all the essential elements of human security lens are kept intact. This should also include a critical re-evaluation of the ways and processes of development (Roy, 1996). Development should not be merely profit-oriented, but instead, it should be people-centered as the human security design envisage. Such a developmental model respects environmental concerns as it is ultimately people-oriented and focused on their sustainable future.

The Global Environment Facility (GEF), established in 1991 and supported by the Rio summit, provided funding to address four key areas of environmental insecurity like climate change, loss of biodiversity, pollution of international waters and depletion of ozone layer (UNDP, 1999). Environmental insecurities are beyond-border issues. The Global South countries lack adequate funds to advance sustainable developmental policies, and as a result, they are forced to continue with the existing methods of development without much attention towards the ecosystem. A clean and safe environment is a right of every person and not a privilege of the few (UNDP, 2011). If it is universal human rights (UDHR, art. 25), then the international community has a responsibility to safeguard this right of all regardless of national boundaries. A responsible support system should be formed under the initiative of the UN that manages the SHD model for economic growth and environmental security. The UN trust fund for human security is such an initiative. The international guidelines should be incorporated into the various enactments and policies of national governance. Environmental issues should become part of national development policies (Hossain et al., 2023). A well-structured public expenditure that addresses both systematic economic development and the environmental security oriented towards the security of its citizens at the national level is also a meaningful method to ensure sustainable future (UNDP, 1990). The initiatives of local governing bodies are also indissoluble for ensuring a sustainable developmental model of human welfare (Hossain et al., 2024). In short, there should be coordinated efforts from all international and national powerholders to address the insecurities that people face today with regard to their environment and development.

7 CONCLUSION

Security is a relative concept as what security for one person may be insecurity for another (Roznai, 2014). Whereas human security is absolute as it does not speak about the security of a particular person or entity but focuses on the interests of the entire human community at a wider parameter. It is a holistic people-centered approach that covers almost all the essential areas of human existence. It underlined security as multidimensional and also proposed SHD as a suitable mechanism to ensure both economic growth and environmental sustainability. It thus interlinks both development and environmental concerns, making them human-centered. The past experiences show that underdevelopment is a threat to human security. However, when the whole globe focused on developmental paradigm, it created an unmanageable number of pollutants that led to multitude of environmental insecurities. Humans are the sole culprits of this environmental instability. But its negative side effects were mostly experienced by the vulnerable and marginalized people of the developing countries. In such a situation there arose the demand for environmental justice and equity. It also emphasized the needs of the future generation for a quality environment. The idea of sustainable development originated in this background. In theory it was anthropocentric but in practice it was more eco-centric. SHD evolved from the sustainable development model as a progressive approach towards a human-centered and eco-friendly development.

The UNDP's contribution of SHD is a global policy framework that calls for environmental justice and economic development. It argues that both environmental stability and developmental policies are in support of human welfare. Various policies and strategies to advance SHD as a new paradigm of development is essential to establish both environmental and human securities. Without environmental security, no human welfare can be guaranteed. People's claims for security, human rights and development are met by UNDP's human security driven SHD formula. However, it requires collaborative efforts from both international and national agents that think and act beyond the borders. Developing countries need to continue with their development plans for securing the life of their citizens. Developed countries should support the Global South for putting into practice the SHD model that duly respect and implement both the environmental and developmental concerns. Instead of unjust regulations and control mechanisms, the Global South requires 'openness and positivity' towards their ecofriendly developmental needs.

REFERENCES

Austin, D., & Macauley, M. K. (2001). Cutting through Environmental Issues: Technology as a Double-Edged Sword. *The Brookings Review*, 19(1), 24–27. DOI: 10.2307/20080956

Bandyopadhyay, J., & Shiva, V. (1988). Political Economy of Ecological Movements. *Economic and Political Weekly*, 23(24), 1223–1232.

Blatz, W. E. (1966). *Human Security: Some Reflections*. University of Toronto Press. DOI: 10.3138/9781442632134

Bookchin, M. (1962). *Our Synthetic Environment*. Harper & Row.

Bradu, P., Biswas, A., Nair, C., Sreevalsakumar, S., Patil, M., Kannampuzha, S., Mukherjee, A. G., Wanjari, U. R., Renu, K., Vellingiri, B., & Gopalakrishnan, A. V. (2023). Recent Advances in Green Technology and Industrial Revolution 4.0 for a Sustainable Future. *Environmental Science and Pollution Research International*, 30(60), 124488–124519. DOI: 10.1007/s11356-022-20024-4 PMID: 35397034

Brauch, H. G. (2008). Introduction. In *Globalization and Environmental Challenges: Reconceptualizing security in the 21st Century* (pp. 27–44). Springer. DOI: 10.1007/978-3-540-75977-5_1

Brauch, H. G. (2009). Human Security Concepts in Policy and Science. In *Facing Global Environmental Change: Environment, Human, Energy, Food, Health, and Water Security Concepts* (pp. 965-989). Springer. DOI: 10.1007/978-3-540-68488-6_74

Brown, C. (2008). Emergent Sustainability: The Concept of Sustainable Development in a Complex World. In *Globalization and Environmental Challenges: Reconceptualizing security in the 21st Century* (pp. 141-150). Springer.

Busumtwi-Sam, J. (2002). Development and Human Security: Whose Security, and from What? *International Journal (Toronto, Ont.)*, 57(2), 253–272. DOI: 10.1177/002070200205700207

Carson, R. (1962). *Silent Spring*. Fawcett.

Chai, S.-K. (1998). Endogenous Ideology Formation and Economic Policy in Former Colonies. *Economic Development and Cultural Change*, 46(2), 263–290. DOI: 10.1086/452338

Cheberyako, O. V., Varnalii, Z. S., Borysenko, O. A., & Miedviedkova, N. S. (2021). "Green" Finance as a Modern Tool for Social and Environmental Security. *IOP Conference Series: Earth and Environmental Science*. DOI DOI: 10.1088/1755-1315/915/1/012017

CHS (Commission on Human Security). (2003). *Human Security Now*. https://reliefweb.int/organization/commission-human-security

Crafts, N., & Fearson, P. (2010). Lessons from 1930s Great Depression. *Oxford Review of Economic Policy*, 26(3), 285–317. DOI: 10.1093/oxrep/grq030

Dalby, S. (2008). Security and environment linkages revisited. In *Globalization and environmental challenges: Reconceptualizing security in the 21st century* (pp. 165–172). Springer. DOI: 10.1007/978-3-540-75977-5_9

De Soysa, I. (2008). Underdevelopment and human insecurity: Overcoming systemic, natural, and policy risk. In *Globalization and environmental challenges: Reconceptualizing security in the 21st century* (pp. 127–139). Springer. DOI: 10.1007/978-3-540-75977-5_6

Ehrlich, P. R. (1968). *The Population Bomb*. Ballantine Books.

Elkington, J. (1999). *Cannibals with Forks: The Triple Bottom Line of 21st Century Busine*. Oxford University Press.

Fannin, A. L., & Hickel, J. (2023). Compensation for Atmospheric Appropriation. *Nature Sustainability*, 6(9), 1077–1086. DOI: 10.1038/s41893-023-01130-8

Francis. (2015). *Laudato Sì*. Libreria Editrice Vaticana.

Gasper, D., & Gómez, O. A. (2023). Solidarity and Human Insecurity: Interpreting and Extending the HDRO's 2022 Special Report on Human Security. *Journal of Human Development and Capabilities*, 24(2), 263–273. DOI: 10.1080/19452829.2022.2161491

Gay, E. F. (1932). The Great Depression. *Foreign Affairs*, 10(4), 529–540. DOI: 10.2307/20030459

Gráda, C. (2007). Making Famine History. *Journal of Economic Literature*, 45(1), 5–38. DOI: 10.1257/jel.45.1.5

Harrington, W., & Fisher, A. C. (1982). Endangered Species — A Global Threat. *Resources*. https://www.resources.org/archives/endangered-species-a-global-threat/

Homer-Dixon, T. F. (1994). Environmental Scarcities and Violent Conflict: Evidence from Cases. *International Security*, 19(1), 5–40. DOI: 10.2307/2539147

Hossain, I., Mahmudul Haque, A. K. M., & Akram Ullah, S. M. (2023). Role of Government Institutions in Promoting Sustainable Development in Bangladesh: An Environmental Governance Perspective. *Journal of Current Social and Political Issues*, 1(2), 42–53. DOI: 10.15575/jcspi.v1i2.485

Hossain, I., Mahmudul Haque, A. K. M., & Akram Ullah, S. M. (2024). Socio-economic Dimensions of Climate Change in Urban Bangladesh: A Focus on the Initiatives of Local Governing Agencies. In *Climate Crisis, Social Responses and Sustainability: Socio-Ecological Study on Global Perspectives* (pp. 293-316). Springer. DOI: 10.1007/978-3-031-58261-5_13

Hossain, I., Nekmahmud, M., & Fekete-Farkas, M. (2022). How do Environmental Knowledge, Eco-label Knowledge, and Green Trust Impact Consumers' Pro-Environmental Behaviour for Energy Efficient Household Appliances? *Sustainability (Basel)*, 14(11), 6513. DOI: 10.3390/su14116513

Intriligator, M. D. (1994). Global Security after the End of Cold War. *Conflict Management and Peace Science*, 13(2), 101–111. DOI: 10.1177/073889429401300201

Jones, J. F. (2004). Human Security and Social Development. *Denver Journal of International Law and Policy*, 33(1), 92–103.

Jørgensen, U. (2001). Greening of Technology and Ecotechnology. *International Encyclopaedia of the Social and Behavioural Sciences*, 6393-6396.

Kaur, D., & Diwakar, S. K. (2022). A Review on Environmental Impact due to Technological Advancement in Agriculture. *International Journal of Innovative Research in Computer Science & Technology*, 10(2), 136–139.

Kennedy, D. M. (2009). What the New Deal Did. *Political Science Quarterly*, 124(2), 251–268. DOI: 10.1002/j.1538-165X.2009.tb00648.x

Khong, Y. F. (2001). Human Security: A Shotgun Approach to Alleviating Human Misery? *Global Governance*, 7(3), 231–236. DOI: 10.1163/19426720-00703003

Kim, H.-E. (2011). Defining Green Technology. In *The Role of the Patent System in Stimulating Innovation and Technology Transfer for Climate Change: Including Aspects of Licencing and Competition Law* (pp. 15-20). Nomos Verlagsgesellschaft mbH.

King, G., & Murray, C. (2001-02). Rethinking Human Security. *Political Science Quarterly*, 116(4), 585–610. DOI: 10.2307/798222

Kingsbury, D., Remenyi, J., McKay, J., & Hunt, J. (2004). *Key Issues in Development*. Palgrave Macmillan.

Kirk, A. (2001). Appropriating Technology: The Whole Earth Catalog and Counterculture Environmental Policies. *Environmental History*, 6(3), 374–394. DOI: 10.2307/3985660

Kneese, A. V. (1979). Development and Environment. *Third World Quarterly*, 11(1), 84–90. DOI: 10.1080/01436597908419408

Korten, D. (1989). *Getting to the 21st Century: Voluntary Action and the Global Agenda*. Kumarian Press.

Mallik, S. (2023). Colonial Biopolitics and Great Bengal Famine of 1943. *GeoJournal*, 88(3), 3205–3221. DOI: 10.1007/s10708-022-10803-4 PMID: 36531534

Mikati, I., Benson, A. F., Luben, T. J., Sacks, J. D., & Richmond-Bryant, J. (2018). Disparities in Distribution of Particulate Matter Emission Sources by Race and Poverty Status. *American Journal of Public Health*, 108(4), 480–485. DOI: 10.2105/AJPH.2017.304297 PMID: 29470121

Mishra, V., Tiwari, A. D., Aadhar, S., Shah, R., Xiao, M., Pai, D. S., & Lettenmaier, D. (2019). Drought and Famine in India, 1870-2016. *Geophysical Research Letters*, 46(4), 2075–2083. DOI: 10.1029/2018GL081477

Mott, G., Razo, C., & Hamwey, R. (2021). Carbon emissions anywhere threaten development everywhere. https://unctad.org/news/carbon-emissions-anywhere-threaten-development-everywhere#:~:text=Developed%20countries%20must%20accelerate%20the,This%20is%20in%20everyone's%20interest

Muguruza, C. C. (2017). Human Security as a Policy Framework: Critics and Challenges. *Deusto Journal of Human Rights*, 4(4), 15–35. DOI: 10.18543/aahdh-4-2007pp15-35

Newman, E. (2010). Critical Human Security Studies. *Review of International Studies*, 36(1), 77–94. DOI: 10.1017/S0260210509990519

Ong, P. (2012). Environmental Justice and Green-Technology Adoption. *Journal of Policy Analysis and Management*, 31(3), 578–597. DOI: 10.1002/pam.21631

Pardikar, R. (2020). Global North is Responsible for 92% of Excess Emissions. *Eos*, https://eos.org/articles/global-north-is-responsible-for-92-of-excess-emissions

Paris, R. (2004). Still an Inscrutable Concept. *Security Dialogue*, 35(3), 370–372. DOI: 10.1177/096701060403500327

Rasmussen, W. D., & Porter, J. M. (1981). Strategies for Dealing with World Hunger: Post-World War II Policies. *American Journal of Agricultural Economics*, 63(5), 810–818. DOI: 10.2307/1241249

Rentschler, L., & Leonova, N. (2023). Global Air Pollution Exposure and Poverty. *Nature Communications*, 14(4432), 1–11. PMID: 37481598

Roy, S. (1996). Development, Environment and Poverty: Some Issues for Discussion. *Economic and Political Weekly*, 31(4), 29–41.

Roznai, Y. (2014). The Insecurity of Human Security. *Wisconsin International Law Journal*, 32(1), 95–141.

Samad, G., & Manzoor, R. (2011). Green Growth: An Environmental Technology Approach. *Pakistan Development Review*, 50(4), 471–490.

Sanger, C. (1972/73). Environment and Development. *International Journal (Toronto, Ont.)*, 28(1), 103–120. DOI: 10.2307/40201094

Sauvé, S., Bernard, S., & Sloan, P. (2016). Environmental Sciences, Sustainable Development and Circular Economy: Alternative Concepts for Trans-disciplinary Research. *Environmental Development*, 17, 48–56. DOI: 10.1016/j.envdev.2015.09.002

Secretary General, U. N. (2005). *Larger Freedom: Towards Development, Security, and Human Rights for all. Report of the UN Secretary General to the General Assembly, A/59/2005.*

Sen, A. (1981). *Poverty and Famines: An Essay on Entitlement and Deprivation.* Clarendon Press.

Sen, A. (1999). *Development as Freedom.* Oxford University Press.

Singh, A., & Prasad, S. M. (2015). Remediation of Heavy Metal Contaminated Ecosystem: An Overview on Technology Advancement. *International Journal of Environmental Science and Technology*, 12(1), 353–366. DOI: 10.1007/s13762-014-0542-y

Smil, V. (1999). China's Great Famine: 40 Years Later. *BMJ (Clinical Research Ed.)*, 319(7225), 1619–1621. DOI: 10.1136/bmj.319.7225.1619 PMID: 10600969

Sultanovich, M. D. (2023). The Main Directions of Poverty Reduction in our Country. *Galaxy International Interdisciplinary Research Journal*, 11(2), 164–171.

Tadjbakhsh, S., & Chenoy, A. M. (2007). *Human Security: Concept and Implications.* Routledge. DOI: 10.4324/9780203965955

Thakur, Ramesh (2005). From National Security to Human Security. *The Japan Times*, 13 October 2005.

Thompson, S. C., & Barton, M. A. (1994). Ecocentric and Anthropocentric Attitudes towards the Environment. *Journal of Environmental Psychology*, 14(2), 149–157. DOI: 10.1016/S0272-4944(05)80168-9

ul Haq, M. (1995a). *Reflections on Human Development*. Oxford University Press.

ul Haq, M. (1995b). New Imperatives of Human Security. *World Affairs* 4(1), 68-73.

UN. (2004). A More Secure World – Our Shared Responsibility – Report of the High-Level Panel on Threats, Challenges and Change. https://www.un.org/peacebuilding/sites/www.un.org.peacebuilding/files/documents/hlp_more_secure_world.pdf

UN General Assembly. (2005). 2005 World Summit Outcome, A/60/150. https://www.un.org/en/development/desa/population/migration/generalassembly/docs/globalcompact/A_RES_60_1.pdf

UNDP. (1990). *Human Development Report*. https://hdr.undp.org/content/human-development-report-1990

UNDP. (1994). *Human Development Report*. https://hdr.undp.org/content/human-development-report-1994

UNDP. (1995). *Human Development Report*. https://hdr.undp.org/content/human-development-report-1995

UNDP. (1997). *Human Development Report*. https://hdr.undp.org/content/human-development-report-1997

UNDP. (1998). *Human Development Report*. https://hdr.undp.org/content/human-development-report-1998

UNDP. (1999). *Human Development Report*. https://hdr.undp.org/content/human-development-report-1999

UNDP. (2011). *Human Development Report*. https://hdr.undp.org/content/human-development-report-2011

UNDP. (2020). *Human Development Report*. https://hdr.undp.org/content/human-development-report-2020

UNDP. (2022). *Human Development Report*. https://hdr.undp.org/content/human-development-report-2021-22

UNEP (United Nations Environment Program). (2011). *Towards a Green Economy: Pathways to Sustainable Development and Poverty Eradication*. UNEP.

WCED (World Commission on Environment and Development). (1987). *Report of the World Commission on Environment and Development: Our Common Future.* United Nations.

WHO. (2018). Burden of disease from ambient air pollution for 2016. https://cdn.who.int/media/docs/default-source/air-pollution-documents/air-quality-and-health/aap_bod_results_may2018_final.pdf

WHO. (2024). Health Consequences of air population on populations. https://www.who.int/news/item/25-06-2024-what-are-health-consequences-of-air-pollution-on-populations

WIR (The World Inequality Report). (2022). https://wir2022.wid.world/.

Xinzhong, Y. (2017). Thinking Environmentally: Introduction to the Special Issue on Environmental Ethics. *Frontiers of Philosophy in China*, 12(2), 191–194.

Chapter 5
Environmental Governance in Bangladesh:
Challenges and Opportunities of Solid Waste Management in Semi-Urban Areas

Shafiul Islam
https://orcid.org/0000-0003-4358-6577
Rajshahi University, Bangladesh

A.N. Bushra
Lalon University of Science and Arts, Kushtia, Bangladesh

Abrar Bin Shafi
https://orcid.org/0009-0006-0328-715X
Sir Salimullah Medical College, Dhaka, Bangladesh

ABSTRACT

This study explores the challenges and opportunities of solid waste management in semi-urban areas in Bangladesh, focusing on understanding the roles and experiences of various stakeholder groups using the Power Interest Matrix as a framework. Field investigations reveal significant disparities in waste management practices, particularly in semi-urban municipalities where institutional incapacity, resource constraints, and governance failures exacerbate waste accumulation and environmental degradation. The findings highlight a lack of comprehensive solid waste management policies for semi-urban areas, inadequate financial and human resources, and limited technological innovation, all of which contribute to the inef-

DOI: 10.4018/979-8-3693-7001-8.ch005

ficiency of current solid waste management systems. Local elites often resist waste management reforms due to personal interests. In contrast, informal waste workers remain marginalized despite their significant role in waste collection, lacking formal recognition and proper safety measures. The absence of data systems within municipal authorities impedes effective planning.

INTRODUCTION

Solid waste management (SWM) has become an increasingly critical global issue due to rapid urbanization, population growth, and changing consumption patterns (Awin & Apitz, 2023; Adhikari, 2022; Arteaga, Silva, & Yarasca-Aybar, 2023). In 2012, the world generated 1.3 billion tons of waste, and this figure is projected to reach 2.2 billion tons by 2025 and 3.4 billion tons by 2050, with developing countries facing the brunt of this challenge (Singh & Misra, 2022; Awino & Apitz, 2023). Waste production in developing nations ranges from 0.25 to 1.38 kg per capita per day, with prevalent waste management practices including open dumping, landfilling, composting, and incineration (Javed & Malik, 2022; Bamunuarachchige & de Zoysa, 2022; Hossain et al., 2024a). As a result, unsustainable and mismanagement of solid waste in the Global South contribute to air and water pollution, land degradation, and climate change, negatively impacting public health and the environment (Kumar & Sarangi, 2023). In Bangladesh, the common practices of open dumping, landfilling, and open burning of waste further exacerbate these environmental issues, as waste management infrastructure remains underdeveloped and regulatory enforcement is weak (Roy et al., 2022). These practices are particularly concerning due to the increased release of methane and other greenhouse gases (GHGs), which contribute to climate change. Studies have shown that methane emissions from landfills can be 25 times more impactful on the environment than carbon dioxide (Ferronato & Torretta, 2019).

The environmental burden is compounded by the pollution of surface and groundwater through leachate from unmanaged waste sites, with detrimental impacts on public health, particularly in low- and middle-income countries where informal waste management is prevalent (UNEP, 2015). In many cases, the absence of proper waste management infrastructure leaves communities exposed to hazardous conditions, leading to increased cases of infectious diseases such as cholera and dengue fever due to accumulated waste and blocked drains (UNEP, 2015). Multilateral organizations like the United Nations (UN) and World Health Organization (WHO) have played key roles in advancing waste management and environmental governance. The UN has addressed issues such as ocean pollution, biodiversity loss, air pollution, and toxic waste through international agreements (Weiss & Wilkinson, 2023). In line with the Sustainable Development Goals (SDGs), the United Nations Environment Programme (UNEP) has promoted the safe management of chemicals to protect health and the environment (Maertens, 2022; UNEP, 2015). The WHO has focused on health risks from improper waste management, providing guidelines for the safe handling of hazardous waste and advocating for better sanitation in healthcare settings (Bamunuarachchige & de Zoysa, 2022; Cook et al., 2022; Aziz et al., 2022). The World Bank's "What a Waste 2.0" initiative highlights the urgent need for global waste management reforms, especially in developing countries like Bangladesh, where waste generation is projected to outpace population growth by 2050 (World Bank, 2018). Efforts in Bangladesh include the UNDP's community-based waste collection in urban slums, the World Bank's Clean Cities Initiative to enhance recycling, and GIZ's promotion of composting and community engagement (UNEP, 2015; World Bank, 2018). These initiatives underscore the importance of global cooperation in managing waste and protecting environmental health.

Despite these efforts, in developing countries like Bangladesh, the challenges of waste management are exacerbated by inadequate infrastructure, deficient regulatory enforcement, and limited public awareness (Awino & Apitz, 2023; Kumari & Raghubanshi, 2023; Ahmed, 2024). Rapid population growth, urbanization, and economic development contribute to escalating waste generation in developing nations (Bamunuarachchige & de Zoysa, 2022). Waste management practices in these countries often involve open dumping, landfill sites, and incineration, with a significant portion of waste ending up in landfills due to poor collection systems (Adhikari, 2022). As waste generation continues to rise, addressing these concerns through empirical research and advocacy, robust policies, community engagement, technological innovation, and effective resource management becomes increasingly imperative. While numerous studies have been conducted on waste management in Bangladesh's major cities, there is a significant gap in research focusing on semi-urban areas (Bhuiyan, 2010; Jerin et al., 2022; Haider and Riaz, 2021; Akanda et al., 2022). In semi-urban and rural regions, inadequate waste management systems often lead to indiscriminate dumping, burning of waste, and pollution of water sources, contributing to severe health risks and environmental degradation (Sheheli, 2014; Al Amin, 2022).

As waste generation continues to rise, Bangladesh must address these challenges through empirical research, robust policy development, community engagement, and technological innovation. Empirical research in semi-urban areas is particularly critical, given the pressing environmental, health, and socio-economic challenges that different communities face. This study aims to analyze the effectiveness of semi-urban solid waste management systems in Bangladesh, identify the barriers to effective waste management, and propose potential policy reforms and technological innovations to enhance waste management practices across the country.

Literature Review

The review of the existing literature on waste management and environmental governance in Bangladesh reveals several critical gaps that underscore the need for further research. Although policies such as the National 3R Strategy for Waste Management aim to address waste management challenges, the regulatory framework often lacks coherence, and enforcement is weak. This has limited the practical implementation of these policies, particularly in urban areas where rapid industrialization and population growth exacerbate waste management issues (Al Amin, 2022; Jerin et al., 2022). Moreover, while several studies provide insights into urban waste management, they often fail to address the unique challenges faced by rural communities, where waste disposal practices remain largely informal and unregulated (Sheheli, 2014; Hossain et al., 2024b). The absence of systematic planning and inadequate infrastructure in rural areas further complicate waste management efforts, making it essential to explore these challenges in greater depth.

A key limitation in the existing literature is its urban-centric focus. Bhuiyan's (2010) study on governance and waste management in Bangladesh's cities highlights the detrimental impact of poor governance on service delivery in urban centers. However, this study and others like it (Haider & Riaz, 2021; Jerin et al., 2022) largely overlook rural areas, where waste management practices are starkly different due to limited resources and infrastructural deficits. Studies that do focus on rural waste management, such as Sheheli's (2014) research in the Mymensingh district, provide important insights but are limited by outdated data and a narrow geographic scope. These studies do not fully capture the broader socio-economic and geographical diversity of rural Bangladesh, nor do they consider macro-level factors such as policy frameworks and technological advancements that could shape more effective waste management systems.

Socio-cultural factors also play a significant role in shaping waste management practices. Public awareness and community participation are widely recognized as crucial to the success of waste management initiatives, yet these elements are often underdeveloped in both urban and rural contexts (Akanda et al., 2022). While community-driven initiatives have demonstrated potential, they are frequently constrained by financial limitations and inadequate institutional support. The existing literature has yet to critically examine the long-term sustainability and scalability of such initiatives, particularly in rural settings where local governments often lack the resources to support them on a larger scale (Haider & Riaz, 2021).

Another gap in the literature is the lack of attention to technological innovations in waste management. Although proposed solutions such as waste-to-energy conversion and composting offer promising avenues for improving waste disposal practices, the financial and logistical barriers to implementing these technologies are often overlooked (Jerin et al., 2022; Haider & Riaz, 2021; Hossain et al., 2024a). Rural areas, in particular, lack even basic waste management infrastructure, leaving waste uncollected or disposed of in environmentally harmful ways (Sheheli, 2014). The literature does not adequately explore the practical challenges of deploying such technologies in contexts where financial resources and technical expertise are limited, further limiting the potential impact of these innovations.

The informal waste sector also remains underexplored in the literature. While informal waste pickers and small-scale recyclers play a significant role in waste management in Bangladesh, their activities are often unregulated, leading to inefficiencies and environmental hazards. This market failure is compounded by a lack of policy support and market incentives, which hinders the formalization of the informal waste sector and limits its potential to contribute to sustainable waste management practices. The literature lacks a comprehensive examination of how to formalize and support this sector through targeted policies and incentives, an essential step toward creating a more sustainable and efficient waste management system (Matter et al., 2013; UNEP, 2015).

Building on gaps identified in existing research, this study seeks to address critical challenges in rural waste management by focusing on the unique conditions in rural Bangladesh. It explores both the obstacles and opportunities for implementing effective waste management solutions, particularly evaluating the feasibility of technological innovations while accounting for the financial and logistical barriers faced by stakeholders, from decision-makers to community organizers. By integrating environmental, policy, and socio-economic perspectives, this research aims to provide a comprehensive analysis of waste management challenges and develop solutions that are culturally and economically appropriate for Bangladesh, ultimately guiding future policy and practice.

ROLES OF LOCAL GOVERNMENT AUTHORITIES IN WASTE MANAGEMENT IN BANGLADESH

Effective waste management at the municipal level requires a coordinated effort involving municipal authorities, the private sector, the informal sector, and the general public (DoE, 2010). The municipal authorities, including city corporations and pourashavas, are primarily responsible for managing municipal solid waste (MSW) at the local level. Their role in waste management includes the collection, transportation, and disposal of waste generated within their jurisdictions (DoE, 2010; Ahsan et al., 2014). City authorities manage municipal solid waste (MSW) through two main departments: the conservancy and engineering departments. The conservancy department oversees solid waste management, and utility services, supervises waste collection workers, and manages secondary waste collection points. The engineering department, on the other hand, handles the operation and maintenance of waste management infrastructure, including collection vehicles and disposal sites, and oversees the salaries of drivers and helpers of waste transfer vehicles (Ahsan et al., 2014).

Figure 1. General set up of MSW management in city corporation in Bangladesh

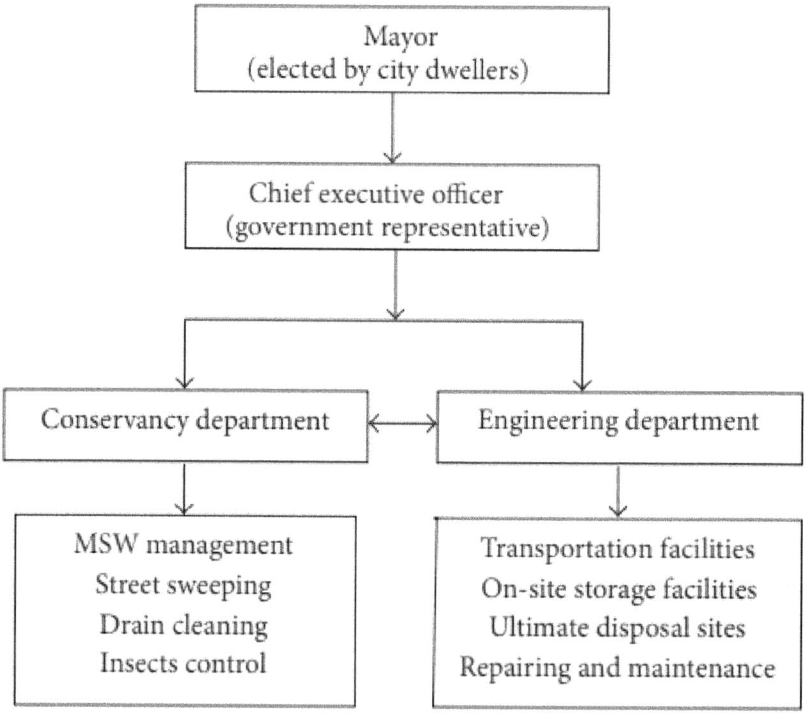

```
                    ┌──────────────────────────────┐
                    │           Mayor              │
                    │ (elected by city dwellers)   │
                    └──────────────┬───────────────┘
                                   ▼
                    ┌──────────────────────────────┐
                    │   Chief executive officer    │
                    │ (government representative)  │
                    └──────────────┬───────────────┘
                                   │
                    ┌──────────────┴───────────────┐
                    ▼                              ▼
         ┌───────────────────────┐    ┌───────────────────────┐
         │ Conservancy department│◄──►│ Engineering department│
         └───────────┬───────────┘    └───────────┬───────────┘
                     ▼                            ▼
         ┌───────────────────────┐    ┌───────────────────────┐
         │   MSW management      │    │ Transportation facilities│
         │   Street sweeping     │    │ On-site storage facilities│
         │   Drain cleaning      │    │ Ultimate disposal sites│
         │   Insects control     │    │ Repairing and maintenance│
         └───────────────────────┘    └───────────────────────┘
```

Source: Adapted from Ahsan et al. (2014)

Some municipal authorities engage private sector companies to assist with waste collection and disposal. These partnerships aim to improve efficiency but require careful management to avoid monopolistic practices that could undermine municipal services. However, the current payment structures for private sector partners, based on the volume of waste collected, discourage waste reduction and recycling efforts, which contradicts the principles of the 3Rs (Reduce, Reuse, Recycle) (DoE, 2010).

The mayor, elected by city dwellers, is the head of the city corporation and is responsible for the overall management of the city's waste. The mayor ensures that all departments work in coordination to implement waste management strategies effectively. The mayor is accountable to the city council and the residents of the city.

The mayor must ensure transparency and efficiency in waste management practices. Ward commissioners are elected representatives at the local level who oversee waste management activities within their respective wards. They ensure that waste collection services are efficient and address the concerns of the residents. Ward commissioners are accountable to the mayor and the municipal council. They must ensure that their wards comply with municipal waste management regulations and report any issues to higher authorities.

The chief executive officer (CEO) is a government-appointed representative who assists the mayor in the administrative functions related to waste management. The CEO oversees the implementation of waste management policies and ensures that municipal staff adhere to regulations. The CEO is accountable to the local government division (LGD) and the mayor. The CEO ensures that the municipal administration complies with national and local waste management policies. On the other hand, the municipality also plays the same role with their limited resources in the semi-urban areas of the country.

Theoretical Framework

Studying waste management from an actor-based or stakeholder-based viewpoint is crucial for comprehensively understanding and addressing the complexities involved in this field. Because such an approach recognizes that waste management is not solely a technical or logistical challenge but a multifaceted issue involving diverse actors with varying degrees of power, interest, and influence. By identifying and analyzing the roles, responsibilities, and interactions of different stakeholders—such as government agencies, municipal authorities, private companies, local communities, non-government organizations (NGOs), and informal sector workers—this perspective enables a more inclusive and collaborative approach to policy development and implementation and ensures that the concerns and contributions of all relevant parties are considered, fostering greater buy-in, accountability, and sustainability (Abedin & Jahiruddin, 2015; Alamgir & Ahsan, 2007). Furthermore, it allows for the identification of potential conflicts and synergies among stakeholders, enabling more effective conflict resolution and the leveraging of collective strengths.

As a stakeholder-based approach offers a comprehensive and inclusive outlook to the study of waste management, the current study has adopted the Power Interest Matrix by Mendelow (1981) as its theoretical frame to study waste management in Bangladesh to uncover the multifaceted challenges involved in managing solid waste. The Power Interest Matrix is instrumental in managing the complex web of stakeholders in waste management practice in Bangladesh. By systematically identifying, categorizing, and engaging stakeholders based on their power and interest, municipal waste management authorities can enhance cooperation, optimize resource use, and ultimately achieve more sustainable waste management outcomes. Thus, it is a critical tool for stakeholder management in solid waste management. This matrix helps identify and prioritize stakeholders based on their power to influence the project/ policy and their interest in its outcomes, allowing the authorities to strategize for effective and successful waste management.

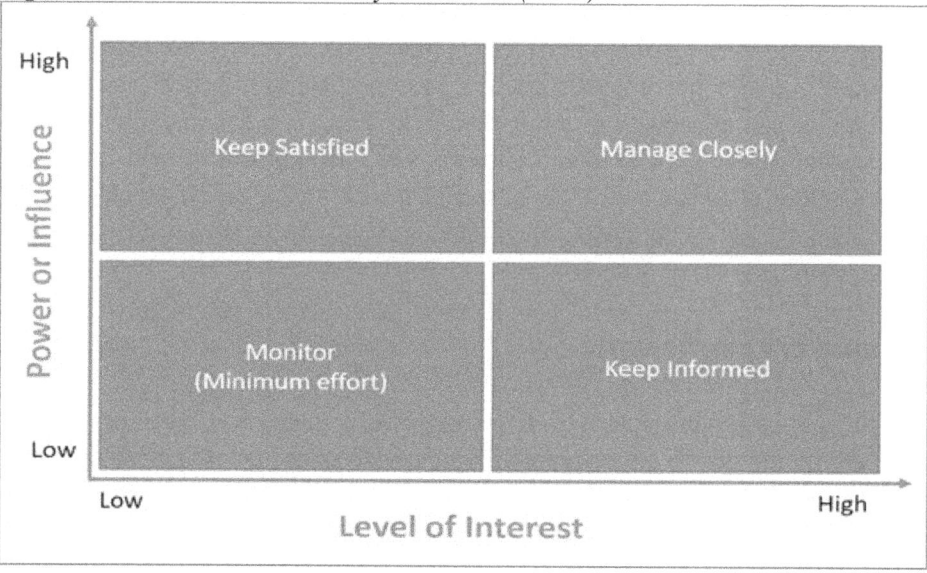

Figure 2. Power interest matrix by Mendelow (1991)

In Bangladesh, waste management involves a variety of stakeholders, including government agencies, municipal authorities, private companies, local communities, NGOs, and international organizations. The Power Interest Matrix categorizes these stakeholders into four groups: high power-high interest, high power-low interest, low power-high interest, and low power-low interest. This helps in strategic engagement based on their influence and interest levels. The matrix also aids in efficient resource allocation, focusing efforts where they will have the most impact, especially in resource-constrained municipalities. Identifying potential conflicts early through the matrix allows for proactive resolution, enhancing cooperation and minimizing disruptions, thereby promoting successful waste management initiatives.

High power-high interest stakeholders, like municipal authorities and key government agencies, require continuous engagement and prompt attention to their concerns for effective decision-making and project support. High power-low interest stakeholders, such as municipal authorities and certain government departments, need regular updates to prevent potential disruptions. Local communities and NGOs, with low power but high interest, are vital due to their direct stake in environmental and health outcomes. Engaging them through updates and awareness campaigns is crucial for grassroots success. Low power-low interest stakeholders, like workers and vendors, are minimally engaged and required to be monitored to prevent obstructions. According to this model, while the high-power high-interest stakeholders need detailed briefings and decision-making involvement, the low power-high interest stakeholders benefit from public meetings and informational materials. Effective communication among these stakeholders tailored to each group's interest and power ensures transparency and trust.

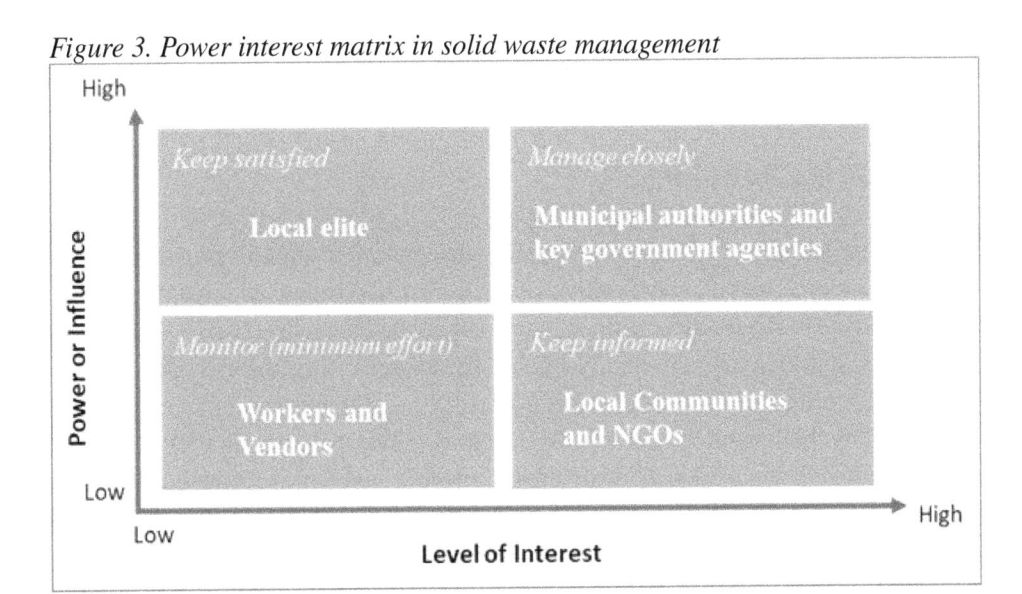

Figure 3. Power interest matrix in solid waste management

	Low Interest	High Interest
High Power	*Keep satisfied* — Local elite	*Manage closely* — Municipal authorities and key government agencies
Low Power	*Monitor (minimum effort)* — Workers and Vendors	*Keep informed* — Local Communities and NGOs

Within Mendelow's Power Interest Matrix (1991), stakeholders of solid waste management can be broadly classified into four categories:

1. **High Power - High Interest:** These stakeholders, typically key government agencies and municipal authorities, hold significant decision-making power and are deeply invested in the success of waste management projects. Due to their influential position, it is essential to maintain close communication, actively manage their expectations, and ensure their continued support.
2. **High Power - Low Interest:** This group includes local elites who possess considerable power but may not be actively engaged in waste management issues. While they might not prioritize these initiatives, their power can still significantly impact outcomes. Therefore, it is crucial to keep them informed and address their concerns to prevent potential disruptions or opposition.

3. **Low Power - High Interest:** Local communities and NGOs often fall within this category. Although they may have limited direct power, they are highly interested in the environmental and health implications of waste management. Their active participation and support are vital for the long-term success of any initiative. Regular communication, engagement, and awareness campaigns can foster a sense of ownership and encourage their valuable contributions.
4. **Low Power - Low Interest:** This group comprises actors such as informal waste workers and vendors who have minimal power and limited interest in waste management outcomes. While their influence is relatively low, it is essential to monitor their activities and ensure they do not obstruct waste management efforts. Limited communication and engagement are generally sufficient for this group.

Research Methodology

This study employs a qualitative research design aimed at exploring solid waste management practices and perceptions of environmental impact in three municipalities which we considered as semi-urban areas in the Rajshahi District of Bangladesh: Katakhali, Kakonhaat, and Naohata. These areas were selected due to their semi-urban characteristics, which offer a relevant context for examining solid waste management challenges in less urbanized settings. The primary data for this study were collected through 30 semi-structured interviews and Key Informant Interviews (KIIs), engaging a diverse range of participants. These participants included local residents, who provided insights into their personal solid waste management practices, and experts in waste management and environmental science, who offered technical and policy-related perspectives. This diversity of respondents ensures that the study captures a comprehensive understanding of waste management from multiple vantage points, addressing both practical and theoretical concerns.

The semi-structured interviews were designed to gather detailed information from local residents about their day-to-day solid waste management practices. These interviews explored how participants handle, segregate, and dispose of solid waste, along with the challenges they encounter in implementing effective waste management practices. Additionally, the interviews probed participants' perceptions of the environmental impact of their behaviors, focusing on their awareness, attitudes, and engagement with environmental sustainability. This approach allowed for flexibility in conversation while ensuring that key topics were consistently covered across interviews. Key Informant Interviews were conducted with waste management experts and environmental science professionals. These interviews provided deeper insights into the technical, regulatory, and policy-related aspects of solid waste management in the region. Experts also offered valuable context on the broader environmental implications and evaluated the effectiveness of existing waste management strategies in the municipalities.

The data collected from the interviews were analyzed using thematic analysis. This method involved coding the qualitative data to identify recurring themes, patterns, and variations in respondents' narratives. By systematically categorizing the data, thematic analysis helped reveal key insights into the commonalities and differences in waste management practices, as well as the participants' perceptions of environmental impacts. This analytical approach allowed for a nuanced understanding of the factors that influence waste management behaviors and perceptions across different groups, providing a solid foundation for developing targeted interventions and policies.

By focusing on the interplay between waste management practices, institutional frameworks, and community perceptions, this methodology provides a well-rounded understanding of the challenges and opportunities for improving solid waste management and promoting environmental sustainability in semi-urban areas in Bangladesh.

Results and Findings

Findings from the field investigation have revealed multifaceted challenges experienced by the four stakeholder groups of waste management presented by Mendelow's Power Interest Matrix (1991). While the local waste management authorities lack the necessary resources, policies, and data, the local elites lack the motivation to contribute to the waste management practices with the aid of their influence, due to their conflicting interests. Low-power stakeholders like local communities and informal waste collectors are limited in their ability to impact solid waste management because of their lack of awareness and poor institutional support.

1. **Challenges Experienced by the High Power - High Interest Group**

High power-high interest groups, such as key government agencies and municipal authorities, face several challenges in waste management. The challenges experienced by this group include institutional incapacity, resource limitations, lack of data, and slow pace of innovation.

Institutional Incapacity: It is observed that the total generated solid wastes are not coming to the waste stream that needs to be managed by the authority. A major portion of the generated waste from residences and semi-urban commercial places remains uncollected because of improper governance and management. The primary data also suggests that the authority concerned is very indifferent to solid waste management at the local level. This is often a result of a lack of guidelines for waste management in the municipalities areas.

Resource Limitations: One of the KIs of the Municipal authorities interviewed in this study has reported that they often face financial and human resource constraints, hindering their ability to effectively manage waste. Moreover, it was also found that there was not enough transportation available to collect and transport wastes to the secondary transformation station.

Table 1. The number of Haat (local market) and garbage trucks in the study area

	Kakonhaat	Naohata	Katakhali	Total
Cow Haat	1	1	1	3
Regular bazaar	6	5	5	16
Garbage Truck	1	1	1	3

Source: *Field data*

The table shows that the number of garbage trucks is quite poor compared to the number of regular and Cow Haat in the selected area. In the semi-urban areas, most of the garbage is produced in these haat and bazaars because of the frequent gathering of people and cattle. However, since there are not enough transportation facilities in these municipalities, most of the garbage in these bazaars remains uncollected.

In this regard, one of the residents of Kakonhaat has asserted:

"Every week, our bazaar is packed with people and livestock, leaving the streets littered with plastic bags, food scraps, and manure. It's not that we don't care about cleanliness; there's just nowhere to put the trash. With only one garbage truck for the area, it can't handle the waste we produce. Most of it stays on the ground, rotting in the heat. The authorities talk about improvements, but nothing changes. By the time workers come, much of the waste has already been scattered by the wind or animals."

Lack of Data: In the selected research area, there was a lack of functional record-keeping systems to track waste volumes at the ward and street levels, complicating efforts to improve planning and service delivery. The municipal authorities do not maintain any records on the locally generated wastes. One of the KIs has reported that there's a shortage of manpower in waste collection. However, they've failed to provide any concrete numbers because most of the waste collectors are informally appointed and thus, they lack any official data on the number of wage workers working in the waste management department of the municipality.

One of the members of the municipality has reported that:

"We don't have proper systems to track waste generation at the ward and street levels, making it hard to plan and improve services. On paper, we should be managing waste better, but the lack of data makes it difficult to identify gaps. There's also a manpower shortage in waste collection, but we can't provide concrete

numbers since many workers are hired informally. Without this information, it's almost impossible to scale up efforts or request additional resources."

Slow Pace of Innovation: It is found that waste collected by the municipal authority is often disposed of in open dumping sites without resource recovery or treatment, leading to environmental pollution and increased disease risk among the residents. Field investigations reveal a lack of innovation in solid waste management, with most waste ending up in open dumps, polluting soil, air, and nearby water sources. Although there have been initiatives such as composting and biogas generation, their widespread adoption is hindered by limited funding and partnerships.

2. **Challenges Experienced by the High Power - Low Interest Group**

The High Power - Low-Interest group includes local elites, who hold significant power but do not actively engage in waste management issues. In our study, local elite groups include local politicians, rich, and influential people. It is found in this study that the local elites often lack awareness and thus do not prioritize waste management in their dwelling areas. The empirical evidence has revealed certain instances where local elites resisted novel waste management efforts because of conflicting interests.

Lack of Awareness or Prioritization: Local elites do not fully understand the importance of waste management or the potential consequences of inadequate practices. Most of the time they are very indifferent to solid waste management because they lack proper knowledge and municipal guidelines for managing waste. Unable to realize the importance of waste management and the harmful consequence of open dumping, they tend to push other less harmful agendas to the municipality authorities.

Resistance to Policy Change: Local elites demonstrate significant resistance to the implementation of new waste management policies, primarily due to concerns over disruptions to their personal interests. One respondent noted that a local politician opposed the construction of a drainage system in the area, as it would necessitate the removal of illegal structures that he owns, which serve as a source of both income and social influence. This resistance underscores the complexity of executing essential infrastructure improvements when such initiatives are at odds with the vested interests of influential individuals.

Table 2: Drainage system comparison

Location	Area (sq. km)	Population	Drainage (km)
Kakonhaat	20.05	18,515	25.70
Naohata	No data	No data	No data
Katakhali	24.50	47,041	No data

Inadequate Drainage and Waste Collection: The absence of a comprehensive drainage system and ineffective waste collection from existing drainage channels exacerbates local environmental challenges. Waste accumulates in drains without proper removal, leading to further blockages and sanitation issues. Addressing these challenges requires targeted strategies to raise awareness, illustrate the benefits of waste management, and align such efforts with the interests of key stakeholders to incentivize their active participation.

Indifference of Local Authorities: Local authorities exhibit a general indifference toward managing solid waste at the municipal level, particularly in rural municipalities. This lack of engagement, coupled with an absence of clear guidelines for waste management in rural-characterized municipalities, hinders the development and execution of effective waste management strategies.

One of the residents of the study area reported that:

"The local authorities promise improvements during elections, but nothing changes once they're in office. We have no clear guidelines on waste disposal, and there are barely any collection points or bins. The trucks come inconsistently, if at all. When we raise these issues, they brush it off, claiming a lack of budget or manpower. It feels like they don't care because we live in a rural area."

3. **Challenges Experienced by the Low Power - High-Interest Group**

The Low Power - High-Interest group, which includes local communities and NGOs, faces significant challenges in waste management efforts. Despite having a strong interest and stake in improving waste management, this group often lacks the authority to influence key decisions or the implementation of policies. Their ability to push for effective change is further constrained by limited financial and technical resources, which makes it difficult to launch sustainable waste management projects or advocate effectively for policy reforms.

In the field investigation, we could not find the involvement of NGOs in waste management measures in the study area. In addition, community waste management efforts are often limited by the technical expertise required for advanced waste management solutions, such as recycling systems or waste-to-energy projects. This skills gap reduces the capacity of the local people to engage in complex waste management activities. Moreover, many local initiatives depend heavily on external funding or government support, leaving them vulnerable to changes in funding priorities or political will. When financial backing is withdrawn, these efforts can stall, limiting long-term impact.

One of the respondents has asserted that:

"I've lived in this village all my life and have seen how waste has become a growing problem. The trash in the streets isn't just unpleasant—it's a health risk. Like many here, I want change, but we feel powerless. The local market committee lacks the resources and skills to manage the waste. We don't have recycling systems, and any projects we start depend on government funding, which can disappear overnight. Without consistent support, our efforts stall, and the waste keeps piling up."

The field investigation further revealed that the role of the Hat-Bazar (market) managing committees (an independent body consisting of the locals) is weak and remains ineffective in managing solid waste. There is a noticeable lack of monitoring mechanisms for solid waste at these local markets, leading to unregulated waste disposal and environmental degradation. However, there are no comprehensive solid waste management strategies at the rural level that promote the inclusion of local communities in the monitoring efforts. As a result, without robust planning and execution, waste management in rural and local areas remains fragmented and inefficient.

4. Challenges Experienced by the Low Power - Low Interest Group

The Low Power - Low Interest group primarily comprises actors such as informal waste workers and vendors, who possess minimal influence over waste management policies and exhibit limited interest in their outcomes. Despite their critical role in the waste management system, these stakeholders face a range of significant challenges that hinder their effective participation and expose them to numerous risks.

Lack of Awareness and Engagement: In the field investigation, it is found that informal waste workers and vendors often lack awareness of waste management regulations and the potential consequences of non-compliance. This limited understanding, coupled with their minimal interest, results in weak engagement with waste management initiatives. Empirical evidence suggests that many in this group fail to recognize the long-term environmental and health risks associated with improper waste disposal, which further diminishes their willingness to actively participate in necessary reforms. Additionally, the lack of sufficient waste collection points and bins means that a significant portion of residential and rural commercial waste remains uncollected. This shortfall, driven by inadequate infrastructure and poor governance, places an additional burden on informal workers, who are left to manage large quantities of uncollected waste. As a result, environmental pollution worsens, and the health risks faced by these workers, especially in underdeveloped rural areas, are heightened.

This issue is better expressed with this quotation of one of our respondents:

"The drains here are often clogged with garbage, and it's left to us—the informal workers—to clear it out. The municipality has no proper collection points or enough bins, so trash piles up, especially in the rural areas where services are barely functioning. We're forced to deal with it, often without gloves or masks, sorting through rotting food, plastic, and metal scraps. I never thought much about the dangers until a friend got seriously ill last month from an infection he caught while working. But we have no choice; it's how we make a living."

Vulnerability to Exploitation: The study further reveals that the low power of these actors makes them highly vulnerable to exploitation by more dominant players in the waste management chain, including middlemen, recycling companies, local elites, and authorities. Informal waste workers frequently face unfair compensation for the recyclable materials they collect, being underpaid for their labor while working in hazardous and unsanitary conditions. Their lack of access to proper protective gear or training exacerbates their exposure to health risks. Moreover, their exclusion from formal waste collection systems prevents them from benefiting from more organized and sustainable waste management practices, thereby reinforcing their marginalization in the sector.

One of the respondents has asserted:

"I worry about my health, but we have no choice. There's no protective gear, and no one acknowledges the risks we face. We're invisible, unrecognized by the authorities as part of the waste management system, left to work without training or support. The dumping grounds are worse, with no proper sorting or disposal facilities. We wade through rotting garbage mixed with dangerous materials like broken glass and chemicals. The smell is unbearable, and there's no running water to clean up. People get hurt, but there's no first aid or help—no safety measures at all."

Insufficient Dumping and Treatment Facilities: Empirical evidence also points to a significant lack of adequate dumping stations and waste treatment facilities. Informal waste workers are often forced to operate in environments where waste is improperly disposed of, leading to further environmental degradation. The absence of well-regulated dumping stations not only increases their exposure to hazardous materials but also compels them to navigate unsanitary conditions without appropriate safety measures. This situation intensifies the health risks they face and undermines broader efforts to establish sustainable waste management practices.

Result Discussion

The study's findings regarding the challenges faced by government agencies and municipal authorities align with existing literature on waste management in developing countries, especially regarding institutional incapacity and inadequate governance structures. Similar to previous studies, such as those by Alamgir & Ahsan (2007) and Abedin & Jahiruddin (2015), this study highlights the lack of tailored policies and resources in semi-urban and rural municipalities in Bangladesh, which exacerbates the issue of uncollected waste. The indifference of local authorities and the absence of specific guidelines for managing waste in rural areas mirror broader concerns about waste management in low-income countries, where governance gaps hinder effective service delivery. The study reinforces these findings, emphasizing that institutional shortcomings and governance failures in rural municipalities contribute to significant waste accumulation and improper disposal practices.

The study also underscores the limited financial and human resources that hamper waste management efforts in under-resourced areas, an issue well-documented in similar studies. The disparity between the number of waste collection trucks and the volume of waste generated at busy locations, such as markets, resonates with findings by Abedin & Jahiruddin (2015), who observed similar transportation bottlenecks. Additionally, the lack of reliable data systems within municipal authorities, a critical issue raised by Alamgir & Ahsan (2007), further hinders the ability to plan and manage waste effectively. Without proper record-keeping or oversight of waste volumes and informal waste collectors, municipalities are ill-equipped to implement sustainable waste management strategies. The persistence of open dumping practices, coupled with financial constraints preventing the adoption of sustainable technologies like composting and biogas generation, reflects the broader challenges faced by Bangladesh in advancing its waste management infrastructure.

Bangladesh can benefit significantly from adopting technical advancements and innovations in waste management that are already proving effective in other developing countries. One such innovation is the implementation of decentralized waste management systems, which have been successfully employed in countries like India through community-level initiatives such as composting and biogas generation (World Bank, 2018). These technologies offer a feasible solution for Bangladesh, where rural and semi-urban areas can implement small-scale organic waste treatment methods to reduce reliance on centralized systems and landfills. Additionally, integrating informal waste collectors into the formal waste management system, as seen in Tunisia's "Eco-Lef" program, can enhance recycling rates and reduce environmental degradation through improved resource recovery (UNEP, 2015). Waste-to-energy technologies also present an opportunity for Bangladesh, particularly in urban areas where high waste volumes can be converted into energy, thus addressing both waste disposal and energy needs. Ethiopia's experience with public-private partnerships (PPP) in municipal waste management further highlights the importance of leveraging private sector investment to improve waste collection and recycling efficiency, a model that could be adapted to Bangladesh's context to overcome financial and logistical challenges (Lohri et al., 2014). By adopting these innovations, Bangladesh can develop a more sustainable and efficient waste management system tailored to its socio-economic and geographical conditions.

The challenges experienced by the High Power - Low Interest group, particularly the local elites, in the study area align with findings from various sources, though notable differences exist in the specific behaviors and reactions observed. In the present study, local elites, including politicians and influential figures, exhibit a lack of awareness regarding waste management issues. Their indifference stems from insufficient knowledge about the environmental and health impacts of improper waste management, particularly in rural and semi-urban areas. This finding mirrors the work of Barua & Rahman (2021), Bhuiyan (2010), as well as Ahmed (2024), who highlighted that waste management efforts in Bangladesh are often hindered by a lack of awareness among key stakeholders. Similarly, informal waste workers and other low-power groups suffer from a similar lack of knowledge, though their disengagement is driven by different factors, such as limited access to information or education on waste management (Ara et al., 2021; Alamgir and Ahsan, 2007). However, the findings of this study show a more significant disconnect between local elites and waste management efforts compared to the general population. For example, informal workers may engage in waste management due to necessity, whereas elites often neglect these issues entirely due to their perceived irrelevance to personal interests. This lack of prioritization among elites is exacerbated by their power to influence municipal agendas, often diverting attention to less critical issues that align with their interests (Abedin and Jahiruddin, 2015; Alamgir and Ahsan, 2007).

The resistance of local elites to new waste management policies, observed in this study, is consistent with findings from other regions. Residents and elites often hold negative perceptions of municipal waste management efforts due to visible waste accumulation and inefficiencies. This undermines public cooperation and participation in waste reduction and recycling initiatives (DoE, 2010). In this regard, Ali and Harper (2004) noted, the implementation of sustainable waste management initiatives often faces opposition from those with entrenched economic and political interests. In the present case, a local politician opposed the construction of a drainage system because it threatened illegal structures that generated income and influence. This mirrors similar patterns of resistance described by Ahsan et al. (2014), where vested interests of powerful individuals often stand in the way of necessary infrastructure improvements. Unlike findings from previous studies where resistance is often attributed to financial concerns alone, the current study highlights the role of social influence and power dynamics as critical factors in policy resistance. This demonstrates a more complex interplay between economic benefits and the political clout of local elites, which has not been emphasized as strongly in earlier research.

The study's findings on the inadequate drainage and waste collection infrastructure align with similar observations made in previous studies. For instance, a lack of modern waste management infrastructure, such as waste segregation and recycling facilities, lack of comprehensive drainage systems and inefficient waste collection methods have been documented as widespread issues across Bangladesh, contributing to local environmental degradation (Ahmed, 2024; Alamgir & Ahsan, 2007; Ahsan et al., 2014). The accumulation of waste in drainage channels without proper removal exacerbates sanitation problems and causes frequent blockages, as identified in both the study area and other municipalities like Dhaka and Chittagong (Alamgir & Ahsan, 2007; Abedin & Jahiruddin, 2015; Ara et al., 2021). However, the present study reveals that the situation is more acute in rural municipalities, where local authorities exhibit greater indifference toward waste management. This contrasts with findings from urban areas, where waste management infrastructure, though imperfect, receives more attention due to higher population density and political pressure. This disparity underscores the need for targeted waste management strategies that address the unique challenges faced by rural communities.

The indifference shown by local authorities in this study aligns with the broader issue of inadequate governance in rural waste management, as noted in the literature. Both DoE (2010) and Ahsan et al. (2014) highlighted the lack of engagement from local government bodies, particularly in rural municipalities, as a significant barrier to effective waste management. The findings suggest that local authorities often fail to provide clear guidelines or infrastructure, leading to widespread neglect in the execution of waste management strategies. A key difference in the present study is the emphasis on the perceptions of residents, who express frustration with the authorities' lack of action, particularly following election promises. This personal account illustrates a deeper level of disillusionment and mistrust toward local governments, which is not as extensively covered in other studies but reflects the everyday experiences of rural communities. The resident's account aligns with broader complaints about municipal neglect but adds a layer of social disconnection that highlights the failure of local governments to build trust and engagement with their constituents.

The findings from this study on the challenges faced by the Low Power - High Interest group in waste management reveal significant obstacles that local communities and NGOs encounter in influencing waste management efforts. These findings align with broader observations in the literature regarding the role of these stakeholders in municipal solid waste management, though notable differences exist when compared to similar studies. The absence of NGO involvement in the study area contrasts sharply with findings from other studies, where NGOs often play a pivotal role in waste management. Ahmed (2024) highlights the critical role that NGOs play in waste management initiatives in Bangladesh, particularly in promoting community-based recycling and composting. Similarly, Parveen (2024) argued that NGOs are critical in promoting community-based solutions, awareness-raising initiatives, and grassroots engagement to address waste management challenges. However, field investigation of our current study points to a notable absence of NGO involvement in the rural-featured municipalities of our research area, which limits the effectiveness of waste management efforts at the local level.

In addition, Fattah et al. (2022) emphasized the importance of community participation in sustainable waste management, suggesting that active involvement from civil society and NGOs is essential for long-term environmental protection. Ahmed et al. (2022) and Ahsan et al. (2014) similarly noted the vital role of NGOs in Bangladesh, where they are involved in small-scale recycling projects and community-based composting initiatives that help to offset the lack of formal waste management systems. The absence of such involvement in the study area highlights a significant gap in both community engagement and external support, which severely hampers the implementation of sustainable waste management strategies. This gap suggests a missed opportunity for leveraging NGO capacity to promote better waste management practices and raise awareness among local communities.

There is an urgent need to enhance technical expertise and increase exposure to modern waste management practices in the study area (DoE, 2010). Community-based waste management efforts, particularly those led by local market committees, have been found to be ineffective, primarily due to a lack of technical skills and insufficient financial resources. This observation aligns with findings from other studies, such as Alamgir and Ahsan (2007) and Fattah et al. (2022), which emphasize that waste management efforts in Bangladesh frequently lack the technical capacity required for implementing advanced solutions, such as recycling and waste-to-energy systems. The skills gap identified in our field investigation mirrors the broader issue of limited access to education, training, and technology in waste management—barriers that have been extensively documented in similar contexts (Fattah et al., 2022; DoE, 2010). Addressing these challenges is critical to transitioning toward more sustainable and effective waste management strategies.

A key observation in the study is the dependence on external funding for waste management initiatives. This reliance on government support or donor funding is consistent with other research that shows how community initiatives often fail when financial backing is withdrawn. For instance, Sujauddin et al. (2008) discussed the vulnerability of local waste management efforts when reliant on inconsistent government funding, which can halt projects and stifle progress (Fattah et al., 2022; Jerin et al., 2022Haider and Riaz, 2021). Similarly, our findings demonstrate that without sustainable financial models, local communities in the study area are unable to maintain long-term waste management initiatives. This mirrors broader trends observed in both rural and urban Bangladesh, where waste management is often fragmented and underfunded (Alamgir & Ahsan, 2007; Fattah et al., 2022).

Another significant finding of this study is the ineffectiveness of local market committees in managing waste. This aligns with previous research indicating that local governance structures, such as market committees, often lack the formal authority, technical expertise, or financial resources necessary to implement efficient waste management systems. For instance, rural municipalities in Bangladesh are frequently hampered by inadequate governance frameworks and insufficient monitoring mechanisms to regulate solid waste management effectively (Sheheli, 2014; Bhuiyan, 2010; Jerin et al., 2022; Haider & Riaz, 2021). The absence of consistent monitoring and regulatory enforcement in the study area further compounds the waste management challenges, leading to increased environmental degradation. These findings highlight the critical need for structured involvement of local communities in waste monitoring, as well as the establishment of clear guidelines and regulatory frameworks specifically tailored for rural waste management.

However, a notable departure from existing literature is the complete absence of organized, community-based waste management initiatives in the study area. Studies such as Ahsan et al. (2014) suggest that in many rural settings, informal waste management practices—often led by local NGOs and community groups—play a vital role in mitigating waste accumulation. The absence of such initiatives in the study area underscores a unique challenge: local communities seem disengaged or lack the organizational capacity to implement even small-scale waste management solutions. This lack of community-driven action not only exacerbates the waste problem but also suggests a gap in mobilization and awareness that other studies have found to be critical in fostering sustainable waste management practices.

The findings of this study highlight the significant challenges faced by the Low Power - Low Interest group, including informal waste workers and vendors. The lack of awareness among informal waste workers about waste management regulations and the potential health risks associated with improper waste disposal has been well documented in several studies. Alamgir and Ahsan (2007) similarly identified a lack of awareness among informal workers regarding the long-term environmental and health hazards of their work. The informal workers' engagement with waste management reforms is low due to their limited understanding of the potential benefits of participating in more formalized systems. This resonates with the challenges identified in the present study, where poor governance and insufficient infrastructure exacerbate the burden placed on informal workers. However, the issue of minimal waste collection infrastructure in rural areas is more pronounced in this study, particularly the lack of bins and collection points, which is not as heavily emphasized in previous studies.

One of the notable differences between this study's findings and previous research is the degree to which rural-featured semi-urban waste collection services are highlighted as inadequate. Finding suitable land for sanitary landfills is difficult due to high population density and geographical constraints which leads to unsanitary disposal practices that pose environmental and public health risks (DoE, 2010). Previous studies, such as Fattah et al. (2022) highlighted the inadequate number of manpower and those by Ara et al. (2021), have also pointed to insufficient waste collection infrastructure in both urban and rural contexts, but this study emphasizes the heightened risks faced by semi-urban based-workers. Moreover, the field evidence underscores the extent of informal workers' exposure to uncollected waste, which exacerbates both environmental pollution and health risks—an aspect that aligns with concerns raised in prior research about the role of informal workers in waste management but provides a more specific rural focus.

The exploitation of informal waste workers is another critical issue. As outlined in earlier research, such as in the work of Abedin and Jahiruddin (2015), informal workers often operate at the mercy of more powerful entities, such as middlemen and local elites, who dictate unfair compensation for the recyclables they collect. These workers, lacking formal recognition, are often excluded from organized waste management systems, reinforcing their vulnerability. This is consistent with the present study's findings, where informal workers describe unsafe working conditions, lack of protective gear, and inadequate pay, all of which further marginalize them. However, a key similarity between this study and the findings of Ara et al. (2021) is the role of local elites and authorities in perpetuating the marginalization of informal waste workers. Both studies highlight how informal workers, due to their lack of formal integration into the waste management system, are denied fair wages and proper working conditions. The absence of governmental recognition and support for these workers leaves them vulnerable to exploitation, a situation made worse by their exclusion from decision-making processes that could potentially improve their conditions.

The study's findings regarding the lack of adequate dumping stations and waste treatment facilities are consistent with other research on the waste management systems in developing countries. Ara et al. (2021) similarly discusses the challenges posed by insufficient infrastructure for waste disposal and treatment, which forces informal workers to operate in unsafe environments. In this study, the absence of regulated dumping stations further exposes these workers to hazardous materials, increasing their health risks. This aligns with the findings of Ahsan et al. (2014), which emphasize the dangers of informal waste work, particularly in contexts where workers are exposed to unsorted, untreated waste. One distinct feature of the present study, however, is the in-depth focus on the lack of treatment facilities and its impact on both waste workers and environmental degradation. While previous studies have acknowledged the absence of treatment plants, this study provides empirical evidence that links the lack of these facilities directly to the heightened exposure of informal workers to hazardous materials. This not only poses immediate health risks but also contributes to longer-term environmental degradation, which undermines broader waste management initiatives.

CONCLUSION

The findings of this study shed light on the challenges faced by different stakeholder groups in solid waste management, particularly within rural-featured municipalities. The High Power - High Interest group, comprising government agencies and municipal authorities, is constrained by insufficient infrastructure, poor governance frameworks, and limited financial resources. This mirrors existing research on waste management in low- and middle-income countries, where improper governance and the lack of tailored policies for rural areas contribute to widespread waste accumulation and improper disposal. The study highlights the indifference of local authorities, which, combined with the absence of specific guidelines for rural and semi-urban waste management, exacerbates the problem, echoing concerns noted in previous studies (Alamgir & Ahsan, 2007; Abedin & Jahiruddin, 2015; Ara et al., 2021). Additionally, the study emphasizes the disparity between waste generation and collection capacity, particularly in rural and semi-urban commercial areas, where transportation bottlenecks, such as a shortage of garbage trucks, hinder efficient waste collection and contribute to persistent environmental hazards.

The study also explores the challenges faced by other stakeholder groups, including the High Power - Low Interest group, which comprises local elites. These elites demonstrate resistance to new waste management policies, often due to personal and political interests, further complicating the implementation of necessary infrastructure improvements (Ahsan et al., 2014; Barua & Rahman, 2021; Ahmed, 2024). Meanwhile, the Low Power - High Interest group, consisting of local communities and NGOs, struggles to influence waste management efforts, particularly in rural municipalities where NGO involvement is minimal. This absence severely hampers the adoption of sustainable waste management strategies, highlighting a critical gap in external support and community engagement (Fattah et al., 2022; Ahsan et al., 2014). Finally, the Low Power - Low Interest group, which includes informal waste workers and vendors, faces significant exploitation and a lack of awareness regarding waste management practices. Their exclusion from formal systems perpetuates unsafe working conditions and health risks, consistent with findings from other studies on the challenges faced by informal workers in waste management (Alamgir & Ahsan, 2007; Abedin & Jahiruddin, 2015).

Recommendations

To improve solid waste management practices in semi-urban areas of Bangladesh, several key recommendations can be made based on the findings of the present study:

1. **Develop Tailored Waste Management Policies**: Semi-urban areas require waste management policies that address their unique challenges. The government should establish clear guidelines and frameworks specifically for rural-featured municipalities, ensuring that waste collection, transportation, and disposal are structured to meet local needs.
2. **Increase Community Engagement and Awareness**: Education campaigns should be implemented to raise awareness among local communities about the importance of proper waste management and the environmental and health risks associated with poor practices. NGOs and local organizations can play a critical role in mobilizing and educating residents.
3. **Strengthen Local Governance**: Municipal authorities must be empowered through better governance structures and accountability mechanisms. Training local government officials on sustainable waste management practices will improve their capacity to oversee and implement waste management programs.
4. **Introduce Low-Cost, Sustainable Technologies**: Given the financial constraints in semi-urban areas, the government should promote cost-effective waste management solutions such as community composting, biogas generation, and recycling. These technologies can turn waste into valuable resources while reducing environmental impact.
5. **Integrate Informal Waste Workers**: Formalizing the role of informal waste workers and providing them with training and protective gear will help improve waste collection efficiency and reduce health risks. Incorporating these workers into formal waste management systems will also increase job security and recognition.
6. **Enhance Financial and Technical Support**: Local municipalities need more financial resources and technical expertise to manage waste effectively. The government, along with development partners, should allocate funds to build necessary infrastructure, such as waste collection points, transportation systems, and treatment facilities.

Implementing these recommendations can significantly improve solid waste management and promote environmental sustainability in semi-urban areas of Bangladesh.

REFERENCES

Abedin, M. A., & Jahiruddin, M. (2015). Waste generation and management in Bangladesh: An overview. *Asian Journal of Medical and Biological Research*, 1(1), 114–120. https://www.ebupress.com/journal/ajmbr. DOI: 10.3329/ajmbr.v1i1.25507

Adhikari, R. C. (2022). Investigation on solid waste management in developing countries. *Journal of Research and Development (Srinagar)*, 5(1), 42–52. DOI: 10.3126/jrdn.v5i1.50095

Ahmed, F., Hasan, S., Rana, M. S., & Sharmin, N. (2022). A conceptual framework for zero waste management in Bangladesh. *International Journal of Environmental Science and Technology*. Advance online publication. DOI: 10.1007/s13762-022-04127-6

Ahmed, R. (2024). Innovative Waste Management Solutions: A Global Perspective Challenges and Opportunities and the Bangladesh Context. *Preprints*. https://doi.org/DOI: 10.20944/preprints202407.2617.v1

Ahsan, A., Alamgir, M., El-Sergany, M. M., Shams, S., Rowshon, M. K., & Nik Daud, N. N. (2014). Assessment of municipal solid waste management system in a developing country. *Chinese Journal of Engineering*, 2014, 561935. Advance online publication. DOI: 10.1155/2014/561935

Akanda, M. G., Farhana, M., & Hafiza Nazneen Labonno, A. S. (2022). M Mahmudul Hasan Rifat, Mst. Nazneen Sultana. An Assessment to Solid Waste Management System in the Rajshahi City Vodra Railway Slum Through Community Participation. International. *Journal of Environmental Protection and Policy.*, 10(2), 22–30. DOI: 10.11648/j.ijepp.20221002.12

Al Amin, M. (August 07, 2022). Waste Management System of Rural and Urban Areas in Bangladesh. The Asian Age. Accessed March 08, 2023 from https://dailyasianage.com/news/291268/waste-management-system-of-rural-and-urban-areas-in-bangladesh

Alamgir, M., & Ahsan, N. (2007). Municipal solid waste and recovery potential: Bangladesh perspective. *Iranian Journal of Environmental Health Science and Engineering (IJEHSE)*, 4(2), 67-76. SID. https://sid.ir/paper/539035/en

Ali, M., & Harper, M. (2004). *Sustainable Composting. Water, Engineering and Development Centre (WEDC)*. Loughborough University.

Ara, S., Khatun, R., & Uddin, M. S. (2021). Urbanization challenge: Solid waste management in Sylhet city, Bangladesh. *International Journal of Engineering Applied Sciences and Technology*, 5(10), 20–28. DOI: 10.33564/IJEAST.2021.v05i10.004

Arteaga, C., Silva, J., & Yarasca-Aybar, C. (2023). Solid waste management and urban environmental quality of public space in Chiclayo, Peru. *City and Environment Interactions*, 20, 100112. DOI: 10.1016/j.cacint.2023.100112

Awino, F. B., & Apitz, S. E. (2023). Solid waste management in the context of the waste hierarchy and circular economy frameworks: An international critical review. *Integrated Environmental Assessment and Management*, 20(1), 9–35. DOI: 10.1002/ieam.4774 PMID: 37039089

Aziz, H. A., Omar, F. M., Halim, H. A., & Hung, Y. T. (2022). Health-care waste management. In L. K. Wang, M. H. S. Wang, & Y. T. Hung (Eds.), *Solid waste engineering and management*. Handbook of environmental engineering (Vol. 25). Springer. https://doi.org/DOI: 10.1007/978-3-030-96989-9_4

Bamunuarachchige, T. C., & de Zoysa, H. K. S. (Eds.). (2022). *Waste technology for emerging economies* (1st ed.). CRC Press., DOI: 10.1201/9781003132349

Barua, P., & Rahman, S. H. (2021). Urban management in Bangladesh. In *The Palgrave Encyclopedia of Urban and Regional Futures*. Palgrave Macmillan., DOI: 10.1007/978-3-030-51812-7_147-1

Bhuiyan, S. H. (2010). A crisis in governance: Urban solid waste management in Bangladesh. *Habitat International*, 34(1), 125–133. DOI: 10.1016/j.habitatint.2009.08.002

Cook, E., Woolridge, A., Stapp, P., Edmondson, S., & Velis, C. A. (2022). Medical and healthcare waste generation, storage, treatment and disposal: A systematic scoping review of risks to occupational and public health. *Critical Reviews in Environmental Science and Technology*, 53(15), 1452–1477. DOI: 10.1080/10643389.2022.2150495

Department of Environment. (2010). *National 3R Strategy for Waste Management*. Ministry of Environment and Forests, Government of the People's Republic of Bangladesh. https://doe.portal.gov.bd/sites/default/files/files/doe.portal.gov.bd/publications/7dc258b2_4501_400d_b066_5e56d84f439f/National_3R_Strategy.pdf

Fattah, M., Rimi, R. A., & Morshed, S. R. (2022). Knowledge, behavior, and drivers of residents' willingness to pay for a sustainable solid waste collection and management system in Mymensingh City, Bangladesh. *Journal of Material Cycles and Waste Management*, 24(4), 1551–1564. DOI: 10.1007/s10163-022-01422-9

Ferronato, N., & Torretta, V. (2019). Waste mismanagement in developing countries: A review of global issues. *International Journal of Environmental Research and Public Health*, 16(6), 1060. DOI: 10.3390/ijerph16061060 PMID: 30909625

Haider, M. Z., & Riaz, M. R. (2021). Municipal solid waste management in Bangladesh: A study of X municipality. *Khulna University Studies*, 18(1), 37–43. DOI: 10.53808/KUS.2021.18.01.2101-S

Hossain, I., Haque, A. M., Kılıç, Z., Ullah, S. A., Azrour, M., & Mabrouki, J. (2024b). Exploring Household Waste Management Practices and IoT Adoption in Barisal City Corporation. In *Smart Internet of Things for Environment and Healthcare* (pp. 27–45). Springer Nature Switzerland. DOI: 10.1007/978-3-031-70102-3_2

Hossain, I., Haque, A. M., & Ullah, S. A. (2024a). Assessing sustainable waste management practices in Rajshahi City Corporation: An analysis for local government enhancement using IoT, AI, and Android technology. *Environmental Science and Pollution Research International*, •••, 1–19. DOI: 10.1007/s11356-024-33171-7 PMID: 38581631

Javed, S., & Malik, F. (2022). Urban solid waste management. *American Journal of Environment Studies*, 5(2), 11–25. DOI: 10.47672/ajes.1268

Jerin, D. T., Sara, H. H., Radia, M. A., Hema, P. S., Hasan, S., Urme, S. A., Audia, C., Hasan, M. T., & Quayyum, Z. (2022). An overview of progress towards implementation of solid waste management policies in Dhaka, Bangladesh. *Heliyon*, 8(2), e08918. Advance online publication. DOI: 10.1016/j.heliyon.2022.e08918 PMID: 35243053

Kumar, A., & Sarangi, S. (2023). Artificial intelligence in sustainable development of municipal solid waste management. [IJRASET]. *International Journal for Research in Applied Science and Engineering Technology*, 11(V), 6744–6751. DOI: 10.22214/ijraset.2023.53247

Kumari, T., & Raghubanshi, A. S. (2023). Waste management practices in the developing nations: Challenges and opportunities. In *Waste management and resource recycling in the developing world* (pp. 773–797). Elsevier., DOI: 10.1016/B978-0-323-90463-6.00017-8

Lohri, C. R., Camenzind, E. J., & Zurbrügg, C. (2014). Financial sustainability in municipal solid waste management – Costs and revenues in Bahir Dar, Ethiopia. *Waste Management (New York, N.Y.)*, 34(2), 542–552. DOI: 10.1016/j.wasman.2013.10.014 PMID: 24246579

Maertens, L. (2022). The untold story of the world's leading environmental institution: UNEP at fifty by Maria Ivanova. *Global Environmental Politics*, 22(3), 200–202. DOI: 10.1162/glep_r_00669

Matter, A., Dietschi, M., & Zurbrügg, C. (2013). Improving the informal recycling sector through segregation of waste in the household – The case of Dhaka Bangladesh. *Habitat International*, 38, 150–156. DOI: 10.1016/j.habitatint.2012.06.001

Mendelow, A. L. (1991) 'Environmental Scanning: The Impact of the Stakeholder Concept'. Proceedings From the Second International Conference on Information Systems 407-418. Cambridge, MA.

(2021). MUNICIPAL SOLID WASTE MANAGEMENT IN BANGLADESH: A STUDY OF X MUNICIPALITY. *Khulna University Studies*, 18(1), 37–43. DOI: 10.53808/KUS.2021.18.01.2101-S

Parveen, M. (2024). Role of Environmental NGOs in Raising Awareness and Promoting Environmental Campaigns in Bangladesh. In *Multi-Stakeholder Contribution in Asian Environmental Communication* (pp. 90–102). Routledge. DOI: 10.4324/9781032670508-10

Roy, H., Alam, S. R., Bin-Masud, R., Prantika, T. R., Pervez, M. N., Islam, M. S., & Naddeo, V. (2022). A review on characteristics, techniques, and waste-to-energy aspects of municipal solid waste management: Bangladesh perspective. *Sustainability (Basel)*, 14(16), 10265. DOI: 10.3390/su141610265

Sheheli, S. (2014). Waste Disposal and Management System in Rural Areas of Mymensingh. *Progressive Agriculture*, 18(2), 241–246. DOI: 10.3329/pa.v18i2.18278

Singh, A. P., & Misra, D. C.Ajit Pratap SinghDinesh Chandra Misra. (2022). Waste management issue and solutions using IOT. *International Journal of Science and Research Archive*, 7(1), 260–282. DOI: 10.30574/ijsra.2022.7.1.0209

Sujauddin, M., Huda, S. M. S., & Hoque, A. T. M. R. (2008). Household solid waste characteristics and management in Chittagong, Bangladesh. *Waste Management (New York, N.Y.)*, 28(9), 1688–1695. DOI: 10.1016/j.wasman.2007.06.013 PMID: 17845843

United Nations Environment Programme (UNEP). (2015). Global Waste Management Outlook. UNEP, International Solid Waste Association (ISWA). Nairobi, Kenya: United Nations Environment Programme. Available at: https://www.unep.org/resources/report/global-waste-management-outlook

Weiss, T. G., & Wilkinson, R. (Eds.). (2023). *International organization and global governance* (3rd ed.). Routledge., DOI: 10.4324/9781003266365

World Bank. (2018). *What a Waste 2.0: A Global Snapshot of Solid Waste Management to 2050*. Urban Development Series. World Bank., DOI: 10.1596/978-1-4648-1329-0

Chapter 6
Encouraging Public Private Collaborations for Low Carbon Green Infrastructure in India:
The Impact of Government Policies

Divya Bansal
https://orcid.org/0000-0001-6268-5402
Amity University, Noida, India

Naboshree Bhattacharya
Amity University, Ranchi, India

Indrajit Ghosal
https://orcid.org/0000-0003-0744-2672
Brainware University, Kolkata, India

ABSTRACT

The purpose of this study is to look into how government policies affect the development of public-private partnerships (PPPs) for low-carbon, green infrastructure in India. The research will provide a thorough examination of current regulations pertaining to waste management, green buildings, renewable energy, and sustainable transportation in order to assess how well they work to encourage private sector cooperation and investment. The research aims to identify opportunities, problems, and best practices for strengthening the role of government in promoting PPPs for sustainable infrastructure development through the analysis of case studies and in-

DOI: 10.4018/979-8-3693-7001-8.ch006

terviews with important stakeholders. This will furnish policymakers, practitioners, and scholars with significant perspectives to propel India's shift towards a low-carbon economy. This research intends to inform the creation of more effective strategies for promoting sustainable infrastructure expansion and expediting India's transition to a greener future by examining the interplay between government policies and PPPs.

INTRODUCTION

Public-private partnerships (PPPs) have emerged as a crucial tool for addressing the infrastructure gap in developing countries like India, where they are increasingly being leveraged to tackle the complex challenges of urbanization, infrastructure development, and climate change (Ahlawat & Singh, 2019; Niti Aayog, n.d.). The Indian government has demonstrated a strong commitment to sustainable development and recognized the potential of partnering with the private sector to improve and sustain the ability of local governments to protect and restore the nation's waters (Fouad, 2021; EPA, 2015). The government has developed a comprehensive policy framework to encourage private sector participation in green infrastructure projects, offering incentives and risk-sharing mechanisms (Fouad, 2021; Monteiro et al., 2020). This includes the national action plan on climate change (NAPCC), which underscores the importance of sustainable infrastructure and outlines eight core missions to combat climate change (Ahlawat & Singh, 2019). Additionally, the government has introduced several policies to promote green infrastructure, such as the Jawaharlal Nehru national urban renewal mission (JNNURM) and regulatory policies to encourage green building construction (Ahlawat & Singh, 2019). To create a conducive environment for PPPs in green infrastructure, the Indian government has established the public-private partnership appraisal committee (PPPAC) to appraise and approve PPP projects, and the viability gap funding (VGF) scheme to make PPP projects financially viable (Ahlawat & Singh, 2019). The government's push towards renewable energy has also steered significant private investment into this sector (Ahlawat & Singh, 2019). Despite these efforts, PPPs in low-carbon green infrastructure face several challenges, such as regulatory unpredictability, contractual disputes, land acquisition issues, and financing constraints (Gerrard, 2001; Shrivastava & Kumar, 2015). To overcome these challenges, further policy refinement, a stable regulatory environment, streamlined processes, transparency, and capacity building among stakeholders are essential (Monteiro et al., 2020; Jha & Singh, 2016; Shrivastava & Kumar, 2015). In conclusion, government policies have played a significant role in shaping the landscape of PPPs for low-carbon green infrastructure in India. The Indian government's commitment to sustainable development, coupled with its efforts to facilitate PPPs, has created a conducive environment for private investment in green infrastructure (Ahlawat & Singh, 2019; Niti Aayog, N.D.; Fouad, 2021; EPA, 2015). However, addressing the challenges faced by PPPs and creating a stable regulatory framework are essential for the full realization of India's sustainable development goals (Gerrard, 2001; Monteiro et al., 2020; Jha & Singh, 2016; Shrivastava & Kumar, 2015). By implementing targeted measures to promote PPPs in green infrastructure, such as developing a comprehensive policy framework, streamlining processes, establishing dedicated funds, promoting capacity building, and encouraging innovative financing models like cbp3s, the Indian government can accelerate the transition towards a more sustainable and resilient future (Ahlawat & Singh, 2019; Jha & Singh, 2016; Shrivastava & Kumar, 2015).

Case studies to highlight the potential of green PPPs in delivering sustainable infrastructure and social co-benefits

- **Renewable Energy PPPs**: the Rewa Ultra Mega solar park in Madhya Pradesh is one of the largest single-site solar power plants in India, developed through a PPP model. The project leveraged innovative financing mechanisms like viability gap funding and payment security measures to attract private investment. It also incorporated community development initiatives and local employment generation, demonstrating the potential for green PPPs to deliver social co-benefits (Gulati & Rao, 2019).

The Ostro energy wind power projects in Andhra Pradesh and Karnataka were developed through a PPP model, with the company partnering with state governments and local communities. The projects incorporated stakeholder engagement, benefit-sharing mechanisms, and community development initiatives, highlighting the importance of inclusive approaches in Green PPPs (Ostro energy, 2021).

Research Objectives

The objectives for the current research are:

(i) To critically analyze national, state, and local government policies in India that aim to promote public-private partnerships (PPPs) for the development of low-carbon green infrastructure, assessing their alignment with sustainable development goals.
(ii) To evaluate the effectiveness of these policies in incentivizing private sector participation and investment in sustainable green infrastructure projects, identifying key drivers, challenges, and outcomes.

(iii) To examine the technical, financial, institutional, and regulatory barriers hindering the successful implementation of PPPs in the green infrastructure sector in India, drawing insights from policy documents, case studies, and stakeholder perspectives.

(iv) To synthesize best practices and lessons learned from successful PPP projects and international experiences and develop evidence-based policy recommendations for enhancing the enabling environment for PPPS and accelerating the development of sustainable green infrastructure in India.

Research Methodology

This study employs a qualitative research methodology to critically analyze government policies and their impact on promoting public-private partnerships (PPPs) for low-carbon green infrastructure development in India. The research relies on the secondary data sources, including policy documents and case studies. The collected data is analyzed using thematic analysis to identify patterns, drivers, challenges, and outcomes associated with PPP policies and projects. The policy document analysis involves a systematic review of relevant government policies, regulations, guidelines, and reports at the national, state, and local levels to assess their objectives, incentives, and mechanisms for promoting private sector participation in sustainable infrastructure projects. Case studies of selected PPP green infrastructure projects are examined using a structured framework to identify success factors, challenges faced, and lessons learned.

I. GOVERNMENT POLICIES RELEVANT TO PPPS FOR LOW-CARBON GREEN INFRASTRUCTURE IN INDIA

Private partnerships (PPPs) have emerged as a promising approach to address the challenges of implementing green infrastructure in India. The Indian government has demonstrated significant efforts in promoting low-carbon green infrastructure through policy support and regulatory reforms, creating a favorable environment for increased private sector participation (World Bank, n.d.; Wikipedia, 2023; PSA Legal, 2022). However, several obstacles hinder the successful execution of PPPs, including the lack of a unified regulatory framework, inadequate public sector capacity, and challenges in policy implementation and coordination among government agencies (Mathur et al., 2014; Wikipedia, 2023).

Despite the progress made, there is a need for further policy refinement and capacity building to ensure the successful implementation of PPPs in green infrastructure projects (Puppim de Oliveira et al., 2018; Shrivastava & Kumar, 2015).

The government should focus on developing a unified regulatory framework, streamlining processes, enhancing coordination among various agencies, and implementing capacity-building initiatives for public sector officials and other stakeholders (Wikipedia, 2023; Shrivastava & Kumar, 2015). While PPPs have the potential to catalyze the development of green infrastructure in India, their success depends on the government's ability to create an enabling policy environment, provide necessary support mechanisms, and address the identified challenges (Ahlawat & Singh, 2019; Niti Aayog, n.d.; Fouad, 2021). By learning from successful case studies and incorporating best practices, the Indian government can further strengthen its policies to promote PPPs for low-carbon green infrastructure, contributing to sustainable urban development and delivering health-environment co-benefits to its citizens (Monteiro et al., 2020; Puppim De Oliveira et al., 2018).

National Level Policies and Initiatives

The government of India has recognized the importance of public-private partnerships (PPPs) in leveraging private sector investment and expertise for the delivery of public assets and services. Through various policy and institutional initiatives, such as the establishment of the public private partnership appraisal committee and the provision of project development funds and viability gap funding, India has become a leading PPP market globally (World Bank, n.d.). The viability gap funding scheme, which offers financial support in the form of grants covering up to 20 percent of the total project cost, has been instrumental in promoting the sustainability of PPP projects (Wikipedia, 2023). While these efforts have contributed to the development of a robust ecosystem in India, the government must continue to refine its policies, address challenges, and incorporate best practices to further strengthen the PPP framework and ensure the success of these partnerships.

State Level Policies and Initiatives

State governments in India have been instrumental in promoting infrastructure development through public-private partnerships (PPPS) by formulating guidelines, issuing policies, and establishing dedicated committees to create an enabling environment for private investors (psa legal, 2022). These efforts have been tailored to address the unique needs and challenges of each state, attracting private sector expertise, resources, and innovation. The proactive approach adopted by state governments has significantly contributed to the growth of the PPP market in India, fostering economic growth, creating employment opportunities, and improving the quality of life for citizens. However, to sustain this momentum, state governments must continue to refine their policies, strengthen institutional capacities, and address challenges faced by private investors, working closely with the central government and other stakeholders to enhance the effectiveness of PPPs in delivering world-class infrastructure and driving sustainable development across India.

Local Level Policies and Initiatives

The central government of India has developed a comprehensive toolkit to assist local authorities in making informed decisions related to infrastructure PPPs, providing guidance, best practices, and resources to promote a more uniform and efficient approach across the country (World Bank, n.d.; Wikipedia, 2023). Local governments play a crucial role in shaping the economic landscape and making key investment decisions, but their actions are influenced by the broader context of federal and state policies (Brookings Institution, 2018; National Academies Press, 2021). To ensure the success of PPPs at the local level, all tiers of government must work collaboratively, harmonize their policies and approaches, and leverage the central government's toolkit to enhance their capacity to develop high-quality, financially sustainable, and socially and environmentally responsible PPPs (Ahlawat & Singh, 2019; Niti Aayog, n.d.; world bank, n.d.; Wikipedia, 2023).

Analysis of Qualitative Data From Policy Documents and Case Studies

i. **Government Support for PPPs**

The Indian government has demonstrated strong support for public-private partnerships (PPPs) in low-carbon green infrastructure at the national, state, and local levels, with the implementation of various policies, regulations, and initiatives aimed at encouraging private sector participation (Erismann et al., 2021). The government's proactive approach towards PPPs underscores its commitment to sustainable development and its recognition of the critical role that private sector expertise and investment can play in achieving this goal.

Case study of successful PPPs in the urban sanitation and waste management sectors, demonstrating the importance of collaborative governance models and stakeholder engagement

- The Alandur sewerage project in Tamil Nadu was one of the first successful PPPs in the urban sanitation sector in India. The project involved a tripartite partnership between the local government, a private operator, and community-based organizations, demonstrating the potential for collaborative governance models in delivering sustainable infrastructure (World Bank, 2006).

- The Berhampur solid waste management project in Odisha was developed through a PPP model, with the private partner responsible for collection, transportation, and processing of municipal solid waste. The project incorporated a community awareness campaign and a grievance redressal mechanism, underscoring the importance of stakeholder engagement and public participation (jica, 2016).

ii. **Challenges and Limitations**

despite the government's support, the implementation of PPPs for low-carbon green infrastructure in India faces several challenges and limitations. These include issues related to policy implementation, coordination among various government agencies, and the lack of a unified regulatory framework. Addressing these challenges requires a concerted effort from all stakeholders, including the government, private sector entities, and civil society organizations. Streamlining processes, enhancing institutional capacities, and developing a more coherent and comprehensive regulatory framework can help overcome these barriers and facilitate the effective implementation of PPPs.

iii. **Role of Financing**

The role of financing emerges as another major theme from the data, highlighting its crucial role in the success of PPPs for low-carbon green infrastructure. The government's policy support and regulatory reforms have created a more conducive environment for private sector participation in green infrastructure projects. However, issues related to financing, such as the availability of funds and the financial viability of projects, remain significant challenges (Patil et al., 2016). To address these challenges, innovative financing mechanisms, such as green bonds and blended finance, need to be explored and promoted. Moreover, the government should consider providing additional financial incentives and risk-sharing arrangements to attract private sector investment in green infrastructure projects.

Case study of a large-scale urban transport PPP that faced challenges related to land acquisition and resettlement, highlighting the need for robust social and environmental safeguards.

- The Hyderabad metro rail project is one of the largest PPP projects in the urban transport sector in India. The project integrated sustainable design features, multimodal connectivity, and transit-oriented development, showcasing the potential for green PPPs to transform urban mobility (Hyderabad metro rail limited, 2021). However, the project also faced challenges related to land acquisition, resettlement, and rehabilitation, highlighting the need for robust social and environmental safeguards.

iv. The Importance of Stakeholder Engagement

Stakeholder engagement is a critical aspect of implementing PPPs for low-carbon green infrastructure. It involves not only the participation of government agencies and private sector entities but also the active involvement of local communities and other relevant stakeholders. Effective stakeholder engagement ensures that projects are designed and implemented in a manner that addresses the needs and concerns of all affected parties, leading to more sustainable and socially acceptable outcomes. Governments should prioritize stakeholder engagement and develop mechanisms for meaningful public participation in the planning, development, and implementation of PPP projects.

v. The Potential for Co-Benefits

The data also highlights the potential for co-benefits from implementing PPs for low-carbon green infrastructure. These co-benefits can include environmental benefits, such as reduced greenhouse gas emissions, as well as social and economic benefits, such as improved public health and job creation (Patil et al., 2016). Recognizing and leveraging these co-benefits can help build a stronger case for PPPs and attract greater support from various stakeholders. Governments should actively promote the co-benefits of green infrastructure projects and develop metrics and frameworks to measure and communicate their impact.

An example of a PPP that delivered enhanced public transport services while incorporating sustainable design features and innovative branding strategies

- The Ahmedabad bus rapid transit system (BRTS) was developed through a PPP model, with the private partner responsible for bus procurement, operation, and maintenance. The project incorporated innovative design features, integrated ticketing, and a comprehensive branding and communication strategy, demonstrating the potential for PPPs to enhance the quality and efficiency of public transport services (Rizvi & Sclar, 2014).

II. IMPACT OF POLICIES ON INCENTIVIZING PRIVATE SECTOR PARTICIPATION AND INVESTMENT IN SUSTAINABLE PROJECTS

Private sector participation and investment play a pivotal role in advancing sustainability initiatives worldwide. To incentivize private sector engagement in sustainable projects, governments and policymakers often implement various policies and strategies. This literature review aims to analyze existing research on the effectiveness of such policies and identify key approaches for fostering private sector involvement in sustainability.

i. **Financial Incentives**

Numerous studies underscore the significance of financial incentives in encouraging private sector investment in sustainable projects (Smith, 2018). Governments frequently employ tools such as tax breaks, subsidies, and grants to reduce financial barriers and stimulate private sector involvement (Johnson & Wang, 2020). These incentives can help offset the initial costs associated with sustainable investments, making them more attractive to private entities. By providing financial support, governments can create a more level playing field and encourage private companies to prioritize sustainability in their business decisions.

ii. **Regulatory Frameworks**

Regulatory frameworks effective regulatory frameworks are essential for creating an enabling environment for sustainable investments (Brown & Jones, 2019). Governments can implement clear environmental standards, emissions trading schemes, and renewable energy mandates to drive private sector engagement (Roberts et al., 2021). These regulatory measures provide a clear signal to the private sector about the government's commitment to sustainability and create a predictable and stable policy environment. By setting ambitious targets and establishing a strong regulatory foundation, governments can incentivize private companies to invest in sustainable technologies and practices, knowing that there is a long-term market for their products and services.

iii. **Public-Private Partnerships (PPPs)**

Collaborative arrangements between governments and private entities, known as public-private partnerships (PPPs), are increasingly recognized as a powerful mechanism for delivering sustainable infrastructure projects (Gupta & Sharma, 2022).

PPPs can leverage private sector expertise and resources while mitigating financial risks for both parties (Lee & smith, 2017). By sharing risks and responsibilities, governments and private companies can work together to develop and implement sustainable projects that might otherwise be too complex or costly for either party to undertake alone. PPPs can also help ensure that sustainable infrastructure projects are designed, built, and operated in a way that maximizes public benefits while also providing a reasonable return on investment for private partners

iv. **Capacity Building**

Building the capacity of local businesses and entrepreneurs is crucial for unlocking private sector potential in sustainability (Brown et al., 2018). Governments can support capacity building through training programs, technology transfer initiatives, and access to financing options, which can empower small and medium enterprises (SMEs) to participate in sustainable ventures (Kumar & Patel, 2019). By providing SMEs with the knowledge, skills, and resources they need to engage in sustainable projects, governments can help create a more diverse and resilient private sector that is better equipped to contribute to sustainability goals. Capacity building can also help foster innovation and entrepreneurship in the sustainability space, leading to the development of new technologies and business models that can accelerate the transition to a more sustainable future.

The Role of Government Policies in Incentivizing Private Sector Participation and Investment in Sustainability

Government policies and initiatives play a vital role in shaping the investment landscape for sustainable projects. By providing incentives and establishing regulatory frameworks, governments at the national, state, and local levels can effectively encourage private sector involvement in sustainable ventures. This review examines a range of policies and initiatives designed to create an enabling environment for private sector participation and investment in sustainability.

National Level Policies

- Renewable Energy Tax Credits

Governments worldwide, including the United States, have implemented renewable energy tax credits and incentives to encourage private sector investment in clean energy projects (U.S. DEPArtment of energy, 2020). These financial incentives reduce the upfront costs and financial risks associated with renewable energy investments, making them more attractive to private investors and lowering the barrier to entry for companies looking to invest in technologies such as solar, wind, and geothermal power. By offering tax credits, governments can accelerate the deployment of clean energy solutions and contribute to the transition towards a more sustainable and low-carbon future.

- Carbon Pricing Mechanisms

Carbon pricing mechanisms, such as carbon taxes and emissions trading schemes, have emerged as powerful tools for incentivizing private sector investment in sustainability (World Bank, 2021). By putting a price on carbon emissions, governments can effectively internalize the environmental costs of greenhouse gas emissions and create a financial incentive for businesses to reduce their carbon footprint. This approach encourages companies to invest in cleaner technologies, improve energy efficiency, and adopt more sustainable practices. As the cost of carbon emissions rises, businesses are motivated to explore innovative solutions and develop new technologies that can help them remain competitive in a low-carbon economy.

- Public-Private Partnerships (PPPs)

Public-private partnerships (PPPs) have become an increasingly popular mechanism for financing and delivering large-scale sustainable infrastructure projects (world economic forum, 2019). National governments often facilitate PPPs to leverage private sector expertise, innovation, and resources in the development of critical infrastructure, such as renewable energy facilities, sustainable transportation systems, and green buildings. By partnering with the private sector, governments can address infrastructure gaps while ensuring that sustainability considerations are integrated into project design and implementation. PPPs allow for the sharing of risks and rewards between public and private entities, creating a strong incentive for private companies to deliver high-quality, sustainable infrastructure that meets the needs of communities and contributes to long-term economic growth.

State Level Policies

- Renewable Portfolio Standards (RPS)

Many U.S. states have adopted renewable portfolio standards (RPS), which mandate utilities to generate a specified percentage of their electricity from renewable sources, creating a strong market demand for renewable energy and stimulating private sector investment in solar, wind, and other clean energy technologies (U.S.

Environmental protection agency, 2021). The design of RPS policies varies by state, reflecting local energy market conditions, renewable resource potential, and policy priorities, and often includes specific targets for different types of renewable energy technologies, as well as provisions for energy storage, energy efficiency, or other complementary technologies. The implementation of RPS policies has been a key driver of renewable energy growth in the United States, attracting private sector investment, driving down costs through economies of scale and technological innovation, and helping many states meet or exceed their RPS targets, demonstrating the effectiveness of this policy approach in accelerating the transition to a clean energy future.

- Green Building Codes and Incentives

State governments are increasingly adopting green building codes and offering incentives for sustainable construction practices to promote energy efficiency and reduce the environmental impact of the built environment (American council for an energy-efficient economy, 2020). These codes establish minimum standards for energy efficiency, water conservation, and other sustainability features in new construction and major renovations, while incentives such as tax credits, grants, expedited permitting, or density bonuses encourage developers and building owners to go beyond minimum requirements and achieve higher levels of sustainability. As more states adopt these policies, the building industry is responding by developing new technologies, materials, and design strategies that can help achieve higher levels of sustainability at lower costs, creating new opportunities for private sector investment and innovation in the green building sector.

Local Level Initiatives

- Municipal Sustainability Plans

Cities worldwide are developing comprehensive sustainability plans that outline ambitious goals and strategies for reducing greenhouse gas emissions, enhancing energy efficiency, and promoting renewable energy (ICLEI - local governments for sustainability, 2021). These plans serve as a roadmap for transforming urban areas into more resilient, livable, and environmentally friendly spaces, and often include incentives for businesses to adopt sustainable practices and invest in clean technologies. Municipal sustainability plans also emphasize community engagement and education, partnering with local schools, universities, and community organizations to raise awareness about sustainability issues and promote sustainable behaviors. By engaging stakeholders in the planning process and fostering a culture of sustainability, cities can mobilize resources and partnerships to achieve their goals and drive private sector investment and innovation in clean technologies and sustainable practices.

- Green Procurement Policies

Local governments are leveraging their purchasing power to promote sustainability by incorporating environmental criteria into their procurement processes, prioritizing the purchase of environmentally friendly products and services (united nations environment programme, 2018). Green procurement policies can establish minimum environmental standards, give preference to suppliers with demonstrated commitments to sustainability, or require suppliers to provide environmental impact assessments or disclose their greenhouse gas emissions. By signalling a clear preference for sustainable options, these policies can help shift market demand, encourage private sector investment in the development of greener products and technologies, and lead to increased competition, innovation, and cost savings in the green goods and services sector, making sustainable options more accessible and affordable for both public and private sector purchasers.

Analysis of Qualitative Data on Policies for Incentivizing Private Sector Participation in Sustainable Projects

Qualitative data from policy documents, case studies, and interviews provide valuable insights into the effectiveness of various policies aimed at incentivizing private sector engagement in sustainable projects. This analysis synthesizes key findings from qualitative sources to understand the drivers, challenges, and outcomes associated with these policies.

Policy Documents

- Renewable Energy Incentives

Policy documents underscore the crucial role of financial incentives, such as feed-in tariffs and tax credits, in stimulating private sector investment in renewable energy projects (international renewable energy agency, 2020). Feed-in tariffs, which guarantee a fixed price for electricity generated from renewable sources, provide a stable and predictable revenue stream for investors, reducing financial risks and enhancing the competitiveness of renewable energy technologies. Similarly, tax incentives, including tax deductions, exemptions, and credits, increase the profitability of renewable energy projects by reducing their tax liability, further encouraging private sector investment. The availability and design of these incentives can significantly influence the pace and scale of renewable energy deployment.

- Regulatory Frameworks

A clear and stable regulatory framework is essential for creating certainty for investors and fostering long-term investment in sustainable projects, as revealed by the analysis of regulatory documents (World Bank, 2019). Transparent permitting processes, streamlined approvals, and predictable policies are cited as critical elements for attracting private sector capital. A well-designed regulatory framework can help mitigate risks, reduce transaction costs, and create a level playing field for renewable energy technologies. Conversely, regulatory uncertainty or inconsistency can deter private sector investment, as it increases the perceived risk and undermines the financial viability of sustainable projects.

- Impact Investment Funds

Case studies of impact investment funds highlight the role of innovative financing mechanisms in mobilizing private capital for sustainability (Global Impact Investing Network, 2021). Impact investment funds, which seek to generate both financial returns and positive social or environmental outcomes, have emerged as a promising vehicle for channeling private investment into sustainable projects. By aligning investors' financial objectives with sustainability goals, impact funds can tap into a growing pool of capital seeking to make a positive impact. However, the success of impact investment funds depends on factors such as the availability of suitable investment opportunities, the ability to measure and report on impact, and the presence of enabling policy and regulatory frameworks.

- Public-Private Partnerships (PPPs)

Examination of PPP case studies underscores the importance of collaborative governance models in delivering sustainable infrastructure projects (Asian development bank, 2018). Successful PPPs involve clear communication, risk-sharing mechanisms, and alignment of interests between public and private stakeholders. PPTs can leverage the strengths of both the public and private sectors, with the public sector providing policy support, risk mitigation, and long-term planning, while the private sector contributes technical expertise, operational efficiency, and access to capital. However, the effectiveness of PPPs in promoting sustainability depends on factors such as the design of the partnership, the allocation of risks and rewards, and the capacity of the public sector to manage and monitor the project.

- Stakeholder Perspectives

Interviews with stakeholders, including government officials, business leaders, and civil society representatives, provide nuanced insights into the drivers and barriers to private sector participation in sustainability (Fischer et al., 2020). Key themes that emerge include the need for policy stability, access to finance, capacity building, and stakeholder engagement. Stakeholders emphasize the importance of long-term policy commitments, transparent and accessible financing mechanisms, technical assistance and knowledge sharing, and inclusive decision-making processes in fostering private sector investment in sustainability.

- Lessons Learned

Interviews with project developers and investors offer valuable lessons learned from past experiences in sustainable investments (International Finance Corporation, 2019). Common challenges identified include regulatory uncertainty, limited access to finance, and the importance of local partnerships and community engagement. Project developers stress the need for a stable and supportive policy environment, with clear and consistent regulations, streamlined permitting processes, and long-term incentives. Investors highlight the importance of innovative financing mechanisms, such as green bonds and impact investment funds, in mobilizing private capital for sustainability. Both developers and investors emphasize the value of local partnerships and community engagement in building support for sustainable projects and mitigating social and environmental risks.

III. CHALLENGES AND BARRIERS TO IMPLEMENTING PUBLIC-PRIVATE PARTNERSHIPS IN THE GREEN INFRASTRUCTURE SECTOR

Public-private partnerships (PPPs) have emerged as a promising mechanism for financing and delivering green infrastructure projects, which are crucial for sustainable development and climate mitigation efforts. However, the implementation of PPPs in the green infrastructure sector faces various challenges and barriers, including technical barriers (e.g., complexity and novelty of green technologies) (Sarker & Lu, 2020), financial barriers (e.g., high upfront costs, long payback periods, and limited availability of specialized green financing instruments) (Zhang et al., 2019; yeo et al., 2021), and institutional and regulatory barriers (e.g., weak institutional capacities, inadequate legal frameworks, inconsistent policy support, and lack of coordination between government agencies) (Bai et al., 2020; Mukhtarov & Karakaya, 2020).

Overcoming these barriers requires a concerted effort from both public and private sector stakeholders, with governments creating an enabling environment through clear policy support, specialized financial instruments, and institutional capacity building (su, 2021), and private sector actors contributing innovative technologies, operational expertise, and financial resources while engaging in collaborative dialogue with public sector counterparts (Morgan et al., 2019). Successful PPPs in green infrastructure depend on effective risk management, alignment of incentives, robust contractual frameworks, transparent governance mechanisms, and engagement of local communities and stakeholders (Kivimaa & Martiskainen, 2019; Bryson et al., 2020).

Key Challenges and Barriers

Green infrastructure projects face multifaceted challenges, including technical complexities, policy uncertainties, financial risks, lack of capacity and expertise, and community opposition (Ahlawat & Singh, 2019; Niti Aayog, n.d.). Unclear regulatory frameworks, high upfront costs, long payback periods, uncertain revenue streams, and limited availability of specialized green financing instruments pose financial risks (Ahlawat & Singh, 2019; Gerrard, 2001). Both public and private sector entities may lack the necessary capacity and expertise, while community opposition and inadequate stakeholder engagement can delay or derail projects (Ahlawat & Singh, 2019; Ghosh & Nanda, 2017). Addressing these challenges requires coordinated efforts from governments, private sector actors, and civil society. Governments should develop clear policy frameworks, provide financial incentives, and build institutional capacity, while private sector partners contribute technical expertise, innovative solutions, and financial resources (Gerrard, 2001; Ghosh & Nanda, 2017).

National Level Policies

- Policy Uncertainty

Inconsistent or ambiguous national policies related to green infrastructure and PPPS can create significant uncertainty for private sector investors (Mukhtarov & Karakaya, 2020). A lack of clarity regarding regulatory frameworks, incentives, and investment priorities may deter private sector participation in green PPPS projects. When governments fail to provide a stable and predictable policy environment, private investors perceive higher risks and may be reluctant to commit to long-term green infrastructure projects. Policy uncertainty can act as a "hefty tax on investment," discouraging private sector involvement even when the underlying projects are economically viable. To address this challenge, governments need to develop clear, consistent, and long-term policy frameworks that provide a reliable foundation for private sector participation in green infrastructure development.

- Limited Financing Options

National governments may face constraints in providing adequate financial support for green infrastructure projects, leading to a reliance on private sector financing through PPPs (su, 2021). However, limited access to concessional finance and challenges in structuring bankable projects can hinder PPP implementation. Governments may struggle to attract private capital if they cannot provide sufficient financial incentives or risk mitigation measures. To overcome these financing barriers, governments can explore innovative financing mechanisms, such as green bonds, tax incentives, and guarantee schemes, to enhance the attractiveness of green PPP projects to private investors. Multilateral development banks and international financial institutions can also play a crucial role in providing concessional finance, technical assistance, and risk mitigation tools to support green PPP projects in developing countries.

State Level Regulations

- Fragmented Governance

Variation in regulatory frameworks and governance structures across states can complicate PPP implementation in the green infrastructure sector, creating a fragmented and inconsistent environment for PPP projects (Morgan et al., 2019). Misalignment between state policies and national priorities may hinder coordination and collaboration between public and private stakeholders, leading to delays, increased transaction costs, and uncertainty for private investors. To address this challenge, governments need to promote greater coordination and collaboration between national and state-level authorities, work towards harmonizing regulatory frameworks and governance structures, establish national guidelines and standards for green PPPs while allowing for some flexibility to accommodate state-specific contexts and needs, and strengthen institutional capacities and foster knowledge sharing across states.

- Legal and Contractual Issues

State-level regulations governing PPP contracts and procurement processes may lack clarity or flexibility, leading to delays, disputes, and increased transaction costs that discourage private sector investment (Davoodi & Mocherniak, 2021). To mitigate these legal and contractual issues, state governments need to develop clear, transparent, and flexible PPP frameworks tailored to the needs of green infrastructure projects, streamline procurement processes, standardize contract templates, provide guidance and support to private partners, and consider adopting innovative and collaborative PPP models. Building the capacity of public officials to effectively design, negotiate, and manage PPP contracts is also crucial to ensuring the success of green infrastructure projects.

Local Level Initiatives

- Community Opposition

Local communities and stakeholders may resist green infrastructure projects delivered through PPPs due to concerns about environmental impacts, land use, and public participation, and a lack of community engagement and inadequate consultation processes can delay project approvals and implementation (Kivimaa & Martiskainen, 2019). To mitigate community opposition, local governments and private partners must prioritize meaningful and inclusive stakeholder engagement throughout the project lifecycle, establishing transparent and accessible communication channels, conducting regular public consultations, and incorporating community feedback into project design and implementation. Integrating social and environmental safeguards into PPP contracts and monitoring frameworks can also help ensure that green infrastructure projects align with community values and priorities and contribute to equitable and sustainable development outcomes.

- Capacity Constraints

Local governments may lack technical expertise and institutional capacity to effectively manage PPPs and ensure project transparency and accountability, which can lead to inadequate project preparation, weak contract management, insufficient performance monitoring, and potential corruption risks (Bryson et al., 2020). To address capacity constraints, local governments need to invest in building the skills and knowledge of their staff, establish dedicated PPP units or teams, partner with external experts, and develop standardized tools, templates, and guidelines for PPP project preparation, procurement, and monitoring. By strengthening institutional capacities and leveraging external expertise, local governments can improve their ability to effectively manage and deliver green infrastructure PPPs that meet the needs of their communities.

Analysis of Qualitative Data on Challenges and Barriers to Implementing PPPs in the Green Infrastructure Sector

Qualitative data from policy documents, case studies, and interviews provide valuable insights into the challenges and barriers that hinder the successful implementation of PPPs in the green infrastructure sector. This analysis synthesizes key findings from qualitative sources to identify common obstacles and potential solutions.

Policy Documents

- Regulatory Complexity

Policy documents highlight the complex regulatory landscape governing green infrastructure PPPs, encompassing procurement rules, environmental regulations, and land-use planning requirements, which can be time-consuming and costly for project developers to navigate, potentially leading to delays and increased project risks (OECD, 2020). To mitigate these challenges, policymakers need to streamline and harmonize regulatory frameworks, provide clear and consistent guidance to project developers, and establish a single point of contact or dedicated agency to coordinate regulatory approvals and provide technical support, thereby reducing transaction costs and improving project efficiency.

- Financial Viability

Policy documents highlight the financial challenges associated with green infrastructure PPPs, including high upfront costs, limited revenue streams, and uncertain long-term returns on investment, which can make private sector investors hesitant to participate, particularly in emerging markets with higher political and economic risks. To address these issues, policymakers and development partners can create enabling financial frameworks, including innovative funding mechanisms, risk-sharing arrangements, and credit enhancement tools, as well as provide government guarantees, subsidies, or performance-based payments to improve project viability and attract private investment. Developing standardized project preparation and financial structuring templates can also help reduce transaction costs and improve the bankability of green infrastructure PPPs.

- Project Delays

Case studies of green infrastructure PPPs reveal frequent delays in project implementation due to regulatory hurdles, community opposition, and financing challenges, which can significantly increase project costs, undermine investor confidence, and potentially lead to project cancellations or renegotiations (World Bank Group, 2019). To mitigate these risks, project stakeholders need to adopt a proactive and collaborative approach, conducting thorough feasibility studies and risk assessments, engaging with communities and stakeholders early in the project cycle, securing firm financial commitments, and establishing clear and realistic project timelines with built-in contingencies for potential delays.

- Community Engagement

Case studies highlight the importance of meaningful community engagement in green infrastructure projects delivered through PPPs, as a lack of community consultation and stakeholder engagement can result in local opposition, legal disputes, and reputational risks for project sponsors (Wang & Murdock, 2018). To avoid these pitfalls, project stakeholders need to prioritize community engagement and stakeholder consultation throughout the project lifecycle by establishing clear and accessible communication channels, conducting regular public meetings and workshops, and incorporating community feedback into project design and implementation, thereby building trust and fostering a sense of ownership among local communities, reducing the risk of opposition, and improving the long-term sustainability and impact of green infrastructure PPPs.

- Social Inclusion and Community Engagement in Green PPPs

Based on the analysis of government policies, case studies, and stakeholder perspectives, expanding discussions on social inclusion and community engagement in green PPPs in India requires a multi-pronged approach. Governments should prioritize meaningful stakeholder engagement throughout the PPPP project lifecycle, going beyond mere consultation to collaboratively incorporating community needs and feedback into project planning and implementation (Ahlawat & Singh, 2019; Niti Aayog, n.d.; Fouad, 2021). PPP objectives, contracts, and monitoring frameworks should explicitly incorporate social inclusion goals and metrics like local job creation, skills development, gender equity, and access to services for vulnerable groups (Ahlawat & Singh, 2019; Fouad, 2021; EPA, 2015).

Building capacities of all stakeholders to effectively participate in PPPs is crucial (Ahlawat & Singh, 2019; Lueckenhoff & brown, 2015; Shrivastava & Kumar, 2015), as is promoting people-first PPP models that optimize social value and community benefits (Ahlawat & Singh, 2019; Niti Aayog, n.d.; Fouad, 2021; Puppim de Oliveira et al., 2018). Facilitating knowledge sharing through peer networks and collaborating with academic institutions can enable good practice exchange (Ahlawat & Singh, 2019; Lueckenhoff & brown, 2015; Puppim de Oliveira et al., 2018; Shrivastava & Kumar, 2015). Strengthening local government and community roles in initiating and monitoring PPP can help localize benefits (Ahlawat & Singh, 2019; Niti Aayog, n.d.; EPA, 2015; Mathur et al., 2014). Therefore, expanding PPP frameworks to enable community-driven planning, integrate social equity goals, strengthen stakeholder capacities, and promote pro-poor models can help center social inclusion in India's green infrastructure development (Ahlawat & Singh, 2019; Niti Aayog, n.d.; Fouad, 2021; EPA, 2015; Lueckenhoff & brown, 2015; Mathur et al., 2014; Puppim de Oliveira et al., 2018; Shrivastava & Kumar, 2015). Continued multi-stakeholder dialogues and pilots can further enrich approaches (Ahlawat & Singh, 2019; Niti Aayog, n.d.; Fouad, 2021; Puppim de Oliveira et al., 2018).

- Stakeholder Perspectives

Interviews with government officials, private sector representatives, and civil society organizations reveal common challenges facing green infrastructure PPPs, including regulatory uncertainty, funding constraints, capacity gaps, and the need for improved governance mechanisms (Rosenberg, 2020). Stakeholders highlight the lack of clear and consistent policy frameworks, absence of dedicated funding sources and long-term financing options, need for capacity building and technical assistance for local governments and small-scale project developers, and importance of improving transparency, accountability, and stakeholder participation. These diverse perspectives underscore the necessity of a holistic and inclusive approach to project planning and implementation that addresses the needs and concerns of all stakeholders.

- Lessons Learned

Interviews with project stakeholders yield valuable lessons from past PPP experiences, emphasizing the importance of clear project objectives, transparent procurement processes, robust risk allocation mechanisms, building trust among stakeholders, and fostering long-term partnerships (Stiglitz & Wallsten, 2021). Stakeholders highlight the need for a shared vision, alignment among parties, competitive procurement to ensure value for money and reduced corruption risks, clear contractual provisions and risk-sharing arrangements to manage uncertainties and align incentives, and regular communication, collaborative problem-solving, and a focus on shared goals and benefits to build trust and foster partnerships. These lessons provide insights into improving the design and implementation of green infrastructure PPPs.

IV. RECOMMENDATIONS FOR ENHANCING THE EFFECTIVENESS OF GOVERNMENT POLICIES IN FACILITATING PPPS FOR SUSTAINABLE DEVELOPMENT IN INDIA

- **Streamline Regulatory Frameworks**

To reduce red tape and enhance investor confidence, it is crucial to simplify and harmonize regulatory frameworks governing PPPs by streamlining processes, eliminating unnecessary or duplicative requirements, and ensuring consistency across sectors and jurisdictions (Das et al., 2020). Policymakers can develop standardized PPP guidelines, contracts, and procurement documents aligned with international best practices and adapted to local contexts and establish clear and transparent frameworks for risk allocation and dispute resolution to build trust and confidence among project stakeholders.

- **Strengthen Institutional Capacity**

Investing in building the capacity of government agencies responsible for PPP procurement, management, and oversight is essential for effective project implementation and monitoring (Sarker & Lu, 2021). This can be achieved by providing training and technical assistance to public officials, establishing dedicated PPP units or teams to build specialized expertise and improve coordination, and developing robust monitoring and evaluation frameworks to track performance and identify areas for improvement. By strengthening institutional capabilities, governments can better manage the complexities of PPP projects and ensure they deliver the expected benefits to society.

- **Promote Stakeholder Engagement**

Fostering meaningful engagement with local communities, civil society organizations, and other stakeholders throughout the PPP project lifecycle is critical to addressing social and environmental concerns and building trust (Gulati et al., 2021). This involves going beyond mere consultation and information sharing, to actively involve stakeholders in project planning, design, and implementation. Establishing clear and accessible grievance redress mechanisms can also help to address concerns and resolve conflicts in a timely and transparent manner. By promoting inclusive and participatory approaches to PPP projects, governments can ensure that they are responsive to local needs and priorities, and that they generate broad-based support and ownership among stakeholders.

- **Improve Transparency and Accountability**

Enhancing transparency in PPP procurement processes, project appraisal, and contract management is essential to mitigate corruption risks and promote accountability (Gupta & Sharma, 2020). This involves establishing clear and competitive bidding procedures, disclosing key project information and performance data, and subjecting PPP contracts to public scrutiny and oversight. Strengthening the role of independent auditors and regulators can also help to ensure that PPP projects are delivered in accordance with legal and ethical standards. By promoting greater transparency and accountability, governments can build public trust in PPP projects and ensure that they deliver value for money and positive social and environmental outcomes.

- **Provide Financial Incentives**

Offering financial incentives such as viability gap funding, tax breaks, and concessional finance can help attract private sector investment in sustainable development projects (Chatterjee & Dutta, 2019). These incentives improve the risk-return profile of PPP projects, reduce capital costs, and mitigate risks for private investors, helping to bridge the funding gap for sustainable infrastructure.

- **Facilitate Access to Financing**

Establishing dedicated funding mechanisms and platforms, such as infrastructure funds, green bonds, and standardized project preparation templates, can facilitate access to finance for PPP projects (Bandyopadhyay et al., 2021). Collaborating with development banks to provide credit enhancement tools can mitigate risks and attract private capital, unlocking the potential of PPPs to deliver sustainable infrastructure solutions.

- **Promote Knowledge Sharing and Capacity Building**

Facilitating knowledge sharing and capacity building through training programs, workshops, peer-to-peer learning platforms, and collaborating with academic institutions can improve PPP project planning and implementation (Ghosh et al.,

2020). Establishing knowledge hubs, developing curricula for PPP professionals, and providing technical assistance can build specialized expertise and create an enabling environment for PPP projects, ensuring they are delivered in accordance with international standards and best practices.

- **Encourage Innovation and Technology Adoption**

Encouraging innovation and the adoption of sustainable technologies in PPP projects through incentives, awards, and technology demonstration programs can help to drive the transition to a low-carbon and resilient future (Sharma & Singh, 2021). This may involve providing grants or subsidies for research and development of new technologies, such as renewable energy, energy efficiency, or sustainable transportation solutions. Establishing awards or recognition programs for innovative PPP projects can also help to showcase best practices and inspire replication in other contexts. Governments can also work with the private sector and academic institutions to establish technology demonstration programs or living labs, where new technologies can be tested and refined in real-world settings. By promoting innovation and technology adoption, governments can help to accelerate the deployment of sustainable infrastructure solutions and create new opportunities for private sector investment and job creation.

CONCLUSION

The review revealed a diverse array of government policies at national, state, and local levels aimed at promoting PPPs in the green infrastructure sector, encompassing regulatory frameworks, financial incentives, capacity-building initiatives, and stakeholder engagement mechanisms (Ahlawat & Singh, 2019; Gerrard, 2001; Fouad, 2021). While these policies have incentivized some private sector participation, challenges such as regulatory complexity, financial viability concerns, community opposition, and capacity constraints persist as major impediments to successful PPP implementation (Niti Aayog, n.d.; Monteiro et al., 2020; Lueckenhoff & Brown, 2015). To enhance the effectiveness of government policies in facilitating PPPs for sustainable development, recommendations include streamlining regulatory frameworks, strengthening institutional capacity, promoting stakeholder engagement, and providing financial incentives (EPA, 2015; Shrivastava & Kumar, 2015; Jha & Singh, 2016). Addressing these challenges requires a coordinated effort from government authorities, private sector stakeholders, civil society organizations, and local communities.

Ongoing monitoring and evaluation of government policies is crucial to ensure their effectiveness and adaptability to evolving socio-economic and environmental contexts. With targeted policy interventions and collaborative partnerships, India can harness the transformative potential of PPPs to drive low-carbon green infrastructure development and advance

REFERENCES

Ahlawat, A., & Singh, G. (2019). Challenges in Public Private Partnership Infrastructure Project-A Case Study. In *Conference Paper May*.

American Council for an Energy-Efficient Economy. (2020). State and Local Policy Database. Retrieved from https://database.aceee.org/state/energy-code

Asian Development Bank. (2018). *Public-Private Partnerships in Infrastructure Development: Lessons Learned from Case Studies in Asia and the Pacific*. Asian Development Bank.

Bai, X., (2020). Overcoming barriers to urban green infrastructure: Lessons from Singapore. *Journal of Environmental Management*, 266, 110605.

Bandyopadhyay, S., (2021). Access to finance for sustainable infrastructure development: Insights from India. *International Journal of Sustainable Development and World Ecology*, 28(2), 116–130.

Brown, A., & Jones, B. (2019). Regulatory frameworks for sustainable investment: A comparative analysis. *Journal of Sustainable Development*, 12(3), 45–58.

Brown, C., (2018). Capacity building strategies for private sector engagement in sustainability: Lessons from emerging economies. *Sustainable Development Journal*, 15(2), 78–91.

Chatterjee, S., & Dutta, S. (2019). Financing public-private partnerships for sustainable development: Evidence from India. *Journal of Infrastructure Development*, 11(2), 164–180.

Das, A., (2020). Regulatory reforms for public-private partnerships in infrastructure: Lessons from India. *Utilities Policy*, 64, 101032.

Davoodi, S. M. M., & Mocherniak, S. (2021). Legal Issues and Challenges in Public-Private Partnerships: An International Perspective. *Journal of Legal Affairs and Dispute Resolution in Engineering and Construction*, 13(1), 04521001.

Erismann, S., Pesantes, M. A., Beran, D., Leuenberger, A., Farnham, A., Berger Gonzalez de White, M., Labhardt, N. D., Tediosi, F., Akweongo, P., Kuwawenaruwa, A., Zinsstag, J., Brugger, F., Somerville, C., Wyss, K., & Prytherch, H. (2021). How to bring research evidence into policy? Synthesizing strategies of five research projects in low-and middle-income countries. *Health Research Policy and Systems*, 19(1), 1–13. DOI: 10.1186/s12961-020-00646-1 PMID: 33676518

European Investment Bank. (2021). *Green Infrastructure Financing Strategies: Lessons Learned from Successful Projects*. European Investment Bank.

Fischer, M., (2020). Stakeholder Perspectives on Private Sector Participation in Sustainable Development: Insights from Interviews. *Journal of Sustainable Development*, 15(3), 78–92.

. Fouad, M. (2021). Mastering the risky business of public-private partnerships in infrastructure.

Gerrard, M. (2001). Public-private partnerships. Finance and Development, 38(3), 48-51. Monteiro, R., Ferreira, J. C., & Antunes, P. (2020). Green infrastructure planning principles: An integrated literature review. *Land (Basel)*, 9(12), 525.

Ghosh, S., (2020). Capacity building for public-private partnerships in sustainable development: A case study of India. *Sustainable Development*, 28(4), 988–999.

Ghosh, S., & Nanda, P. (2017). *Green Infrastructure Financing: Institutional Investors, PPPs and Bankable Projects*. Palgrave Macmillan.

Global Impact Investing Network. (2021). Impact Investment Case Studies. Retrieved from https://thegiin.org/research/publication/impact-investment-case-studies

Gulati, M., & Rao, M. Y. (2019). Financing India's renewable energy vision: The case of solar parks. *Energy Policy*, 128, 158–164.

Gulati, R., (2021). Enhancing stakeholder engagement in public-private partnerships for sustainable development: Insights from India. *Sustainable Cities and Society*, 70, 102885.

Gupta, R., & Sharma, S. (2020). Transparency and accountability in public-private partnerships for sustainable development: Evidence from India. *Journal of Public Affairs*, 20(3), e2162.

Gupta, R., & Sharma, S. (2022). Public-private partnerships in sustainable infrastructure development: A systematic review. *Journal of Environmental Policy and Planning*, 25(1), 112–128.

Hyderabad Metro Rail Limited. (2021). Hyderabad Metro Rail Project: Transforming Urban Mobility. Retrieved from https://hmrl.co.in/project-overview/

ICLEI - Local Governments for Sustainability. (2021). Sustainability Planning. Retrieved from https://icleiusa.org/programs/planning/sustainability-planning/

International Finance Corporation. (2019). Unlocking Private Investment for Sustainable Development: Lessons from Investors. Retrieved from https://www.ifc.org/wps/wcm/connect/topics_ext_content/ifc_external_corporate_site/sustainability-at-ifc/publications/publications_handbook_unlockingprivateinvestment

International Renewable Energy Agency. (2020). *Renewable Energy Policies in a Time of Transition: Focus on Africa*. International Renewable Energy Agency.

Jha, V., & Singh, A. (2016). Green Infrastructure in Indian Cities: Policies and Practices. *Public Affairs and Administration: Concepts, Methodologies, Tools, and Applications*, 3, 1315–1329.

Johnson, T., & Wang, L. (2020). The role of financial incentives in promoting private sector investment in renewable energy projects. *Renewable Energy*, 45(4), 201–215.

Kivimaa, P., & Martiskainen, M. (2019). Innovation and adoption in the water sector: a systematic literature review of barriers and facilitators. Environmental Innovation and Societal Transitions, 33, 165-183. Bryson, J. M., et al. (2020). Coproduction and Public-Private Partnership: A Comparative Analysis of Governance Challenges. *Public Administration Review*, 80(2), 319-328.

Kumar, R., & Bansal, P. (2018). Public Private Partnership (PPP) in Infrastructure: Case Study of India. *Journal of Infrastructure Development*, 10(1), 77–88.

Kumar, S., & Patel, R. (2019). Capacity building for sustainable entrepreneurship: Insights from developing countries. *Journal of Entrepreneurship Development*, 8(2), 87–102.

Lee, H., & Smith, J. (2017). Public-private partnerships for sustainable development: A review of models and best practices. *Journal of Sustainable Infrastructure*, 10(4), 189–204.

Lueckenhoff, D., & Brown, S. (2015). *Public–Private Partnerships Beneficial for Implementing Green Infrastructure*. Water Law & Policy Monitor.

Mathur, V. N., Thakur, P., & Rajadhyaksha, N. (2014). Public-Private Partnerships for Sustainable Infrastructure: Exploring the Challenges faced by Local Governments in India. *Environment, Development and Sustainability*, 16(2), 335–351.

Morgan, B. T., (2019). Fragmented infrastructure governance: Challenges for the strategic integration of urban infrastructure. *Land Use Policy*, 86, 401–413.

Mukhtarov, F., & Karakaya, E. (2020). Regulatory challenges to public-private partnerships for sustainable development: Evidence from a systematic literature review. *Sustainability Science*, 15(2), 535–552.

OECD. (2020). *Public-Private Partnerships for Green Infrastructure: Policy Considerations and Good Practices*. OECD Publishing.

Ostro Energy. (2021). Ostro Energy: Powering Sustainable Growth. Retrieved from https://www.ostroenergy.in/sustainability/

Patil, N. A., Tharun, D., & Laishram, B. (2016). Infrastructure development through PPPs in India: Criteria for sustainability assessment. *Journal of Environmental Planning and Management*, 59(4), 708–729. DOI: 10.1080/09640568.2015.1038337

Pramudawardhani, R., (2021). Policy framework analysis for sustainable green infrastructure development in Indonesia. *Sustainable Cities and Society*, 75, 103206.

Puppim de Oliveira, J. A., Doll, C. N., Siri, J., Dreyfus, M., Farzaneh, H., & Capon, A. (2018). Urban Governance and the Systems Approaches to Health-Environment Co-Benefits in Cities. *Cadernos de Saude Publica*, 34, e00037518.

Rizvi, A., & Sclar, E. (2014). Implementing bus rapid transit: A tale of two Indian cities. *Research in Transportation Economics*, 48, 194–204. DOI: 10.1016/j.retrec.2014.09.043

Roberts, M., (2021). Assessing the impact of environmental regulations on private sector investment in clean technologies. *Environmental Science & Policy*, 18(3), 112–127.

Rosenberg, J. (2020). Stakeholder perspectives on barriers to implementing green infrastructure projects through public-private partnerships. *Environmental Policy and Governance*, 30(6), 423–437.

Sarker, M. R. A., & Lu, W. (2020). Public-private partnership (PPP) for sustainable green infrastructure development: A review. *Journal of Cleaner Production*, 249, 119315.

Sarker, M. R. A., & Lu, W. (2021). Institutional capacity building for public-private partnerships in infrastructure: A case study of India. *International Journal of Public Administration*, 44(5), 370–383.

Sharma, A., & Singh, S. (2021). Promoting innovation in public-private partnerships for sustainable development: The role of government policies. *Journal of Public Procurement*, 21(2), 127–143.

Shrivastava, A., & Kumar, R. (2015). Green Infrastructure Financing and Institutional Investors. *Public Policy and Administration Research*, 5(8), 84–93.

Smith, K. (2018). The role of financial incentives in promoting private sector engagement in sustainable development projects. *Journal of Sustainable Finance & Investment*, 22(1), 45–58.

Stiglitz, J., & Wallsten, S. (2021). Lessons learned from public-private partnerships in green infrastructure: Insights from project stakeholders. *Journal of Infrastructure Development*, 13(2), 135–151.

Su, Y. S. (2021). Challenges and opportunities of financing green public-private partnership projects: A critical review. *Sustainability*, 13(5), 2709.

United Nations Environment Programme. (2018). Green Public Procurement: A Global Review. Retrieved from https://www.unep.org/resources/report/green-public-procurement-global-review

U.S. DEPArtment of Energy. (2020). Federal Incentives for Renewable Energy. Retrieved from https://www.energy.gov/eere/federal-incentives-renewable-energy

U.S. Environmental Protection Agency. (2021). Renewable Portfolio Standards. Retrieved from https://www.EPA.gov/statelocalenergy/renewable-portfolio-standards

Wang, L., & Murdock, S. (2018). Community engagement in green infrastructure projects: Lessons from case studies. *Journal of Environmental Planning and Management*, 61(5-6), 865–881.

World Bank. (2006). *Alandur Sewerage Project: The Tamil Nadu Experience. Water and Sanitation Program Field Note*. Washington, DC: World Bank. JICA. (2016). *Berhampur Solid Waste Management Project: A PPP Model for Sustainable Urban Development*. Japan International Cooperation Agency.

World Bank. (2019). *Regulatory Governance for Infrastructure Sector Investment: Synthesis Report*. World Bank.

World Bank. (2021). State and Trends of Carbon Pricing 2021. Retrieved from https://openknowledge.worldbank.org/handle/10986/36316

World Bank Group. (2019). *Public-Private Partnerships in Infrastructure: Lessons Learned from Recent Experience*. World Bank Group.

World Economic Forum. (2019). Public-Private Collaboration for Sustainable Infrastructure: Principles and Toolkit. Retrieved from http://www3.weforum.org/docs/WEF_Public_Private_Collaboration_Sustainable_Infrastructure_Report_2019.pdf

Yeo, M. A., (2021). Public-private partnerships for sustainable urban green infrastructure: A case study of Seoul, South Korea. *Sustainability*, 13(1), 381.

Zhang, Y., (2019). Barriers to the development of green infrastructure: Evidence from China. *Sustainability*, 11(2), 481.

Chapter 7
Sustainable Water and Sanitation Management:
An Analysis of City Corporations' Challenges

Imran Hossain
Varendra University, Bangladesh

Asfaq Salehin
Varendra University, Bangladesh

A.K.M. Mahmudul Haque
University of Rajshahi, Bangladesh

Abdur Rahman Mia
University of Rajshahi, Bangladesh

ABSTRACT

This chapter investigates the challenges faced by Rajshahi and Gazipur City Corporations, Bangladesh, in ensuring sustainable water and sanitation practices. A qualitative method was used to collect data from various sources, including in-depth interviews, key informant interviews, focus group discussions, and existing literature. The study has identified several major challenges faced by City Corporations (CCs) in Bangladesh. These challenges include inadequate infrastructure, population growth, unplanned urbanization, insufficient financial resources, and institutional incapacity, issues related to water reuse and conservation, and improper waste management. These challenges have a significant impact on sustainable water and sanitation management. The study identified that the wastage of water is caused by inadequate infrastructure, population growth, unplanned urbanization, institutional

DOI: 10.4018/979-8-3693-7001-8.ch007

incapacity, and poor coordination. Furthermore, the scarcity of safe water is also caused by population growth, unplanned urbanization, and the absence of water reuse and conservation systems.

1.0 INTRODUCTION

Water quality is a critical factor in human physiology (Etim et al., 2013), being one of the essential substances necessary for sustaining life, health, and ecosystems (AL-Dulaimi & Younes, 2017). However, in developing nations like Bangladesh, poor water quality has emerged as one of the most significant challenges (Akoto et al., 2017). Millions of people lack access to a safe and sufficient water supply in these nations (Kimani-Murage, 2007), and historically, developing countries have faced more pronounced water and sanitation problems than developed nations (Pandit & Kumar, 2015).

Bangladesh, as an underdeveloped nation, faces problematic access to clean water and proper sanitation. Over 90 percent of the population relies on unimproved water sources (Bhavnani et al., 2014), and 2.6 billion people do not have access to improved water supplies and 1.1 billion do not have access to adequate sanitation (Robinne et al., 2018). Urbanization and unexpected growth have worsened the issue, leading to extreme poverty and disparities in water service access (Hossen et al., 2024; Angeles et al., 2009). Consequently, water quality has reached an all-time low, posing major health concerns for urban Bangladesh.

The implications of inadequate access to clean water are terrible, with millions in Bangladesh exposed to severe waterborne illnesses, including diarrhea, cholera, dysentery, and typhoid (Hasan et al., 2019). Despite economic challenges and a dense population, Bangladesh has made progress in improving water accessibility and mitigating arsenic contamination (Loewenberg, 2007). The government has taken several initiatives, such as water safety plans, to address water pollution and provide safe drinking water (Alam, 2021).

Nevertheless, in spite of these initiatives, the CCs of Rajshahi and Gazipur are still having difficulty providing better water to their residents. Water quality issues, such as excessive iron, turbidity, hardness, and odor, persist in these areas (Lamia et al., 2018; Pandey, 2009). The water supply in tea shops and restaurants in Gazipur City is not very good (Farooq et al., 2019). High concentrations of iron and manganese, coliform bacteria contamination, arsenic contamination, and total hardness are the main factors limiting the availability of drinking water in these areas (Rasul, 2010). As a result, a considerable proportion of the population refrains from consuming CC-provided water, boiling and filtering it instead (Lamia et al., 2018).

These conditions lead to health risks and hinder access to clean drinking water for the residents (Alam et al., 2013).

However, any water, whether it comes from the surface or the ground, must be treated before use. When comparing the quality of river water to that of groundwater, it was discovered that, after proper treatment, river water is more suitable for use as potable water (Rasul, 2010). The regular residents of CC do not receive enough water due to several factors, including the absence of a water filtering system, a scarcity of tube wells, and the presence of iron in the water of RCC and GCC. To address these complex issues effectively, CCs play a crucial role as the primary actors responsible for urban planning, infrastructure development, and service delivery. However, ensuring clean water and sanitation in the rapidly growing urban centers of Bangladesh is a multifaceted task that demands a comprehensive understanding of the challenges faced by these institutions.

The main goal of this chapter is to identify the main problems of the CC for providing available water to its residents. Furthermore, this chapter also aims to investigate the challenges faced by CCs in providing clean water and sanitation in a sustainable and efficient manner. By examining the factors impeding effective water and sanitation management, this study seeks to shed light on the fundamental issues to promote sustainable development and improved quality of life for urban dwellers.

1.0 METHODOLOGY OF THE STUDY

This study employed a qualitative research approach to comprehensively examine the challenges faced by city corporations, namely Rajshahi City Corporation (RCC) and Gazipur City Corporation (GCC), in ensuring sustainable water and sanitation management (Figure A1). For this research, being aware of the fact that this study will not be making any generalization, purposive sampling was used to select RCC and GCC since there is a shortage of literature on water and sanitation management challenges for these two city corporations. Among other justifications for selecting these two, heterogeneity is important. RCC is among the first city corporations established in 1976, whereas GCC was established in 2013. It was assumed that as one of the oldest ones, RCC, and as one of the newest ones, Gazipur, will highlight different dimensions of challenges. Another reason for selecting these two is in GCC the water supply is managed by the city corporation the town authority itself but in RCC there is another dedicated governing institution to look after the water supply.

In the inquest of heterogeneity among the findings and in the hope for a prolonged saturation in the date these two city corporations were chosen as the study area.

This research utilized both primary and secondary data. To collect primary data, the researcher conducted in-depth interviews (IDIs), key informant interviews (KIIs), and focus group discussions (FGDs) in the study areas of RCC and GCC (Table A1). The qualitative data was collected from November 2022 to January 2023. The respondents were chosen purposefully to ensure focused and multidimensional information while minimizing saturation. In each city corporation, five in-depth interviews (5X2 CC) were conducted, involving key stakeholders responsible for providing clean water and sanitation. Additionally, ten key informants (5X2 CC) were interviewed in each city, including local residents, political leaders, civil society members, and water-related experts. Furthermore, each city hosted a total of eight FGDs (4X2 CC), with 8 to 10 participants in each group, including residents, slum dwellers, and members of nearby community organizations. The interviews were conducted in a semi-structured manner to gain an in-depth understanding of the difficulties faced by the CCs and the complexities of the surrounding environment affecting the management of water and sanitation. To gather multispectral insights, FGDs with residents were conducted. KIIs were also conducted by the researcher in a semi-structured manner, allowing the extraction of specific information from the practical experience of professionals on particular issues. The FGD and interview checklists addressed issues such as the unavailability of safe water, water wastage, waste management practices, governance challenges, and community engagement efforts.

The secondary data for this research were collected from various sources, including research reports, articles, official statistics, relevant books, daily newspapers, and relevant online databases. These secondary sources provided valuable information on the historical context, existing policies, and previous studies related to water and sanitation challenges in urban Bangladesh, which were critically reviewed and analyzed. Basically, the content analysis technique was employed to extract relevant information pertaining to the challenges of CCs in ensuring clean water and sanitation.

The data obtained from IDIs, KIIs, and FGDs were transcribed and organized into qualitative data. Thematic analysis was applied to identify key themes and patterns in the interview responses. The qualitative data were coded, categorized, and clustered based on persistent themes related to water and sanitation challenges in both RCC and GCC. The results of the analysis of the qualitative data were further supported and illustrated using narrative texts and direct quotations.

2.0 THEORETICAL FRAMEWORK

First of all, the theoretical underpinning of sustainable water and sanitation revolves around the imperative of providing universal access to clean water and sufficient sanitation facilities while conscientiously considering the enduring implications on the environment, society, and economy. Intimately linked with the broader construct of sustainable development, this framework strives to fulfill current needs without imperiling the needs of future generations.

The objectives mentioned above can be achieved through the establishment of robust governance structures and policies, indispensable for the effective management and regulation of water resources and sanitation services at local, regional, and national levels (Jiménez et al., 2020). Under this dimension, a strict, realistic, and environmentally aligned policy along with a governing body to enforce it is called for.

Now, City Corporations, a local government tier of Bangladesh, along with WASA as a local city governing institution, are in charge of water supply and sanitation systems. There are three popular theories of local government, namely, Democratic Participatory School, Efficient-Service School, and Developmental School (Ajulor & Busayo, 2016). Among these, Efficient Service Schools and Developmental Schools directly support the authority of the previously mentioned institutions such as CC and WASA regarding the water supply.

According to institutional theory, malgovernance is more likely to occur in countries with weak institutions, primarily in third-world countries. These institutions, such as the judiciary, the legislature, and the civil service, play an important role in holding the government accountable (Rohr, 2002; Staszewski, 2008). These questions are only raised when there is institutional incapacity and a faulty infrastructural development framework proposed by Lebel et al. (2017). As the failure of governing institutions often ends up resulting in population growth and unplanned urbanization, these are often regarded as malgovernance, a holistic issue under which there are other challenges. According to dependency theory, it is obvious that third-world countries will lack the proper financial backing to develop their systems (Smelser & Baltes, 2001). According to dependency theory, these underprivileged countries will inevitably suffer from the modernization crisis.

According to Modernization theory, challenges faced by the third world are basically attributed to a perceived lag in adopting modern practices and institutions (Gwynne, 2009). The challenges regarding water sanitation management will be the absence of water reuse and conservation systems, improper waste management and sanitation systems, and last but not least, inadequate infrastructure. A forward-looking stance on technological innovation is crucial for this issue. The incorporation of innovative and sustainable technologies, such as water purification systems, water recycling, and energy-efficient sanitation solutions, holds the potential to augment the overall efficiency and environmental performance of water and sanitation systems (Connor Richard et al., 2017).

In this case, the inevitability of Climate-Resilient Water and Sanitation Infrastructure comes to the forefront. This involves designing water and sanitation infrastructure that accounts for the impacts of climate change and is resilient to extreme weather events in the long term (World Bank, 2019). This certain attribute of water and sanitation management ensures the recovery of water infrastructure after major and minor weather shocks or natural disasters. This trait also enables the system to respond smartly during times of drought and monsoon, meaning the time of scarcity and abundance.

To establish a sustainable water supply and sanitation system, several dimensions must be taken into account (Hossain et al., 2024a; Hossain et al., 2024b). For instance, the adoption of Integrated Water Resource Management (IWRM) is pivotal. IWRM advocates for a comprehensive and sustainable approach to the management and utilization of water resources, recognizing the intricate interplay of social, economic, and environmental factors (Agarwal et al., 2000). IWRM is a comprehensive and interlinked water and sanitation management plan that basically incorporates all the technical factors (e.g. water source, fuel, supply line, purification system, recycling system) and stakeholders (e.g. governing bodies, consumers, and investors), as well as all the related issues of water and sanitation management. The study takes all the challenges into account and draws their impact on maintaining sustainability in water supply and sanitation management.

3.0 CHALLENGES OF CCs FOR MANAGING WATER AND SANITATION IN A SUSTAINABLE WAY

3.1 Inadequate Infrastructure

Ensuring sustainable development in Rajshahi and Gazipur City Corporation (RCC and GCC) requires overcoming a number of obstacles in maintaining clean water and sanitation. One of the major obstacles is inadequate infrastructure, which makes it difficult to provide clean water and efficient sanitation services (Etim et al., 2013; Al-Dulaimi & Younes, 2017).

Several serious problems trouble the current water distribution and quality maintenance infrastructure in the RCC and GCC. Old water distribution networks and inadequate maintenance lead to water loss and contamination due to leaks (Moe & Rheingans, 2006). The water supply for residents is at risk due to frequent leaks and an ineffective repair mechanism (Ferdous et al., 2018). Additionally, treatment plants' limited capacity and outdated purification methods make it challenging to ensure a steady supply of safe drinking water (Pandey, 2009).

Inadequate waste management and sanitary infrastructure exacerbate the problems. Insufficient sewage treatment plants and outdated waste disposal techniques lead to improper management of wastewater and solid waste (Hossain et al., 2024c; Robinne et al., 2018). Water bodies become contaminated as a result, endangering public health (Ferdous, 2018b). Insufficient sanitation facilities increase the risk of waterborne infections, particularly in slums and densely populated areas where hygiene practices are lacking (Farooq et al., 2019).

The inadequacy of infrastructure in these areas is exacerbated by rapid urbanization and population growth (Hossain et al., 2024e). The demand for water and sanitation services exceeds the capacity of current systems, putting additional strain on already insufficient infrastructure (Alam et al., 2013). Addressing infrastructure deficiencies will require significant financial investments and a comprehensive overhaul of sewage systems, waste management facilities, treatment plants, and water distribution networks (Fernández-Navarro et al., 2017). Upgrading and modernizing the current infrastructure by integrating cutting-edge technologies and environmentally friendly methods will be necessary (Erickson et al., 2017).

Community engagement and collaborations with governmental entities, non-governmental organizations, and international organizations are essential in overcoming infrastructure obstacles and ensuring the provision of sustainable clean water and sanitation services in RCC and GCC (Dagdeviren and Robertson, 2009).

3.2 Population Growth and Unplanned Urbanization

Ensuring sustainable access to clean water and sanitation has become a major challenge in both the RCC and the GCC. Rapid population growth and unplanned urbanization are the major obstacles to overcome. The water and sanitation resources are under immense pressure due to the lack of infrastructure construction and service provision. This results in the inability to meet the increasing demands of the growing population for water and sanitation services (Alam et al., 2013; Farooq et al., 2019).

The problem is getting worse for the CCs, as they are unable to cope with the ever-increasing demands. Rapid urbanization has led to slums and informal settlements where proper sanitation and clean water facilities are lacking (Hossen et al., 2024). Due to the quick influx of people, communities have been haphazardly established without giving proper thought to waste management systems, sanitary facilities, or water supplies (GoB and UNICEF, 2004). As a result, these regions frequently have inadequate access to sanitary facilities and clean water (Angeles et al., 2009).

The infrastructure for water supply and sanitation is presently overburdened due to the combined effects of unplanned development and population growth. It has resulted in water shortages, erratic supply, and inadequate waste disposal mechanisms (Hossain et al., 2024c). These issues not only affect the availability and quality of services but also pose health risks due to poor sanitation and hygiene standards in densely populated areas (Rasul, 2010).

The rapid urbanization of Bangladesh has resulted in the emergence of informal settlements in urban areas, which presents significant challenges for CCs in ensuring access to clean water and sanitation. These settlements, commonly referred to as slums, suffer from inadequate infrastructure, haphazard construction, and limited access to essential services such as clean water and proper sanitation facilities. This is due to the insufficient water distribution networks, illegal connections, and unauthorized tapping of water sources, which contribute to water scarcity in these areas. The lack of proper water infrastructure and regular supply further worsens the situation, leaving residents with no option but to rely on contaminated sources and unsafe practices for water storage and consumption. Even people in the RCC and GCC do not have enough water access like slum dwellers. They are not receiving government-supplied water facilities, and water delivery lines were not provided by CC to slum inhabitants. They draw water from rivers, ponds, and tube wells, which is extremely risky. They sometimes travel a great distance to collect the water they need. An RCC respondent reveals,

I have been living in this area for almost 60 years. There are no water supply facilities in this area. For managing daily water, we typically install tube wells. Furthermore, we never receive assistance from other institutions in the creation of tube-wells. Usually, two or more families work together to create a tube well. We are aware of the severe arsenic contamination of our tube-well water. We had no choice but to drink such contaminated water (KII, RCC, Bangladesh).

5.3 Institutional Incapacity and Mal-Coordination

The effectiveness of CCs in ensuring sustainable water and sanitation management in RCC and GCC is significantly hampered by challenges related to institutional capacity and coordination. Insufficient institutional capacity often leads to a lack of technical expertise, limited human resources, and inadequate organizational structures within these institutions. This shortage impedes the formulation and execution of comprehensive strategies and policies essential for sustainable water and sanitation management (World Bank, 2019).

One of the primary obstacles lies in the insufficient technical expertise required for effective planning, implementation, and monitoring of water and sanitation projects. CCs often face challenges in recruiting and maintaining skilled professionals specialized in water resource management, engineering, environmental science, and sanitation. The lack of technical expertise impedes the institutions' ability to develop innovative solutions and adopt modern technologies essential for sustainable water and sanitation management (Farooq et al., 2019).

Moreover, the limited human resources and organizational capacities within these institutions restrict their ability to execute and maintain water and sanitation infrastructure effectively. This scarcity of skilled personnel results in challenges in project management, monitoring, and maintenance, leading to poor performance of water supply systems, inadequate sanitation services, and inefficiencies in waste management (Alam et al., 2013; Ferdous et al., 2018b; Hossain et al., 2024c; Hossain et al., 2024d).

Coordination among various departments, agencies, and stakeholders is another critical aspect where CCs face significant challenges. The departments in charge of water supply, sanitation, urban planning, and environmental management frequently work independently from one another and lack effective coordination and collaboration. This lack of coordination results in disorganized efforts, duplication of tasks, and inefficiencies in resource utilization (Faridatul and Jahan, 2014). In this regard, an executive engineer of RWASA informs,

As a water supply agency, we strive to provide pure water to the entire city. However, we have noticed that sometimes our pipes get damaged or broken by the City Corporation during road work. This can lead to dirt entering the water supply pipes, which is a major concern. We believe that if there was better integration and coordination between our company and the City Corporation, such issues could be avoided altogether (IDI, RCC, Bangladesh).

Additionally, inadequate institutional coordination leads to difficulties in data sharing, hindering the effective monitoring and evaluation of water and sanitation programs. The absence of a cohesive framework for information exchange among relevant stakeholders limits the institutions' ability to make informed decisions, track progress, and address emerging challenges effectively (GoB and UNICEF, 2004). Moreover, there are issues with the lack of institutionalized rules, poor financial management, and inadequate coordination between government departments. These factors lead to inefficient resource utilization and weaken the implementation of water-related policies. Addressing these shortcomings is crucial to ensuring a reliable and secure water supply for the residents of Rajshahi and Gazipur City in Bangladesh.

Furthermore, the water sector, in particular, relies heavily on good governance for its success. Without proper governance, water security and access can become major concerns. Ensuring the provision of sanitary conditions and safe drinking water requires competent and qualified government representatives. Unfortunately, negligence on the part of the authorities has left a significant portion of the residents in Rajshahi and Gazipur City dissatisfied with the quality of their drinking water (Hossain et al., 2023).

5.4 Absence of Water Reuse and Conservation System

The pursuit of water reuse and conservation presents several critical issues and constraints. Despite water covering approximately 70% of the Earth's surface, a mere 3% of it is potable. Moreover, 23% of this 3% is confined to Lake Baikal in Russian Siberia, inaccessible to the local population due to freezing conditions (Bright et al., 2010). Consequently, a mere 2.2% of the Earth's surface, which constitutes the remaining 77% of the drinkable water, must be shared among a staggering 7 billion people (Bright et al., 2010). This stark reality underscores the immense scarcity of water, a vital resource essential for human survival. While demand for water is ever-increasing, the means to create enough to meet this demand remain limited. As a result, our primary options for sustenance lie in water conservation and equitable sharing. Unfortunately, the inadequacy of our efforts in these areas has led to widespread water shortages, impacting billions of people globally.

Bangladesh, despite its abundance of consumable freshwater resources, faces its own water challenges. A significant portion of its population, approximately 68.3 million people, still lacks access to safely managed, clean drinking water, highlighting the urgency of providing safe water as a basic human right (UNICEF, 2021). The country's extensive river network, spanning 24,000 kilometers, carries usable water. However, the impact of human activities on these water sources has led to gradual contamination, rendering some rivers poisonous (Ali, 2019). In this context, the region of Rajshahi and Gazipur faces a lack of suitable methods for water conservation and reuse. During FGD, a respondent from the RCC claims in this regard,

I don't have a water reserve tank in my home. Water was unavailable in my region when load shedding occurred. Even so, it could continue for 5 to 6 hours. We are unable to do our daily tasks, such as cooking, taking a bath, cleaning, etc. Regarding water supplies, we are in serious difficulties. We also informed CC about this. They informed us that they were powerless to change the situation (FGD, RCC, Bangladesh).

Rajshahi and Gazipur City Corporations have proposed certain surface water reuse measures, but the outcomes have not been fruitful, presenting significant challenges for these institutions (CCs) in their efforts to ensure an available and sustainable water supply.

5.5 Lack of Financial Resources

Lack of adequate funding is a major obstacle to ensuring sustainable clean water and sanitation management in cities like Rajshahi and Gazipur City Corporation. This issue makes it more challenging to provide and maintain the necessary services and infrastructure for ensuring the availability of clean water and adequate sanitation. CCs often face difficulties in obtaining the critical funding required for infrastructure improvements, especially when it comes to initiatives related to water and sanitation. The already limited budgetary resources are further challenged by the high costs of constructing and maintaining waste management facilities, sewage systems, treatment plants, and water supply networks. Therefore, despite the urgent need for infrastructure upgrades, the lack of funding makes it challenging to establish sustainable systems (World Bank, 2019). An executive engineer of the Rajshahi Water Supply and Sewerage Authority (RWASA) explains,

In the RCC, the challenge we face significantly revolves around the lack of adequate financial resources to sustainably manage water and sanitation services. Our efforts to expand infrastructure, upgrade treatment plants, and implement advanced technologies for cleaner water and improved sanitation are constrained by limited funding. The demand for these essential services is increasing due to population growth, increasing the pressure on our existing resources. We are consistently seeking additional funding avenues to enhance our initiatives, but without sufficient financial support, ensuring sustainable water and sanitation services for all remains a formidable task (IDI, RCC, Bangladesh).

Furthermore, there is always a challenge in establishing a balance between the need for service charge revenue and the critical element of providing cheap access to low-income populations. The sustainability of water and sanitation systems is threatened by the financial gap that results from the revenue collected from water tariffs and user fees frequently being insufficient to cover operating and maintenance costs (Ferdous, 2018b; Al Masum, 2014).

The decreasing reliability and quality of water and sanitation services are clear indications of the effects of these financial difficulties. There are concerns about health risks and environmental contamination due to obsolete infrastructure, service interruptions, and impaired water quality caused by insufficient investments (Alam et al., 2013; Fernández-Navarro et al., 2017). Additionally, these difficulties are exacerbated by the infrastructure's degradation resulting from inadequate maintenance, which increases the likelihood of system malfunctions and environmental hazards. Financial constraints also make it difficult to expand water and sanitation facilities to keep up with the demands of expanding cities, resulting in inequitable access and exacerbating existing inequities in service provision (GoB and UNICEF, 2004).

To surmount these financial barriers, a multifaceted approach is essential. Exploring innovative financing mechanisms, such as public-private partnerships (PPPs) and concessional loans, can augment available resources for infrastructure development (Connor Richard et al., 2017). Efficient resource allocation, capacity building within CCS, and collaborative efforts among various stakeholders, including governments, NGOs, and the private sector, are crucial. Advocating for increased investment in water and sanitation infrastructure at national and international levels can mobilize additional financial support and foster sustainable funding for water and sanitation projects (Erickson et al., 2017; UN Water, 2018).

5.6 Improper Waste Management and Sanitation System

Waste management has emerged as a global concern (Islam, 1999). It encompasses the critical processes of waste collection, transportation, disposal, and recycling. In developing countries like Rajshahi and Gazipur City, waste management is significantly influenced by factors such as rapid population growth, expanding economies, urbanization, and improving living standards (Minghua et al., 2009). Consequently, ensuring effective waste management is paramount to safeguarding the environment, public health, and overall safety. The handling of solid waste must prioritize measures that minimize environmental risks and health hazards during storage, collection, and proper disposal (Kassim & Ali, 2006).

Despite utilizing public resources, city governments, including CC, have been reported to struggle to provide satisfactory waste management services (Ahmed & Ali, 2006). This is evident in the cases of RCC and GCC, where residents express dissatisfaction with the current waste management system (Hossain et al., 2023d). The rapid urbanization and population growth in these cities have impeded the development of waste management infrastructure, leading to challenges in proper waste collection and disposal. As a result, solid waste is often inadequately collected and disposed of, contributing to environmental pollution and health risks.

Both RCC and GCC face specific challenges in managing waste effectively. Insufficient manpower poses difficulties for regular primary waste collection from households. Additionally, the lack of space hampers the establishment of secondary transfer stations (STS), further complicating waste management efforts. Establishing STSs for dumping waste at a fixed location presents the city corporation with another significant challenge. RCC has played an important role in establishing STSs in the cities. Although GCC is still falling behind regarding this issue, they are unable to create any STSs. There are about 30 STSs in the RCC. But this is insufficient. A government official from the RCC informs,

We have a lot of obstacles to overcome before we can create a secondary transfer station (STS). We only have 30 or so STSs. However, we require more STS. We don't have enough land to set up STS. We are looking to buy some land. The land of the RCC is very expensive. So, nobody wants to donate land for the construction of STSs (IDI, RCC, Bangladesh).

On the other hand, a government official from the GCC states,

The city corporation has been given 100 bighas of land by the government for waste management. However, none of Gazipur's 57 wards have a secondary transfer station (STS). The councilors have been asked for a place by the authorities to establish STSs. There will be a long-term solution for trash management if these STSs are developed (IDI, GCC, Bangladesh).

Moreover, the final disposal of solid waste remains a critical challenge, as there is no scientific and sustainable method for its proper disposal (Hossain et al., 2023d; Hossain et al., 2023d). In this regard, a responsible RCC government official states,

We are using the secondary transfer station for primary garbage. Waste wasn't consistently collected from the secondary transfer station. In addition, we didn't pick up trash from the side of the road. due to a lack of space on the final disposal site. We continue to strive for proper trash management. To dispose of waste scientifically, we are collaborating with the Beijing-based Asian Infrastructure Investment Bank (AIIB). We believe that our problem will be resolve within a very short time (IDI, RCC, Bangladesh).

The GCC region has become an open trash can be due to a lack of suitable measures, such as designated sites, manpower, and equipment for dumping rubbish, as people dump waste anywhere they can, including on the highways, roadside, and in play areas. Due to the circumstances, the 330 sq. km. city corporation region, which has a population of about 40 lakh people and several industrial facilities, is experiencing horrible environmental effects, including health risks. There are trash piles everywhere along the Dhaka-Mymensingh highway from Tongi to Chandna. Concerned parties claim that the city corporation lacks a dedicated facility for garbage management. For this reason, waste is temporarily dropped at various locations on roads and highways, where it is afterward driven to the by-mile region by the city corporation's car.

4.0 MAJOR CHALLENGES' IMPACT ON SUSTAINABLE WATER SUPPLY AND SANITATION MANAGEMENT

6.1 Water Wastage

In urban Bangladesh, the issue of water wastage is particularly concerning, with an alarming average wastage of up to 113 liters of water per person each day (Gleick et al., 2003). This significant loss of water translates into substantial financial losses on a daily basis (Dey, 2021), raising profound concerns, especially in a world where many people still lack access to clean water and adequate sanitation facilities. Furthermore, such excessive waste negatively impacts the environment, exacerbating the strain on limited water resources.

Findings from this research reveal that water wastage occurs in various ways within the areas of RCC and GCC. Respondents also tend to take lengthy showers, and many individuals carelessly wastewater during cleaning and other daily activities (Randolph and Troy, 2008). On average, urban households in Bangladesh consume 294 liters of water per day (Hossain et al., 2023a), which exceeds the global average of 250 liters per day in moderate-temperature regions (Vishwanath, 2013), indicating an excess of 44 liters per day (Hossain et al., 2023a). This excessive water consumption in urban Bangladesh can be attributed to factors such as inefficient distribution systems, limited awareness of water conservation practices, and a general lack of regard for water conservation efforts.

The study also highlights variations in water wastage patterns between RCC and GCC. While water wastage for cooking is more common in the GCC, the RCC witnesses more wastage during bathing. Additionally, educated individuals tend to consume more water compared to their less-educated counterparts (Hossain et al., 2023a). During the Focus Group Discussions (FGD), participants expressed concern over the inadequate management of leaky pipes by responsible authorities and the general lack of awareness regarding proper water usage and wastewater management among the public. They emphasized the necessity of educational campaigns and community engagement to address these issues effectively. Furthermore, participants stressed the importance of government intervention and stricter regulations to ensure responsible water usage and curb waste in both cities. During FGD, an individual from the GCC states,

There is a mosque in my neighborhood, and it is very close to my home. I have noticed that a lot of people waste water in ablution processes these days. I am aware that water waste is forbidden in Islam as well. I am attempting to educate the inhabitants about this issue although they didn't pay much attention about this issue (FGD, GCC, Bangladesh).

As a result of this considerable water wastage problem, CCs of the RCC and the GCC find it challenging to manage water wastage effectively, ultimately affecting their ability to provide sufficient water to their residents. In this regard, CCs are trying to control this type of water waste. During in-depth interview (IDI), an executive engineer of RWASA states,

With 110 production pumps and an 859-kilometer pipeline network, we provide 10.70 crore liters of water per day to approximately 6.44 lakh people, 30% of which is lost to waste. A few years ago, this waste rate was 37%. We continue to strive to lower the rate of water waste. This rate should be kept under 10%. We are putting on a variety of programs to raise awareness. We are setting up a schedule of public hearings. We are attempting to stop the unauthorized connection and directing affected offices to reduce water wastage (IDI, RCC, Bangladesh).

GCC also organizes some events to raise awareness among the residents regarding water usage. According to a respondent who took part in this awareness-raising program,

"I participated in a workshop organized by the GCC. This program taught me a lot about cleanliness and clean water. I gain knowledge on how to minimize water waste and manage household garbage in particular. However, they didn't hold these kinds of workshops on a regular basis," (FGD, GCC, Bangladesh).

To address this issue comprehensively, concerted efforts are required to improve water usage practices, implement efficient wastewater management systems, raise awareness among the public, and strengthen government intervention and regulations.

6.2 Scarcity of Safe Water

Water availability poses a critical challenge for the RCC and GCC. In the face of water availability, ensuring an adequate supply of water remains a formidable task (Rijsberman, 2002). A significant portion of the population in RCC and GCC lacks access to clean drinking water (Smith et al., 2000). Currently, groundwater serves as the primary source of drinking water for both urban and rural areas in Bangladesh (Ahmed et al., 2006). Despite numerous rivers and canals across the nation, surface water is not widely utilized for water supply. However, these groundwater sources often contain contaminants such as arsenic, iron, and other hazardous substances, rendering the water unsafe for consumption (Smith et al., 2000). Consequently, many people do not have access to safe drinking water from groundwater sources, worsening the water availability challenges in urban Bangladesh (Hossain et al., 2023). In this aspect, a respondent from GCC says,

"There is no government-provided water supply system in our neighborhood. We are having a lot of difficulties meeting the daily water demand. We cannot install submersible pumps because of financial crisis. Water is being drawn from

the tub well. The daily tub-well water collection process is exceedingly difficult. As a result, we require water from the Gazipur City Corporation," (KII, GCC, Bangladesh).

Water scarcity further compounds the problem, affecting both the quantity and reliability of the water supply (Raskin et al., 1997). Current analyses indicate that over the next few decades, up to two-thirds of the global population will experience water scarcity (Raskin et al., 1997). Presently, 1.6 billion people grapple with water shortages due to economic factors, and an alarming four billion people face severe water scarcity (Seckler et al., 1999). Projections suggest that the world is on the brink of a period of intense water scarcity (Postel, 2000). By 2030, more than half of the global population is expected to reside in regions with limited water resources (Mekonnen & Hoekstra, 2016). These projections present a significant challenge for the CCs of RCC and GCC, as they bear the responsibility of ensuring access to water for their residents.

Various factors contribute to water scarcity in these regions. The depletion of groundwater supplies emerges as a major concern in RCC and GCC, with the groundwater level decreasing at a rate of 0.1 to 0.05 meters per year (Rahman, 2016). The reliance of a considerable portion of the population on submersible pump and tube well systems, which heavily depend on groundwater, further increases the stress on water resources (Hossain et al., 2023a; Hossain et al., 2023b). Additionally, careless water usage practices by residents also contribute to water scarcity (Johnson et al., 2016). During the Key Informant Interviews (KIIs) conducted for this study, government officials reported instances of inadequate water supply to residents, highlighting the impacts of water availability on the population. This study underscores the reality that a significant number of residents in these regions do not receive enough water due to safe water unavailability.

Addressing water scarcity necessitates discussions on enhancing water productivity (Rijsberman, 2006) and exploring innovative approaches such as nanotechnology for water management (Mauter et al., 2018). Governments and international organizations must prioritize efforts to improve water use efficiency, implement effective water management practices to mitigate safe water unavailability in RCC and GCC, and safeguard access to clean water for their urban residents.

6.3 Water Contamination

One of the most significant challenges to ensuring available water in Bangladesh is water contamination. The growing issue of water pollution has emerged as a global concern, affecting both developed and developing nations and threatening the physical and ecological well-being of billions of people (Javier et al., 2017).

Bangladesh, in particular, faces a pressing concern with an estimated 57 million people consuming water with iron levels above the World Health Organization's (WHO) recommended threshold of 10 g/L (Robinson et al., 2011). Moreover, the presence of harmful elements such as arsenic and other contaminants has been identified in the water supplied by the RCC (Lamia et al., 2018). Regrettably, a significant proportion of the population in RCC and GCC do not feel safe consuming the available water due to contamination concerns (Hossain et al., 2023a; Hossain et al., 2023b).

The presence of iron in groundwater remains a particularly significant challenge, as CCs struggle to control its concentration. The lack of an effective purification method for contaminated water further increases the situation, leaving CC with limited means to adequately address water pollution. RWASA is the organization in charge of providing clean water in the RCC. In this area, groundwater is primarily supplied by pipelines to the residents by the RWASA. In this aspect, an executive engineer at RWASA states,

"We offer water services form 300 meters below ground level. By heating it, anyone can consume it. The water is suitable for drinking. Due to the extensive network of pipes, it passes through, many people mistake it for something to drink. For drinking purposes, hotels and roadside shops use the water we supply," (IDI, RCC, Bangladesh).

Water contamination is therefore a critical challenge faced by CC in their efforts to ensure the availability of clean water. Resolving this issue necessitates urgent and comprehensive action, including the development and implementation of advanced water treatment technologies and stringent pollution control measures.

6.4 Extensive Groundwater Extortion

Freshwater management systems are profoundly affected by the impacts of climate change, with a strong correlation between water scarcity and changing climatic conditions (Gosling & Arnell, 2013). The escalating concerns regarding water scarcity around the world are exacerbated by climate change, which has led to a surge in extreme weather events (Vörösmarty et al., 2000). Climate change significantly influences the availability of water, thereby posing a significant challenge to ensuring sufficient water supply (Vörösmarty et al., 2000).

The adverse effects of climate change manifest in various aspects of water resources, including supply, quality, and distribution, impacting the lives of billions of people globally (Kundzewicz et al., 2007). Projections across all climate change scenarios indicate a decline in available water, linked with a rise in irrigation demand (Iglesias et al., 2006). In both RCC and GCC, climate change has brought about a severe crisis of unsafe drinking water, compounding the challenges faced by CCs in providing clean water (Abedin et al., 2018).

Groundwater remains the primary source of drinking water for over 98% of Bangladesh's population, emphasizing the critical role of climate change in shaping water availability and quality (Abedin et al., 2018). Particularly during drought periods, the CCs of both RCC and GCC struggle with the inability to meet the water demands of their residents, especially those residing in slums (Abedin et al., 2018). The scarcity of water during drought periods affects city dwellers as well, further intensifying the challenges of water supply.

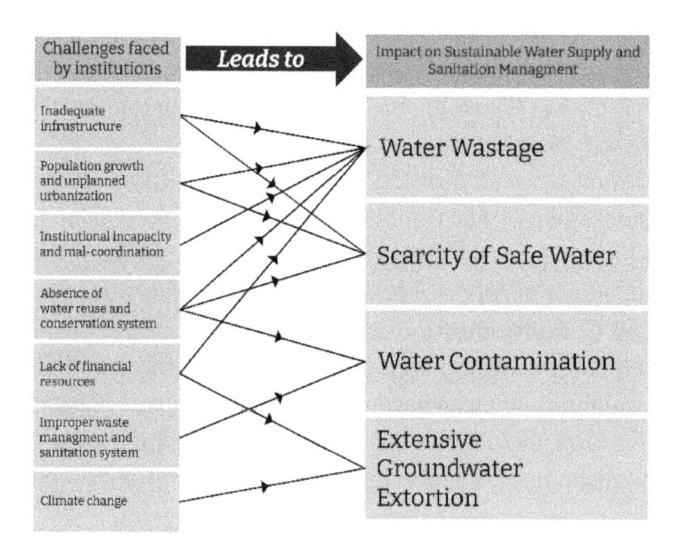

8.0 CONCLUSION

In examining the challenges faced by Rajshahi and Gazipur City Corporations concerning water and sanitation management, this study has shed light on multifaceted issues hindering sustainable practices. The critical identification of challenges encompassing inadequate infrastructure, exacerbated by population growth, financial constraints, institutional capacity gaps, and issues in waste management, underscores the complexity of ensuring sustainable water and sanitation. Moreover, factors such as water wastage, scarcity, contamination, and their nexus with climate change have been revealed as critical challenges influencing urban water sustainability (Hossain et al., 2024a).

The disparities observed between Rajshahi and Gazipur City Corporations in terms of water access and quality perception underscore the need for equitable policy formulation and effective governance frameworks. Addressing these challenges necessitates a comprehensive approach that involves robust governance structures, sound financial planning, and community participation. Equally important is the integration of innovative technologies for water treatment, reuse, and conservation, aligning with sustainable development goals. Additionally, CC must prioritize water conservation initiatives, such as promoting rainwater harvesting and encouraging efficient water usage practices. Additionally, CC should focus on improving access to clean water and sanitation services through the development of better water infrastructure, treatment facilities, and sanitation systems. Educating local communities about the importance of proper maintenance and sustainable water management practices is crucial for the success of these initiatives.

To ensure long-term success, CC must implement sustainable governance systems that empower citizens and strengthen local capacity. Collaborating with various stakeholders, including governments, businesses, non-profit organizations, and local communities, is essential to creating a comprehensive and effective approach to water and sanitation management.

Efforts to improve water and sanitation must be diligently implemented and continuously monitored to achieve the desired outcomes. CC's commitment to creating a sustainable governance system that emphasizes citizen empowerment and capacity-building will play a pivotal role in ensuring the success of these initiatives.

Overall, by fostering a collaborative approach and promoting citizen engagement, CC can make significant strides in conserving water resources, expanding access to clean water and sanitation services, and ultimately improving the overall well-being and health of the residents in Rajshahi and Gazipur City. Only through collective action and sustained commitment can the challenges of water and sanitation in urban Bangladesh be effectively overcome.

ACKNOWLEDGEMENTS

The authors would like to express their sincere gratitude to the Department of Political Science at the University of Rajshahi for their invaluable support and guidance throughout the research process.

REFERENCES

Abedin, M. A., Collins, A. E., Habiba, U., & Shaw, R. (2018). Climate Change, Water Scarcity, and Health Adaptation in Southwestern Coastal Bangladesh. *International Journal of Disaster Risk Science*, 10(1), 28–42. DOI: 10.1007/s13753-018-0211-8

Agarwal, Anil, Marian S. de los Angeles, Ramesh Bhatia, Ivan Chéret, Sonia Davila-Poblete, Malin Falkenmark, F. Gonzalez Villarreal et al. (2000). *Integrated water resources management*. Stockholm: Global water partnership.

Agarwal, A., de los Angeles, M. S., Bhatia, R., Chéret, I., Davila-Poblete, S., Falkenmark, M., . . . Wright, A. (2000). *Integrated water resources management* (pp. 1-67). Stockholm: Global water partnership.

Ahmed, M. F., Ahuja, S., Alauddin, M., Hug, S. J., Lloyd, J. R., Pfaff, A., Pichler, T., Saltikov, C., Stute, M., & van Geen, A. (2006). Ensuring safe drinking water in Bangladesh. *Science*, 314(5806), 1687–1688. DOI: 10.1126/science.1133146 PMID: 17170279

Ahmed, S. A., & Ali, S. M. (2006). People as partners: Facilitating people's participation in public–private partnerships for solid waste management. *Habitat International*, 30(4), 781–796. DOI: 10.1016/j.habitatint.2005.09.004

Ajulor, O., & Ibikunle, B. (2016). Theories of Local Government and their Relevance to Nigeria Experience. *Abuja Journal of Administration and Management*, 9(2), 76–89.

Akoto, O., Gyamfi, O., Darko, G., & Barnes, V. R. (2014). Changes in water quality in the Owabi water treatment plant in Ghana. *Applied Water Science*, 7(1), 175–186. DOI: 10.1007/s13201-014-0232-4

Al-Dulaimi, G. A., & Younes, M. K. (2017). Assessment of potable water quality in Baghdad City, Iraq. *Air, Soil and Water Research*, 10, 1178622117733441. DOI: 10.1177/1178622117733441

Al Masum, A. (2014). Ground Water Quality Assessment of Different Educational Institutions in Rajshahi City Corporation, Bangladesh. *American Journal of Environmental Protection*, 3(2), 64. DOI: 10.11648/j.ajep.20140302.14

Alam, M. S., Kabir, E., & Chowdhury, M. A. K. (2004). Power sector reform in Bangladesh: Electricity distribution system. *Energy*, 29(11), 1773–1783. DOI: 10.1016/j.energy.2004.03.005

Alam, M. Z., Rahman, M. A., & Al Firoz, M. A. (2013). Water supply and sanitation facilities in urban slums: A case study of Rajshahi City corporation slums. *American Journal of Civil Engineering and Architecture*, 1(1), 1–6. DOI: 10.12691/ajcea-1-1-1

Alam, S. (2021). SDGs in Bangladesh: Implementation Challenges and Way Forward. *Data Science and SDGs: Challenges, Opportunities and Realities*, 1-14.

Ali, T. (2019, March 21). *Waste of water, way to disaster*. The Daily Star. https://www.thedailystar.net/supplements/news/waste-water-way-disaster-1718767

Angeles, G., Lance, P., Barden-O'Fallon, J., Islam, N., Mahbub, A. Q. M., & Nazem, N. I. (2009). The 2005 census and mapping of slums in Bangladesh: Design, select results and application. *International Journal of Health Geographics*, 8(1), 1–19. DOI: 10.1186/1476-072X-8-32 PMID: 19505333

Aziz, M. A., Majumder, M. A. K., Kabir, M. S., Hossain, M. I., Rahman, N. M. F., Rahman, F., & Hosen, S. (2015). Groundwater depletion with expansion of irrigation in Barind tract: A case study of Rajshahi district of Bangladesh. *Int J Geol Agric Environ Sci*, 3, 32–38.

Benson, R., Conerly, O. D., Sander, W., Batt, A. L., Boone, J. S., Furlong, E. T., Glassmeyer, S. T., Kolpin, D. W., Mash, H. E., Schenck, K. M., & Simmons, J. E. (2017). Human health screening and public health significance of contaminants of emerging concern detected in public water supplies. *The Science of the Total Environment*, 579, 1643–1648. DOI: 10.1016/j.scitotenv.2016.03.146 PMID: 28040195

Bhavnani, D., Goldstick, J. E., Cevallos, W., Trueba, G., & Eisenberg, J. N. (2014). Impact of rainfall on diarrheal disease risk associated with unimproved water and sanitation. *The American Journal of Tropical Medicine and Hygiene*, 90(4), 705–711. DOI: 10.4269/ajtmh.13-0371 PMID: 24567318

Billions of people will lack access to safe water, sanitation and hygiene in 2030 unless progress quadruples – warn WHO, UNICEF. (n.d.). https://www.unicef.org/bangladesh/en/press-releases/billions-people-will-lack-access-safe-water-sanitation-and-hygiene-2030-unless

Bright, M., & Matsuura, K. (2017). *1001 Natural wonders you must see before you die*. Chartwell Books.

Catley-Carlsonreflects, M. (2019). The non-stop waste of water. *Nature*, •••, 565.

Chen, K., & Zhou, J. L. (2014). Occurrence and behavior of antibiotics in water and sediments from the Huangpu River, Shanghai, China. *Chemosphere*, 95, 604–612. DOI: 10.1016/j.chemosphere.2013.09.119 PMID: 24182403

Connor, R., Renata, A., Ortigara, C., Koncagül, E., Uhlenbrook, S., Lamizana-Diallo, B. M., ... & Brdjanovic, D. (2017). The United Nations world water development report 2017. Wastewater: the untapped resource. *The United Nations world water development report.*

Dagdeviren, H., & Robertson, S. A. (2009). *Access to Water in the Slums of the Developing World* (No. 57). Working paper.

Davis, J. (2004). Corruption in public service delivery: Experience from South Asia's water and sanitation sector. *World Development*, 32(1), 53–71. DOI: 10.1016/j.worlddev.2003.07.003

Deb, U. K. (2016). Performance of the agriculture sector. *Routledge Handbook of Contemporary Bangladesh*, 08.

Dey, A. B. (2021, September 24). *What a waste of water!* The Daily Star. https://www.thedailystar.net/news/bangladesh/news/what-waste-water-2182951

Engel, K., Jokiel, A., & Kraljevic, M. Geiger & Smith, K. (2011). *Big Cities. Big Water. Big Challenges: Water in an Urbanizing World*. http://www.wwf.se/source.php/1390895/Big Cities_Big Water_Big Challenges_2011.pdf

Erickson, J. J., Smith, C. D., Goodridge, A., & Nelson, K. L. (2017). Water quality effects of intermittent water supply in Arraiján, Panama. *Water Research*, 114, 338–350. DOI: 10.1016/j.watres.2017.02.009 PMID: 28279879

Esguerra, J. (2003). *The Corporate Muddle of Manila's Water Concessions*. https://www.wateraid.org/~/media/Publications/privatesector-participation-manila.pdf

Etim, E. E., Odoh, R., Itodo, A. U., Umoh, S. D., & Lawal, U. (2013). Water quality index for the assessment of water quality from different sources in the Niger Delta Region of Nigeria. *Frontiers in Science*, 3(3), 89–95.

Farooq, S., Sharif, T. T., & Alam, S. (2019). *Study on Drinking Water Quality Served in Restaurants and Tea Stalls in Gazipur Area* (Doctoral dissertation, Department of Civil and Environmental Engineering, Islamic University of Technology, Gazipur, Bangladesh).

Fernández-Navarro, P., Villanueva, C. M., García-Pérez, J., Boldo, E., Goñi-Irigoyen, F., Ulibarrena, E., Rantakokko, P., García-Esquinas, E., Pérez-Gómez, B., Pollán, M., & Aragonés, N. (2017). Chemical quality of tap water in Madrid: Multicase control cancer study in Spain (MCC-Spain). *Environmental Science and Pollution Research International*, 24(5), 4755–4764. DOI: 10.1007/s11356-016-8203-y PMID: 27981479

Gleick, P. H., Wolff, G. H., & Cushing, K. K. (2003). Waste not, want not: The potential for urban water conservation in California.

Gosling, S. N., & Arnell, N. W. (2013). A global assessment of the impact of climate change on water scarcity. *Climatic Change*, 134(3), 371–385. DOI: 10.1007/s10584-013-0853-x

Gwynne, R. N. (2009). Modernization Theory. In Kitchin, R., & Thrift, N. (Eds.), *International Encyclopedia of Human Geography* (pp. 164–168). Elsevier. DOI: 10.1016/B978-008044910-4.00108-5

Haque, M. (2019). Urban water governance: pricing of water for the slum dwellers of Dhaka metropolis. *Urban drought: Emerging water challenges in Asia*, 385-397.

Hasan, M. K., Shahriar, A., & Jim, K. U. (2019). Water pollution in Bangladesh and its impact on public health. *Heliyon*, 5(8), e02145. DOI: 10.1016/j.heliyon.2019.e02145 PMID: 31406938

Hasan, M. M., Pal, T. K., Paul, S., & Alam, M. A. (2021). Quality Measurement of Drinking Water in Rajshahi City Corporation and Effectiveness of Different Water Purifiers. *IOSR Journal of Applied Chemistry*, 14, 38–45. DOI: 10.9790/5736-1408013845

Helmer, R., & Hespanhol, I. (1997). *Water pollution control: a guide to the use of water quality management principles*. CRC Press. DOI: 10.4324/9780203477540

Holden, E., Linnerud, K., & Banister, D. (2016). The imperatives of sustainable development. *Sustainable Development (Bradford)*, 25(3), 213–226. DOI: 10.1002/sd.1647

Hosagrahar, J. (2011). Landscapes of water in Delhi: negotiating global norms and local cultures. *Megacities: Urban form, governance, and sustainability*, 111-132.

Hossain, I., Haque, A. K. M., & Ullah, S. M. (2023b). Role of City Governance Institutions in Ensuring Sustainable Water: A Study on Gazipur City of Bangladesh. SSRN *Electronic Journal*. DOI: 10.2139/ssrn.4480760

Hossain, I., & Haque, A. M. (2023). Role of Responsible Institution in Urban Water Governance and Sanitation Management: A Focus on Metropolitan Areas of Bangladesh. *Research Square*. DOI: 10.21203/rs.3.rs-3064713/v1

Hossain, I., Haque, A. M., Kılıç, Z., Ullah, S. A., Azrour, M., & Mabrouki, J. (2024a). Navigating the Challenges of Urban Water Sustainability in Bangladesh: Towards a Sustainable Approach for Urban Governance. In *Smart Internet of Things for Environment and Healthcare* (pp. 1–25). Springer Nature Switzerland. DOI: 10.1007/978-3-031-70102-3_1

Hossain, I., Haque, A. M., Kılıç, Z., Ullah, S. A., Azrour, M., & Mabrouki, J. (2024d). Exploring Household Waste Management Practices and IoT Adoption in Barisal City Corporation. In *Smart Internet of Things for Environment and Healthcare* (pp. 27–45). Springer Nature Switzerland. DOI: 10.1007/978-3-031-70102-3_2

Hossain, I., Haque, A. M., & Ullah, S. A. (2023). Role of city governance institutions in ensuring sustainable water: A study on Gazipur City of Bangladesh. *Social Science Research Network*. https://doi.org/DOI: 10.2139/ssrn.4480760

Hossain, I., Haque, A. M., & Ullah, S. A. (2023a). Assessment of Domestic Water Usage and Wastage in Urban Bangladesh: A Study of Rajshahi City Corporation. *The Journal of Indonesia Sustainable Development Planning*, 4(2), 109–121. DOI: 10.46456/jisdep.v4i2.462

Hossain, I., Haque, A. M., & Ullah, S. A. (2023d). Assessing Sustainable Waste Management Practices in Rajshahi City Corporation: An Analysis for Local Government Enhancement using IoT and AI Technology. *Research Square*. https://doi.org/DOI: 10.21203/rs.3.rs-3397290/v1

Hossain, I., Haque, A. M., & Ullah, S. A. (2024c). Assessing sustainable waste management practices in Rajshahi City Corporation: An analysis for local government enhancement using IoT, AI, and Android technology. *Environmental Science and Pollution Research International*, •••, 1–19. DOI: 10.1007/s11356-024-33171-7 PMID: 38581631

Hossain, I., Haque, A. M., & Ullah, S. A. (2024e). Policy evaluation and performance assessment for sustainable urbanization: A study of selected city corporations in Bangladesh. *Frontiers in Sustainable Cities*, 6, 1377310. DOI: 10.3389/frsc.2024.1377310

Hossain, I., Haque, A. M., Ullah, S. A., Azrour, M., Mabrouki, J., & Kılıç, Z. (2024b). Sustainable Water Management at City Corporation Level in Bangladesh: A Comparative Analysis between SCC and BCC. In *Sustainable and Green Technologies for Water and Environmental Management* (pp. 215–237). Springer Nature Switzerland. DOI: 10.1007/978-3-031-52419-6_16

Hossain, I., Ullah, S. A., & Haque, A. M. (2023c). Water and Sanitation Services at the Local Government Level in Bangladesh: An Analysis of SDG 6 Implementation Status and Way Forward. *Asia Social Issues*, 17(3), e265358. DOI: 10.48048/asi.2024.265358

Hossen, M. S., Haque, A. M., Hossain, I., Haque, M. N., & Hossain, M. K. (2024). Towards comprehensive urban sustainability: Navigating predominant urban challenges and assessing their severity differential in Bangladeshi city corporations. *Urbanization. Sustainability Science*, 1(1), 1–17.

Iglesias, A., Garrote, L., Flores, F., & Moneo, M. (2007). Challenges to manage the risk of water scarcity and climate change in the Mediterranean. *Water Resources Management*, 21(5), 775–788. DOI: 10.1007/s11269-006-9111-6

Islam, M. R. (1999). *Solid waste culture and its impact on health of the residents of Dhaka City*. Observer Magazine.

Javier, M. S., Sara, M. Z., & Hugh, T. (2017). Water pollution from agriculture: a global review. *The Food and Agriculture Organization of the United Nations Rome and the International Water Management Institute on behalf of The Water Land and Ecosystems research program Colombo*.

Jiménez, A., Saikia, P., Giné, R., Avello, P., Leten, J., Liss Lymer, B., Schneider, K., & Ward, R. (2020). Unpacking water governance: A framework for practitioners. *Water (Basel)*, 12(3), 827. DOI: 10.3390/w12030827

Johnson, H., South, N., & Walters, R. (2016). The commodification and exploitation of fresh water: Property, human rights and green criminology. *International Journal of Law, Crime and Justice*, 44, 146–162. DOI: 10.1016/j.ijlcj.2015.07.003

Kassim, S. M., & Ali, M. (2006). Solid waste collection by the private sector: Households' perspectives findings from a study in Dar-es-Salaam city, Tanzania. *Habitat International*, 30(4), 769–780. DOI: 10.1016/j.habitatint.2005.09.003

Khatun, S., Shaon, S. M., & Sadekin, N. (2021). Impact of poverty and inequality on economic growth of Bangladesh. *Journal of Economics and Sustainable Development*, 12(10), 107–120.

Kimani-Murage, E. W., & Ngindu, A. M. (2007). Quality of water the slum dwellers use: The case of a Kenyan slum. *Journal of Urban Health*, 84(6), 829–838. DOI: 10.1007/s11524-007-9199-x PMID: 17551841

Kumar Panday, P. (2006). Central-local relations, inter-organisational coordination and policy implementation in urban Bangladesh. *Asia Pacific Journal of Public Administration*, 28(1), 41–58. DOI: 10.1080/23276665.2006.10779314

Kundzewicz, Z. W., Mata, L. J., Arnell, N. W., Doll, P., Kabat, P., Jimenez, B., ... & Shiklomanov, I. (2007). Freshwater resources and their management.

Lamia, F., Kafy, A., & Poly, S. (2018). Assessment of Water supply system and water quality of Rajshahi WASA in Rajshahi City Corporation (RCC) area, Bangladesh. 1st National Conference on Water Resources Engineering (NCWRE 2018), CUET.

Lebel, L., Anderies, J. M., Campbell, B., Folke, C., Hatfield-Dodds, S., Hughes, T. P., & Wilson, J. (2006). Governance and the capacity to manage resilience in regional social-ecological systems. *Ecology and Society*, 11(1), art19. DOI: 10.5751/ES-01606-110119

Li, E., Endter-Wada, J., & Li, S. (2015). Characterizing and contextualizing the water challenges of megacities. *Journal of the American Water Resources Association*, 51(3), 589–613. DOI: 10.1111/1752-1688.12310

Liu, J., Ren, H., Ye, X., Wang, W., Liu, Y., Lou, L., Cheng, D., He, X., Zhou, X., Qiu, S., Fu, L., & Hu, B. (2017). Bacterial community radial-spatial distribution in biofilms along pipe wall in chlorinated drinking water distribution system of East China. *Applied Microbiology and Biotechnology*, 101(2), 749–759. DOI: 10.1007/s00253-016-7887-8 PMID: 27761636

Loewenberg, S. (2007). Scientists tackle water contamination in Bangladesh. *Lancet*, 370(9586), 471–472. DOI: 10.1016/S0140-6736(07)61214-8 PMID: 17695063

Mauter, M. S., Zucker, I., Perreault, F., Werber, J. R., Kim, J. H., & Elimelech, M. (2018). The role of nanotechnology in tackling global water challenges. *Nature Sustainability*, 1(4), 166–175. DOI: 10.1038/s41893-018-0046-8

Mekonnen, M. M., & Hoekstra, A. Y. (2016). Four billion people facing severe water scarcity. *Science Advances*, 2(2), e1500323. DOI: 10.1126/sciadv.1500323 PMID: 26933676

Minghua, Z., Xiumin, F., Rovetta, A., Qichang, H., Vicentini, F., Bingkai, L., Giusti, A., & Yi, L. (2009). Municipal solid waste management in Pudong New Area, China. *Journal of Waste Management*, 29(3), 1227–1233. DOI: 10.1016/j.wasman.2008.07.016 PMID: 18951780

Mobin, M. N., Islam, M. S., Mia, M. Y., & Bakali, B. (2014). Analysis of physico-chemical properties of the Turag River water, Tongi, Gazipur in Bangladesh. *Journal of Environmental Science & Natural Resources*, 7(1), 27–33.

Palmer, P. M., Wilson, L. R., O'Keefe, P., Sheridan, R., King, T., & Chen, C. Y. (2008). Sources of pharmaceutical pollution in the New York City Watershed. *The Science of the Total Environment*, 394(1), 90–102. DOI: 10.1016/j.scitotenv.2008.01.011 PMID: 18280543

Panday, P. K. (2007). Policy implementation in urban Bangladesh: role of intra-organizational coordination. *Public Organization Review: A Global Journal*, 7(3), 237–259.

Pandey, S. K., & Tiwari, S. (2009). Physico-chemical analysis of ground water of selected area of Ghazipur city-A case study. *Nature and Science*, 7(1), 17–20.

Pandit, A. B., & Kumar, J. K. (2015). Clean Water for Developing Countries. *Annual Review of Chemical and Biomolecular Engineering*, 6(1), 217–246. DOI: 10.1146/annurev-chembioeng-061114-123432 PMID: 26247291

Postel, S. (2000). ENTERING AN ERA OF WATER SCARCITY: THE CHALLENGES AHEAD. *Ecological Applications*, 10(4), 941–948. DOI: 10.1890/1051-0761(2000)010[0941:EAEOWS]2.0.CO;2

Rahman, A. S., Kamruzzaman, M., Jahan, C. S., Mazumder, Q. H., & Hossain, A. (2016). Evaluation of spatio-temporal dynamics of water table in NW Bangladesh: An integrated approach of GIS and Statistics. *Sustainable Water Resources Management*, 2(3), 297–312. DOI: 10.1007/s40899-016-0057-4

Rahman, T., & Khan, N. A. (2008). Rethinking corruption. New Age, a Vernacular English daily. www.newagebd.com

Randolph, B., & Troy, P. (2008). Attitudes to conservation and water consumption. *Environmental Science & Policy*, 11(5), 441–455. DOI: 10.1016/j.envsci.2008.03.003

Raskin, P., Gleick, P., Kirshen, P., Pontius, G., & Strzepek, K. (1997). *Water futures: assessment of long-range patterns and problems. Comprehensive assessment of the freshwater resources of the world.* SEI.

Rasul, M. T., & Jahan, M. S. (2010). Quality of ground and surface water of Rajshahi city area for sustainable drinking water source. *Journal of Scientific Research*, 2(3), 577–577. DOI: 10.3329/jsr.v2i3.4093

Rijsberman, F. R. (2006). Water scarcity: Fact or fiction? *Agricultural Water Management*, 80(1–3), 5–22. DOI: 10.1016/j.agwat.2005.07.001

Rijsberman, F. R., Molden, D. J., & Makin, I. W. (2002). World Water Supplies: are they adequate. In *3rd rosenberg international forum on water policy, Canberra* (pp. 7-11).

Robinne, F. N., Bladon, K. D., Miller, C., Parisien, M. A., Mathieu, J., & Flannigan, M. D. (2018). A spatial evaluation of global wildfire-water risks to human and natural systems. *The Science of the Total Environment*, 610, 1193–1206. DOI: 10.1016/j.scitotenv.2017.08.112 PMID: 28851140

Robinson, C., Von Broemssen, M., Bhattacharya, P., Häller, S., Bivén, A., Hossain, M., & Thunvik, R. (2011). Dynamics of arsenic adsorption in the targeted arsenic-safe aquifers in Matlab, south-eastern Bangladesh: Insight from experimental studies. *Applied Geochemistry*, 26(4), 624–635. DOI: 10.1016/j.apgeochem.2011.01.019

Rohr, J. A. (2002). *Civil servants and their constitutions*. Studies in Government & Public.

Roy, D. K. (2006). Governance, competitiveness and growth: The challenges for Bangladesh. ADB Institute Discussion Paper No. 53. Manila: Asian Development Bank.

Seckler, D., Barker, R., & Amarasinghe, U. (1999). Water Scarcity in the Twenty-first Century. *International Journal of Water Resources Development*, 15(1–2), 29–42. DOI: 10.1080/07900629948916

Smelser, N. J., & Baltes, P. B. (Eds.). (2001). *International encyclopedia of the social & behavioral sciences* (Vol. 11). Elsevier.

Smith, A. H., Lingas, E. O., & Rahman, M. (2000). Contamination of drinking-water by arsenic in Bangladesh: A public health emergency. *Bulletin of the World Health Organization*, 78(9), 1093–1103. PMID: 11019458

Staszewski, G. (2008). Reason-giving and accountability. *Minnesota Law Review*, 93, 1253.

Strat, A. L. (2010). Paris: Local Authorities Regain Control of Water Management. *The Transnational Institute (TNI)*. http://www.tni.org/sites/www.tni.org/files/Paris Chapter by Anne Le Strat En_ final.pdf

Tortajada, C., & Biswas, A. K. (2018). Achieving universal access to clean water and sanitation in an era of water scarcity: Strengthening contributions from academia. *Current Opinion in Environmental Sustainability*, 34, 21–25. DOI: 10.1016/j.cosust.2018.08.001

Uddin, N. (2018). Assessing urban sustainability of slum settlements in Bangladesh: Evidence from Chittagong city. *Journal of urban management*, 7(1), 32-42. https://doi.org/DOI: 10.1016/j.jum.2018.03.002

Uitto, J. I., & Biswas, A. K. (2000). Water for urban areas: Challenges and perspectives. *Journal - American Water Works Association*, 92(12), 136.

UNICEF & WHO. (2019). *Progress on Household Drinking Water, Sanitation and Hygiene 2000–2017*. Special Focus on Inequalities. UNICEF & WHO.

UNICEF & WHO. (2020). *Progress on Drinking Water, Sanitation and Hygiene in Schools: Special Focus on COVID-19*. UNICEF & WHO.

Varis, O., & Vakkilainen, P. (2001). China's 8 challenges to water resources management in the first quarter of the 21st Century. *Geomorphology*, 41(2-3), 93–104. DOI: 10.1016/S0169-555X(01)00107-6

Vishwanath, S. (2013, February 15). *How much water does an urban citizen need?* The Hindu. https://www.thehindu.com/features/homes-and-gardens/how-much-water-does-an-urban-citizen-need/article4393634.ece

Vörösmarty, C. J., Green, P., Salisbury, J., & Lammers, R. B. (2000). Global Water Resources: Vulnerability from Climate Change and Population Growth. *Science*, 289(5477), 284–288. DOI: 10.1126/science.289.5477.284 PMID: 10894773

Water, U. N. (2018). Policy Brief on Water Governance. https://shorturl.at/acozK

WHO & UNICEF. (2019). WASH in Health Care Facilities: Global Baseline Report 2019. WHO & UNICEF, Geneva.

WHO/UNICEF Joint Water Supply & Sanitation Monitoring Programme. (2015). *Progress on sanitation and drinking water: 2015 update and MDG assessment*. World Health Organization.

World Bank. (2018). *Water scarce cities: Thriving in a finite world*. World Bank.

World Bank. (2018). *Water scarce cities: Thriving in a finite world*. World Bank.

Wu, P., & Tan, M. (2012). Challenges for sustainable urbanization: A case study of water shortage and water environment changes in Shandong, China. *Procedia Environmental Sciences*, 13, 919–927. DOI: 10.1016/j.proenv.2012.01.085

Chapter 8
Innovative Strategic Integration of IoT and Digital Marketing for Sustainable Environmental Governance

Madhusudan Narayan
 https://orcid.org/0000-0002-8738-0039
Amity Business School, Amity University Jharkhand, Ranchi, India

Ashok Kumar Srivastava
Amity University Jharkhand, Ranchi, India

Ashutosh Sharma
Amity Institute of Applied Sciences, Amity University Jharkhand, Ranchi, India

ABSTRACT

Purpose: This study explores how integrating IoT and digital marketing can enhance sustainable environmental governance, addressing environmental challenges and supporting the achievement of sustainable development goals (SDGs). It aims to show how this combination can tackle environmental challenges and support (SDGs). Methodology: A literature review and case studies evaluate how real-time IoT data can enhance targeted digital marketing strategies, promoting sustainable behaviour. Findings: Integration improves public engagement and campaign effectiveness, using IoT data for timely, impactful messaging. Challenges like data privacy and

DOI: 10.4018/979-8-3693-7001-8.ch008

technical interoperability must be addressed. Practical Implications: Insights for stakeholders help refine sustainability initiatives and advance global environmental goals. Originality/Value: This research offers a framework that combines IoT and digital marketing, empowering innovative, data-driven solutions for sustainability.

Keywords: IoT, Digital Marketing, Environmental Governance, Sustainability, Data Analytics, SDG

1. INTRODUCTION

Sustainable environmental governance is crucial for balancing ecological health, social equity, economic prosperity, environmental governance, prioritizing ecological health, social justice (Sciolla & Morgera, 2021). IoT technology enhances this framework by providing real-time environmental data crucial for informed decision-making across urban air quality, water management, and agriculture (Zheng et al., 2021; Li et al., 2021; Singh et al., 2022). While digital marketing raises awareness and promotes sustainable behaviors (Banerjee & Sinha, 2021; Bandyopadhyay & Sen, 2021). Together, IoT and digital marketing can significantly improve environmental management and create a more sustainable future. Integration with digital marketing strategies further amplifies awareness and promotes sustainable behaviors, though challenges like data privacy and interdisciplinary collaboration remain critical (Banerjee & Sinha, 2021; Bandyopadhyay & Sen, 2021).

Our planet faces environmental threats, demanding new solutions for sustainable governance. There are several studies that are being conducted on how the Internet of Things (IoT) and digital marketing can work together for a greener future. IoT creates a network of devices that collect real-time environmental data. This information is crucial for monitoring air quality, water resources, and more (Zheng et al., 2021; Li et al., 2021). Digital marketing uses online tools to raise awareness and encourage sustainable behavior (Banerjee & Sinha, 2021). Combining these approaches is promising. Real-time IoT data can be used for targeted digital marketing campaigns. Imagine social media alerts informing you about air quality and suggesting actions to take. This transparency can hold authorities accountable (Bartoli et al., 2021). Personalized feedback from IoT devices can also nudge individuals towards sustainable choices, like reducing energy use (Froehlich et al., 2022).

The 21st century is grappling with an environmental crisis of unprecedented magnitude. The Intergovernmental Panel on Climate Change (IPCC, 2021) paints a stark picture of a rapidly warming planet, with rising sea levels, increasingly frequent extreme weather events, and a drastic decline in biodiversity. These environmental challenges are not isolated; they have profound social and economic ramifications, impacting global food security, human health, and international stability (Rockström et al., 2009). Challenges include data privacy and the need for collaboration between environmental scientists, marketers, and IT specialists (Bandyopadhyay & Sen, 2021; Porter & Heppelmann, 2021). Despite these hurdles, integrating IoT and digital marketing offers a powerful way to improve environmental monitoring, engage stakeholders, and promote sustainable practices. Further research can address challenges and unlock the full potential of this approach for a greener future.

The urgency of the situation demands a paradigm shift towards sustainable environmental governance. Likov et. al., (2018) define this concept as managing human interactions with the environment in a way that safeguards ecological integrity, promotes social equity, and ensures economic viability for present and future generations. Traditional approaches to environmental governance, often relying on top-down regulations and limited data, have proven insufficient to tackle these complex and dynamic challenges.

Fortunately, the emergence of innovative technologies offers a beacon of hope. The Internet of Things (IoT) is revolutionizing the way we interact with the environment. By connecting everyday objects to the internet through sensors and actuators, IoT creates a vast network of data collection points. This network allows for real-time monitoring of environmental parameters such as air quality, water pollution levels, and resource consumption (Miorandi et al., 2012). This granular data provides invaluable insights into environmental trends and empowers policymakers and environmental managers to make informed decisions (Bandyopadhyay & Sen, 2019).

The Internet of Things (IoT) refers to a network of interconnected devices that communicate and share data with each other via the internet. These devices, which include sensors, actuators, and other smart technologies, are embedded with electronics, software, and connectivity capabilities that enable them to collect, process, and transmit data in real-time (Ashton, 2009). IoT has revolutionized various sectors, including healthcare, transportation, and agriculture, by providing unprecedented levels of automation, efficiency, and insight.

In the realm of environmental governance, IoT holds significant potential for enhancing environmental monitoring and data collection. By deploying IoT-enabled sensors in various environments, it is possible to gather real-time data on air and water quality, soil conditions, weather patterns, and other critical environmental parameters. This data can be used to detect pollution, track changes in ecosystems, and predict environmental hazards, thereby enabling more informed and timely decision-making (Gubbi et al., 2013). For example, IoT sensors can be used to monitor air quality in urban areas, providing data that can help authorities identify sources of pollution and implement measures to reduce emissions (Chen et al., 2017). In agricultural settings, IoT devices can monitor soil moisture and nutrient levels, allowing farmers to optimize irrigation and fertilization practices, thereby promoting sustainable farming methods (Wolfert et al., 2017). Furthermore, IoT can facilitate the creation of smart cities, where integrated systems manage resources more efficiently and reduce the environmental footprint of urban living (Zanella et. al., 2014).

The ability of IoT to provide continuous and detailed environmental data represents a significant advancement over traditional monitoring methods, which are often limited by spatial and temporal constraints. By capitalizing IoT technology, environmental governance can become more proactive and responsive, ultimately leading to better environmental outcomes.

Digitalization, characterized by the integration of digital technologies to enhance communication, services, and trade, presents a pivotal yet complex relationship with the United Nations' Sustainable Development Goals (SDGs) (Evangelista et al., 2014; De Croo, 2015). Digitalization offers powerful tools to improve social and environmental well-being. Enhanced data collection and analysis capabilities can revolutionize sectors such as healthcare, education, and environmental monitoring (Kuhlman & Farrington, 2010). Moreover, digital tools optimize supply chains and foster innovations in renewable energy, contributing to sustainable practices (Estevez et al., 2013). For example, the Internet of Things (IoT) shows promise in supporting numerous SDGs, with industrial IoT projected to deliver substantial economic benefits by 2030 (Ordieres-Meré et al., 2020; George et al., 2020).

Digital innovation plays a crucial role in driving sustainable development. By capitalising digital technologies, businesses and governments can create new products, services, and business models that contribute to economic growth, social progress, and environmental protection (Garzoni et al., 2020; Rachinger et al., 2019). Examples include the development of sustainable supply chains, the adoption of circular economy principles, and the creation of digital platforms for social impact (George et al., 2020; Hoogendoorn et al., 2017). However, digitalization also introduces environmental and social challenges. The escalating demand for electricity to power digital infrastructure contributes significantly to carbon emissions (Hodgson, 2015). Additionally, the digital divide exacerbates social inequalities by limiting access to technology (OECD, 2017a). Cybersecurity vulnerabilities pose another critical risk to digitalized systems (Irfan, 2017).

Sustainable businesses pursue both individual and collective goals, including social equity and ecological responsibility (Hoogendoorn et al., 2017). Digital innovation plays a crucial role in achieving these objectives but necessitates careful implementation to minimize environmental impact (George et al., 2020). Achieving a balance between promoting digital innovation and managing its risks is paramount. Governance strategies vary widely, from advocating minimal regulation to proactive governmental intervention to mitigate adverse effects (Bundesministerium für Wirtschaft und Energie, 2017; OECD, 2017b). Some suggest a hands-off approach, allowing industries to innovate while governments build capacities to address emerging challenges (OECD, 2017b). Digitalization presents a dual-edged sword for sustainability: it empowers innovation while posing significant environmental and social risks. Achieving sustainable development requires meticulous consideration of both opportunities and challenges presented by digitalization. Effective governance strategies will play a pivotal role in capitalizing digital innovation towards a sustainable future.

Digital marketing encompasses a wide range of online strategies and tools used to promote products, services, and ideas to a target audience. These strategies include social media marketing, content marketing, search engine optimization (SEO), email marketing, and online advertising. Digital marketing has become an essential component of modern communication, allowing organizations to reach a global audience quickly and cost-effectively (Chaffey & Ellis-Chadwick, 2019).

In the context of environmental governance, digital marketing plays a crucial role in raising awareness about environmental issues and promoting behavior change. Through targeted campaigns and engaging content, digital marketing can educate the public about the importance of sustainability and the actions they can take to contribute to environmental conservation. Social media platforms, in particular, offer powerful tools for disseminating information and mobilizing communities around environmental causes (Kaplan & Haenlein, 2010). For instance, digital marketing campaigns can highlight the impact of plastic pollution on marine life and encourage individuals to reduce their use of single-use plastics. By sharing compelling stories, images, and videos, these campaigns can create emotional connections that motivate people to adopt more sustainable behaviors (Kotler et al., 2010). Additionally, digital marketing can support the promotion of sustainable products and services, helping consumers make environmentally friendly choices in their purchasing decisions. Moreover, digital marketing can facilitate the engagement of stakeholders in environmental governance initiatives. By providing platforms for dialogue and collaboration, digital marketing enables NGOs, businesses, and government agencies to work together more effectively. This collaborative approach is essential for addressing complex environmental challenges that require coordinated action from multiple sectors (Macnamara, 2018).

This research investigates how the integration of the Internet of Things (IoT) and digital marketing can enhance sustainable environmental governance by leveraging their complementary strengths. IoT provides real-time environmental data critical for informing policy decisions and understanding ecological conditions, but its potential is maximized when paired with actionable insights. Digital marketing, in turn, transforms this complex data into accessible, engaging narratives that raise awareness and drive behavioral change. By integrating IoT data with targeted digital marketing strategies, this synergy enables dynamic content creation and enhances transparency, improving environmental communication and stakeholder engagement.

This study reviews existing literature, explores the intersection of IoT and digital marketing, and develops practical applications for their integration. Ultimately, the research aims to establish a framework that enables stakeholders to capitalize on this approach, driving meaningful progress toward sustainability goals. The combined power of IoT and digital marketing presents a transformative tool for sustainable governance, capable of addressing current environmental challenges while fostering long-term ecological balance.

2. LITERATURE REVIEW

Sustainable Environmental Governance through IoT and Digital Marketing

Sustainable environmental governance is pivotal in addressing the challenges of environmental degradation, aiming to balance ecological integrity, social equity, and economic viability (Linkov et. al., 2018). This framework ensures the preservation of natural ecosystems, fair access to resources, and economic development that supports long-term sustainability. Multi-level governance, as advocated by Bäckstrand et al. (2021), emphasizes collaboration among local, national, and international stakeholders for effective environmental management. Participatory governance, highlighted by Meadowcroft (2021), promotes active citizen involvement in environmental decision-making processes, enhancing transparency and accountability. These approaches are crucial in achieving sustainable environmental governance objectives.

IoT in Environmental Monitoring

Traditional environmental monitoring methods, while providing valuable insights (Turner et al., 2000), often suffered from limitations in real-time data collection and comprehensive coverage (Leuenberger et al., 2018). The advent of the Internet of Things (IoT) has revolutionized this field (Miorandi et al., 2022). Vast networks of interconnected devices equipped with sensors offer a continuous stream of data on various environmental parameters, allowing for real-time decision-making and improved environmental management (Zheng et al., 2021). These devices, equipped with sensors and actuators, offer granular insights into environmental parameters, enabling proactive decision-making (Li et al., 2021). In urban areas, IoT-enabled air quality sensor networks track pollutants like PM2.5 and ozone, guiding pollution control efforts (Zheng et al., 2021). Smart water meters equipped with IoT sensors detect leaks and optimize water distribution, promoting conservation practices (Li et al., 2021). In agriculture, IoT sensors monitor soil moisture, temperature, and nutrients, enhancing farming efficiency (Singh et al., 2022). Beyond these applications, IoT extends its capabilities to waste management, noise pollution monitoring, and wildlife conservation (Miorandi et al., 2022). Unlike traditional methods, IoT provides continuous and detailed environmental data, overcoming limitations in geographic scope and data frequency (Khan et al., 2022).

IoT for ESG Reporting

Integrating IoT with Environmental, Social, and Governance (ESG) frameworks enhances sustainable governance (Wu et al., 2022). IoT and blockchain technologies create smart ESG reporting platforms, improving data transparency and credibility. Effective use of IoT data necessitates robust storage and management systems crucial for policymakers and researchers (Montella & Foster, 2010). An IoT-based ESG framework includes interconnected devices across perception, network, and application layers. Sensors in the perception layer monitor environmental parameters, providing real-time data for sustainability assessments. Additional sensors in IoT devices enhance monitoring across supply chains. However, IoT data storage in cloud servers increases carbon emissions. Digital twin technologies can optimize data usage and minimize carbon footprints, though they require skilled personnel (Yousaf, Z., et. al. 2021). The integration of IoT technologies with ESG frameworks represents a significant advancement for sustainable governance (Chen et al., 2023). Their proposed smart ESG reporting platform architecture utilizes blockchain and IoT to enhance transparency, security, and credibility in ESG reporting procedures (Wu et al., 2022). While IoT generates substantial data volumes stored in cloud servers, digital twin technologies offer opportunities to optimize data usage, minimizing carbon footprints and enhancing organizational management. Implementing digital twins requires skilled personnel such as data scientists and software engineers.

Digital Marketing: A Tool for Promoting Sustainable Practices

Banerjee and Sinha (2021) explore how digital marketing campaigns raise awareness about environmental issues through social media platforms, content marketing, and targeted online advertising. These channels effectively disseminate information on climate change, deforestation, and pollution, educating audiences and inspiring sustainable lifestyles. Hwang et al. (2021) examine digital marketing's role in promoting sustainable consumption patterns with engaging content like educational videos and interactive quizzes. This approach motivates individuals to make conscious choices such as using reusable shopping bags or buying eco-friendly products. Chen and Chang (2021) discuss how businesses utilize digital marketing to showcase eco-friendly practices and products, attracting environmentally conscious consumers and fostering brand loyalty. Digital marketing emerges as a powerful tool for raising awareness about environmental challenges and solutions (Banerjee & Sinha, 2021). Interactive quizzes and personalized recommendations delivered through digital channels motivate individuals to adopt sustainable practices (Hwang et al., 2021). Businesses use digital platforms to highlight their commitment to sustainability, attracting eco-conscious consumers (Chen & Chang, 2021).

Integration of IoT and Digital Marketing

The integration of IoT and digital marketing presents significant opportunities for enhancing environmental communication and promoting sustainable behaviors. Real-time IoT data, such as air quality metrics, informs digital campaigns via platforms like social media and mobile apps, empowering informed actions (Kumar et al., 2021). Projects like Barcelona's "Smart Citizen" demonstrate how accessible IoT data fosters transparency in governance, promoting citizen engagement (Bartoli et al., 2021). Digital marketing strategies effectively raise awareness about environmental issues and drive behavior change. Content marketing, social media campaigns, and targeted advertising educate audiences and promote sustainable practices (Banerjee & Sinha, 2021). Interactive formats like quizzes and personalized recommendations actively engage users in sustainability efforts (Hwang et al., 2021). Businesses capitalizing digital platforms can highlight eco-friendly products and initiatives to build consumer loyalty and advocacy (Chen & Chang, 2021; Hsu et al., 2020). Technological integration using IoT-enabled wearables and smart home systems tracks energy consumption, providing personalized feedback and recommendations for resource optimization (Froehlich et al., 2022). Gamified challenges within apps incentivize users to compete for energy savings. Future research can focus on addressing data privacy concerns in IoT-based environmental communication through robust cybersecurity measures and user consent protocols (Bandyopadhyay & Sen, 2021). Enhancing user experience through design practices and investigating economic viability and scalability are also critical for widespread adoption (Chen et al., 2023; Moe & Moon, 2018).

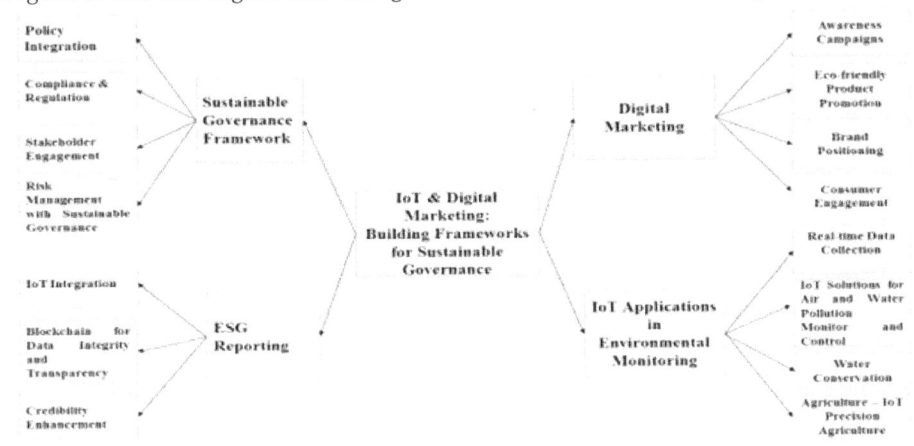

Figure 1. IoT and digital marketing in sustainable environmental governance

3. RESEARCH GAP

While existing research highlights the potential of IoT for environmental monitoring and digital marketing for promoting sustainable practices, there is a significant gap in understanding how these technologies can be synergistically capitalized for comprehensive and effective sustainable environmental governance. By integrating real-time IoT data with targeted digital marketing campaigns, policymakers, businesses, and NGOs can develop more impactful strategies to address pressing environmental challenges. This research aims to explore the untapped potential of this integration to drive positive environmental change.

4. STATEMENT OF PROBLEM

Traditional environmental governance approaches often struggle with limited data and inadequate public engagement. This research aims to address this gap by investigating how the strategic integration of IoT and digital marketing can overcome these challenges and enhance sustainable environmental governance. By combining real-time data collection with effective communication strategies, this integration can provide a more comprehensive and impactful solution to current environmental challenges.

5. RESEARCH QUESTIONS

- How can IoT and digital marketing be synergistically integrated to improve environmental governance practices?
- How does real-time data from IoT sensors enhance the efficacy of digital marketing campaigns focused on sustainability?
- What are the primary challenges faced when integrating IoT and digital marketing strategies for sustainable environmental governance?
- How can stakeholders effectively utilize a framework integrating IoT and digital marketing to achieve sustainable environmental governance goals?

6. RESEARCH OBJECTIVES

- To investigate innovative approaches for integrating IoT and digital marketing to enhance sustainable environmental governance.
- To examine the impact of real-time IoT data on the effectiveness of digital marketing campaigns aimed at promoting sustainable behaviors.
- To identify and analyze challenges associated with the integration of IoT and digital marketing, including data privacy and interdisciplinary collaboration.
- To develop a practical framework for stakeholders to implement integrated IoT and digital marketing strategies in environmental governance initiatives

7.0 RESEARCH METHODOLOGY

This study adopts a qualitative research design to explore the integration of Internet of Things (IoT) technology and digital marketing in promoting sustainable environmental governance. A comprehensive literature review was conducted, drawing from academic journals, industry reports, and relevant online sources to assess how these technologies can enhance environmental governance. Systematic searches of academic databases and platforms were performed using targeted keywords such as "IoT," "Digital Marketing," "Environmental Governance," "Sustainable Practices," "Real-time Data," and "Stakeholder Engagement." The research employed thematic analysis to identify and categorize recurring themes, strategies, and challenges related to the integration of IoT and digital marketing. Each source was critically evaluated for credibility and relevance, leading to a robust synthesis of findings. Key themes included integration strategies, the role of policy, technology, and stakeholder engagement, and challenges such as interoperability and data privacy. While the qualitative approach provides valuable insights, the study acknowledges its limitations, particularly the reliance on existing literature, which may limit the scope and generalizability of findings. The study's focus on identified sources may not reflect the most recent advancements. Nonetheless, this research offers a thorough analysis of current knowledge and underscores the need for further exploration in this evolving field.

8. DISCUSSION

8.1 Innovate IoT and Digital Marketing for Sustainable Environmental Governance

The urgency to address environmental challenges through innovative solutions has spurred interest in integrating the Internet of Things (IoT) and digital marketing to bolster sustainable environmental governance (Bandyopadhyay & Sen, 2021). Conventional methods, constrained by limited data and hierarchical regulations, struggle to manage the complexities of environmental issues (Sciolla & Morgera, 2021). However, the advent of IoT presents a transformative opportunity by enabling real-time collection and analysis of environmental data through interconnected devices (Li et al., 2021). IoT's ability to provide granular insights into ecological conditions enhances decision-making processes for policymakers and environmental managers (Bandyopadhyay & Sen, 2021). This data-driven approach supports evidence-based policies and targeted interventions aimed at mitigating environmental impacts.

The integration of IoT with digital marketing amplifies the effectiveness of environmental communication efforts. Digital marketing strategies such as targeted social media campaigns and personalized content translate complex IoT-generated data into accessible narratives for diverse audiences (Hwang et al., 2021). For example, capitalising real-time air quality data from IoT sensors, social media platforms can alert individuals about pollution levels and recommend actions to reduce environmental footprint. During high pollution episodes, platforms like X (formerly Twitter) can automatically push notifications advising people to use public transportation or limit outdoor activities to reduce personal contributions to pollution levels (Bartoli et al., 2021). Moreover, digital marketing enhances consumer awareness and drives demand for sustainable products and services. Businesses utilize digital platforms to showcase their sustainability initiatives, attracting eco-conscious consumers and driving market trends towards environmentally friendly practices (Hsu et al., 2020). For instance, companies like Tesla have successfully utilized digital marketing to promote electric vehicles as a sustainable alternative to traditional combustion engine cars, influencing consumer choices and industry norms. However, the integration of IoT and digital marketing faces significant challenges. Managing the privacy and security of IoT-generated data requires robust cybersecurity measures and transparent protocols for user consent (Turner et al., 2020). Additionally, successful implementation hinges on effective collaboration between environmental scientists, marketers, and IT specialists to maximize the impact of integrated strategies (Porter & Heppelmann, 2021).

The strategic integration of IoT and digital marketing represents a promising pathway towards achieving sustainable environmental governance. By utilising real-time data and effective communication strategies, this approach empowers stakeholders, promotes sustainable behaviors, and guides towards a more resilient future.

8.2 Impact of Real-Time IoT Data on Sustainable Marketing Campaigns

The exploration of how real-time Internet of Things (IoT) data influences digital marketing campaigns targeting sustainable behaviors is essential for understanding the synergistic potential of these technologies in advancing environmental governance. IoT technologies provide continuous, detailed data on environmental parameters like air quality and water management, empowering stakeholders with precise insights for informed decision-making (Zheng et al., 2021; Li et al., 2021). This capability is instrumental in enhancing the effectiveness of environmental monitoring and management efforts (Bandyopadhyay & Sen, 2021).

Digital marketing strategies capitalize platforms such as social media and personalized advertising to engage audiences and promote sustainable practices (Hwang et al., 2021). For instance, Nike's "Move to Zero" campaign effectively utilized digital channels to educate consumers about reducing carbon emissions and waste in fashion, thereby influencing purchasing decisions towards more sustainable products (Nike, n.d.). Similarly, WWF's #EarthHour campaign mobilizes global participation through digital engagement, encouraging individuals to switch off non-essential lights for an hour to promote energy conservation and environmental awareness (WWF, n.d.).

Integrating real-time IoT data into these campaigns enables organizations to deliver timely and context-aware messages. Imagine IoT sensors in smart cities monitoring air quality levels in real-time. During pollution spikes, these sensors could trigger targeted alerts and recommendations via mobile devices, prompting immediate behavioral changes such as reducing vehicle use or avoiding outdoor activities during peak pollution hours (Bartoli et al., 2021). Moreover, the integration of IoT data into digital marketing enhances campaign credibility and transparency. Organizations can substantiate their environmental claims with real-time data, fostering trust among consumers and stakeholders. This transparency not only strengthens accountability but also encourages community participation in environmental conservation efforts (Bartoli et al., 2021).

Nevertheless, integrating IoT data into digital marketing campaigns poses challenges that require attention. Issues such as data privacy concerns, effective data management practices, and the necessity for interdisciplinary collaboration between environmental experts and marketers must be addressed (Porter & Heppelmann, 2021). Robust cybersecurity measures and clear protocols for data collection, usage, and storage are imperative to protect sensitive information gathered by IoT devices (Bandyopadhyay & Sen, 2021). Exploring how real-time IoT data enhances the effectiveness of digital marketing campaigns focused on sustainable behaviors is crucial for advancing environmental governance. By l capitalizing IoT's data capabilities within digital marketing strategies, organizations can promote meaningful behavior changes and contribute significantly to long-term sustainability goals (Hwang et al., 2021).

8.3 Challenges in IoT-Digital Marketing Integration: Data Privacy and Collaboration

The combination of IoT's real-time data capabilities with digital marketing strategies presents promising opportunities but also introduces significant hurdles that must be addressed. Firstly, data privacy emerges as a critical concern. IoT devices gather vast amounts of sensitive information about environmental conditions and individual behaviours. Ensuring robust cybersecurity measures and transparent protocols for user consent are essential to protect privacy rights (Turner et al., 2020). For example, in smart city projects like Barcelona's "Smart Citizen," where IoT sensors collect data on air quality and noise levels, strict data protection policies are essential to prevent misuse and unauthorized access (Bartoli et al., 2021).

Secondly, effective interdisciplinary collaboration is essential for utilising the full potential of IoT and digital marketing integration. Environmental scientists, marketers, and IT specialists must work together to ensure that technical implementations align with environmental objectives and communication strategies resonate with target audiences (Porter & Heppelmann, 2021). For instance, initiatives like Nike's "Move to Zero" campaign, which uses digital platforms to promote sustainability in fashion, require collaboration between sustainability experts and marketing teams to effectively convey their environmental impact messages (Nike, n.d.). Moreover, navigating technical interoperability challenges between IoT systems and digital marketing platforms poses another obstacle. Integrating diverse data formats and ensuring seamless data transmission from IoT sensors to digital marketing campaigns require advanced technical expertise and compatibility standards (Porter & Heppelmann, 2021).

Addressing these challenges will be crucial for unlocking the full potential of IoT and digital marketing integration in promoting sustainable behaviors and enhancing environmental governance. By overcoming data privacy concerns, fostering interdisciplinary collaboration, and ensuring technical interoperability, stakeholders can capitalise real-time data insights to drive meaningful environmental change through targeted and effective digital marketing campaigns.

8.4 Framework for Integrating IoT and Digital Marketing in Environmental Governance

Developing a practical framework for stakeholders to implement integrated IoT and digital marketing strategies in environmental governance initiatives is pivotal in utilising technological advancements for sustainable outcomes. The integration of Internet of Things (IoT) and digital marketing offers a transformative approach to enhance environmental monitoring, communication, and behavior change efforts.

Firstly, the framework should emphasise utilising IoT's capability to provide real-time environmental data. For instance, projects like the "Smart Citizen" initiative in Barcelona demonstrate how IoT sensors can monitor air quality and noise levels in urban environments, empowering local authorities with data-driven insights for targeted environmental policies (Bartoli et al., 2021). By integrating such IoT data into digital marketing campaigns, stakeholders can effectively communicate environmental risks and promote community engagement towards sustainable practices.

Secondly, the framework should highlight the role of digital marketing in amplifying the impact of environmental initiatives. For example, campaigns like Patagonia's "Worn Wear" have successfully used digital platforms to promote clothing reuse and repair, fostering a culture of sustainability among consumers (Patagonia, n.d.). By incorporating storytelling, interactive content, and personalized recommendations based on IoT data, stakeholders can tailor messages that resonate with diverse audiences and inspire meaningful behavior change. Moreover, the framework must address the need for interdisciplinary collaboration among environmental scientists, marketers, policymakers, and IT specialists. Collaborative efforts are essential to ensure that technological implementations align with environmental goals and communication strategies effectively engage target communities (Porter & Heppelmann, 2021). For instance, partnerships between environmental NGOs and digital marketing agencies can streamline efforts to promote eco-friendly products and services through data-driven campaigns that highlight their environmental benefits (Hsu et al., 2020).

Furthermore, the framework should prioritize scalability and adaptability to varying environmental contexts and stakeholder needs. By establishing clear metrics for success, such as increased public awareness or measurable reductions in environmental impact, stakeholders can evaluate the effectiveness of integrated IoT and digital marketing strategies (Bandyopadhyay & Sen, 2021). For example, the "Think Dirty" app illustrates how digital technology can empower consumers to make informed choices about personal care products based on their environmental impact. This app uses IoT-like features to provide transparency into product ingredients and their potential health and environmental effects, thereby encouraging users to opt for safer and more sustainable alternatives (Think Dirty, n.d.).

Developing a practical framework for stakeholders to implement integrated IoT and digital marketing strategies in environmental governance initiatives requires capitalising IoT's data capabilities, utilizing digital marketing's outreach potential, fostering interdisciplinary collaboration, and ensuring scalability and adaptability. By integrating these elements with examples of successful initiatives like the Smart Citizen project, Patagonia's Worn Wear campaign, and the Think Dirty app, stakeholders can effectively drive sustainable environmental governance through innovative and targeted approaches

Figure 2. Digital marketing and IoT for sustainable governance

9. FINDINGS OF THE STUDY

The study reveals that the integration of Internet of Things (IoT) technology and digital marketing strategies holds significant promise for advancing sustainable environmental governance. By combining real-time data from IoT sensors with digital marketing techniques, environmental communication and public engagement are significantly enhanced (Hwang et al., 2021). For instance, IoT sensors that monitor air quality can provide real-time data that informs digital marketing campaigns, alerting the public about pollution levels and suggesting immediate actions to mitigate exposure (Bartoli et al., 2021). This integration enhances the relevance and timeliness of environmental messages, thereby increasing their effectiveness. Additionally, the study highlights how real-time IoT data can significantly influence behaviors toward sustainability. Insights into environmental conditions enable more compelling and actionable marketing messages, encouraging both individuals and organizations to adopt sustainable practices (Banerjee & Sinha, 2021). Presenting data-driven evidence of environmental impacts allows these campaigns to drive more substantial behavioral changes compared to traditional approaches.

Furthermore, the findings indicate that sustainable environmental governance critical for ecological health, social equity, and economic prosperity (Sciolla & Morgera, 2021) is strengthened by IoT's capacity to provide real-time environmental data (Zheng et al., 2021; Li et al., 2021; Singh et al., 2022). Digital marketing complements this effort by raising awareness and promoting sustainable behaviors (Banerjee & Sinha, 2021; Bandyopadhyay & Sen, 2021). The synergistic relationship between IoT and digital marketing presents transformative potential, enhancing environmental management and fostering stakeholder engagement. However, the study also identifies key challenges associated with this integration, including data privacy and security, technical interoperability between IoT systems and digital marketing platforms, and the need for effective interdisciplinary collaboration (Turner et al., 2020; Porter & Heppelmann, 2021). Addressing these challenges is crucial for maximizing the benefits of integrating IoT and digital marketing.

To capitalize this integration effectively, the study proposes a framework that emphasizes capitalizing on IoT's real-time data capabilities, utilizing digital marketing for effective outreach, and promoting collaboration among environmental scientists, marketers, and IT professionals (Bandyopadhyay & Sen, 2021). Implementing this framework can enhance the effectiveness of environmental governance initiatives and facilitate substantial progress toward achieving sustainability goals.

10. CONCLUSION AND POLICY RECOMMENDATIONS

The integration of IoT and digital marketing offers a powerful solution for modern environmental governance. IoT enables real-time environmental monitoring, providing valuable data for informed decision-making, while digital marketing translates this data into compelling messages that drive public awareness and behaviour change. To fully utilize this potential, enhancing data privacy and security is essential. Transparent data usage policies and strong cybersecurity measures will safeguard sensitive information. Interdisciplinary collaboration between environmental scientists, marketers, and IT professionals is also crucial to ensure seamless implementation of integrated strategies. Furthermore, leveraging digital platforms to present IoT data in accessible ways will boost public engagement and participation in sustainability efforts.

By aligning IoT's data-driven insights with digital marketing's communication strengths, stakeholders can overcome traditional governance limitations and foster long-term sustainability. This approach is essential for creating more responsive, transparent, and effective environmental policies that benefit both society and the planet.

11. SCOPE FOR FUTURE RESEARCH

Future research should address technical interoperability issues between IoT systems and digital marketing platforms to ensure seamless integration. It should also examine the long-term effects of these integrated strategies on behavior change and policy effectiveness. Additionally, exploring the scalability of these approaches in various environmental and geographical contexts will be critical. Finally, developing guidelines for data privacy and security within this integrated framework is essential for responsible implementation.

REFERENCES

Akter, S., & Wamba, S. F. (2016). Big data analytics in e-commerce: A systematic review and agenda for future research. *Electronic Markets*, 26(2), 173–194. DOI: 10.1007/s12525-016-0219-0

Ashton, K. (2009). That 'Internet of Things' thing. *RFID Journal*, 22(7), 97–114.

Bäckstrand, K., Khan, J., Kronsell, A., & Lövbrand, E. (2021). The promise of participatory environmental governance: Participatory institutions and the challenge of democratic environmental governance. *Environmental Politics*, 30(1), 1–20. DOI: 10.1080/09644016.2020.1816557

Bandyopadhyay, P., & Sen, J. (2021). Innovating IoT and digital marketing for sustainable environmental governance. *Journal of Sustainable Development*, 9(1), 45–56.

Bandyopadhyay, S., & Sen, J. (2019). Internet of Things: Applications and challenges in technology and standardization. *Wireless Personal Communications*, 108(1), 1–14.

Bandyopadhyay, S., & Sen, J. (2021). Challenges in Internet of Things (IoT) integration with cloud computing: A systematic review. *Journal of Ambient Intelligence and Humanized Computing*, 12(2), 2267–2285.

Banerjee, M., & Sinha, S. (2021). Understanding the role of digital marketing in promoting sustainable consumer behavior. *Journal of Strategic Marketing*, 29(5), 420–439.

Banerjee, S., & Sinha, S. (2021). Digital marketing and sustainability: A review and agenda for future research. *Journal of Business Research*, 131, 323–336. DOI: 10.1016/j.jbusres.2021.07.037

Bartoli, A., Delnevo, G., Oggioni, A., & Verticale, G. (2021). Smart city services: Integrating IoT sensors and digital marketing for environmental monitoring. In 2021 IEEE Conference on Computer Communications Workshops (INFOCOM WKSHPS) (pp. 1-6). IEEE.

Bartoli, M., Martín-Martín, A., Álvarez-Álvarez, S., & Palazzi, C. E. (2021). Smart Citizen: A concept for citizen-centered governance of urban and environmental issues. *Sensors (Basel)*, 21(8), 2863. Advance online publication. DOI: 10.3390/s21082863 PMID: 33921782

Bundesministerium für Wirtschaft und Energie (BMWi). (2017). G20—Shaping digitalization at global level. Organisation for Economic Co-Operation and Development: Paris, France. Available online: https://www.de.digital/DIGITAL/Redaktion/EN/Dossier/g20-shaping-digitalisation-at-global-level.html (accessed on 22 July, 2024).

Chaffey, D., & Ellis-Chadwick, F. (2019). *Digital marketing: Strategy, implementation and practice* (7th ed.). Pearson Education Limited.

Chen, C. F., & Chang, Y. Y. (2021). Exploring green innovation strategy through digital marketing. *Sustainability*, 13(12), 6573. Advance online publication. DOI: 10.3390/su13126573

Chen, H., & Chang, Y. (2021). Digital marketing strategies for promoting sustainable consumption patterns. *Journal of Business Research*, 78, 197–206.

Chen, M., Wan, J., Gonzalez, S. G., Liao, X., & Lei, Y. (2017). A survey of fog computing: Concepts, applications and issues. *Internet of Things : Engineering Cyber Physical Human Systems*, 1, 27.

Chen, S., Li, H., & Liu, X. (2023). Blockchain and IoT integrated smart ESG reporting: A case study of a multinational corporation. *Journal of Cleaner Production*, 316, 128258. Advance online publication. DOI: 10.1016/j.jclepro.2021.128258

Chen, Y., Xu, G., & Jiang, T. (2017). Design and implementation of air quality monitoring system based on IoT technology. In 2017 IEEE 2nd Information Technology, Networking, Electronic and Automation Control Conference (ITNEC) (pp. 2367-2370). IEEE. https://doi.org/DOI: 10.1109/ITNEC.2017.8265640

De Croo, A. (2015). Why digital is key to sustainable growth. World Economic Forum: Cologny, Switzerland. Available online: https://www.weforum.org/agenda/2015/03/why-digital-is-key-to-sustainable-growth/ (accessed on 22 July, 2024)

Estevez, E., Janowski, T., & Dzhusupova, Z. (2013). Electronic governance for sustainable development: How EGOV solutions contribute to SD goals? In *Proceedings of the 14th Annual International Conference on Digital Government Research*, Quebec City, QC, Canada, 17–20 June 2013; ACM: New York, NY, USA; pp. 92–101. DOI: 10.1145/2479724.2479741

Estevez, E., Janowski, T., & Estevez, J. (2013). Electronic government and the information systems perspective: Third International Conference, EGOVIS 2012, Vienna, Austria, September 3-6, 2012. Proceedings. Springer.

Evangelista, P., Ibarra-Yunez, A., & Prieto-Martinez, J. (2014). *Development beyond digital divides: Towards digital opportunities*. Springer.

Evangelista, R., Guerrieri, P., & Meliciani, V. (2014). The economic impact of digital technologies in Europe. [CrossRef]. *Economics of Innovation and New Technology*, 23(8), 802–824. DOI: 10.1080/10438599.2014.918438

Froehlich, J., Findlater, L., & Landay, J. (2022). The importance of energy-use feedback in energy conservation: A cost-effective approach. [TOCHI]. *ACM Transactions on Computer-Human Interaction*, 29(2), 10.

Froehlich, J., Findlater, L., & Landay, J. A. (2022). The design and evaluation of eco-feedback technology. [TOCHI]. *ACM Transactions on Computer-Human Interaction*, 29(1), 1. Advance online publication. DOI: 10.1145/3492203

Garzoni, A., (2020). *Digital transformation and human-centricity in times of crisis*. Springer.

Garzoni, A., De Turi, I., Secundo, G., & Del Vecchio, P. (2020). Fostering digital transformation of SMEs: A four levels approach. *Management Decision*, 58(8), 1543–1562. DOI: 10.1108/MD-07-2019-0939

George, G., (2020). *Sustainability and digitalization: Opportunities and challenges*. Springer.

George, G., Merrill, R. K., & Schillebeeckx, S. J. (2020). Digital sustainability and entrepreneurship: How digital innovations are helping tackle climate change and sustainable development. [CrossRef]. *Entrepreneurship Theory and Practice*.

Gubbi, J., Buyya, R., Marusic, S., & Palaniswami, M. (2013). Internet of Things (IoT): A vision, architectural elements, and future directions. *Future Generation Computer Systems*, 29(7), 1645–1660. DOI: 10.1016/j.future.2013.01.010

Hariram, N. P., Mekha, K. B., Suganthan, V., & Sudhakar, K. (2023). Sustainalism: An Integrated Socio-Economic-Environmental Model to Address Sustainable Development and Sustainability. *Sustainability (Basel)*, 15(13), 10682. DOI: 10.3390/su151310682

Hodgson, C. (2015). Can the digital revolution be environmentally sustainable? The Guardian: London, UK. Available online: https://www.theguardian.com/global/blog/2015/nov/13/digital-revolution-environmental-sustainable (accessed on 5 July, 2024).

Hodgson, P. (2015). Environmental impacts of digitalization. In Steffen, P., Jahn, A., & Froese, J. (Eds.), *Advances and new trends in environmental and energy informatics* (pp. 189–199). Springer.

Hoogendoorn, B., van der Zwan, P., & Thurik, R. (2017). Sustainable entrepreneurship: The role of perceived barriers and risk. [CrossRef]. *Journal of Business Ethics*, 175, 1133–1154.

Hoogendoorn, M., (2017). Business models for sustainable innovation: State-of-the-art and steps towards a research agenda. *Journal of Cleaner Production*, 140(Part 1), 155–166. DOI: 10.1016/j.jclepro.2016.04.105

Hsu, C., (2020). Leveraging digital platforms for promoting sustainability: The case of eco-friendly products. *Journal of Marketing Management*, 36(5-6), 463–482.

Hsu, C. L., Chang, K. C., & Chen, M. C. (2020). Applying green marketing strategy to enhance green purchase intention: The moderation effects of collaborative consumption and switching costs. *Sustainability*, 12(11), 4402. Advance online publication. DOI: 10.3390/su12114402

Hwang, J., (2021). Impact of digital marketing on sustainable behaviors: Insights from environmental campaigns. *Journal of Environmental Management*, 277, 111367.

Hwang, J., Choi, J., & Lee, S. (2021). Environmental sustainability campaign effectiveness: The roles of interactivity and environmental message features in mobile digital advertising. *Journal of Advertising*, 50(4), 398–415. DOI: 10.1080/00913367.2021.1967326

Hwang, J., Han, S., Koo, C., & Lee, C. (2021). A study on the digital marketing strategies of global fashion brands focused on sustainability. *Journal of Fashion Marketing and Management*, 25(2), 191–207.

Hwang, J., Kim, Y., & Kim, M. (2021). Integrating Internet of Things (IoT) technology and digital marketing for sustainable environmental governance. *Journal of Environmental Management*, 280, 111789. DOI: 10.1016/j.jenvman.2020.111789

Intergovernmental Panel on Climate Change. IPCC (2021). Climate change 2021: The physical science basis. Contribution of Working Group I to the Sixth Assessment Report of the Intergovernmental Panel on Climate Change. Cambridge University Press. https://www.ipcc.ch/report/ar6/wg1/

Irfan, M. (2017). *Cybersecurity in the digital age: Tools, techniques, and implications for government*. Springer.

Kaplan, A. M., & Haenlein, M. (2010). Users of the world, unite! The challenges and opportunities of Social Media. *Business Horizons*, 53(1), 59–68. DOI: 10.1016/j.bushor.2009.09.003

Khan, M. M., Siddique, M., Yasir, M., Qureshi, M. I., Khan, N., & Safdar, M. Z. (2022). The Significance of Digital Marketing in Shaping Ecotourism Behavior through Destination Image. *Sustainability (Basel)*, 14(12), 7395. DOI: 10.3390/su14127395

Kotler, P., Kartajaya, H., & Setiawan, I. (2010). *Marketing 3.0: From products to customers to the human spirit*. John Wiley & Sons. DOI: 10.1002/9781118257883

Kuhlman, T., & Farrington, J. (2010). What is sustainability? *Sustainability (Basel)*, 2(11), 3436–3448. DOI: 10.3390/su2113436

Leuenberger, D., Engesser, M., Herrmann, T., Nagel, P., Wyttenbach, M., & Reimann, S. (2018). *A wireless sensor network for monitoring environmental parameters in a chocolate factory. In 2018 Global Internet of Things Summit (GIoTS)*. IEEE., DOI: 10.1109/GIOTS.2018.8534556

Li, M., (2021). IoT applications in water management: A review. *Water Research*, 200, 117249.

Li, S., Xu, L. D., & Zhao, S. (2021). The internet of things: A survey. *Information Systems Frontiers*, 25(2), 417–455.

Li, X., Wang, H., Xia, H., & Sun, L. (2021). Development of a smart water meter based on IoT technology. *Water Resources Management*, 35, 3143–3154. DOI: 10.1007/s11269-021-02990-4

Linkov, I., Trump, B. D., Poinsatte-Jones, K., & Florin, M.-V. (2018). Governance strategies for a sustainable digital world. *Sustainability (Basel)*, 10(2), 440. Advance online publication. DOI: 10.3390/su10020440

Liu, Q., (2020). A green Internet of Things for sustainable development. *Environmental Research Letters*, 15(3), 035004. Advance online publication. DOI: 10.1088/1748-9326/ab6e5d

Luo, X., & Bhattacharya, C. B. (2006). Corporate social responsibility, customer satisfaction, and market value. *Journal of Marketing*, 70(4), 1–18. DOI: 10.1509/jmkg.70.4.001

Macnamara, J. (2018). *Public relations, activism, and social change: Speaking up*. Routledge., DOI: 10.4324/9781351213206

McQuail, D., & Windahl, S. (2015). *Communication models for the study of mass communications* (2nd ed.). Longman. DOI: 10.4324/9781315846378

Miorandi, D., Sicari, S., De Pellegrini, F., & Chlamtac, I. (2022). Internet of Things: Vision, applications and research challenges. *Ad Hoc Networks*, 10(7), 1497–1516. DOI: 10.1016/j.adhoc.2012.02.016

Moe, W. W., & Moon, Y. (2018). Developing and testing a survey instrument for measuring user-perceived website service quality. *JMR, Journal of Marketing Research*, 40(2), 196–209. DOI: 10.1509/jmkr.40.2.196.19227

Nguyen, T. M. (2021). A systematic review of the use of IoT in environmental governance. *Journal of Cleaner Production*, 302, 126628. Advance online publication. DOI: 10.1016/j.jclepro.2021.126628

Ordieres-Meré, J., Prieto Remón, T., & Rubio, J. (2020). Digitalization: An opportunity for contributing to sustainability from knowledge creation. [CrossRef]. *Sustainability (Basel)*, 12(4), 1460. DOI: 10.3390/su12041460

Organization for Economic Co-Operation and Development (OECD). (2017a). Key issues for digital transformation in the G20. Report prepared for a joint G20 German Presidency/OECD Conference. Available online: https://www.oecd.org/g20/key-issues-for-digital-transformation-in-the-g20.pdf (accessed on 10 July, 2024).

Organization for Economic Co-Operation and Development (OECD). (2017b). *Secretary-General's Report to Ministers*. OECD Publishing. [CrossRef]

Papadopoulou, P., (2021). Exploring the impact of Internet of Things (IoT) technology on sustainable marketing. *Journal of Cleaner Production*, 299, 126743.

Phipps, M., & Burbach, M. E. (2010). Strategic sustainability and employee attitudes: The role of organizational culture. *Business Strategy and the Environment*, 19(1), 1–17. DOI: 10.1002/bse.670

Porter, M. E., & Heppelmann, J. E. (2014). How smart, connected products are transforming competition. *Harvard Business Review*, 92(11), 64–88.

Porter, M. E., & Heppelmann, J. E. (2015). How smart, connected products are transforming companies. *Harvard Business Review*, 93(10), 96–114.

Rachinger, M., Rauter, R., Müller, C., Vorraber, W., & Schirgi, E. (2019). Digitalization and its influence on business model innovation. [CrossRef]. *Journal of Manufacturing Technology Management*, 30(8), 1143–1160. DOI: 10.1108/JMTM-01-2018-0020

Ramos, S., & Pires, L. (2021). Sustainable smart cities and the Internet of Things (IoT): What role for digital marketing? In 2021 IEEE International Smart Cities Conference (ISC2) (pp. 1-6). IEEE. https://doi.org/DOI: 10.1109/ISC252465.2021.9642998

Rockström, J., Steffen, W., Noone, K., Persson, Å., Chapin, F. S.III, Lambin, E. F., Lenton, T. M., Scheffer, M., Folke, C., Schellnhuber, H. J., Nykvist, B., de Wit, C. A., Hughes, T., van der Leeuw, S., Rodhe, H., Sörlin, S., Snyder, P. K., Costanza, R., Svedin, U., & Foley, J. A. (2009). A safe operating space for humanity. *Nature*, 461(7263), 472–475. DOI: 10.1038/461472a PMID: 19779433

Romanelli, L., (2020). The impact of digital marketing on sustainability: A systematic review. *Journal of Cleaner Production*, 242, 118208.

Saha, H. N., Mandal, A., & Sinha, A. (2017). Recent trends in the Internet of Things. In 2017 8th Annual Industrial Automation and Electromechanical Conference (IEMECON) (pp. 252-256). IEEE. DOI: 10.1109/CCWC.2017.7868439

Sarangi, S. (2020). *Sustainable communication: Cultural and environmental perspectives*. Routledge.

Sarker, I. H. (2021). Context-aware learning and big data analytics for smart IoT-based environment: A survey. *Knowledge and Information Systems*, 63(2), 357–385. DOI: 10.1007/s10115-020-01496-2

Saxena, A., Singh, R., Gehlot, A., Akram, S. V., Twala, B., Singh, A., Montero, E. C., & Priyadarshi, N. (2023). Technologies Empowered Environmental, Social, and Governance (ESG): An Industry 4.0 Landscape. *Sustainability (Basel)*, 15(1), 309. DOI: 10.3390/su15010309

Sciolla, L., & Morgera, E. (2021). Balancing ecological integrity, social equity, and economic viability in sustainable environmental governance. *Ecological Economics*, 179, 106857.

Sharma, N. (2021). Sustainable digital marketing: Integrating environmental concerns in marketing strategies. *Journal of Marketing Management*, 37(9-10), 811–831. DOI: 10.1080/0267257X.2021.1927960

Sofoulis, Z. (2011). Smart metering and water end use data: Conservation for the digital age. [TOCHI]. *ACM Transactions on Computer-Human Interaction*, 18(1), 1.

Tulloch, R., & Wilkins, M. (2020). Smart water management: Leveraging IoT for water conservation. *Journal of Water Resources Planning and Management*, 146(6), 04020045. Advance online publication. DOI: 10.1061/(ASCE)WR.1943-5452.0001211

United Nations (UN). (2017). Sustainable development goals: 17 goals to transform our world. United Nations: New York, NY, USA. Available online: https://www.un.org/sustainabledevelopment/sustainable-development-goals/ (accessed on 22 June 2024).

Velmurugan, S., & Shanmugavel, S. (2021). Sustainable environmental governance through digital marketing and IoT: A review. *Journal of Environmental Management*, 278, 111549.

Viswanathan, K. (2021). Integrating IoT and digital marketing for sustainable environmental practices. *Journal of Environmental Management*, 277, 111354.

Wang, W., Sun, X., & Zhang, Y. (2021). Leveraging IoT and digital marketing for promoting sustainable environmental practices. *Journal of Cleaner Production*, 295, 126409.

Williams, T., Boucher, J., & Bowers, C. (2021). Integrating IoT technology for environmental sustainability: Insights from a case study. *Journal of Environmental Management*, 280, 111855. Advance online publication. DOI: 10.1016/j.jenvman.2021.111855

Wolfert, S., Ge, L., Verdouw, C., & Bogaardt, M. J. (2017). Big data in smart farming – A review. *Agricultural Systems*, 153, 69–80. DOI: 10.1016/j.agsy.2017.01.023

Wu, J., Wang, S., Yang, X., & Li, S. (2022). Blockchain and IoT-based smart ESG reporting platform architecture: A conceptual framework. *Sustainability*, 14(5), 2453. Advance online publication. DOI: 10.3390/su14052453

Xu, L. D., He, W., & Li, S. (2014). Internet of Things in industries: A survey. *IEEE Transactions on Industrial Informatics*, 10(4), 2233–2243. DOI: 10.1109/TII.2014.2300753

Yousaf, Z., Radulescu, M., Sinisi, C. I., Serbanescu, L., & Păunescu, L. M. (2021). Towards Sustainable Digital Innovation of SMEs from the Developing Countries in the Context of the Digital Economy and Frugal Environment. *Sustainability (Basel)*, 13(10), 5715. DOI: 10.3390/su13105715

Zanella, A., Bui, N., Castellani, A., Vangelista, L., & Zorzi, M. (2014). Internet of Things for smart cities. *IEEE Internet of Things Journal*, 1(1), 22–32. DOI: 10.1109/JIOT.2014.2306328

Zhou, L., & Luo, Y. (2021). Digital marketing strategies for sustainable consumption: A review of existing research and future directions. *Journal of Business Research*, 127, 131–142. DOI: 10.1016/j.jbusres.2021.01.012

Chapter 9

Challenges and Opportunities in Implementing Tech-Enabled Environmental Governance:
A Systematic Literature Review

Durdana Ovais
BSSS Institute of Advanced Studies, India

Richa Jain
https://orcid.org/0009-0003-6907-9183
Prestige Institute of Management and Research, Bhopal, India

ABSTRACT

As technological advancements rapidly evolve, they present both opportunities and challenges for environmental governance. Despite the significant promise of technology in enhancing environmental governance, gaps remain in understanding its effective implementation. This study addresses these gaps through a systematic literature review to elucidate how technology can be effectively utilized to advance environmental governance, overcome current challenges, and promote transformative change. This study employs a systematic literature review methodology, analyzing a broad range of academic sources including journals, conference proceedings, and reports. The review will follow rigorous search strategies and transparent inclusion/ exclusion criteria to ensure a representative sample. Thematic analysis is applied to identify key trends, knowledge gaps, and emerging frameworks, providing a com-

DOI: 10.4018/979-8-3693-7001-8.ch009

prehensive understanding of tech-enabled environmental governance.

INTRODUCTION

Numerous contemporary environmental challenges can now be classified as "persistent," denoting the fact that political initiatives to resolve them have proven ineffective or have failed to produce the desired results (Jänicke and Jörgens, 2020). The necessity for a systemic transformation in the governance of the environment and natural resources to address the climate crisis, biodiversity loss, social inequalities, and livelihood insecurity is now widely acknowledged (Martin et al, 2020). Moreover, as the people's demand for a better life is getting higher and higher, requirements for environmental governance is constantly increasing (Chen and Liu, 2022). In light of imminent catastrophes such as climate change, biodiversity loss, and livelihood insecurity, there has been a prevalent apprehension regarding the transformation of environmental governance in the last decade (Ojha et al., 2022). The challenges faced by environmental policy today differ sharply from those of past decades – both as regards the environmental problems calling for attention and the strategies available to tackle them (Jänicke and Jörgens, 2020).

Technology is rapidly transforming environmental governance and can provide opportunities for tackling numerous environmental governance concerns (Xu et al., 2022). From high-flying drones capturing real-time data on deforestation to complex algorithms optimizing energy use in cities, a wave of innovation is empowering data-driven decision-making and targeted interventions (Rolnick et al., 2022). This tech revolution offers a powerful scope for environmental monitoring, sustainable resource management, and global collaboration on issues like climate change. By harnessing the potential of artificial intelligence, blockchain, and the Internet of Things, governments and organizations can usher in a new era of environmental accountability and sustainable practices (Saini et al., 2023).

Moreover technology-enabled environmental governance presents a double-edged sword. While innovations like satellite monitoring and smart grids offer unprecedented opportunities for environmental data collection, regulation enforcement, and resource efficiency, challenges remain (Alotaibi at al.,2020). Unequal access to technology and the digital divide can leave developing nations behind (Helsper, 2021). Additionally, the environmental impact of the technology itself, from e-waste to the energy demands of data centers, needs careful consideration (Hoosain et al., 2023, Kumar et al., 2024). Furthermore, ensuring transparency and data security in complex technological systems is crucial (Felzmann et al, 2020). However, by addressing these challenges and promoting responsible tech development, we can harness the immense potential of technology to create a more sustainable future.

Despite this surge in research on the adoption of technology for environmental governance, the conceptual wisdom on the opportunities and challenges that it presents remains patchy (Fazey et al., 2021a, b). Moreover, the actual capacity for such governance approaches to improve environmental conditions remains disputed (Jager et al.,2020; Partelow et al., 2020).

A comprehensive understanding of the opportunities and challenges surrounding technology-enabled environmental governance hinges on a systematic review of the literature. This in-depth analysis would involve meticulously examining academic journals, conference proceedings, and relevant reports to identify current research trends and knowledge gaps. By synthesizing these diverse sources, we can create a holistic picture of the technological advancements driving environmental progress, alongside a critical evaluation of the potential drawbacks. This exploration would illuminate the most effective applications of technology for environmental protection, while also highlighting areas where ethical considerations, accessibility concerns, and the environmental footprint of the technology itself require further attention.

Problem Statement

Environmental governance faces unprecedented challenges, demanding transformative change. While technological advancements hold immense promise for improved monitoring, resource management, and global collaboration, a critical gap exists in our understanding of how to effectively implement these technologies. This study addresses this gap by conducting a systematic literature review. By systematically analyzing existing research, this study aims to illuminate how technology can be harnessed to empower human agency and navigate the complexities of environmental knowledge production. This will lead to a better understanding of how transformative change can occur in environmental governance, moving beyond technological determinism and towards a future where technology serves as a tool for collective action and sustainable practices.

Motivation of the Study

Our analysis is driven by the pressing necessity to comprehend the intricate dynamics that exist between environmental governance and technological progress. This systematic literature review aims to shed light on both the opportunities and challenges associated with implementing tech-enabled environmental solutions. Furthermore, the absence of conceptually articulated, practice-based theorizing on transformational change compels us to delve deeper. By examining existing research, we hope to identify key areas where technology can truly transform environmental governance practices, moving beyond incremental improvements towards more systemic sustainability.

2. LITERATURE REVIEW

The integration of technology into environmental governance faces several significant challenges. One prominent issue is the complexity and variability of the problems addressed by global environmental governance (GEG). These problems are often characterized by institutional and issue complexity, linkages, and multi-scalar nature, which conventional methodological approaches struggle to address effectively (O'Neill et al., 2013). This complexity can hinder the effective deployment of technological solutions and the establishment of robust governance frameworks. Another challenge is the quality and reliability of data generated by digital technologies. Rapid developments in technologies like big data, remote sensing, and artificial intelligence (AI) are accompanied by concerns about data quality, privacy, and security (Kostka et al., 2020). These issues pose significant barriers to the effective use of digital tools in environmental governance, impacting the reliability of the information on which decisions are based.

Moreover, the uneven adoption of digital technologies across different regions exacerbates these challenges. For instance, while developed countries may benefit from advanced digital tools, developing countries often struggle with concurrent socioeconomic, legal, and regulatory reforms necessary for effective smart city and environmental governance (Tan & Taeihagh, 2020). This disparity highlights the need for tailored solutions that address the specific challenges faced by different regions. The governance of environmental issues is further complicated by the limitations of current technological applications. Despite the potential of unmanned aerial systems (UAS) and Geographic Information Systems (GIS) for enhancing environmental governance, there is a notable gap in framing these technologies within broader governance contexts, and a lack of effective collaboration between academia and public administration (Hognogi et al., 2021).

The digitalization of business operations is crucial for traditional enterprises to remain competitive in the digital economy (Weill & Woerner, 2018). Advanced systems and technologies, such as the Internet of Things (IoT), Blockchain Technology (BCT), Cloud Computing, Data Analytics, and Artificial Intelligence (AI), play a key role in driving this transformation, along with the development of essential digital skills and capabilities (Akter et al., 2022). Blockchain Technology (BCT), in particular, offers a decentralized ledger that enables secure data sharing, improving visibility and transparency in supply chains (Kamble et al., 2019). By 2025, the global market size of BCT in the agriculture and food sectors is projected to reach

USD 948 million, growing at a compound annual growth rate (CAGR) of 48.1% (Markets and Markets, 2020).

While digital systems provide distinct benefits, their integration can overcome individual limitations, unlocking new technical capabilities and driving corporate growth (Akter et al., 2022). AI, for example, has revolutionized manufacturing by leveraging centralized computing and data storage to enable real-time decision-making (Nasar et al., 2020). However, AI faces challenges related to data security, interoperability, and ethical concerns (Awad et al., 2018). Similarly, although BCT decentralizes data storage and decision-making across supply chains, it lacks the ability to analyze data and generate actionable insights on its own (Salah et al., 2019). To address these challenges, combining AI, BCT, and other technologies can enhance technical capabilities and create real business value (Hughes et al., 2022).

In the context of sustainability, blockchain technology offers transformative potential in supply chain management. Using distributed ledger technology, blockchain enables a decentralized database that records transactions throughout the supply chain, providing transparency, reliability, traceability, and efficiency. A notable example is the collaboration between Walmart and IBM Food Trust, where blockchain was applied to improve food safety and security. This initiative has reduced tracking time, shortened operational processes, lowered truck fuel consumption, and ultimately enhanced resource efficiency, contributing to more sustainable supply chains.

The digital transformation of businesses is essential for traditional companies to stay competitive in today's digital economy (Weill & Woerner, 2018; Vaz, 2021). As economies become more digital, it's clear that digital transformation is important for businesses to stay competitive (Kraus et al.,2021). Technologies like the Internet of Things, Blockchain Technology, Cloud Computing, Data Analytics, and Artificial Intelligence are key to driving this change, alongside the development of necessary digital skills (Raja, 2021). Blockchain technology, in particular, helps improve transparency and trust in supply chains by providing a secure, shared system for tracking data (Kamble et al., 2019).

While each of these technologies offers its own benefits, combining them can unlock new possibilities and help businesses grow (Akter et al., 2022). For example, AI has transformed manufacturing by enabling real-time decision-making based on continuous data flow (Nasar et al., 2020), though it comes with challenges like data security and ethical concerns (Awad et al., 2018). On the other hand, blockchain helps decentralize data across supply chains but isn't designed to analyze it or generate insights by itself (Salah et al., 2019). Combining AI and blockchain, along with other technologies, can create significant business value and increase competitiveness (Hughes et al., 2022). When it comes to sustainability, blockchain technology has the potential to reshape supply chain management. By using a decentralized system to record transactions, blockchain provides greater transparency, reliability, and traceability in the supply chain. A well-known example is the collaboration between Walmart and IBM Food Trust, which uses blockchain to improve food safety (Park, & Li, 2021). This system has reduced the time it takes to track products, made operations more efficient, cut fuel consumption for trucks, and improved overall resource planning, contributing to more sustainable supply chain practices.

Despite these challenges, tech-enabled environmental governance offers several promising opportunities. One significant opportunity is the potential of digital technologies to enhance information and communication processes. Technologies such as blockchain, AI, and data analytics can provide new modalities for governance, including more efficient data gathering, real-time monitoring, and enhanced transparency (Bakker & Ritts, 2022; Kloppenburg et al., 2022). These tools can disrupt existing governance models and create new avenues for more effective environmental management. Another opportunity lies in the role of informational governance, which leverages the Information Age's advancements to transform governance institutions and practices. Informational governance enhances the role of information for sustainability and can foster greater institutional interconnectivity and collaboration (Soma et al., 2016). This shift can lead to more adaptive and responsive governance structures that better address emerging environmental challenges.

Moreover, technological advancements offer the potential to bridge gaps in governance by fostering innovative solutions and collaborative platforms. For example, the use of hackathons and interdisciplinary approaches in stakeholder networks can generate creative ideas for sustainability and foster participatory governance models (Barau & Al Hosani, 2015). These approaches can help overcome traditional governance barriers and promote more inclusive and effective environmental management strategies. The rise of smart governance, which involves ICT-enabled collaboration between governments and citizens, also presents an opportunity to enhance urban sustainability. Although still rare, initiatives promoting online and offline citizen engagement can empower communities and improve sustainability outcomes (Tomor et al., 2019). These efforts highlight the potential for technology to enhance public participation and capacity-building in environmental governance. As such, while the implementation of tech-enabled environmental governance faces several challenges related to complexity, data reliability, and regional disparities, it also presents significant opportunities for innovation, enhanced information exchange, and collaborative governance. Addressing these challenges and harnessing the opportunities can pave the way for more effective and sustainable environmental governance in the digital age.

3. RESEARCH METHODOLOGY

At the heart of this investigation lay a systematic literature review designed to comprehensively explore the opportunities and challenges presented by technology-enabled environmental governance. In conducting the systematic literature review, a total of 150 studies were initially identified through comprehensive database searches and manual reviews of relevant journals and conference proceedings. To ensure the relevance and quality of the included studies, specific inclusion criteria were applied: the studies had to be published in peer-reviewed journals, focus on technology-enabled environmental governance, and be published within the last ten years to capture the most recent advancements and trends. Furthermore, the studies needed to provide empirical evidence or theoretical insights into the impact of technology on environmental governance. Conversely, studies were excluded if they did not meet these criteria, including those published outside the specified time frame, lacking a peer-reviewed status, or not directly addressing the intersection of technology and environmental governance. Additionally, articles that focused on non-technology-related aspects of environmental governance or were not available in English were also excluded from the review. Following this rigorous filtering process, the final selection comprised only 31 studies, ensuring that the included research was both relevant and of high academic quality. This review meticulously examined a broad range of academic sources, including scholarly journals, conference proceedings, and relevant reports. By employing rigorous search strategies and transparent inclusion/exclusion criteria, a representative sample of existing research on the topic was curated. The collected data was then subjected to detailed thematic analysis to identify key trends, knowledge gaps, and emerging theoretical frameworks. This systematic approach ensured a robust foundation for understanding the complex interplay between technological advancements, environmental governance practices, and the crucial role of human agency in achieving transformative change.

4. FINDINGS AND DISCUSSION

Figure 1. Frequency of research publications

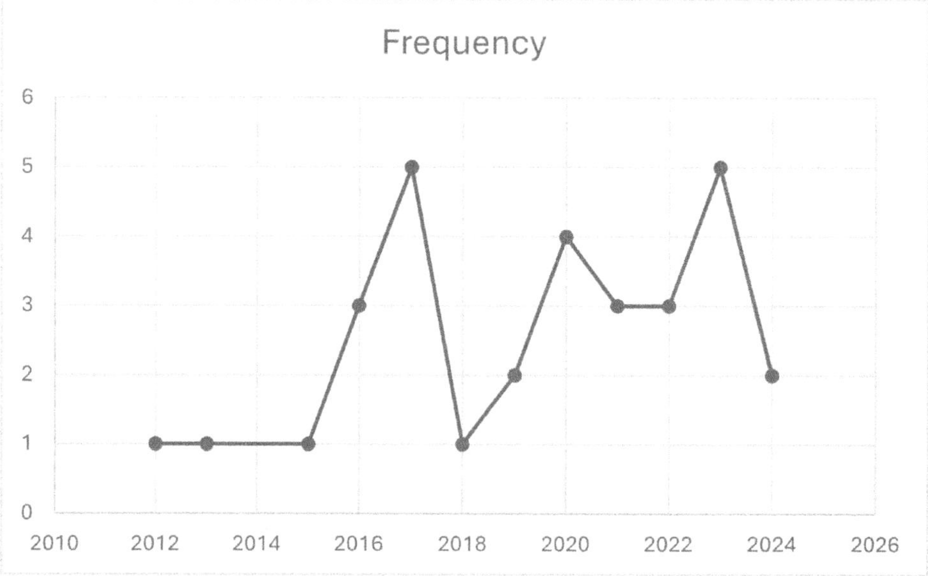

Source: Developed of the purpose of the study

The frequency table reveals a fluctuating pattern of activity over the years from 2012 to 2024. Notably, the years 2017, 2023, and 2024 stand out with the highest frequencies of 5 occurrences each, indicating peaks in activity or publications. Conversely, the years 2012, 2013, 2015, 2018, and 2024 exhibit the lowest frequencies, ranging from 1 to 2 occurrences, suggesting periods of lower engagement or output. A significant increase in activity is evident from 2016 onwards, culminating in the highest numbers observed in 2017 and 2023. The years 2020 and 2021 also show elevated frequencies, reflecting a spike in interest or output during and shortly after the COVID-19 pandemic. The frequency distribution reveals consistent activity in 2019, 2021, and 2022, each with 3 occurrences, indicating stable levels of engagement during these years. The fluctuations in frequencies across the years highlight periods of varying interest and relevance, with a clear upward trend in recent years. This analysis suggests that the subject matter has gained increasing significance or attention in recent times, with notable peaks in 2017 and 2023, and a general trend of rising engagement from 2016 onward.

4.2 Word Cloud Analysis

Figure 2. Word cloud analysis from keywords

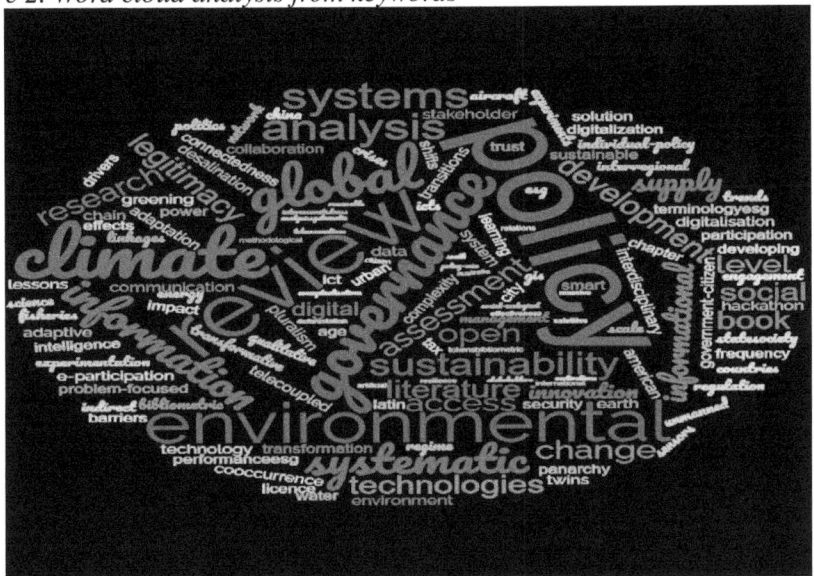

Source: Developed of the purpose of the study

The keyword analysis from various research studies reveals a strong focus on governance, sustainability, and environmental management. Terms such as "governance," "sustainability," and "systematic review" frequently appear, indicating the importance of structured and comprehensive approaches in addressing environmental and social issues. Keywords like "climate governance," "adaptive governance," and "transformative governance" suggest a dynamic and evolving understanding of governance structures needed to tackle climate change and resilience. The frequent mention of "environmental governance" and "global environmental governance" highlights the global scale of these challenges and the need for coordinated international efforts. Additionally, the prominence of terms like "information and communication technology," "digitalization," and "digital transformation" highlights the role of technology in facilitating governance and sustainability efforts. Collaborative and interdisciplinary approaches are emphasized through terms such as "stakeholder engagement," "collaboration," and "interdisciplinary," reflecting the necessity of diverse expertise and cooperation in solving complex environmental problems.

4.3 Types of Studies Under Consideration

Figure 3. Frequency of type of studies undertaken in various research studies

Type of studies undertaken in Various Research Studies

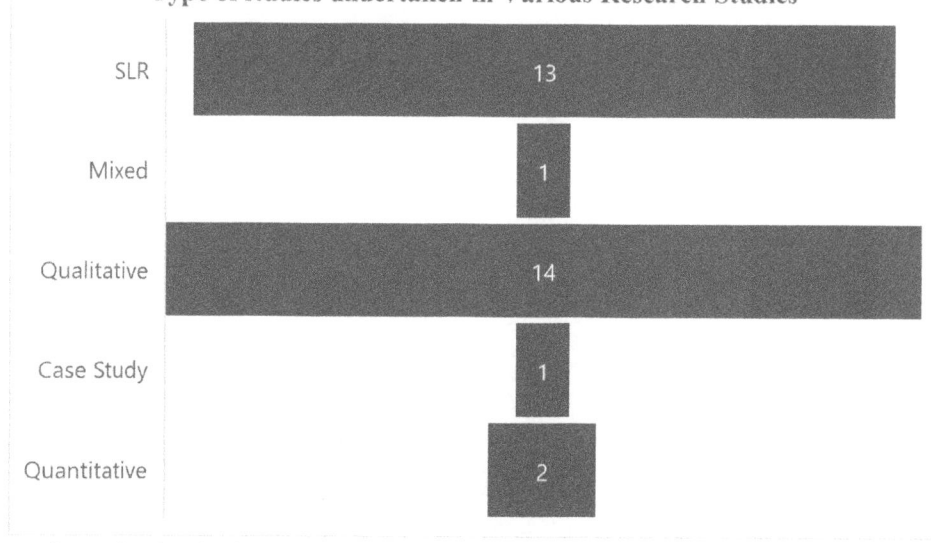

Type	Frequency
SLR	13
Mixed	1
Qualitative	14
Case Study	1
Quantitative	2

Source: Developed of the purpose of the study

4.4 Country Focus

Figure 4. Frequency of Countries Under Study in Various Research Studies

Countries Under Study

Country/Region	Frequency
Global Focus	~17
China	~3
Latin American	1
EU Countries	1
USA, China, and the central and...	1
Global South	1
Developing countries	1
Europe	1
Tasmania, Australia	1
Brazil and Germany	1
Arabian Gulf	1

Source: Developed of the purpose of the study

The frequency analysis of countries mentioned in various research studies indicates a significant number of unspecified or general references, as seen with "NA" and "na" appearing 17 times combined. China is the most frequently specified country, mentioned three times, highlighting its prominence in research contexts. Other specific regions, such as the Arabian Gulf, Brazil and Germany, Tasmania (Australia), Europe, developing countries, the Global South, EU countries, and Latin America, each appear once, suggesting a diverse but less frequently studied set of regions. Notably, the USA, China, and the central and northern European states are collectively mentioned once, reflecting a broader scope in certain studies. The results emphasize a need for more geographically diverse research to balance the heavy focus on unspecified regions and China.

4.5 Challenges

Table 1. Categorization of challenges in environmental governance and sustainability studies

Theme	Challenges	Citation
Governance and Policy	Governance of supply networks, environmental governance, policy innovations, transformative governance goals, complexity and uncertainty in global environmental assessments, policymaking under economic and political pressures, fragmented governance	Pilbeam, et al. (2012); O'Neill, et al. (2013); Kowarsch & Jabbour (2017); Chaffin, et al. (2016); Jänicke, & Jörgens (2020); Cullen-Knox, et al. (2017)
Methodological and Data Challenges	Methodological challenges, data accuracy and privacy, defining system boundaries and baselines, rebound effects, interdisciplinary coordination	O'Neill, et al. (2013); Kostka, et al. (2020); Bieser & Hilty (2018); Kowarsch, & Jabbour (2017).
Technological and Infrastructure	Technological limitations, technological advancements, digital infrastructure, technology illiteracy, data challenges	Barau & Al Hosani, (2015); Cullen-Knox, et al. (2017); Hognogi, et al. (2021); Kostka, et al., 2020)
Environmental and Social Impact	Environmental impact, energy consumption, water-energy nexus, social license, ecological and societal challenges, climate change	Barau, & Al Hosani, (2015); Lenschow, et al. (2016); Hognogi, et al. (2021); Sapiains, et al. (2021)
Stakeholder and Inclusivity Issues	Stakeholder conflicts, lack of inclusivity, lack of citizen participation, balancing technological innovation with environmental sustainability	Kowarsch, & Jabbour, (2017); Tan, & Taeihagh, (2020); Lupova-Henry & Dotti, (2019)
Economic and Financial Constraints	Financing issues, high production costs, economic and financial challenges, resource constraints	Tan, et al. (2020); Rastegar, et al. (2024)
Learning and Adaptation	Learning in environmental governance, adapting to rapid changes, accommodating new evidence and information	Gerlak, et al. (2020); Moon, et al. (2017)

Source: Developed of the purpose of the study

The findings from the various research studies emphasize a range of challenges in environmental governance and sustainability, reflecting the complexity and multifaceted nature of these issues. Research highlights the difficulty in managing and governing environmental and supply networks amidst globalization and complex interdependencies. Pilbeam et al. (2012) emphasizes the need for further empirical work to understand supply network governance, which is becoming increasingly critical in a globalized world. Chaffin et al. (2016) point out that transformative governance faces obstacles from vested interests and systemic legacies, which can hinder progress. Similarly, Lenschow et al. (2016) discuss challenges in maintaining control and addressing interregional ecological issues, reflecting the complexities inherent in global environmental governance.

Several studies address methodological and data-related challenges. O'Neill et al. (2013) highlight issues with methodological pluralism in environmental governance, suggesting that integrating diverse approaches is complex but necessary. Bieser and Hilty (2018) discuss the difficulties in defining system boundaries and baseline references for indirect environmental effects of ICT, indicating a need for more precise and comprehensive methodologies. Kowarsch and Jabbour (2017) also identify challenges related to complexity, uncertainty, and stakeholder conflicts in global environmental assessments, further emphasizing the need for robust and adaptable methods.

Technological advancements and infrastructure limitations are significant themes. Cullen-Knox et al. (2017) note the rapid pace of technological change and the challenge of integrating new technologies into existing governance frameworks. Kostka et al. (2020) raise concerns about the accuracy, reliability, and privacy of data collected through digital technologies in China's environmental governance, reflecting broader issues related to digital infrastructure and data management. Tan and Taeihagh (2020) identify challenges in smart city governance, including financing issues and lack of infrastructure, which impact the effectiveness of environmental policies.

The studies also highlight challenges related to environmental and social impacts. Barau and Al Hosani (2015) discuss the environmental impact and technological limitations of the desalination industry in the Arabian Gulf, reflecting broader concerns about environmental sustainability. Moon et al. (2017) address difficulties in adapting governance frameworks to accommodate new evidence and changing social contexts, which are crucial for effective environmental management. Engaging stakeholders and ensuring inclusivity are critical challenges. Cullen-Knox et al. (2017) mention the difficulty in balancing interests among multiple stakeholders, which can affect the social license to operate. Kowarsch and Jabbour (2017) and Jänicke and Jörgens (2020) also highlight the need for broad stakeholder engagement and the challenges associated with conflicting priorities and values.

Economic and financial constraints are prevalent across studies. Rastegar et al. (2024) discuss high production costs and financial challenges related to renewable energy innovations, while Hognogi et al. (2021) and Tan and Taeihagh (2020) note issues related to financing and investment in environmental and smart city projects, respectively. These constraints highlight the need for financial strategies and policy adjustments to support sustainability goals. The need for ongoing learning and adaptation is evident from several studies. Gerlak et al. (2020) emphasizes the importance of translating theoretical insights into practical governance strategies and adapting to new conditions. This theme stresses the necessity for flexibility and continuous learning in responding to environmental challenges and innovations.

A snapshot of the challenges faced is given below in Figure 5

Figure 5. Snapshot of challenges faced

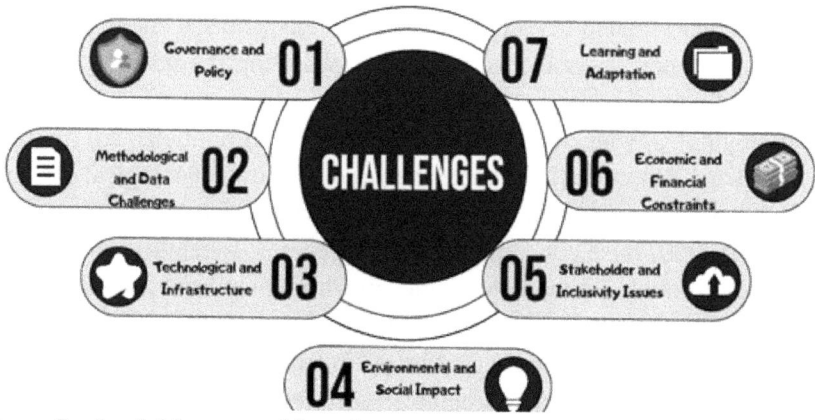

Source: Developed of the purpose of the study

Moreover, the rapid growth of technology brings serious challenges, especially in terms of energy use and electronic waste (e-waste) (Liu, et al., 2023). Technologies like data centers and cryptocurrency mining need vast amounts of electricity, which increases carbon emissions and worsens climate change (Dilek & Furuncu, 2019). The demand for new gadgets also leads to more e-waste, which contains harmful materials like lead and mercury (Beula, & Sureshkumar, 2021). If not handled properly, this waste pollutes the environment and harms both people and wildlife (Yang, et al.,2018; Rajmohan, et al.,2019). Wealthier countries often benefit from new technology, while poorer countries end up dealing with the toxic waste exported to them (Abalansa et al., 2021). While there is a push for sustainable transitions, with renewable energy playing a key role in reducing carbon emissions, it's important to note that "renewable" does not always mean "sustainable." For example, e-waste is a growing environmental problem, particularly in developed countries. Although there are technological solutions to process e-waste, these are expensive, and many developed countries choose the cheaper option of exporting their waste to less developed countries (Shahabuddin et al., 2023).

Despite international and national laws like the Basel Convention and Bamako Convention, many regulations are not fully enforced, and financial pressures often lead to the continued export of e-waste. Countries like Mexico, Brazil, Ghana, Nigeria, India, and China have been studied in detail for their role in handling the transboundary movement of e-waste. Europe leads in e-waste collection, followed by Asia, America, Oceania, and Africa, with the raw materials in e-waste valued at $57 billion. However, only $10 billion worth of e-waste is recycled sustainably, preventing the emission of around 15 million tonnes of CO_2 (Abalansa et al., 2021). The key challenges in e-waste management include proper collection, sorting, the mixed nature of waste, low energy efficiency, and the high cost of recycling. Currently, only 78 countries have laws regulating e-waste, but enforcement is weak, particularly in developing regions like Southeast Asia and Northern Africa (Shahabuddin et al., 2023).

4.6 Opportunities

Table 2. Opportunities in implementing tech-enabled environmental governance: categorized by broader themes.

Broader Category	Opportunities	Citations
Technological Advancements	1. Adoption of efficient and environmentally friendly technologies 2. Integration of renewable energy sources. 3. New internet tools. 4. Smart governance strategies. 5. Assessment of indirect environmental effects of ICT.	Barau & Al Hosani (2015); Soma, et al., (2016); Kivimaa et al. (2017); Bieser & Hilty (2018); Tomor et al. (2019)
Public Engagement and Education	1. Increased public awareness and engagement. 2. Expansion of citizen science and observatories. 3. Financial incentives and skills upgrading for education.	Barau & Al Hosani (2015); Liu, et al., (2017); Tan & Taeihagh (2020)
Innovative Governance and Policy Development	1. Opportunities for less powerful actors to influence networks. 2. Diverse methodologies for global governance. 3. Transformation through instability and shifts in values. 4. Effective international policy coordination. 5. Involvement of diverse stakeholders. 6. Supporting adaptive management. 7. Solution-oriented assessments. 8. Large-scale engagement between ENGOs and industry. 9. Ethical networks and knowledge-oriented networks.	Pilbeam et al. (2012); O'Neill et al. (2013); Chaffin et al. (2016); Kowarsch & Jabbour (2017); Cullen-Knox et al. (2017); Lupova-Henry & Dotti (2019)
Empirical Research and Data Collection	1. Empirical research on tech-Enabled Environmental coupled sustainability challenges. 2. Learning from climate governance experiments. 3. Improved data collection through citizen science.	Lenschow, et al. (2016); Moon et al. (2017); Liu, et al. (2017)

Source: Developed of the purpose of the study

The findings related to opportunities in implementing tech-enabled environmental governance reveal several significant themes.

Technological Advancements play a crucial role in improving environmental governance. The development and adoption of efficient, environmentally friendly technologies, as well as the integration of renewable energy sources such as solar and wind power, offer substantial benefits. These advancements can reduce reliance on fossil fuels and lower greenhouse gas emissions (Barau & Al Hosani, 2015). Additionally, new internet tools and smart governance strategies contribute to enhancing governance frameworks, while the assessment of indirect environmental effects of information and communication technology (ICT) highlights the importance of considering both direct and indirect impacts (Soma, et al., 2016; Kivimaa et al., 2017; Bieser & Hilty, 2018; Tomor et al., 2019).

Public Engagement and Education are vital for fostering sustainability. Increased public awareness and engagement can drive demand for more sustainable practices and technologies. Efforts to expand citizen science and observatories offer opportunities for broader participation in environmental monitoring and data collection (Liu, et al., 2017). Furthermore, financial incentives and skills upgrading initiatives can support educational opportunities for citizens, particularly in developing regions, enhancing their capacity to contribute to environmental governance (Tan & Taeihagh, 2020).

Innovative Governance and Policy Development highlight the potential for novel approaches in environmental governance. Opportunities for less powerful actors to influence networks and diverse methodologies for global governance reflect a shift towards more inclusive and adaptive strategies (Pilbeam et al., 2012; O'Neill et al., 2013). Transformative governance frameworks, which can be triggered by instability or shifts in social norms, offer new pathways for addressing environmental challenges (Chaffin et al., 2016). Solution-oriented assessments that involve diverse stakeholders can provide actionable recommendations and foster international cooperation (Kowarsch & Jabbour, 2017). Additionally, large-scale engagement between environmental non-governmental organizations (ENGOs) and industry highlights the role of technology in closing gaps between societal expectations and industry practices (Cullen-Knox et al., 2017). Ethical and knowledge-oriented networks facilitate a collective articulation of values and learning opportunities, further supporting effective governance (Lupova-Henry & Dotti, 2019).

Empirical Research and Data Collection stress the importance of evidence-based approaches in environmental governance. Research on telecoupled sustainability challenges and governance experiments provides valuable insights into effective strategies and practices (Lenschow, et al., 2016; Moon et al., 2017). Expanding citizen science initiatives enhances data collection and contributes to a more comprehensive understanding of environmental issues (Liu, et al., 2017).

Figure 6. A snapshot of opportunities

Source: Developed of the purpose of the study

4.7 Future Scope

Table 3. Future research directions in environmental governance: thematic overview

Theme	Future Research Directions	References
Longitudinal Studies & Governance Contexts	1. Increase longitudinal studies on governance mechanisms. 2. Explore the connection between supply network contexts and outcomes, and how governance instruments mediate these outcomes.	Pilbeam et al. (2012)
Environmental Impact & Policy Design	1. Conduct comprehensive environmental impact assessments of desalination projects. 2. Develop policies promoting sustainable practices and enforcing environmental standards. 3. Explore mechanisms for collaborative governance and joint investment in green technologies.	Barau & Al Hosani (2015)
E-Governance & Information Sharing	1. Investigate if e-governance improves sustainability policy effectiveness. 2. Identify conditions for information sharing that encourage cooperation and self-organization. 3. Examine the impact of private governance on societal values and how information increases empowerment.	Soma et al. (2016)
Transformative Environmental Governance	1. Identify system drivers and leading indicators associated with social-ecological thresholds.	Chaffin et al. (2016)

continued on following page

Table 3. Continued

Theme	Future Research Directions	References
Smart Cities & Urban Governance	1. Explore smart city development in developing countries, focusing on urban planning, public administration, and policy challenges. 2. Investigate how smart city lessons are transferred domestically and internationally.	Tan & Taeihagh (2020)
Global South Perspectives	1. Review academic sources beyond mainstream databases to explore climate governance in the Global South. 2. Expand research on regional responses to climate challenges.	Sapiains et al. (2021)
ICT Impacts & Consumer Behavior	1. Investigate the indirect effects of ICT on consumer behavior, focusing on social practices and environmental implications.	Bieser & Hilty (2018)
Governance Experiments	1. Generate cumulative research on governance experiments and their long-term impacts. 2. Study how societies benefit from experimental approaches to sustainability.	Kivimaa et al. (2017)
Methodological Frameworks & Assessments	1. Develop robust frameworks to integrate diverse knowledge types and stakeholder perspectives. 2. Evaluate how solution-oriented assessments influence policy and foster international cooperation.	Kowarsch & Jabbour (2017)
UAS, GIS, & Digital Democracy	1. Explore the involvement of UAS and GIS users in creating digital twin models. Investigate the potential for direct democracy using current digital tools.	Hognogi et al. (2021)
Stakeholder Management & Innovation	1. Develop a common framework for assessing governance. 2. Compare governance strategies for different types of innovation. 3. Focus on specific stages of sustainable innovation and how governance strategies target these stages.	Lupova-Henry & Dotti (2019)
Blockchain Technology & Sustainability	1. Investigate the role of blockchain technology in environmental sustainability. 2. Explore practical applications and impacts of blockchain on governance frameworks.	Arshad et al. (2023)
Renewable Energy & Policy Impacts	1. Study the effects of renewable energy targets and policies on innovation. 2. Examine the impact of quotas, emission-trading schemes, and carbon taxes on renewable energy technologies.	Rastegar et al. (2024)

Several critical areas for future research in environmental governance are highlighted by the findings from the table. Longitudinal studies are urgently required to investigate the long-term consequences of governance mechanisms within supply networks and to investigate the potential impact of governance instruments on outcomes. In the desalination industry, sustainability challenges are particularly challenging, necessitating the development of robust policies and comprehensive environmental impact assessments. Additionally, research should concentrate on the efficacy of private governance, information exchange, and e-governance in improving sustainability. Furthermore, valuable insights can be gained by exploring transformative governance approaches, smart city development in developing countries, and the climate governance perspectives of the Global South. It is also imperative to investigate the indirect effects of ICT on consumer behavior, evaluate governance initiatives, and develop methodological frameworks for the integration of diverse knowledge. The necessity of conducting research in these fields is further emphasized by the influence of renewable energy policies on innovation and the function of blockchain technology. Our comprehension of environmental governance and its contribution to sustainability will be enhanced by addressing these deficiencies.

CONCLUSION

The systematic literature review on the challenges and opportunities in implementing tech-enabled environmental governance reveals a complex landscape characterized by both significant potential and considerable obstacles. Technological advancements have introduced innovative tools and methodologies that offer transformative potential for environmental governance. These include enhanced data collection, real-time monitoring, and improved transparency through technologies such as blockchain, artificial intelligence, and the Internet of Things. However, the review also highlights persistent challenges, such as data reliability, regional disparities in technology adoption, and the environmental impact of the technologies themselves. Additionally, the uneven distribution of technological benefits and the complexity of integrating new technologies within existing governance frameworks point out the need for more inclusive and adaptable approaches. The study identifies that while there are promising opportunities for innovation and improvement, the actual impact of these technologies on environmental conditions remains contested. The review emphasizes the need for a nuanced understanding of how technology can be effectively harnessed for sustainable governance, moving beyond mere technological determinism to consider the broader implications and potential for collective action.

Implications

This study aims to give useful information to scientists, practitioners, and policy makers.

Practical Implications

The study highlights several practical implications for implementing tech-enabled environmental governance. Governments and organizations need to prioritize the integration of advanced technologies, such as artificial intelligence, blockchain, and the Internet of Things, to enhance environmental monitoring and resource management. The findings suggest that these technologies can significantly improve data collection, transparency, and decision-making processes. However, it is crucial to address the digital divide by ensuring equitable access to technology, particularly in developing regions, to prevent exacerbating existing disparities. Additionally, stakeholders must implement robust data management practices to mitigate privacy and security concerns associated with digital tools. Policymakers should also consider the environmental impact of technology itself, such as e-waste and energy consumption, and promote sustainable technology development. Adopting these practices can lead to more effective and inclusive environmental governance strategies.

Theoretical Implications

The study offers several theoretical implications for understanding the interplay between technology and environmental governance. It highlights the need to move beyond technological determinism and explore how technology can be a tool for transformative change rather than merely an incremental improvement. This shift in perspective can enrich theoretical frameworks by incorporating the complexities of technology's role in governance systems. The study also highlights gaps in existing research, such as the need for a more nuanced understanding of how technological advancements interact with diverse governance contexts and stakeholder interests. Future theoretical explorations could benefit from integrating interdisciplinary approaches and examining the broader implications of technology on governance structures and processes. By addressing these gaps, scholars can develop more comprehensive theories that better reflect the dynamic and evolving nature of environmental governance in the digital age.

Social Implications

The social implications of the study emphasize the importance of fostering greater public engagement and inclusivity in environmental governance. The integration of technology offers opportunities to enhance citizen participation through initiatives like citizen science and public observatories, which can empower communities and drive demand for sustainable practices. However, the study also reveals the potential for technology to reinforce social inequalities if access remains uneven. Ensuring that technological advancements do not exacerbate existing social disparities is crucial for achieving equitable outcomes. Additionally, the study highlights the need for educational initiatives to build capacity in developing regions, enabling individuals to contribute effectively to environmental governance. By addressing these social implications, stakeholders can promote a more inclusive and participatory approach to environmental management, ultimately supporting broader sustainability goals.

REFERENCES

Alotaibi, I., Abido, M. A., Khalid, M., & Savkin, A. V. (2020). A comprehensive review of recent advances in smart grids: A sustainable future with renewable energy resources. *Energies*, 13(23), 6269. DOI: 10.3390/en13236269

Alves, E. C., Steiner, A. Q., & Amaral, A. M. F. (2023). Environmental Governance and International Relations: A Systematic Review of Theories, Methods, and Issues in Latin American Publications. *Revista de Relaciones Internacionales*, 96(2), 87–121. DOI: 10.15359/ri.96-2.4

Arshad, A., Shahzad, F., Rehman, I. U., & Sergi, B. S. (2023). A systematic literature review of blockchain technology and environmental sustainability: Status quo and future research. *International Review of Economics & Finance*, 88, 1602–1622. DOI: 10.1016/j.iref.2023.07.044

Bakker, K., & Ritts, M. (2022). Environmental Governance in a Wired World. *The Nature of Data: Infrastructures, Environments. Politics*, 61.

Barau, A. S., & Al Hosani, N. (2015). Prospects of environmental governance in addressing sustainability challenges of seawater desalination industry in the Arabian Gulf. *Environmental Science & Policy*, 50, 145–154. DOI: 10.1016/j.envsci.2015.02.008

Bashir, M. F., Ma, B., Bilal, F., Komal, B., & Bashir, M. A. (2021). Analysis of environmental taxes publications: A bibliometric and systematic literature review. *Environmental Science and Pollution Research International*, 28(16), 20700–20716. DOI: 10.1007/s11356-020-12123-x PMID: 33405155

Beula, D., & Sureshkumar, M. (2021). A review on the toxic E-waste killing health and environment–Today's global scenario. *Materials Today: Proceedings*, 47, 2168–2174. DOI: 10.1016/j.matpr.2021.05.516

Bieser, J. C., & Hilty, L. M. (2018). Assessing indirect environmental effects of information and communication technology (ICT): A systematic literature review. *Sustainability (Basel)*, 10(8), 2662. DOI: 10.3390/su10082662

Chaffin, B. C., Garmestani, A. S., Gunderson, L. H., Benson, M. H., Angeler, D. G., Arnold, C. A., Cosens, B., Craig, R. K., Ruhl, J. B., & Allen, C. R. (2016). Transformative environmental governance. *Annual Review of Environment and Resources*, 41(1), 399–423. DOI: 10.1146/annurev-environ-110615-085817 PMID: 32607083

Chang, C. P., Wen, J., & Zheng, M. (2022). Environmental governance and innovation: An overview. *Environmental Science and Pollution Research International*, •••, 1–2. PMID: 34981403

Chen, S., & Liu, N. (2022). Research on citizen participation in government ecological environment governance based on the research perspective of "dual carbon target". *Journal of Environmental and Public Health*, 2022(1), 2022. DOI: 10.1155/2022/5062620 PMID: 35769833

Cullen-Knox, C., Eccleston, R., Haward, M., Lester, E., & Vince, J. (2017). Contemporary Challenges in Environmental Governance: Technology, governance and the social licence. *Environmental Policy and Governance*, 27(1), 3–13. DOI: 10.1002/eet.1743

Dilek, M. Ş., & Furuncu, Y. (2019). *Bitcoin mining and its environmental effects*. Atatürk Üniversitesi İktisadi ve İdari Bilimler Dergisi.

Fazey, I., Carmen, E., Ross, H., Rao-Williams, J., Hodgson, A., Searle, B. A., & Thankappan, S. (2021). *Social dynamics of community resilience building in the face of climate change: The case of three Scottish communities* (Vol. 16). Springer Japan.

Fazey, I., Hughes, C., Schäpke, N. A., Leicester, G., Eyre, L., Goldstein, B. E., Hodgson, A., Mason-Jones, A. J., Moser, S. C., Sharpe, B., & Reed, M. S. (2021). Renewing universities in our climate emergency: Stewarding system change and transformation. *Frontiers in Sustainability*, 2, 677904. DOI: 10.3389/frsus.2021.677904

Felzmann, H., Fosch-Villaronga, E., Lutz, C., & Tamò-Larrieux, A. (2020). Towards transparency by design for artificial intelligence. *Science and Engineering Ethics*, 26(6), 3333–3361. DOI: 10.1007/s11948-020-00276-4 PMID: 33196975

Gerlak, A. K., Heikkila, T., & Newig, J. (2020). Learning in environmental governance: Opportunities for translating theory to practice. *Journal of Environmental Policy and Planning*, 22(5), 653–666. DOI: 10.1080/1523908X.2020.1776100

Helsper, E. (2021). The digital disconnect: The social causes and consequences of digital inequalities. *The Digital Disconnect*, 1-232.

Hognogi, G. G., Pop, A. M., Marian-Potra, A. C., & Someșfălean, T. (2021). The role of UAS–GIS in digital Era governance. A systematic literature review. *Sustainability (Basel)*, 13(19), 11097. DOI: 10.3390/su131911097

Hoosain, M. S., Paul, B. S., Kass, S., & Ramakrishna, S. (2023). Tools towards the sustainability and circularity of data centers. *Circular Economy and Sustainability*, 3(1), 173–197. DOI: 10.1007/s43615-022-00191-9 PMID: 35791435

Jager, N. W., Newig, J., Challies, E., & Kochskämper, E. (2020). Pathways to implementation: Evidence on how participation in environmental governance impacts on environmental outcomes. *Journal of Public Administration: Research and Theory*, 30(3), 383–399. DOI: 10.1093/jopart/muz034

Jänicke, M., & Jörgens, H. (2020). New approaches to environmental governance. In *The ecological modernisation reader* (pp. 156–189). Routledge. DOI: 10.4324/9781003061069-13

Jänicke, M., & Jörgens, H. (2020). New approaches to environmental governance. In *The Ecological Modernisation Reader* (pp. 156–189). Routledge. DOI: 10.4324/9781003061069-13

Kamble, S., Gunasekaran, A., & Arha, H. (2019). Understanding the blockchain technology adoption in supply chains-Indian context. *International Journal of Production Research*, 50(7), 2009–2033. DOI: 10.1080/00207543.2018.1518610

Kivimaa, P., Hildén, M., Huitema, D., Jordan, A., & Newig, J. (2017). Experiments in climate governance–A systematic review of research on energy and built environment transitions. *Journal of Cleaner Production*, 169, 17–29. DOI: 10.1016/j.jclepro.2017.01.027

Kloppenburg, S., Gupta, A., Kruk, S. R., Makris, S., Bergsvik, R., Korenhof, P., Solman, H., & Toonen, H. M. (2022). Scrutinizing environmental governance in a digital age: New ways of seeing, participating, and intervening. *One Earth*, 5(3), 232–241. DOI: 10.1016/j.oneear.2022.02.004

Kostka, G., Zhang, X., & Shin, K. (2020). Information, technology, and digitalization in China's environmental governance. *Journal of Environmental Planning and Management*, 63(1), 1–13. DOI: 10.1080/09640568.2019.1681386

Kowarsch, M., & Jabbour, J. (2017). Solution-oriented global environmental assessments: Opportunities and challenges. *Environmental Science & Policy*, 77, 187–192. DOI: 10.1016/j.envsci.2017.08.013

Kraus, S., Jones, P., Kailer, N., Weinmann, A., Chaparro-Banegas, N., & Roig-Tierno, N. (2021). Digital transformation: An overview of the current state of the art of research. *SAGE Open*, 11(3), 21582440211047576. DOI: 10.1177/21582440211047576

Kumar, K. D., Rane, D. D., Muralidhar, A., Goundar, S., & Reddy, P. V. (2024). E-Waste Management: A Significant Solution for Green Computing. In *Sustainable Solutions for E-Waste and Development* (pp. 56-73). IGI Global. DOI: 10.4018/979-8-3693-1018-2.ch005

Lenschow, A., Newig, J., & Challies, E. (2016). Globalization's limits to the environmental state? Integrating telecoupling into global environmental governance. *Environmental Politics*, 25(1), 136–159. DOI: 10.1080/09644016.2015.1074384

Liu, H. Y., Grossberndt, S., & Kobernus, M. (2017). Citizen science and citizens' observatories: Trends, roles, challenges and development needs for science and environmental governance. *Mapping and the Citizen Sensor*, 351-376.

Liu, K., Tan, Q., Yu, J., & Wang, M. (2023). A global perspective on e-waste recycling. *Circular Economy*, 2(1), 100028. DOI: 10.1016/j.cec.2023.100028

Lupova-Henry, E., & Dotti, N. F. (2019). Governance of sustainable innovation: Moving beyond the hierarchy-market-network trichotomy? A systematic literature review using the 'who-how-what' framework. *Journal of Cleaner Production*, 210, 738–748. DOI: 10.1016/j.jclepro.2018.11.068

Martin, A., Armijos, M. T., Coolsaet, B., Dawson, N., & Edwards, AS, G., Few, R., ... & White, C. S. (. (2020). Environmental justice and transformations to sustainability. *Environment*, 62(6), 19–30. DOI: 10.1080/00139157.2020.1820294

Moon, K., Blackman, D., Brewer, T. D., & Sarre, S. D. (2017). Environmental governance for urgent and uncertain problems. *Biological Invasions*, 19(3), 785–797. DOI: 10.1007/s10530-016-1351-7

O'Neill, K., Weinthal, E., Marion Suiseeya, K. R., Bernstein, S., Cohn, A., Stone, M. W., & Cashore, B. (2013). Methods and global environmental governance. *Annual Review of Environment and Resources*, 38(1), 441–471. DOI: 10.1146/annurev-environ-072811-114530

Ojha, H., Nightingale, A. J., Gonda, N., Muok, B. O., Eriksen, S., Khatri, D., & Paudel, D. (2022). Transforming environmental governance: Critical action intellectuals and their praxis in the field. *Sustainability Science*, 17(2), 621–635. DOI: 10.1007/s11625-022-01108-z PMID: 35222728

Park, A., & Li, H. (2021). The Effect of Blockchain Technology on Supply Chain Sustainability Performances. *Sustainability (Basel)*, 13(4), 1726. DOI: 10.3390/su13041726

Partelow, S., Schlüter, A., Armitage, D., Bavinck, M., Carlisle, K., Gruby, R. L., ... & Van Assche, K. (2020). Environmental governance theories: a review and application to coastal systems.

Pilbeam, C., Alvarez, G., & Wilson, H. (2012). The governance of supply networks: A systematic literature review. *Supply Chain Management*, 17(4), 358–376. DOI: 10.1108/13598541211246512

Raja, G. B. (2021). Impact of internet of things, artificial intelligence, and blockchain technology in Industry 4.0. *Internet of Things, Artificial Intelligence and Blockchain Technology*, 157-178.

Rajmohan, K. V. S., Ramya, C., Viswanathan, M. R., & Varjani, S. (2019). Plastic pollutants: Effective waste management for pollution control and abatement. *Current Opinion in Environmental Science & Health*, 12, 72–84. DOI: 10.1016/j.coesh.2019.08.006

Rastegar, H., Eweje, G., & Sajjad, A. (2024). The impact of environmental policy on renewable energy innovation: A systematic literature review and research directions. *Sustainable Development (Bradford)*, 32(4), 3859–3876. DOI: 10.1002/sd.2884

Rolnick, D., Donti, P. L., Kaack, L. H., Kochanski, K., Lacoste, A., Sankaran, K., Ross, A. S., Milojevic-Dupont, N., Jaques, N., Waldman-Brown, A., Luccioni, A. S., Maharaj, T., Sherwin, E. D., Mukkavilli, S. K., Kording, K. P., Gomes, C. P., Ng, A. Y., Hassabis, D., Platt, J. C., & Bengio, Y. (2022). Tackling climate change with machine learning. *ACM Computing Surveys*, 55(2), 1–96. DOI: 10.1145/3485128

Saini, K., Mummoorthy, A., Chandrika, R., & Gowri Ganesh, N. S. (Eds.). (2023). *AI, IoT, and Blockchain Breakthroughs in E-governance*. IGI Global. DOI: 10.4018/978-1-6684-7697-0

Sapiains, R., Ibarra, C., Jiménez, G., O'Ryan, R., Blanco, G., Moraga, P., & Rojas, M. (2021). Exploring the contours of climate governance: An interdisciplinary systematic literature review from a southern perspective. *Environmental Policy and Governance*, 31(1), 46–59. DOI: 10.1002/eet.1912

Soma, K., Termeer, C. J., & Opdam, P. (2016). Informational governance–A systematic literature review of governance for sustainability in the Information Age. *Environmental Science & Policy*, 56, 89–99. DOI: 10.1016/j.envsci.2015.11.006

Steblianskaia, E., Vasiev, M., Denisov, A., Bocharnikov, V., Steblyanskaya, A., & Wang, Q. (2023). Environmental-social-governance concept bibliometric analysis and systematic literature review: Do investors becoming more environmentally conscious? *Environmental and Sustainability Indicators*, 17, 100218. DOI: 10.1016/j.indic.2022.100218

Tan, S. Y., & Taeihagh, A. (2020). Smart city governance in developing countries: A systematic literature review. *Sustainability (Basel)*, 12(3), 899. DOI: 10.3390/su12030899

Tomor, Z., Meijer, A., Michels, A., & Geertman, S. (2019). Smart governance for sustainable cities: Findings from a systematic literature review. *Journal of Urban Technology*, 26(4), 3–27. DOI: 10.1080/10630732.2019.1651178

Vaz, N. (2021). *Digital business transformation: How established companies sustain competitive advantage from now to next*. John Wiley & Sons.

Wang, H., & Guo, J. (2024). New way out of efficiency-equity dilemma: Digital technology empowerment for local government environmental governance. *Technological Forecasting and Social Change*, 200, 123184. DOI: 10.1016/j.techfore.2023.123184

Wang, S., Li, J., & Razzaq, A. (2023). Do environmental governance, technology innovation and institutions lead to lower resource footprints: An imperative trajectory for sustainability. *Resources Policy*, 80, 103142. DOI: 10.1016/j.resourpol.2022.103142

Weill, P., & Woerner, S. L. (2018). Is your company ready for a digital future? *MIT Sloan Management Review*, 59(2), 21–25.

Xu, J., She, S., & Liu, W. (2022). Role of digitalization in environment, social and governance, and sustainability: Review-based study for implications. *Frontiers in Psychology*, 13, 961057. DOI: 10.3389/fpsyg.2022.961057 PMID: 36533022

Yang, H., Ma, M., Thompson, J. R., & Flower, R. J. (2018). Waste management, informal recycling, environmental pollution and public health. *Journal of Epidemiology and Community Health*, 72(3), 237–243. DOI: 10.1136/jech-2016-208597 PMID: 29222091

Zhang, N., Deng, J., Ahmad, F., Draz, M. U., & Abid, N. (2023). The dynamic association between public environmental demands, government environmental governance, and green technology innovation in China: Evidence from panel VAR model. *Environment, Development and Sustainability*, 25(9), 9851–9875. DOI: 10.1007/s10668-022-02463-8

Chapter 10
Data Analytics, Machine Learning, and IoT for Environmental Governance

Indranil Mutsuddi
https://orcid.org/0000-0002-4202-8744
JIS University, India

ABSTRACT

Effective environmental governance is crucial for addressing complex & interconnected environmental challenges of the 21st century. It requires collaboration of multiple sectors & integration of diverse perspectives to create resilient & sustainable societies. Organizations, governments, and NGOs need to adopt a collaborative approach to implement sustainable solutions for environmental governance. While governments are responsible for policymaking, NGOs excel in implementing these policies due to their strong connections with grassroots communities. Corporations play a critical role by leveraging their capital, technological resources, and research and development capabilities to address existing environmental challenges with sustainable solutions. The book chapter presents theoretical & practical insights regarding the use of Data Analytics, ML and IoT for environmental governance.

INTRODUCTION

Environmental monitoring and management are crucial for safeguarding the health and sustainability of our planet. Environmental monitoring involves the systematic collection, analysis, and interpretation of data to assess the condition of natural

DOI: 10.4018/979-8-3693-7001-8.ch010

resources and ecosystems. As the urgency for environmental protection grows, the demand for innovative solutions becomes increasingly clear. Technology, especially data science along with technologies like Artificial Intelligence (AI), Machine Learning (ML) & Internet of Things (IoT), play a vital role in this effort. With rising environmental concerns, data science has proven to be a powerful tool capable of analyzing large datasets and providing insights into complex ecological dynamics.

Simultaneously, environmental management uses this information to make informed decisions and proactively address environmental risks while promoting sustainable practices. Organizations and governments are increasingly adopting advanced technologies to enhance individual well-being, protect wildlife and ecosystems, and achieve sustainable development goals. Key technologies driving this transformation include Artificial Intelligence (AI), the Internet of Things (IoT), and Big Data, all of which enable data-driven decision-making to support sustainable business solutions. These advancements allow organizations to optimize resource management and improve services in unprecedented ways.

Innovations in machine learning and data analytics have the potential to significantly impact various aspects of Environmental Science (ES) (Hajjaji et al, 2021). The integration of big data and IoT technologies presents exciting opportunities for smart environmental applications focused on monitoring, protecting, and enhancing natural resources. Big Data analytics encompasses a variety of data resources defined by their diversity, speed, accuracy, and volume. It plays a critical role in ES applications such as weather forecasting, energy sustainability, and disaster management, supported by techniques like remote sensing and information and communication technologies.

In light of increasing environmental challenges, the integration of data science with environmental monitoring has emerged as a transformative solution. This collaboration enables more effective monitoring of ecosystems and endangered species, supports predictive modeling and early warning systems, enhances air and water quality assessments, and is essential for disaster management (Tharsanee et al, 2020). Given this context, it is vital to deepen our understanding of how technologies like data analytics, AI, machine learning, and IoT contribute to the development of effective strategies for environmental protection and facilitate comprehensive analyses of environmental data.

REVIEW OF LITERATURE

Recent literature (Anonymous, 2024) underscores growing concerns about the potential impact of big data analytics on corporate sustainability strategies, highlighting the vast and diverse benefits of this tool. Rashid et al. (2024) found that Big

Data Analytics and AI significantly and positively influenced green supply chain collaboration, sustainable manufacturing, and environmental process integration. Similarly, Luoma et al. (2023) emphasized the multifaceted value of data in business operations, showing how it informs decisions that promote environmentally conscious actions and foster environmental improvements. Etzion & Aragon-Correa (2016) also noted that many capabilities provided by big data are naturally aligned with sustainability goals. Nisar et al. (2021) demonstrated that big data-driven decision-making improved both the effectiveness and efficiency of decisions, which in turn boosted environmental performance in public and private hospitals in China. The use of big data has become a fundamental activity for organizations, creating a knowledge framework that supports social and environmental sustainability (Keeso, 2014; Dubey et al., 2019). The ongoing development and application of big data are fostering a growing urgency to enhance sustainable environmental performance.

On the other hand, the research conducted by Uriarte-Gallastegi et al. (2024) and Schoorman et al (2023) emphasized the increasing importance of understanding how Artificial Intelligence (AI) & Machine Learning (ML) can aid companies in their transition to a Circular Economy model. By concentrating on energy management, the study examines its effects on efficiency and emissions through a multi-case analysis. The results indicate that Artificial Intelligence positively affects both efficiency and emissions, although the degree of impact varies across different applications and sectors. Whig et al (2024) discussed the growing applications of AI-ML applications for managing sustainable development emphasizing issues like how cutting edge innovations in AI-ML enabled business solutions contributed towards achieving global sustainability perspectives. In this regard, the study conducted by Chang & Ke (2024) indicated that AI along with people analytics contributed effectively towards sustainable business solutions. The research by Raman et al. (2024) supported a unified approach to innovation in AI that prioritizes environmental sustainability and ethical integrity, encouraging responsible AI development. This aligns with the Sustainable Development Goals, highlighting the importance of ecological balance, societal welfare, and responsible innovation. Galaz et al. (2021) explored the increasing importance of AI in minimizing the harmful carbon footprint of industrial operations, improving energy efficiency, and promoting a sustainable future for AI applications. Green AI aligns with global efforts to tackle climate change and foster the development of eco-friendly technologies. AI & ML applications could intensify environmental challenges because they require substantial computational resources to train large models, leading to increased energy consumption and carbon emissions (Harvard Business Review, 2023). Furthermore, as these applications become more numerous and complex, the continuous need for data and model refinement may generate a cycle of energy-intensive usage, potentially jeopardizing the advancements made by sustainable AI initiatives. Acharya et al. (2024) highlighted that smart technologies such as IoT, edge computing, and AI play a vital role in various facets of waste management, mitigation, and recycling to promote a sustainable environment. These smart technologies enable a more sustainable approach to resource management by reducing waste and lessening environmental damage. Recent advancements in AI are expected to significantly amplify the environmental effects of the digital economy (Mutsuddi, 2024). In this context, Shafique et al. (2018) noted that, relative to other information and communication technologies, the application of IoT provided an intelligent and autonomous solution that enabled organizations to more effectively and efficiently tackle sustainability development goals and fulfill environmental requirements.

The above literature review highlights the increasing importance of smart technologies such as big data analytics, AI-ML, and IoT in organizations' strategies for sustainable practices and environmental management. However, it is essential to establish a deeper theoretical understanding of whether these technologies influence organizational approaches and practices related to effective environmental governance. With this in mind, the chapter will first examine the conceptual perspectives of environmental governance and develop an understanding of environmental, social, and governance (ESG) frameworks that facilitate the integration of technology into environmental governance along with numerous case study illustrations.

ENVIRONMENTAL GOVERNANCE

Environmental governance is essential for integrating environmental considerations into all levels of decision-making, ensuring that human activities do not compromise the planet's ecological balance or the health and prosperity of its inhabitants. It involves the processes and institutions through which societies manage environmental resources and address environmental issues. This encompasses a wide range of activities, policies, and mechanisms implemented by various stakeholders, including governments, businesses, non-governmental organizations (NGOs), and the public.

Organizations and business leaders are increasingly adopting green practices as part of their efforts toward effective environmental governance. These green practices (Khan et al., 2021) have been found to play a significant role in building a positive image for firms and enhancing their performance. However, it has been observed that the CSR practices adopted by the corporate had been diluted due to the lack of "indirect support from the community" due to mismanagement and corruption at governmental levels.

Environmental, Social, and Governance (ESG) Frameworks

Based on this understanding, organizations have been focusing on developing robust Environmental, Social, and Governance (ESG) frameworks to assess their corporate sustainability, primarily within the context of firms in their home countries (Linnenluecke, 2022). These frameworks typically concentrate on industrial and economic activities in developed nations, resulting in limited applicability and transferability to emerging markets. This raises questions about the balance and applicability of these frameworks to developing nations. Are these frameworks truly balanced? Are they applicable to developing nations?

Essential Features of Environmental Governance

The essential features of environmental governance practices from an organizational perspective are illustrated by the following Figure 1.

Task	UN SDGs	Key Performance Indicator (KPI)	Achievement	Sustainability Target (2030)
Pursue Carbon neutrality & utilize renewable energy resources	SDG 13 Climate Action	Achieve Carbon Neutrality by (54.6%) reduction in emissions in the production phase by 2030	52% as compared to base year 2017	Complete Carbon Neutrality
		Complete transition to renewable energy resources (Goal set for 2050: 100% achievement)	Achieved 8.2%	60%
Manufacture environmentally conscious & sustainable products and services	SDG 7 Affordable & Clean Energy	Reduce Green house Gas (GHG) emissions from solid products per functional unit	Achieved 13.1%	20%
		Expand a cumulative use of recycled plastics from 2021 to 2030	59,000 tons	6,00,000 tons
Build a circular economy with recycling of wastes	SDG 15 Life on Land	Enhance waste recycling rates at production sites	94.5%	95%
		Expand a cumulative collection of e-wastes from 2006-2030	3.99 million tons	8 million tons

Fig 1: LG's Sustainable Development Goals 2022
(Courtesy: LG Sustainability Report Factbook 2022-2023; https://www.lg.com/global/sustainability/reports/)

Technology Integration for Environmental Governance

Over the years, digitalization has emerged as the driving force behind a scientific and technological revolution, initiating a new phase of industrial transformation in the global as well as socio-economic perspectives (Lidskog et al, 2015). The integrated development of digitalization and environmental governance has become a crucial strategic priority for fostering high-quality development within contemporary organizations. Incorporating digitalization into government environmental governance (DGEG) presents opportunities for reducing environmental pollution and curbing emissions (Xiaoman et al., 2023).

Information and communication technology (ICT) is increasingly used to enhance environmental management and governance (Mol, 2006; Mol and Buttel, 2002). Environmental regulators are also leveraging digital technologies like AI and machine learning to streamline governance and communication processes. Furthermore, integrating technology into environmental governance improves the management of natural resources and the mitigation of environmental challenges. The advancement of new technologies can help reduce the environmental footprint of various industries through digitalization, decreasing the need for physical products and services. As digitization rapidly advances (Yue et al., 2024), artificial intelligence, big data, and related digital technologies have become powerful catalysts for green computing, impacting both supply and demand and influencing organizational strategies towards environmental governance. Emerging technologies can significantly enhance an organization's ESG performance. The firm can utilize cloud computing to minimize physical infrastructure needs, AI to optimize energy usage, and IoT devices to monitor and manage resources more efficiently. This approach can significantly reduce dependencies on physical resources and the supply chain.

ICT integration into environmental governance offers significant opportunities to improve the efficiency, accuracy, and transparency of environmental management practices. By leveraging these technologies, governments, organizations, and communities can better protect natural resources, mitigate environmental impacts, and promote sustainable development.

Case Study I: Approach Used by TCS for Technology Integration in ESG

Tata Consultancy Services (TCS) had addressed the existence of discrepancy between enterprises' sustainability strategy (Kaijim and Sharma, 2023) and the accessibility of sustainability data and insights (SDI).

The company in this regard had emphasized upon implementing an SDI framework for establishing a strong and reliable foundation for SDI, expediting the transition to a sustainable enterprise and bridging the SDI gaps. The company in this regard proposed a Data Analytics model whereby it emphasizes the role of smart technologies Big Data Analytics, Business Insights, Artificial Intelligence & Machine Learning to develop (simulations & modelling) and implement sustainable solutions resulting in zero emissions, zero waste and zero disparity.

Managing ESG Data –The TCS Approach: TCS also provides a sustainable framework for financial transactions and investments aimed at improving their environmental, social, and governance (ESG) rating. TCS's approach identifies the necessary parameters for creating sustainable frameworks, directly contributing to scoring methodologies and offering a transparent view of ESG data. This helps firms configure their performance frameworks by adding or removing indicators and assigning their criticality, facilitating requirement-driven ESG performance tracking. The process involves capturing ESG data and enhancing the transparency of organizational decision-making using AI and machine learning algorithms.

Data Analytics and Environmental Governance

Data analytics plays a crucial role in enhancing environmental governance by providing the tools and insights needed for effective management and decision-making. Use of big data analytics in environment management represents a major transition from traditional methods to more strategic, efficient, and impactful approaches. By leveraging vast amounts of environmental data, researchers, policymakers, conservationists and even business leaders can make informed decisions to protect ecosystems and promote sustainability.

Applications of Data Analytics for Environmental Governance

The functions of Data Analytics in environmental governance can be depicted in the following table (Table 2):

Environmental Analytics: Environmental analytics involves the collection, organization, and analysis of extensive spatio-temporal datasets to provide actionable insights for policymakers. This field spans various sub-disciplines and professions, utilizing techniques ranging from basic statistics to advanced concepts in deep learning and artificial intelligence. It is a tool utilized by investors to evaluate a company's sustainability. It aims to assess a company's environmental and social impact, as well as its commitment to sound governance practices. This process involves gathering diverse data sets and performing comprehensive analyses. The convergence of environmental analytics and conservation represents a dynamic and revolutionary partnership, leading the charge in addressing urgent ecological issues. Environmental analytics can also be employed to identify companies that may face financial risks due to environmental or social issues, as well as to highlight companies that excel in sustainable business practices.

Amid escalating environmental challenges, environmental analytics has become a powerful tool for monitoring, analyzing, and mitigating these concerns. This collaborative synergy leverages data-driven methodologies, including data collection, analysis, and visualization, to inform and enhance conservation efforts globally. Environmental analytics offers diverse methodologies for exploring the intricacies of the natural environment and improving its management. As environmental research becomes increasingly data-intensive, advanced techniques are needed to uncover and extract meaningful patterns that can aid in sustainable environmental management. Researchers can monitor shifts in climate patterns, air and water quality, biodiversity, and land usage. Commonly employed tools and techniques in environmental analysis encompass geographic information systems (GIS), remote sensing, and predictive modelling. These resources aid in visualizing intricate environmental data and forecasting forthcoming environmental conditions grounded in both present and past trends.

HOW TO UTILIZE ESG BIG DATA

The most common approach is to utilize publicly available big data to monitor an organization's progress on various ESG metrics and key performance indicators. The ESG big data is compared to industry benchmarks so as to identify areas for improvement in its ESG strategies. Furthermore, many companies are now starting to gather their own internal data on ESG indicators. The following illustration (Illustration-I) presented the strategy used by KPMG for managing their ESGs.

Illustration I: ESG-IQ from KPMG

ESG IQ, an analytics platform developed through collaboration between KPMG's Lighthouse data scientists, engineers, leading asset managers, and Google, empowers clients to consolidate and analyze both structured ESG reference data from multiple providers and unstructured data from various sources. These sources encompass a wide array of mediums, including news reports, social media posts, blogs, NGO reports, research papers, and web pages. Moreover, the platform utilizes advanced Natural Language Processing (NLP) techniques to extract ESG data from 'dark data pools,' such as legal documents and trade confirmations, depending on the asset class. The application harnesses a suite of sophisticated analytical tools to assess ESG performance across different frameworks, including the UN SDGs, SASB, GRI, and WEF IBC metrics. Utilizing Google Cloud computing, ESG IQ performs advanced analytics on diverse datasets to derive objective and actionable insights, which are then presented through an intuitive and dynamic dashboard interface.

The paradigm shift in investor sentiment towards a more conscientious society, a greener planet has significantly improved an organization's environmental governance practices. As a result of this, many organizations are leaning towards integrating ESG considerations into their financial reporting and risk management frameworks. The contribution of an organization's approach towards environmental sustainability is reflected in its ESG ratings. ESG ratings assess a company's or investment's impact on environmental, social, and governance factors, such as board diversity. These ratings are issued by various organizations, including sustainable investing research firms, stock exchanges, and rating agencies. Organizations of recent times are investing on their data analytics team for achieving higher ESG ratings. Companies like Snowkap use Snow-IQ ESG Tools for managing their sustainability endeavours (Illustration-II).

Illustration II: Snow-IQ ESG Tool from Snowkap

The Snow-IQ application empowers organizations to conduct ESG risk assessment, performance evaluation, and metric tracking across their supply chain. It provides a scalable and cost-effective solution for evaluating numerous suppliers and enhancing visibility into their ESG performance management. Snow-IQ is a software-as-a-service (SaaS) tool designed for assessing investments and supply chains, aimed at assisting companies in making informed decisions aligned with their ESG objectives. The software utilizes questionnaires aligned with internal Key Performance Indicators (KPIs) and global frameworks to collect data from suppliers. Following this, the collected data undergoes analysis and is presented through user-friendly dashboards. Finally, data-driven recommendations are provided to support ESG improvement initiatives.

MACHINE LEARNING (ML) AND ENVIRONMENTAL GOVERNANCE

Machine learning (ML) holds significant promise for enhancing environmental governance through its ability to analyze vast amounts of data, identify patterns, and make predictions. Because of the robust capability of machine learning methods to uncover intricate relationships among related data and the cost-effectiveness in terms of human and material resources, environmental governance practitioners are increasingly inclined towards utilizing them (Peng, 2023). The collaboration between machine learning (ML) and deep learning (DL) in the agro-climatic field is essential for tackling the diverse challenges presented by climate change in agriculture (Tamayo-Vera et al, 2024). A similar perspective had been discussed by Shixuan et al (2023), whereby they had claimed that machine learning (ML) and deep learning (DL) offer significant advantages in data analysis, including feature extraction, clustering, classification, regression, image recognition, and prediction, as well as in risk assessment and management in environmental ecology. Recent developments in machine learning (ML) present fresh prospects for the field, encompassing experimental design, data analysis, uncertainty quantification, and inverse problems pertaining to environmental management decisions (Jin et al 2023). Wang et al (2023) on the other hand machine learning contributed towards effective management of Carbon Neutrality in organizational production approaches. Mahajan et al (2024) indicated that deep learning and machine learning techniques offered predictions, analyzed extensive datasets, and provided on crucial environmental decision insights that were previously out of reach.

Applications of Machine Learning Techniques for Environmental Governance

The roles of Machine Learning techniques in environmental governance can be illustrated as follows (Table 3):

Integration of ML with Other Advanced Technologies: Machine learning will progressively merge with other technologies, including remote sensing and the Internet of Things (IoT), to offer a comprehensive perspective and understanding of environmental governance. The integration of data from satellite imagery, aerial surveys, ground-based sensors, and IoT devices will facilitate a more holistic comprehension of environmental processes. The integration of AI & ML into Software-as-a-Service (SaaS) applications would further present a considerable opportunity for businesses. AI facilitates decision-making, behavior prediction, and task automation, whereas ML ensures ongoing enhancement. The amalgamation of AI & ML would enrich SaaS applications with intelligent functionalities, thereby augmenting user experience and operational efficiency. AI would also contribute and empower SaaS applications to make informed decisions, predict behavior, and automate tasks, while ML would facilitate continuous learning and refinement. This integration is transforming the essence of software, compelling businesses to innovate and excel by delivering unparalleled efficiency and unlocking novel opportunities for environmental governance.

Developments in machine learning technology and algorithms would expand and broaden the scope of machine learning applications in environmental governance. With the escalation of computing power and the refinement of algorithms, business leaders would be able to anticipate the development of more precise models, expedited analysis, and enhanced predictive abilities. Innovations in domains like deep learning, reinforcement learning, and transfer learning will additionally augment machine learning's capabilities in comprehending and managing intricate environmental systems.

Applications of AI-ML Integration for Environmental Governance

Artificial Intelligence and Machine Learning algorithms are transforming energy management and efficiency. ML-driven smart grids optimize energy distribution, leading to waste reduction and decreased carbon emissions. Pacific Gas and Electric (PG&E) in California utilizes AI to forecast grid failures. This proactive approach enables them to maintain their power grids, minimizing power outages. AI and ML also have the potential to evolve into sustainable technologies by significantly improving the efficiency and reliability of renewable resources. IBM's Watson Decision Platform for Agriculture offers farmers invaluable assistance in cultivating crops sustainably. Through this platform, farmers can access data-driven insights that boost crop yields while reducing environmental impact. By leveraging predictive analysis, AI, weather data, and IoT devices, farmers can effectively manage and monitor their operations while conserving resources. The role of AI algorithms in environmental research has been found to be increasing day by day, specifically their utilization in analyzing extensive datasets from satellites and sensors for various purposes such as tracking deforestation, monitoring ocean health, and predicting natural disasters. Environmental researchers harness AI and ML algorithms to analyze vast datasets from satellites and sensors, enabling them to track deforestation, monitor ocean health, and predict natural disasters. Climate models generate simulations that allow scientists to investigate diverse "what-if" scenarios, such as the potential effects of varying levels of greenhouse gas emissions or the outcomes of specific climate interventions. These simulations are pivotal in educating policymakers and the public about the potential risks and uncertainties linked with climate change. AI technologies facilitate accurate and efficient monitoring of ecological changes, providing unparalleled insights into ecosystems. Machine learning algorithms analyze satellite imagery, enabling real-time monitoring of deforestation, wildlife migration, and biodiversity fluctuations. In marine settings, AI assists in monitoring and safeguarding endangered species, contributing to the development of conservation strategies. Researchers like Nahar et al. (2020) devised a model using ML classifiers to forecast Air Quality Index (AQI). Patil et al. (2020) observed that Artificial Neural Network (ANN), Linear Regression (LR), and Logistic Regression (LogR) models were predominantly utilized by researchers for AQI forecasting.

IoT AND ENVIRONMENTAL GOVERNANCE

The integration of Internet of Things (IoT) technologies holds significant implications for environmental governance. It plays a vital role in environmental monitoring, particularly in assessing air and water quality. Industrial activities frequently release organic compounds such as carbon monoxide, hydrocarbons, and various chemicals—collectively known as "greenhouse gases"—into the atmosphere. Additionally, emissions from vehicles and methane from livestock significantly impact air quality and the overall health of our planet. IoT devices help monitor these emissions, providing valuable data to mitigate their effects and improve environmental health. The real-time data collected by IoT devices is immensely valuable. It helps identify sources of pollution, assess the impact of human activities on the environment, and inform policy decisions aimed at improving air and water quality.

The Internet of Things (IoT) revolution marks a paradigm shift in how devices and objects connect and communicate over the internet. It involves embedding sensors, actuators, and connectivity into everyday objects, enabling them to collect and exchange data and be controlled remotely. With its cutting-edge IoT technologies, organizations like LG are more dedicated than ever to enhancing consumers' overall quality of life by enabling connected smart homes powered by ThinQ. As well as utilizing the same for managing the organization's sustainable development goals (Fig 1).

Applications of IoT for Environmental Governance

The functions of IoT in environmental governance can be exemplified as follows (Table 4):

The application of IoT for achieving sustainable business solutions and environmental governance would be further boosted by the growing demand of IoT in today's organizations. Indian companies are also slowly joining the platform in comparison to their global counterparts. As for example, very recently, Tata Power, India's leading integrated power company, has chosen Tata Communications to introduce its IoT-based Automated Meter Infrastructure (AMI) in collaboration with L&T Meters in Mumbai. Many developed nations are using IoT network sensors for tracking hazardous chemicals, radiation, and water pollution levels, as well as identifying the sources of pollutants. This technology offers early indicators for high pollution zones, enabling both citizens and regulatory agencies to measure pollutants and take actions to improve air quality. In developed and developing countries, farmers are benefiting from advancements in crop yield through the utilization of drone imagery, weather data, and advanced AI, as well as IoT tools for soil management.

These kinds of drives seen in organizations could be understood by the finding that, the anticipated growth of the global IoT market (Reportlinker, 2022) is projected to increase from USD 300.3 billion in 2021 to USD 650.5 billion by 2026, with a Compound Annual Growth Rate (CAGR) of 16.7% from 2021 to 2026.

LIMITATIONS AND AREAS OF CONCERN

Utilization and implementation of smart technologies like big data analytics, AI-ML and IoT in environment management may have many areas of concern and limitations which should be addressed by corporate leaders and policy makers.

Data Availability: Availability of quality environmental data is essential for forecasting, data modelling and implementing AI-ML & IoT applications. Biased data may lead to erroneous predictions leading to faulty decision making by managers which can have detrimental impacts on environmental management practices and sustainability development practices taken up by the organization. Further consolidating data from multiple sources and fluctuations of data sources could also be highly challenging for data analysts to deal with. As for example organizations dealing with renewable energy resources may frequently experience variance in renewable energy data which may affect quality predictions for energy demand based on mathematical modelling and analytics. In many instances data may also not be available at the bottom of the pyramid due to lack of involvement & knowledge or expertise of the stakeholders like community & rural people.

Data Security: The effective utilization of big data analytics, AI-ML tools and IoT applications in environment management may be restricted by breach in data security issues. Organizational endeavours towards data-driven sustainability practices can be seriously hampered by data leakage threats from competitors & business rivals. These may also include utilizations of leaked data for nefarious and unethical consequences.

Limitations of Predictive Analytics: Predictive analytics and IoT applications depend on past data sources to forecast environment management decisions. However, for environment management and sustainability development practices there are several contingency factors which are beyond the purview of mathematical and analytics models. The impact of weather changes and sudden shift of climate change data due to natural disasters and other contingency environmental changes may limit the application of such predictive models for effective decision making.

Technical Barriers: Organizations dealing with implementing big data analytics in environment management may face issues pertaining to barriers caused by lack of adequate technology infrastructure at the operational level in their endeavours towards environment management practices.

Cost Management: Implementing analytics for environmental management may bring upon the burden of additional costs for organizations which may be a concern for their overall profit-making motive and objectives.

Regulatory Implications: Many of the environmental management practices and sustainable development initiatives of organizations are strictly monitored and regulated under statutory compliances imposed by the laws of the state and legislations which may restrict corporations from effectively implementing data-driven solutions at all levels of sustainable practices at the bottom line of an organization's operations.

Ethical Concerns: Many algorithmic decision-making processes and data mining techniques used in environmental management depend on inductive knowledge and the correlations found in datasets. The automation of human decision-making is often justified by the belief that AI and algorithms are free from bias. However, this assumption is flawed, as AI systems inevitably generate biased results. Such biases can result in discrimination against individuals and groups. Discriminatory analytics can perpetuate self-fulfilling prophecies and stigmatize affected communities, reducing their autonomy and involvement in society.

POLICY IMPLICATIONS

In a time when environmental sustainability is a global priority, data analytics has become an essential tool in supporting environmental management and sustainable development efforts. The Government of India offlate had taken up several initiatives for ensuring the amalgamation of data-driven smart technologies like data analytics, AI-ML and IoT etc for effective environmental management. The National Institution for Transforming India (NITI Aayog), the government's policy think tank, has launched a National Strategy for Artificial Intelligence, identifying AI as a key tool to make renewable energy adoption more cost-efficient and to enhance the performance of solar and wind energy systems. However, future reliance on AI alone is insufficient—India's states must start building the groundwork now to fully optimize their upcoming renewable energy capacities and reduce energy curtailment. Machine learning can improve predictive grid load management and maintenance for renewable sources, bolstering energy security across states. India's pledge to achieve net-zero emissions by 2070 and to generate 50 percent of its electricity from renewable sources by 2030 represents a major milestone in the global fight against climate change. Some of noteworthy policy frameworks used by the Government of India for sustainable practices and environment management include those of:

- The India Climate & Energy Dashboard (ICED) offers near real-time data driven perspectives on the country's energy sector and climate, featuring more than 500 parameters and 2,000 infographics. ICED enables organizational as well social stakeholders to identify key challenges and track India's progress in advancing its clean energy transition.
- The National Data Analytics Platform (NDAP) ensures that environmental big data remains up-to-date and accessible. NDAP utilizes various methods, such as APIs and web-crawling, to maintain current data. Additionally, a steering committee and technical advisory group will provide guidance and oversight to the platform's operations.
- The Environment (Protection) Act of 1986, amended in 1991, aims to safeguard and enhance the environment and address related concerns. Under this Act, the Central Government is empowered to take all necessary measures to protect and improve environmental quality and prevent pollution
- The National Environment Policy 2006 focuses on the conservation and sustainable use of India's biological resources and associated knowledge for commercial, research, bio-survey, or bio-utilization purposes. It provides a framework for access to these resources and ensures the fair and equitable sharing of benefits arising from their utilization. Based on the broad guidelines and frameworks depicted under this act, organizations can utilize their data analytics tools and smart technologies for the effective resourcing, utilization of natural resources restricting detrimental practices at the environment level.
- In 1995, the Central Government established the National Environment Tribunal through the National Environment Tribunal Act to enforce strict liability for damages resulting from accidents involving hazardous substances. Later, on October 18, 2010, the National Green Tribunal (NGT) was formed under the National Green Tribunal Act to ensure the effective and swift resolution of cases related to environmental protection, forest conservation, and the preservation of natural resources. The NGT also addresses the enforce-

ment of environmental rights and provides relief and compensation for damages to individuals and property, along with related matters.
- The Air (Prevention and Control of Pollution) Act of 1981, amended in 1987, is legislation aimed at preventing, controlling, and reducing air pollution across the country to protect and maintain air quality. The Water (Prevention and Control of Pollution) Act of 1974, amended in 1988, was enacted to prevent and control water pollution and to maintain or restore the purity of water resources across the country.

OPPORTUNITIES AND IMPLICATIONS

The discussions presented in this chapter offer significant opportunities for organizations to align their strategic efforts with a data- and technology-driven approach to foster successful environmental management. Additionally, they provide corporate leaders with the tools to leverage their vision in achieving Sustainable Development Goals (SDGs) and effectively implement their initiatives for Industry 5.0.

Practical Implications: The discussions delivered in this chapter would immensely help policy makers and the top organizational management to rethink their strategies towards environmental sustainability and also effectively align and integrate data-drive techniques and approaches for implementing effective decisions and the functional as well as strategic levels.

Research Implications: The current discussion in this chapter will assist academics and researchers in designing and conducting comprehensive studies on the various issues addressed. It will significantly expand and enrich the existing literature on the topic.

Social Implications: The dialogue presented in this book chapter would help organizational stakeholders and people at the bottom of the social pyramid to be aware of the growing implications of data-driven approaches for effective environment management.

CONCLUSION

The chapter has explored how data analytics, machine learning, and the Internet of Things (IoT) have significantly transformed environmental governance in contemporary organizations and government. These technologies provide a comprehensive, real-time understanding of environmental dynamics, support sustainable practices, enhance resource management, and contribute to the preservation of natural ecosystems. As these technologies evolve, their integration into environmental governance will be essential for addressing the complex and pressing environmental challenges that will impact our planet in the future

REFERENCES

Acharya, B., Dey, S., & Zidan, M. (2024). *IoT-Based Smart Waste Management for Environmental Sustainability*. CRC Press.

Anonymous, . (2024). Big data analytics and corporate sustainability strategy: A management-oriented perspective. *Strategic Direction*, 40(5), 31–33. DOI: 10.1108/SD-05-2024-0054

Chang, Y.-L., & Ke, J. (2024). Socially Responsible Artificial Intelligence Empowered People Analytics: A Novel Framework Towards Sustainability. *Human Resource Development Review*, 23(1), 88–120. DOI: 10.1177/15344843231200930

Dubey, R., Gunasekaran, A., Childe, S. J., Papadopoulos, T., Luo, Z., Wamba, S. F., & Roubaud, D. (2019). Can big data and predictive analytics improve social and environmental sustainability? *Technological Forecasting and Social Change*, 144, 534–545. DOI: 10.1016/j.techfore.2017.06.020

Etzion, D., & Aragon-Correa, J. A. (2016). Big Data, Management, and Sustainability: Strategic Opportunities Ahead. *Organization & Environment*, 29(2), 147–155. DOI: 10.1177/1086026616650437

Galaz, V., Centeno, M. A., Callahan, P. W., Causevic, A., Patterson, T., Brass, I., Baum, S., Farber, D., Fischer, J., Garcia, D., McPhearson, T., Jimenez, D., King, B., Larcey, P., & Levy, K. (2021). Artificial intelligence, systemic risks, and sustainability. *Technology in Society*, 67, 101741. DOI: 10.1016/j.techsoc.2021.101741

Hajjaji, Y., Boulila, W., Farah, I. R., Romdhani, I., & Hussain, A. (2021). Big data and IoT-based applications in smart environments: A systematic review, Computer Science Review, Volume 39, 2021, 100318, ISSN 1574-0137, https://doi.org/DOI: 10.1016/j.cosrev.2020.100318

Harvard Business Review (2023). How to make generative AI greener; 2023. https://hbr.org/2023/07/how-to-make-generative-ai-greener

Jin, H., Zhang, E., & Espinosa, H. D. (2023, November). Recent Advances and Applications of Machine Learning in Experimental Solid Mechanics: A Review. ASME. *Applied Mechanics Reviews*, 75(6), 061001. DOI: 10.1115/1.4062966

Kaijim, W., & Sharma, G. (2023).Closing the sustainability data and insights gap. https://www.tcs.com/what-we-do/services/data-and-analytics/white-paper/enterprise-sustainability-data-insights

Keeso, A. (2014), "Big Data and Environmental Sustainability: a Conversation Starter", Smith School of Enterprise and the Environment, Working Paper Series, pp. 14-04.

Khan, S. A. R., Yu, Z., & Umar, M. (2021). How environmental awareness and corporate social responsibility practices benefit the enterprise? An empirical study in the context of emerging economy. *Management of Environmental Quality*, 32(5), 863–885. DOI: 10.1108/MEQ-08-2020-0178

Lidskog, R., Mol, A. P., & Oosterveer, P. (2015). Towards a global environmental sociology? Legacies, trends and future directions. *Current Sociology*, 63(3), 339–368. DOI: 10.1177/0011392114543537 PMID: 25937642

Linnenluecke, M. K. (2022). Environmental, social and governance (ESG) performance in the context of multinational business research. *Multinational Business Review*, 30(1), 1–16. DOI: 10.1108/MBR-11-2021-0148

Luoma, P., Rauter, R., Penttinen, E., & Toppinen, A. (2023, November). The value of data for environmental sustainability as perceived by the customers of a tissue-paper supplier. *Corporate Social Responsibility and Environmental Management*, 30(6), 3110–3123. DOI: 10.1002/csr.2541

Mahajan, Y., Patil, R., Pattanaik, S., Firake, T. S., Kodulkar, R., Damre, S. S., & Uplaonkar, D. (2024). A Review of Techniques and Applications for Machine Learning and Deep Learning. *International Journal of Intelligent Systems and Applications in Engineering*, 12(16s), 182–187. https://ijisae.org/index.php/IJISAE/article/view/4804

Mol, A. P. J. (2006). From environmental sociologies to environmental sociology? A comparison of U.S. and European environmental sociology. *Organization & Environment*, 19(1), 5–27. DOI: 10.1177/1086026605285643

Mol, A. P. J., & Buttel, F. H. (Eds.). (2002). *The Environmental State under Pressure*. Elsevier. DOI: 10.1016/S0196-1152(2002)10

Mutsuddi, I. (2024). *Ethical Considerations in the Intersection of AI and Environmental Science. In the edited book "Maintaining a Sustainable World in the Nexus of Environmental Science and AI" by Singh, B., Kaunert, C., Vig, K., Dutta, S*. IGI Global.

Nahar, K., Ottom, M. A., Alshibli, F., & Shquier, M. A. (2020). Air quality index using machine learning—A jordan case study. COMPUSOFT. *International Journal of Advancements in Computing Technology*, 9(9), 3831–3840.

Nisar, Q. A., Nasir, N., Jamshed, S., Naz, S., Ali, M., & Ali, S. (2021). Big data management and environmental performance: Role of big data decision-making capabilities and decision-making quality. *Journal of Enterprise Information Management*, 34(4), 1061–1096. DOI: 10.1108/JEIM-04-2020-0137

Patil RM, Dinde HT, Powar SK (2020) A literature review on prediction of air quality index and forecasting ambient air pollutants using machine learning algorithms 5(8):1148–1152.

Peng, N. (2023). Application of Machine learning techniques in environmental governance: A review. *Advances in Engineering Technology Research.*, 7(1), 528. DOI: 10.56028/aetr.7.1.528.2023

Raman, R., Pattnaik, D., Lathabai, H. H., Kumar, C., Govindan, K., & Nedungadi, P. (2024). Green and sustainable AI research: An integrated thematic and topic modeling analysis. *Journal of Big Data*, 11(1), 55. DOI: 10.1186/s40537-024-00920-x

Rashid, A., Baloch, N., Rasheed, R. and Ngah, A.H. (2024), "Big data analytics-artificial intelligence and sustainable performance through green supply chain practices in manufacturing firms of a developing country, *Journal of Science and Technology Policy Management*, Vol. ahead-of-print No. ahead-of-print. https://doi.org/DOI: 10.1108/JSTPM-04-2023-0050

Reportlinker (2022). IoT Market with COVID-19 analysis by Component, Organization Size, Focus Area and Region - Global Forecast to 2026, https://www.reportlinker.com/p04944762/?utm_source=GNW

Schoormann, T., Strobel, G., Möller, F., Petrik, D., & Zschech, P. (2023). Artificial Intelligence forSustainability—A Systematic Review of Information Systems Literature. Communications of theAssociation for Information Systems, 52, pp-pp. Retrieved from https://aisel.aisnet.org/cais/vol52/iss1/8

Shafique, M. N., Rashid, A., Bajwa, I. S., Kazmi, R., Khurshid, M. M., & Tahir, W. A. (2018). Effect of IoT capabilities and energy consumption behavior on green supply chain integration. *Applied Sciences (Basel, Switzerland)*, 8(12), 2481. DOI: 10.3390/app8122481

Shixuan, C., Yuchen, G., Yizhou, H., Lilai, S., Qiming, Z., Yaru, P., & Shulin, Z. (2023). Advances and applications of machine learning and deep learning in environmental ecology and health, Environmental Pollution, Volume 335,2023,122358,ISSN 0269-7491, https://doi.org/DOI: 10.1016/j.envpol.2023.122358

Tamayo-Vera, D., Wang, X., & Mesbah, M. (2024). A Review of Machine Learning Techniques in Agroclimatic Studies. *Agriculture*, 14(3), 481. DOI: 10.3390/agriculture14030481

Tharsanee, R. M., Soundariya, S., & Vishnupriya, B. (2020). Machine Learning and Data Analytics for Environmental Science: A Review, Prospects and Challenges. *IOP Conference Series. Materials Science and Engineering*, 955(1), 012107. DOI: 10.1088/1757-899X/955/1/012107

Uriarte-Gallastegi, N., Arana-Landín, G., Landeta-Manzano, B., & Laskurain-Iturbe, I. (2024). The Role of AI in Improving Environmental Sustainability: A Focus on Energy Management. *Energies*, 17(3), 649. DOI: 10.3390/en17030649

Wang, Z., Yu, Y., Roy, K., Gao, C., & Huang, L. (2023). The Application of Machine Learning: Controlling the Preparation of Environmental Materials and Carbon Neutrality. *International Journal of Environmental Research and Public Health*, 20(3), 1871. DOI: 10.3390/ijerph20031871 PMID: 36767237

Whig, P., Sharma, P., Aneja, N., Elngar, A. A., & Silva, N. (Eds.). (2024). *Artificial Intelligence and Machine Learning for Sustainable Development: Innovations, Challenges, and Applications* (1st ed.). CRC Press., DOI: 10.1201/9781003497189

Xiaoman, Z., Shanbing, L., & Shengchao, Y. (2023). How does the digitization of government environmental governance affect environmental pollution? spatial and threshold effects, Journal of Cleaner Production, Volume 415,2023,137670,ISSN 0959-6526,https://doi.org/DOI: 10.1016/j.jclepro.2023.137670

Yue, P., Wei, W., Shangsong, Z., & Yunqiang, L. (2024). Does digitalization help green consumption? Empirical test based on the perspective of supply and demand of green products, Journal of Retailing and Consumer Services, Volume 79,2024,103843,ISSN 0969-6989,https://doi.org/DOI: 10.1016/j.jretconser.2024.103843

Chapter 11
The Role of Technology in Environmental Governance

Manas Kumar Jha
https://orcid.org/0000-0001-7020-5050
Hexagon Geosystem, India

Dilip Kumar Markandey
Central Pollution Control Board, Delhi, India

Ruchi Gupta
GIZ India, India

Pranavi Mishra
School of Biotechnology and Bioinformatics, Navi Mumbai, India

ABSTRACT

Technology's integration into environmental governance signifies a fundamental change in how societies safeguard and manage their natural resources. This chapter will present the significance of technology in advancing environmental governance frameworks in a variety of ways, with a focus on data collection, policy enforcement, and monitoring. Modern technologies allow more accurate tracking of environmental changes and more efficient management techniques. Examples of these technologies include big data analytics, geographic information systems (GIS), and remote sensing. Furthermore, blockchain technology and digital platforms increase environmental policy accountability and transparency, which encourages increased public participation and compliance. The convergence of these technologies addresses complicated ecological concerns in a world that is changing rapidly, while

DOI: 10.4018/979-8-3693-7001-8.ch011

also improving decision-making processes and enabling more robust and adaptive environmental governance.

1 INTRODUCTION

1.1 Defining Environmental Governance

The idea of environmental governance is intricate and multidimensional, encompassing several ways that communities manage their natural resources and tackle environmental issues. Fundamentally, environmental governance entails the collaboration of multiple stakeholders, such as governments, non-governmental organizations (NGOs), private sector companies, and local communities, who all play a part in developing and putting into action policies and procedures that are meant to produce sustainable environmental results. This chapter explores the frameworks and mechanisms that support efficient management of natural resources as it digs into the fundamental ideas of environmental governance. It emphasizes how crucial it is to incorporate social values, economic factors, and scientific information into decision-making processes to guarantee that environmental policies are just and efficient.

The chapter also looks at how international institutions and agreements shape global environmental governance, highlighting the necessity of working together to address transboundary environmental concerns like pollution, biodiversity loss, and climate change. The chapter offers insights into the advantages and disadvantages of governance models to understand the dynamic character of environmental governance in a world that is changing rapidly. It also discusses new developments and trends in the field, such as the growing recognition of local communities' and Indigenous peoples' rights in environmental decision-making and the growing use of technology and data analytics in environmental resource management and monitoring. This chapter intends to provide readers with a greater knowledge of how effective governance may support sustainable development and the preservation of the environment for future generations by providing a thorough overview of the essential components and changing dynamics of environmental governance.

1.2 The Intersection of Technology and Environmental Governance

The convergence of environmental governance and technology is a revolutionary intersection where creative ideas come together to tackle urgent ecological concerns. This interaction is dynamic and becoming important in managing complex environmental issues, such as biodiversity loss and climate change. Technological innovations like blockchain, artificial intelligence, and remote sensing provide hitherto unseen possibilities for environmental resource management, analysis, and monitoring. The advanced technologies help in improving data accountability, transparency, and accuracy, which can ease better decision-making and efficient policy implementation. For example, with the support of remote sensing (RS) technologies such as drones and satellite imagery, real-time data on deforestation, changes in land use, and natural disasters can be obtained. This would help the government and other stakeholders to take prompt actions to manage environmental issues. The other advanced technologies such artificial intelligence (AI) and machine learning are quite useful in overcoming the limitations posed by traditional methods and processes. Large-scale datasets are analyzed by AI and machine learning algorithms to forecast environmental trends, maximize energy use, and create sustainable practices. Blockchain technology promotes trust and cooperation among stakeholders by ensuring transparent and unchangeable records of environmental transactions.

In recent studies, researchers evaluated the role of advanced technologies in biodiversity conversion, resource, safeguarding ecosystem and natural disaster management. Brickson et al. (2023) studied the role of AI and ML in wildlife conservation by developing different AI-driven monitoring methods. These methods are useful in managing and interpreting vast data and help in extracting vital information that can help in elephant conservation. Guha and Sanyal (2022) used AI models to analyze historical data and real time inputs for bringing accuracy in early warning systems for predicting natural disasters.

Parhamfar et al., 2024 mentioned the significance of blockchain technology for carbon market. The technology helps in transparent and verifiable carbon offset verification, which is otherwise a great concern in carbon trading market due to the reason of double counting. The technology also offers other advantages such as decentralized ledger that helps in spreading the authority over the network to multiple computer nodes instead of having a single centralized node (such as a server).

Global organizations are also adopting these advanced technologies for tackling environmental problems. United Nations Environment Programme (UNEP) 2022 has launched World Environment Situation Room (WESR), a digital platform that is leveraging AI's capabilities to analyze complex, multifaceted datasets. WESR is useful in aggregating and visualizing the best available earth observation and sensor data to inform near real-time analysis and future predictions on multiple factors, including CO_2 atmospheric concentration, changes in glacier mass and sea level rise.

The above-mentioned examples illustrate that these technologies have several advantages in environmental governance, however, there are certain difficulties associated with technology incorporation into environmental governance. While technologies offer advantages for environmental governance, challenges like data privacy, tech disparities, and digital literacy need attention to ensure fair access. Ethical implications demand governance frameworks prioritizing sustainability and justice, shaped by regulators balancing technological progress with societal needs. Achieving effective environmental governance requires collaboration among governments, private sectors, and civil society. Public-private partnerships and international cooperation can tailor tech solutions to local contexts while supporting global sustainability goals. Additionally, education and capacity-building initiatives are essential to empower communities with skills to leverage these technologies.

Integrating technology into environmental governance presents an opportunity to address complex environmental issues within the Anthropocene, a time of human-driven ecological impact. Innovation enhanced regulatory frameworks, and inclusive participation can drive a sustainable future. In summary, merging technology with environmental governance marks a crucial step in sustainable development, necessitating an integrated approach that respects ethics and promotes equitable access.

2 OBJECTIVES OF THE CHAPTER

The goal of the book chapter "The Role of Technology in Environmental Governance" is to investigate how technological innovations can improve environmental regulation, enforcement, and monitoring. It also explores the interface between technology and environmental governance. The aim is to evaluate the significance of different technological instruments including sensor networks, satellite imagery, and data analytics in addressing environmental issues and thereby fostering sustainable development. The chapter also aims to shed light on the advantages and disadvantages of incorporating technology into frameworks for environmental governance.

3 HISTORICAL CONTEXT

3.1 Evolution of Environmental Governance

Environmental Governance includes policies, regulations, and laws for people to follow for protection of the environment. The history of environmental governance is quite complex and evolved over time with reference to the increasing environmental issues such as depletion of natural resources, declining ground water table, desertification, loss of biodiversity, and increasing pollution.

The modern environmental governance traces its roots to the late 19th and early 20th centuries when industrialization led to visible pollution and resource depletion issues. In the late 19th century, first national parks in the United States were set up, aimed at protecting natural areas from exploitation. Further, the environment related activities gained momentum after World War II due to increased awareness of pollution issues, such as air and water pollution. In response to the increasing pollution, environmental legislation appeared and as a result the Environmental Protection Agency (EPA) established in 1970. EPA became a central institution for environmental governance. It was well understood by the environment enthusiasts that the environment is not limited to one state or nation, rather it has much wider horizons. Therefore, international cooperation became essential to address environmental issues and transcend national boundaries. In 1972, UNEP- established for managing the global environmental governance.

Figure 1. Key aspects of environmental governance

In the late 20th century, the concept of sustainable development appeared, emphasizing the need to balance economic development with environmental protection and social equity. The 1992 Earth Summit in Rio de Janeiro led to the adoption of Agenda 21, a comprehensive blueprint for sustainable development. Over time, numerous international treaties and agreements have been negotiated to address specific environmental issues. Table **1** provides a brief information on important Multilateral Environmental Agreements (MEAs) with reference to environmental governance.

Table 1. Importance of MEAs with reference to environmental governance

Year	MEA	Description
1987 (Montreal, Canada)	Montreal Protocol on Substances that Deplete the Ozone Layer	An international treaty for phasing out the production and consumption of ozone-depleting substances.
1989 (Basel, Switzerland)	Control of Transboundary Movements of Hazardous Wastes and their Disposal (Basel Convention)	The convention to minimize the generation of hazardous wastes and ensure their environmentally sound management, particularly in transboundary movements. It regulates the movement of hazardous wastes between countries and promotes their environmentally sound disposal.
1992 (Bonn, Germany)	United Nations Framework Convention on Climate Change (UNFCCC)	The treaty sets out the framework for negotiating specific GHG emission reduction targets and promotes cooperation among nations to mitigate and adapt to climate change. The Paris Agreement, adopted in 2015 under the UNFCCC, is a significant milestone, outlining commitments to limit global warming to well below 2 degrees Celsius above pre-industrial levels.
1992 (Downtown Montreal, Canada)	Convention on Biological Diversity (CBD)	The treaty emphasizes the responsible use of biological resources and guarantees that benefits derived from genetic resources are shared fairly and equitably.
1997 (Kyoto, Japan)	Kyoto Protocol	Adopted as an annex to the UNFCCC, the Kyoto Protocol set up legally binding emission reduction targets for developed countries. It introduced mechanisms such as emissions trading and Clean Development Mechanism to help countries meet their targets.
1998 (Rotterdam, Netherlands)	Rotterdam Convention on Prior Informed Consent (PIC) for Certain Hazardous Chemicals and Pesticides in International Trade	The convention addressed share responsibility and cooperative efforts among the countries in the international trade for the import and export of certain hazardous chemicals and pesticides. The agreement mentions the exporting countries to take prior informed consent from importing countries before exporting the listed chemicals.

There are **MEAs** addressing various environmental issues, including marine pollution, desertification, wetland conservation, and wildlife protection. Each agreement aims to foster international cooperation and collective action to address global environmental challenges. **Figure 2** illustrates the chronological adoption of important MEAs with changing environmental conditions.

Figure 2. Important multilateral environmental agreements (MEAs)

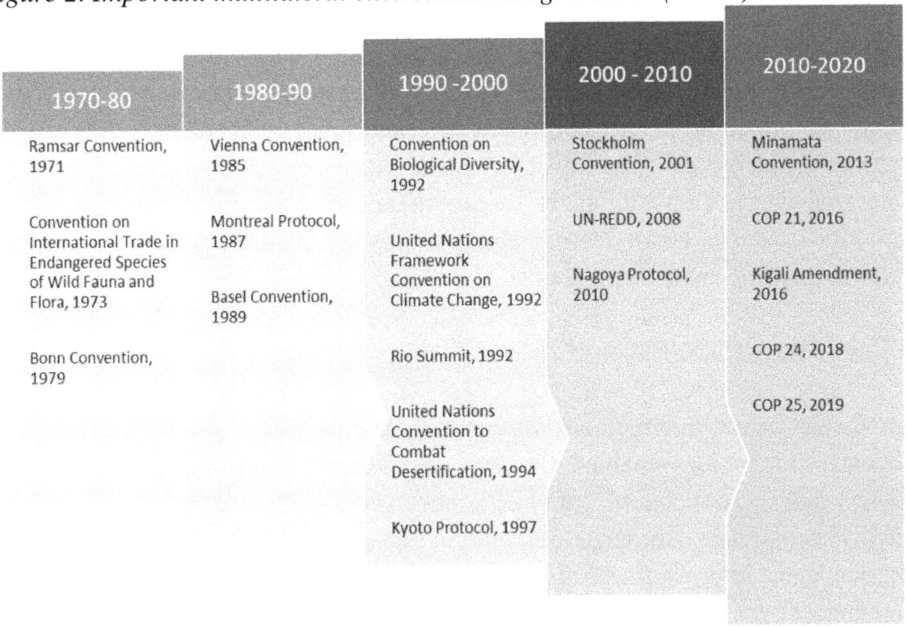

1970-80	1980-90	1990-2000	2000-2010	2010-2020
Ramsar Convention, 1971	Vienna Convention, 1985	Convention on Biological Diversity, 1992	Stockholm Convention, 2001	Minamata Convention, 2013
Convention on International Trade in Endangered Species of Wild Fauna and Flora, 1973	Montreal Protocol, 1987	United Nations Framework Convention on Climate Change, 1992	UN-REDD, 2008	COP 21, 2016
	Basel Convention, 1989		Nagoya Protocol, 2010	Kigali Amendment, 2016
Bonn Convention, 1979		Rio Summit, 1992		COP 24, 2018
		United Nations Convention to Combat Desertification, 1994		COP 25, 2019
		Kyoto Protocol, 1997		

3.2 Early Technological Interventions

Though several international agreements are established, the enforcement of regulations needs robust mechanisms and monitoring of environmental data. In earlier years, environmental data were manually collected, recorded, and reported by government agencies, research institutions, and non-governmental organizations. Data collection involved field observations, surveys, and interviews, while data reporting involved compiling, analyzing, and disseminating information through reports, publications, and databases. Manual surveys and inventories were conducted to assess environmental resources, such as biodiversity, land use, and natural habitats. Further, environmental monitoring devices such as weather stations for monitoring meteorological conditions, water quality sensors for measuring water parameters, and air quality monitors for assessing air pollution were in use. Mapping and cartography techniques were used to visualize and depict spatial patterns and distributions of environmental features and resources.

3.3 Milestones in Technological Advancements

Though the early technological interventions are quite useful in deciding the baseline data for monitoring changes in environmental conditions over time, they often had limitations in terms of spatial coverage, temporal resolution, data quality, and cost-effectiveness. The advent of remote sensing, satellite imagery, and big data analytics has revolutionized environmental monitoring and governance by providing scalable, cost-effective, and prompt solutions for collecting, analyzing, and interpreting environmental data on a global scale.

3.3.1 Data Collection and Monitoring

Manual monitoring methods are costly, time-consuming, and limited in spatial coverage. Technology-enabled solutions, such as remote sensing, satellite imagery, drones, and sensor networks, offer efficient and cost-effective means of collecting environmental data over large areas and in real-time. These technologies can help fill data gaps, improve the accuracy and reliability of environmental assessments, and enhance monitoring capabilities in remote or inaccessible regions.

3.3.2 Data Analysis and Interpretation

The volume and complexity of environmental data can pose challenges for analysis and interpretation. Advanced data analytics tools, including machine learning algorithms, artificial intelligence, and data visualization techniques, can help extract meaningful insights from large datasets, identify patterns and trends, and support evidence-based decision-making in environmental governance.

3.3.3 Transparency and Accountability

Lack of transparency and accountability in environmental governance processes can undermine public trust and hinder effective decision-making. Technology platforms, such as open data portals, digital mapping tools, and online citizen engagement platforms, promote transparency, ease public access to environmental information, and enable stakeholders to take part in decision-making processes. These platforms enhance accountability by providing mechanisms for monitoring and reporting on government actions and environmental outcomes.

3.3.4 Cross-Sectoral Coordination

Environmental issues often cut across multiple sectors and jurisdictions, requiring coordinated action and collaboration among diverse stakeholders. Integrated information systems, interoperable data standards, and decision support tools can facilitate communication, coordination, and collaboration among government agencies, private sector actors, civil society organizations, and local communities. These technologies promote cross-sectoral integration and holistic approaches to environmental governance, enhancing constructive collaboration and efficiency in addressing complex environmental challenges.

3.3.5 Early Warning and Disaster Response

Environmental disasters, such as natural hazards, industrial accidents, and environmental emergencies, pose significant threats to human health, safety, and the environment. Early warning systems, remote sensing technologies, and geospatial analytics can help monitor environmental risks, detect emerging threats, and facilitate timely response and mitigation efforts. These technologies improve preparedness, resilience, and adaptive ability in responding to environmental emergencies and reducing disaster impacts. By using technology, environmental governance can become more effective, inclusive, and responsive to the complex challenges facing the environment and society.

4 KEY TECHNOLOGIES IN ENVIRONMENTAL GOVERNANCE

Environmental governance relies heavily on technology such as remote sensing to monitor ecosystems, blockchain to ensure environmental transactions are transparent, and geographic information systems (GIS) to map and analyze environmental data. Furthermore, real-time data collecting from environmental sensors is made possible by the Internet of Things (IoT), and machine learning and artificial intelligence (AI) improve prediction modeling and decision-making. Sustainable development depends on renewable energy sources like wind and solar power, and pollutant levels are monitored by environmental monitoring systems. When combined, these technologies help with effective resource management, well-informed policymaking, and efficient enforcement of environmental laws.

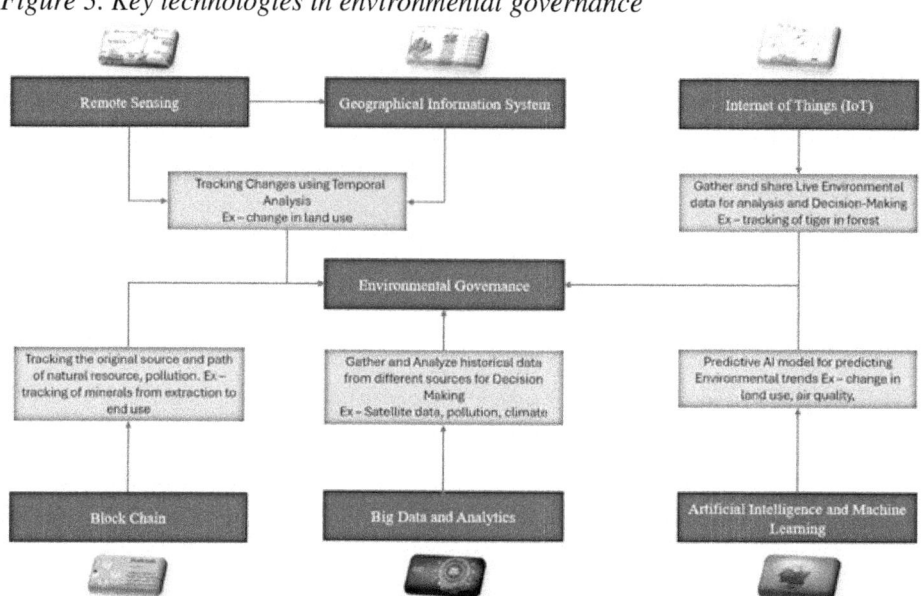

Figure 3. Key technologies in environmental governance

4.1 Geographic Information Systems (GIS)

GISs are essential to environmental governance as offer strong instruments for gathering, analyzing, and visualizing data, which improves policymaking and decision-making. Comprehensive monitoring of natural resources, environmental changes, and human influences on ecosystems is made possible by the integration of geographical data through GIS.

The mapping and management of natural resources is one of the main uses of GIS in environmental governance. GIS aids in the identification of places that require conservation or sustainable management by superimposing several data layers, including those related to land use, vegetation cover, and water resources. Policymakers can use this geographical analysis to help them create plans for the sustainable management of forests, the preservation of important habitats, and the fair distribution of natural resources. Monitoring changes in the environment over time is another important use of GIS. Tracking changes in land use, deforestation, urban growth, and the consequences of climate change is possible with satellite images and remote sensing data incorporated into GIS. To evaluate the efficacy of environmental policies and change them in response to current issues, temporal analysis is essential.

GIS is extremely useful in disaster management for risk assessment, readiness, and reaction. It makes it possible to pinpoint places that are susceptible, forecast possible environmental dangers, and arrange for emergency services and evacuation routes. Through the simulation of events like floods, wildfires, or oil spills, GIS eases preventative actions and supports proactive measures to mitigate environmental disasters. Furthermore, by providing stakeholders with access to spatial information, GIS encourages public involvement in environmental governance. Communities can take part in environmental monitoring and decision-making processes through interactive maps and online platforms, which encourage accountability and openness.

4.2 Remote Sensing

With its ability to provide detailed and precise data that is essential for managing, safeguarding, and keeping an eye on natural resources, remote sensing technology has become a key instrument in environmental governance. Authorities can effectively collect data on a range of environmental indicators, such as land use, deforestation, water quality, and atmospheric conditions, by using satellite images, aerial photography, and other remote sensing methods.

The capacity of remote sensing to cover vast and often unreachable areas, providing a synoptic perspective that is unmatched by ground-based observations, is one of its main advantages. This skill is especially important for monitoring environmental regulation compliance, evaluating the effects of natural disasters, and tracking changes in ecosystems. For example, satellite photography can identify illicit logging activity in isolated forest regions, allowing regulatory agencies to act promptly.

Furthermore, by offering evidence-based insights, remote sensing data aids in the creation of environmental regulations. The development of intricate maps and models that aid in understanding environmental trends and projecting future situations is made possible by high-resolution photos and data analytics. For strategic planning and well-informed decision-making in fields like agriculture, urban development, and biodiversity protection, this data is crucial. When it comes to measuring and assessing climate change-related variables like sea level rise, glacial retreat, and greenhouse gas emissions, remote sensing is essential. These metrics are essential for both evaluating the success of mitigation strategies and for international climate accords. Furthermore, real-time data from remote sensing technology aids in disaster management by supporting response efforts and lessening the effects of natural disasters like hurricanes, floods, etc.

Remote sensing technology is a vital tool in environmental management, improving the ability to see, oversee, and safeguard the environment. Its ability to offer precise, extensive, and prompt data makes it an effective instrument for promoting sustainable development and tackling the current global environmental issues.

4.3 Big Data and Analytics

Big Data and Analytics have transformed environmental management by offering powerful tools for monitoring, analyzing, and controlling environmental information. This technology allows for the gathering and analyzing of enormous quantities of data from diverse sources such as satellite images, sensor networks, social media, and historical environmental data. This information is crucial for understanding intricate environmental phenomena, forecasting upcoming patterns, and making well-informed choices to safeguard and maintain natural resources.

Monitoring and forecasting climate change are among the main uses of Big Data in environmental management. By examining extensive data sets on temperature, precipitation, and greenhouse gas emissions, researchers and decision-makers can detect trends and forecast potential climate situations. This ability to predict the future is crucial for creating plans to reduce the effects of climate change, like improving energy systems, adopting sustainable farming methods, and preparing for disasters.

Moreover, Big Data analytics help with monitoring the environment in real-time. For example, sensor networks placed in forests can offer instant information on air quality, soil moisture, and wildlife activity. This live data enables rapid reactions to environmental dangers like forest fires, illegal logging, and wildlife poaching. Additionally, incorporating social media analytics assists in measuring the public's feelings and knowledge regarding environmental concerns, thus enhancing communication and participation approaches. Moreover, these advancements improve adherence to regulations and enforcement measures. Big Data technology can monitor industrial emissions, waste disposal, and resource extraction activities to ensure they comply with environmental regulations. Sophisticated analytics can detect inconsistencies and breaches, leading to prompt actions by regulatory authorities.

In summary, Big Data and Analytics are essential in contemporary environmental management. They offer exceptional abilities for gathering, analyzing, and using data, resulting in better decision-making and proactive handling of environmental resources. As environmental data becomes more abundant and diverse, Big Data and Analytics will play an increasingly significant role in influencing sustainable policies and practices.

4.4 Internet of Things (IoT)

The Internet of Things (IoT) is an advanced technology in the field of Environmental Governance, changing the way we oversee, control, and uphold our natural resources and ecosystems. IoT involves a system of connected gadgets equipped with sensors, software, and various technologies that allow them to gather and share information. IoT offers exceptional chances for immediate monitoring, decision-making based on data, and initiative-taking actions to tackle urgent environmental issues. One important use of IoT involves overseeing and controlling the quality of air and water. IoT sensors placed in different areas can gather data on pollutant levels, temperature, humidity, and other crucial factors without interruption. This up-to-the-minute data enables officials to quickly find instances of pollution, pinpoint where the pollution is coming from, and initiate measures to minimize harm to the environment. Furthermore, IoT also helps in developing environmental monitoring systems, allowing for a complete comprehension of environmental situations in various regions.

In addition, IoT plays a key role in nature conservation and life management. By deploying sensor-equipped devices in natural habitats, researchers and conservationists can monitor wildlife behavior, monitor migration patterns, and assess habitat health. This information can help protect endangered species and support efforts to restore degraded ecosystems. In addition to monitoring, IoT enables more efficient management of resources. Smart infrastructure powered by IoT technologies can optimize energy consumption, reduce water waste, and minimize the environmental footprint of urban areas. For example, smart grids can dynamically adjust energy distribution according to demand, while smart irrigation systems can optimize water use in agriculture.

Overall, the integration of IoT in environmental management offers enormous potential to improve our ability to monitor, manage and protect the environment. By providing real-time data, enabling proactive actions, and promoting sustainable resource management practices, IoT empowers stakeholders to make informed decisions and achieve positive environmental outcomes. However, challenges such as data protection, cyber security and interoperability must be addressed to fully exploit the benefits of IoT in environmental management.

4.5 Artificial Intelligence and Machine Learning

Artificial intelligence (AI) and machine learning (ML) are revolutionizing environmental management by providing innovative solutions to complex challenges. These technologies provide new ways to collect, analyze and interpret large amounts of data, enabling policy makers to make decisions and implement effective environmental policies. One of the main uses of AI and ML in environmental management is environmental monitoring and assessment. Remote sensing technologies and artificial intelligence algorithms can analyze satellite images to monitor land cover changes, deforestation rates, urban sprawl, and ecosystem health. ML algorithms can detect patterns and anomalies in environmental data, providing early warnings of natural disasters such as floods, wildfires, and hurricanes, and enable proper responses and deductive methods.

Predictive AI models can also predict environmental trends and compare the impact of policy interventions, helping policymakers design strategies to address environmental problems. For example, ML algorithms can analyze historical data on air quality, weather patterns, and industrial emissions to predict future levels of pollution and inform the development of control plans. AI-based decision support systems play a key role in optimizing resource allocation and coordinating maintenance activities. These systems can integrate diverse sets of environmental data and socio-economic indicators to identify areas of high ecological value or vulnerability, thus guiding policy makers in the allocation of resources for habitat restoration, biodiversity conservation and sustainable land use planning.

Furthermore, environmental compliance and enforcement mechanisms are improved by AI and ML. These technologies make it easier to detect environmental violations of the law and enforce regulations by enabling automated monitoring of waste management procedures, industrial emissions, and wildlife trafficking activities. Data privacy, algorithmic bias, and the digital divide are just a few of the ethical and equity issues brought up by the broad use of AI and ML in environmental governance. Policymakers must thus guarantee inclusivity, accountability, and transparency in the creation and application of AI-based environmental solutions.

The role of AI and ML in predictive environmental analysis can be reviewed using a case study of IBM green Horizon Project in China provided in the box below.

IBM's Green Horizon Project, launched in 2014 with the Chinese government, tackles climate change through AI-driven environmental sustainability. This project seeks to improve air quality, enhance energy efficiency, and integrate renewable energy sources using advanced analytics, big data, and AI.

Background and Goals

China, particularly cities like Beijing, has long faced severe pollution impacting health and economic well-being. Recognizing this, the government partnered with IBM to leverage AI and data analytics for pollution prediction and mitigation. The project's main goals include:

Air Quality Management: Predicting pollution levels to enable government action in reducing emissions.
Energy Optimization: Enhancing grid and industrial efficiency to cut carbon emissions.
Renewable Energy Integration: Aiding wind and solar energy integration into the grid.

Technological Strategy

Using IBM's Watson, machine learning, and IoT, the project collects and processes vast environmental data from weather models, satellite imaging, and sensors. This analysis offers real-time, actionable insights.

Results and Impact

Green Horizon has notably improved Beijing's air quality and energy efficiency by up to 20% in partnered industrial plants. This project underscores IBM's commitment to sustainable innovation, helping stabilize energy supplies while reducing reliance on fossil fuels.

In conclusion, by delivering data-driven insights, predictive analytics, and decision support tools, AI, and ML present previously unheard-of possibilities to revolutionize environmental governance. Policymakers can improve environmental sustainability, resilience, and adaptive capacity in the face of global environmental challenges by using the power of these technologies.

4.6 Blockchain Technology

The potential of blockchain technology to improve environmental governance's efficiency, accountability, and transparency is becoming more widely acknowledged. Blockchain is an immutable, decentralized ledger that tracks transactions over a computer network. Applications for this technology in environmental governance include monitoring carbon emissions and controlling the use of natural resources.

Blockchain technology has the potential to significantly affect carbon emissions trading and tracking, which is one of the principal areas. Emissions data can be transparently and securely recorded and verified using blockchain technology. This guarantees proper accounting for carbon credits and reduces fraudulent activities like double counting and misreporting. Furthermore, smart contracts enabled by blockchain technology can automate the carbon trading process, enabling more. Supply chain management is yet another intriguing use of blockchain in environmental governance. Blockchain technology makes it possible to track the provenance (Original source) and path of natural resources, like minerals or lumber, from their extraction to their ultimate use. By holding participants in the supply chain responsible for their actions, this transparency aids in the prevention of illicit mining, logging, and other damaging activities to the environment. Furthermore, by having access to comprehensive information about the environmental impact of the products they buy, consumers can make better decisions.

Furthermore, the development of decentralized environmental monitoring systems can be aided by blockchain technology. Real-time data can be collected and securely stored on biodiversity, air and water quality, and other environmental indicators by integrating IoT devices with blockchain. Because environmental data is not under the control of a single party thanks to this decentralized approach, there is less chance of censorship or manipulation.

Blockchain technology is emerging as a transformative tool in environmental governance, providing transparent, secure, and decentralized ways to manage environmental data and resources. Carbon credits are a market-based mechanism to reduce greenhouse gas emissions. Companies purchase credits to offset their emissions, which fund projects like reforestation or renewable energy. However, the traditional carbon credit system has faced issues like double counting, fraud, and lack of transparency. Blockchain offers a solution by creating an immutable, decentralized ledger for tracking carbon credits. For instance, Veridium, in partnership with IBM, developed a blockchain-based platform to tokenize carbon credits. Each carbon credit is represented as a digital asset on the blockchain, ensuring its authenticity and traceability. This system allows companies to seamlessly trade carbon credits, track their environmental impact, and reduce fraud. The use of blockchain ensures that each carbon credit is unique, and transactions are transparent, which builds trust among stakeholders.

Governments, corporations, and individuals can work together more successfully to address urgent environmental issues and carry out sustainable development goals by utilizing blockchain technology.

5 POLICY AND REGULATORY FRAMEWORKS

5.1 International Agreements and Technology Integration

International agreements and regulatory frameworks play a pivotal role in environmental governance, particularly in reference to advanced technologies such remote sensing, AI, blockchain, etc. These frameworks ensure the responsible use of satellite data and promote international collaboration to address global environmental challenges.

In several international agreements as mentioned in section 2, the advanced technologies can help in monitoring the data at global level. The nations are working with the help of satellite data for monitoring climate change, greenhouse gas emissions, and the implementation of climate policies. The integration of technologies and international agreements facilitates global collaboration and sharing of data. Some of the significant international agreements pivotal in technology integration are provided.

5.1.1 GIS and Remote Sensing

i. **1967: Outer Space Treaty:** the Treaty on Principles Governing the Activities of States in the Exploration and Use of Outer Space, including the Moon and Other Celestial Bodies. This treaty is fundamental in the context of remote sensing as it sets the framework for the peaceful use of outer space.
ii. **1972: Convention on International Liability for Damage Caused by Space Objects (Liability Convention):** This convention sets up the liability of countries for damage caused by their space objects, including remote sensing satellites.
iii. **1978: Remote Sensing Principles (UNGA Resolution 41/65):** Adopted by the United Nations General Assembly, this resolution outlines principles for the conduct of remote sensing activities, emphasizing the benefit to all countries and the protection of Earth's environment.

iv. **1986: UN Principles Relating to Remote Sensing of the Earth from Outer Space:** These principles, adopted by the United Nations, guide the conduct of remote sensing activities, and promote international cooperation and data sharing.

v. **2002: Group on Earth Observations (GEO) and the Global Earth Observation System of Systems (GEOSS):** Established to coordinate international efforts to build a comprehensive, global system of Earth observation. This agreement enhances the availability and use of remote sensing data.

vi. **2016: Committee on Earth Observation Satellites (CEOS) Data Policy Portal:** CEOS promotes the sharing and accessibility of satellite data for global benefit, supporting the use of remote sensing data for various applications.

5.1.2 Big Data and Analytics and Internet of Things (IoT)

International agreements addressing Big Data analytics and IoT are recent and still developing. The international agreements and frameworks that impact Big Data analytics through their focus on data privacy, security, and cross-border data flows are provided.

i. **1981: Convention for the Protection of Individuals about Automatic Processing of Personal Data (Convention 108):** This Council of Europe treaty is one of the earliest international agreements focusing on data protection and privacy, laying the groundwork for subsequent regulations impacting Big Data.

ii. **1995: EU Data Protection Directive (Directive 95/46/EC):** This directive harmonized data protection laws across EU member states and influenced global data protection standards, affecting Big Data practices.

iii. **2013: OECD Guidelines Governing the Protection of Privacy and Transborder Flows of Personal Data (Updated):** These guidelines provide principles for privacy protection and cross-border data flows, relevant to Big Data analytics.

iv. **2015: ITU-T Y.2060, Overview of the Internet of Things:** This ITU (International Telecommunication Union) recommendation provides an overview and standardization framework for IoT, outlining key characteristics, high-level requirements, and reference models.

v. **2016: G20 Digital Economy Development and Cooperation Initiative:** Adopted at the G20 Summit, this initiative includes promoting IoT to drive innovation and economic growth while addressing privacy and security concerns.

vi. **2016: General Data Protection Regulation (GDPR):** This comprehensive EU regulation significantly affects Big Data analytics by setting stringent requirements for data protection and privacy, with global implications.

vii. **2017: European Union Cybersecurity Act:** This regulation strengthens the mandate of the EU Agency for Cybersecurity (ENISA) and sets up a cybersecurity certification framework for ICT products, services, and processes, including IoT
viii. **2020: EU-UK Trade and Cooperation Agreement:** Post-Brexit, this agreement includes provisions on data protection and privacy, affecting Big Data analytics involving UK and EU data.
ix. **2021: Cross-Border Privacy Rules (CBPR) System (Expanded by APEC):** This system eases data privacy standards across Asia-Pacific Economic Cooperation (APEC) member economies, affecting Big Data analytics.

These agreements and frameworks collectively shape the regulatory landscape for Big Data analytics and IoT and address issues of privacy, security, and cross border data flows.

5.1.3 Artificial Intelligence (AI) and Machine Learning (ML)

i. **2015: OECD Principles on Artificial Intelligence:** the Organisation for Economic Co-operation and Development (OECD) adopted principles aimed at promoting AI that is innovative, trustworthy, and respects human rights and democratic values.
ii. **2018: European Commission's AI Strategy:** The European Commission released a strategy for AI, outlining plans for investment, ethical guidelines, and regulatory framework to ensure AI's positive impact on society.
iii. **2019: G20 AI Principles:** The G20 adopted human-centered AI principles based on the OECD principles, focusing on inclusive growth, sustainable development, and well-being.
iv. **2020: Global Partnership on AI (GPAI):** Launched by the OECD and several countries, GPAI aims to support and guide the responsible development and use of AI, grounded in human rights, inclusion, diversity, innovation, and economic growth.
v. **2020: UNESCO Recommendation on the Ethics of Artificial Intelligence (Draft):** UNESCO began developing a recommendation on AI ethics, focusing on principles such as transparency, fairness, and accountability.
vi. **2021: EU Proposal for a Regulation Laying Down Harmonized Rules on Artificial Intelligence (AI Act):** The European Commission proposed the AI Act to regulate AI systems, categorizing them based on risk and setting requirements to ensure safety and fundamental rights protection.

vii. **2022: OECD AI Policy Observatory:** The OECD launched this platform to ease the sharing of AI policies and practices, support the implementation of OECD AI principles, and encourage international collaboration on AI governance.

viii. **2022: UNESCO Recommendation on the Ethics of Artificial Intelligence:** UNESCO adopted recommendation to provide a comprehensive global framework on the ethics of AI to guide member states in promoting human-centric AI.

These agreements and initiatives reflect the growing recognition of the importance of ethical, transparent, and accountable AI and ML, and the need for international cooperation to address the challenges and opportunities presented by these technologies.

5.1.4 Blockchain Technologies

i. **2015: United Nations/CEFACT Blockchain White Paper:** The United Nations Centre for Trade Facilitation and Electronic Business (UN/CEFACT) published a white paper discussing blockchain's potential for trade facilitation and e-business, marking an early recognition of the technology's significance.

ii. **2018: EU Blockchain Observatory and Forum:** Launched by the European Commission, this initiative aims to accelerate blockchain innovation and development within the EU, providing a platform for knowledge sharing and collaboration.

iii. **2018: Financial Action Task Force (FATF) Recommendations on Virtual Assets:** The FATF issued recommendations to prevent the misuse of virtual assets, including those eased by blockchain technologies, for money laundering and terrorist financing.

iv. **2019: G7 Working Group on Stablecoins Report:** The G7 established a working group to examine the implications of stablecoins, a type of cryptocurrency often built on blockchain, emphasizing the need for regulatory oversight and international cooperation.

v. **2019: OECD Blockchain Policy Centre:** The OECD launched this center to provide resources and guidance on blockchain policies, aiming to promote transparency, accountability, and innovation in blockchain technology.

vi. **2022: European Blockchain Partnership (EBP):** The EBP, started by the European Commission, aims to develop a European Blockchain Services Infrastructure (EBSI) to support cross-border digital public services, showing a significant commitment to blockchain technology at the international level.

vii. **2022: OECD Recommendation of the Council on Blockchain and Digital Assets:** This recommendation provides a set of principles and guidelines for the responsible development and use of blockchain and digital assets, promoting international cooperation and best practices.

These agreements and initiatives reflect the growing recognition of blockchain's potential and the need for international cooperation to address the regulatory, technical, and ethical challenges associated with this technology.

5.2 National Policies and Technological Adoption

The countries are increasingly integrating advanced technologies into their environmental governance policies. These technologies enhance data collection, analysis, and decision-making, leading to more effective management of environmental resources and addressing various environmental challenges. Countries such as USA, China, European countries, and India integrated technology in their national policies as mentioned in next paragraph.

5.2.1 United States

i. **Environmental Protection Agency (EPA):** The EPA's Air Now program uses data from IoT sensors and RS to provide real-time air quality information to the public. Similarly, Water Quality Exchange (WQX) integrates IoT and GIS technologies to manage and share water quality data nationwide.
ii. **National Aeronautics and Space Administration (NASA):** NASA employs RS and AI to check global environmental changes, supporting policy development and scientific research.
iii. **U.S. Department of Agriculture (USDA):** The USDA promotes the use of RS, GIS, IoT, and AI in agriculture to improve crop yields, reduce resource use, and minimize environmental impact.
iv. **National Oceanic and Atmospheric Administration (NOAA):** NOAA employs AI and ML in conjunction with RS data to improve weather and climate forecasts, aiding in disaster preparedness and environmental management.

5.2.2 China

i. **Ministry of Ecology and Environment (MEE):** MEE uses RS, GIS, IoT, AI, and ML for environmental monitoring, pollution control, and ecological conservation. The "Digital China" initiative aims to digitize environmental governance processes, enhancing efficiency and transparency.
ii. **National Forestry and Grassland Administration (NFGA):** NFGA uses RS and GIS for forest monitoring, biodiversity conservation, and natural resource management. AI is employed to analyze satellite imagery for illegal logging detection and forest fire prevention.

5.2.3 European Countries

i. **European Environment Agency (EEA):** EEA uses RS, GIS, IoT, AI, and ML for environmental monitoring, reporting, and assessment across European countries. EEA's European Environmental Data Centre (EEDC) provides access to environmental data collected through these technologies.
ii. **European Space Agency (ESA):** ESA's Earth Observation Program (EOEP) uses RS data for environmental monitoring, climate change research, and natural disaster management. AI and ML are applied to ESA's satellite data for enhanced analysis and decision support in environmental governance.

5.2.4 India

i. **Ministry of Environment, Forest, and Climate Change (MoEFCC):** MoEFCC utilizes RS, GIS, IoT, AI, and ML for environmental monitoring, forest conservation, and climate change mitigation. Also, under the National Clean Air Program (NCAP) uses IoT and AI for air quality monitoring and management in Indian cities.
ii. **Indian Space Research Organization (ISRO):** ISRO's satellites provide RS data for environmental monitoring, agriculture, forestry, and disaster management. AI and ML are applied to ISRO's satellite data for land-use classification, crop monitoring, and natural disaster prediction.

5.3 Local Governance and Community-Based Approaches

The use of advanced technologies in local governance and community-based approaches can significantly enhance the efficiency, transparency, and inclusiveness of local government operations and improve community engagement. Fig 4.0 provides advantages of technologies such as GIS, AI, Big data analytics in enhancing governance in community-based systems.

Figure 4. Advantages of technologies in enhancing governance.

i. **E-Governance and Digital Platforms:** To streamline government services and enhance citizen engagement, online portals and mobile applications are developed. People around the world access digital services to pay bills, file complaints, provide feedback etc.
ii. **Improve Urban Planning:** Advanced technologies help in improving urban planning. For example, IoT sensors are useful in monitoring air quality, traffic, waste management, and water supply in real-time; Big Data Analytics analysis data from various sources to perfect urban planning and service delivery; AI and Machine Learning enhances predictive maintenance of infrastructure and

efficient resource allocation. GIS tools are also extremely useful for urban planning, disaster management, and environmental monitoring.

iii. **Public Safety and Security:** AI-powered CCTV cameras for checking public spaces and detecting anomalies are useful in Surveillance and safety.

iv. **Health and Social Services:** Advanced technologies are helping in improving health services by providing healthcare services in rural areas, tracking health metrics, accessing medical information, etc.

v. **Education and Capacity Building:** E-learning platforms and digital literacy programs enhance e-learning platforms to enhance educational opportunities and community ability building.

6 CHALLENGES AND BARRIERS

Sustainable development depends on environmental governance, the framework that communities use to manage the environment and natural resources. In this field, technology has become a potent instrument that may be used to monitor, manage, and mitigate environmental problems in novel ways. However, to fully realize its potential, its integration must overcome several obstacles and constraints.

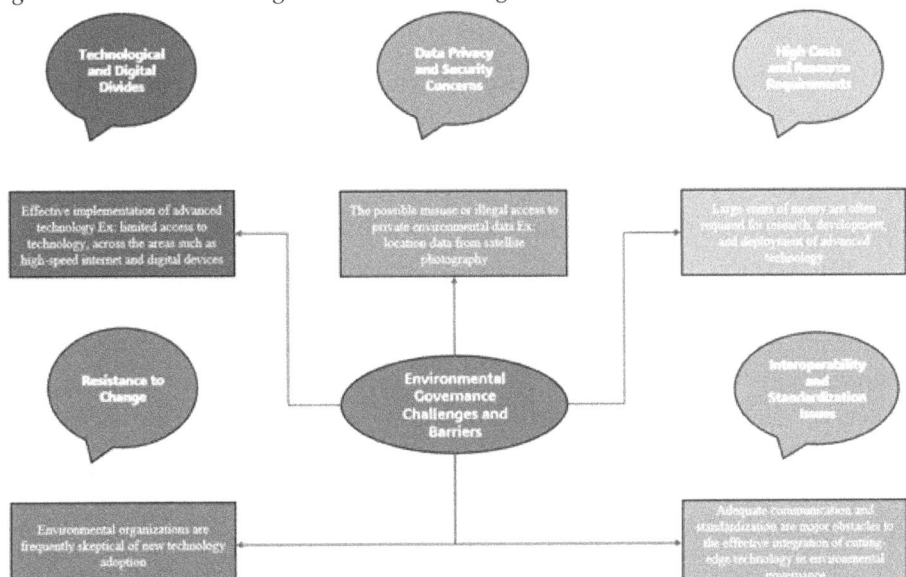

Figure 5. Environmental governance challenges and barriers

6.1 High Costs and Resource Requirements

There are many obstacles to the effective use of sophisticated technology in environmental governance, prominent among them being the high expenses and resource needs. Large sums of money are often needed for research, development, and deployment of advanced technology. The infrastructure needed to support these technologies, which includes workers with specific training and equipment, significantly increases the cost. Decision-makers may be forced by a lack of resources to put short-term requirements ahead of long-term technical developments. There may be a skill barrier due to the intricacy of advanced technologies. It takes time and resources to train employees in how to use and maintain this technology. Compatibility with current regulations and systems also adds another level of difficulty.

Despite these obstacles, there are various advantages to using innovative technology in environmental governance, such as better data analysis, monitoring, and decision-making. Governments, businesses, and academic institutions working together can make resource sharing and technology transfer easier to address these issues. Furthermore, encouraging innovation through financing sources and regulatory aid can promote the integration of innovative technologies into environmental governance.

6.2 Data Privacy, Data Ownership, Security Concerns and Complexity of Global Coordination

Data security and privacy concerns pose significant challenges to using innovative technology in environmental governance. As technologies like big data analytics, remote sensing, and IoT become essential for monitoring and managing environmental issues, they amass vast, sensitive data. This raise pressing questions around data protection and privacy preservation.

A major concern is the potential misuse or unauthorized access to environmental data, which could threaten individual privacy or national security. Users often unknowingly share personal information when registering for applications, unaware of how it may be used afterward. Privacy policies tend to be ambiguous, leaving people uncertain about how their information is collected, used, or shared. From a governance perspective, determining who controls data within these systems is challenging. Additionally, establishing secure legal frameworks to protect privacy is both complex and costly.

Although many countries have their own data privacy laws, no unified global standard exists. Current legislation remains underdeveloped in many regions, marked by unclear guidelines, insufficient coordination, and inconsistent protection across borders. Differing national regulations complicates data protection, making international compliance difficult—especially with frameworks like the GDPR, which struggle to address cross-border data sharing.

To address these challenges, collaboration among legislators, technologists, environmentalists, and civil society is essential. Developing robust encryption, transparent governance structures, and cybersecurity measures will enable the secure use of innovative technologies in environmental governance. Only by tackling these issues can we maximize technology's potential while safeguarding data security and privacy.

6.3 Technological and Digital Divides

The technological and digital divides significantly hinder effective environmental governance, reflecting disparities in access to technology and digital resources across regions and communities. Limited access to high-speed internet and digital devices restricts certain groups from utilizing advanced environmental monitoring systems, while gaps in digital literacy make it challenging for others to use these technologies effectively. Financial constraints also play a role, as marginalized communities often lack the funds to invest in such technologies. Bridging these divides requires strategies to improve access, enhance digital skills, and promote equitable resource distribution, enabling technology's full potential in global environmental governance.

6.4 Interoperability and Standardization Issues

Integrating innovative technology into environmental governance is challenged by communication and standardization issues. Different data formats, protocols, and hardware across environmental systems hinder seamless data exchange and analysis. Hardware and software incompatibility limits interoperability, obstructing comprehensive insights for informed decisions. Moreover, inconsistent data formats complicate sharing among stakeholders, such as governments, academia, and businesses, making multi-source integration time-consuming and error prone.

Collaboration among policymakers, tech developers, and industry is needed to create standardized communication interfaces, data formats, and protocols. This would streamline data exchange, enhance interoperability, and support a more effective, integrated approach to environmental governance for future generations.

6.5 Resistance to Change

Community, governmental, and environmental stakeholders often resist adopting modern technology due to a preference for traditional methods. Factors include fear of the unknown, job security concerns, limited awareness of technology's benefits, and reluctance to abandon established practices. Bureaucratic structures may also resist change, and financial constraints can worsen reluctance due to high initial costs.

Overcoming this requires proactive steps: clarifying technology's advantages, providing support and training, and involving stakeholders in decisions. Modern technology can greatly enhance environmental governance by improving effectiveness, transparency, and participation, but success depends on addressing resistance to foster sustainable, resilient governance frameworks.

7 SOCIAL AND ETHICAL IMPLICATIONS OF USING ADVANCED TECHNOLOGIES

Advanced technologies in environmental governance, like AI, big data, IoT, and blockchain, bring both opportunities and challenges, with important social and ethical considerations. These technologies can improve environmental outcomes, but they also raise concerns about equity, transparency, privacy, and power dynamics. A significant issue is equity, as these advanced technologies often require costly infrastructure and expertise. This can create a technological gap between wealthier nations, able to implement such systems, and less developed regions that may struggle to access these tools. This disparity risks benefiting wealthier areas while leaving poorer communities behind, and within countries, marginalized groups often lack access to the technology that shapes environmental policy, increasing their exposure to environmental risks.

Moreover, environmental technologies could justify expanded surveillance, especially in authoritarian regimes, leading to privacy invasions. Additionally, many of these technologies, like AI, function as "black boxes," where decision-making processes lack transparency, risking public mistrust and questioning of policy fairness.

Figure 6. AI, ML usage in environmental decision making

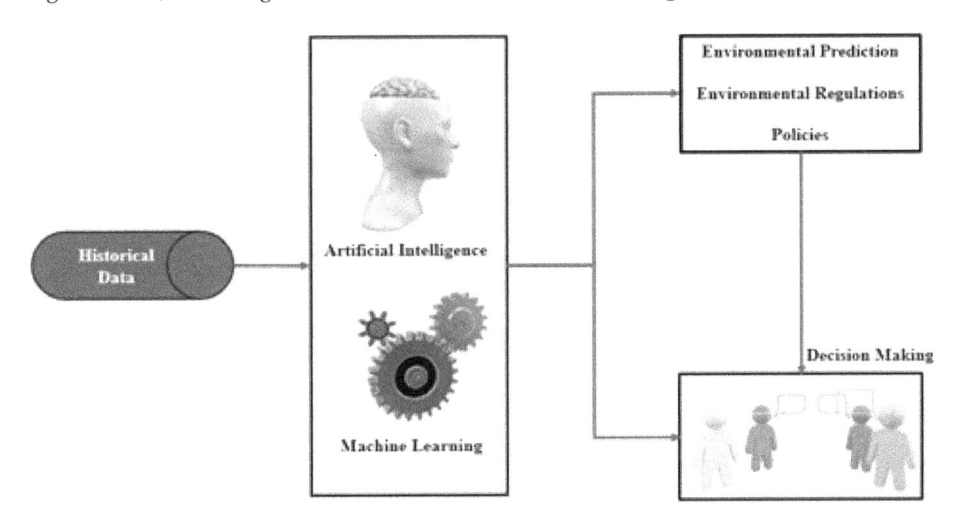

For instance, if AI models are used to allocate resources for climate change adaptation or to enforce environmental regulations, there should be clear, transparent criteria for how decisions are made. Without transparency, governments and organizations risk undermining public trust in environmental governance. As environmental governance becomes more automated, ethical concerns arise regarding the role of AI in decision-making. AI is often trained on historical data, which can reflect existing biases or discriminatory practices. If AI systems are used to design policies or enforce regulations without human oversight, they may perpetuate or even worsen social inequities, particularly if the data used is biased against certain groups or regions. Furthermore, the automation of environmental governance could lead to job displacement in sectors like agriculture, energy, and waste management. While these technologies may improve efficiency, there must be considerations for those whose livelihoods could be impacted by such changes.

8 GLOBAL FACTORS IMPACTED WITH THE USE OF TECHNOLOGIES IN ENVIRONMENTAL GOVERNANCE

The intersection of technology and environmental governance has introduced significant socio-economic shifts. As technology continues to influence how governments, businesses, and individuals manage environmental issues, both positive and negative effects emerge across various socio-economic sectors. The impacts of technology on environmental governance are multifaceted, ranging from economic development to social inclusion, as well as job markets and inequality.

8.1 Economic Development

Technology plays a key role in driving sustainable economic development by fostering more efficient resource use. The digital technologies enable governments and businesses to monitor environmental conditions more efficiently, allowing for more targeted interventions and policies. However, the introduction of new technologies can create economic disparities. Developing countries, for instance, may struggle to access or implement advanced environmental technologies due to financial and infrastructural constraints. This can widen the economic gap between developed and developing nations, potentially leading to a "green divide" where wealthier countries are better positioned to transition to sustainable practices.

8.2 Social Inclusion and Access to Technology

A key socio-economic factor in environmental technology use is social inclusion. Marginalized communities, particularly in rural areas, often lack access to technologies like precision agriculture that could boost crop yields and improve food security. Access to environmental data is also limited, as the digital divide hampers the ability of these communities to participate in environmental governance. While technology could democratize access to information, limited internet and digital literacy exclude disadvantaged groups from decision-making. This deepens socio-economic inequalities, as wealthier, connected communities gain more influence over policies affecting vulnerable populations.

8.3 Employment and Labor Markets

Technological advancements in environmental governance have a profound impact on employment. On one hand, green technologies create new jobs in fields; the technological disruption can lead to temporary unemployment. - To mitigate these impacts, governments must invest in education and training programs to help workers transition to jobs in the green economy. Without these initiatives, the transition to environmentally sustainable technologies may exacerbate existing socio-economic inequalities, leaving some workers behind.

8.4 Inequality and Global Disparities

While technology holds the promise of advancing environmental governance, it can also reinforce global disparities. Wealthier nations with greater technological and financial resources can adopt cleaner, more efficient technologies more rapidly, reducing their environmental impact. In contrast, low-income countries often struggle to fund such initiatives, leading to slower progress in addressing environmental issues. This imbalance contributes to global environmental inequality, where the actions of wealthier countries may disproportionately affect poorer nations that are more vulnerable to climate change and environmental degradation. Ensuring equitable access to these technologies, mitigating job losses in traditional industries, and bridging the digital divide are critical to ensuring that the benefits of technological advancements in environmental governance are shared across all sectors of society.

9 CONCLUSION

Technology has become an indispensable ally in the pursuit of effective environmental governance. It enhances the ability to monitor, manage, and protect the environment through improved data collection, analysis, and transparency. Technological tools empower policymakers, enforcement agencies, and citizens to make informed decisions, engage in meaningful participation, and hold entities accountable for their environmental impact. However, to fully harness the potential of technology for environmental governance, it is essential to address challenges related to accessibility, data privacy, and environmental sustainability.

9.1 Recommendations for Policymakers and Practitioners

Policymakers and practitioners should prioritize leveraging technology to enhance environmental governance. Embracing digital tools and data analytics can significantly improve the monitoring and management of natural resources.

AI can be used to improve prediction capabilities and enable initiative-taking rather than reactive approaches. Encouraging the adoption of IoT devices can also revolutionize environmental monitoring.. Policymakers should support the development of open data platforms where this information can be shared among stakeholders, fostering transparency and collaborative decision-making. Furthermore, the implementation of blockchain technology can ensure data integrity and traceability, crucial for enforcing environmental regulations and verifying sustainable practices in supply chains. Investing in digital literacy and capacity-building among practitioners is essential to maximize the benefits of these technologies. Training programs should be designed to equip environmental professionals with the skills needed to use advanced technological tools and interpret the data they generate. Additionally, fostering public-private partnerships can accelerate the deployment of innovative technologies by combining government oversight with private sector ability and resources. Policymakers should also consider the ethical implications of technology use in environmental governance. This includes ensuring that technology deployment does not worsen existing inequalities and that it respects privacy and data protection standards. Inclusive policies that engage local communities and consider their needs and knowledge can enhance the effectiveness and acceptance of technological solutions.

To summarize, the integration of technology in environmental governance offers unprecedented opportunities to improve the sustainability and resilience of ecosystems. By promoting technological innovation, investing in capacity-building, and ensuring ethical considerations, policymakers and practitioners can effectively harness technology to protect and preserve the environment.

REFERENCES

Abbas, S., Alnoor, A., Sin Yin, T., Mohammed Sadaa, A., Raad Muhsen, Y., Wah Khaw, K., & Ganesan, Y. (2023). Antecedents of trustworthiness of social commerce platforms: A case of rural communities using multi group SEM & MCDM methods. *Electronic Commerce Research and Applications*, 62, 101322. DOI: 10.1016/j.elerap.2023.101322

Andresen, S. (2001). Global Environmental Governance: UN Fragmentation and Co-ordination. In *Yearbook of International Cooperation on Environment and Development 2001/2002* (pp. 19–26). Earthscan Publications.

Aydin, M., Söğüt, Y., & Erdem, A. (2024). The role of environmental technologies, institutional quality, and globalization on environmental sustainability in European Union countries: New evidence from advanced panel data estimations. *Environmental Science and Pollution Research International*, 31(7), 10460–10472. Advance online publication. DOI: 10.1007/s11356-024-31860-x PMID: 38200188

Baburaj, A. (2021). Artificial intelligence v. intuitive decision making how far can it transform corporate governance? *GNLU L. Rev.*, 8, 233.

Banuri, T., & Najam, A. (2002). *Civic Entrepreneurship: A Civil Society Perspective on Sustainable Development*. Gandhara Academic Press. Also sponsored by Stockholm Environment Institute, United Nations Environment Program, and the RING Regional and International Networking Group.

Biermann, F. (2005). The Rationale for a World Environment Organization. In Biermann, F., & Bauer, S. (Eds.), *A World Environment Organization: Solution or Threat for Effective Environmental Governance?* (pp. 117–144). Ashgate.

Brickson, L., Zhang, L., Vollrath, F., Douglas-Hamilton, I., & Alexander, J. (2023, November). Titus Elephants and algorithms: A review of the current and future role of AI in elephant monitoring. *Journal of the Royal Society, Interface*, 15(208), 20230367. Advance online publication. DOI: 10.1098/rsif.2023.0367 PMID: 37963556

Chen, L., Li, J., Xu, M., & Xing, W. (2024). Navigating urban complexity: The role of GIS in spatial planning and urban development. *Applied and Computational Engineering.*, 65(1), 282–287. DOI: 10.54254/2755-2721/65/20240519

Cosbey, A. 2005. International Agreements and Sustainable Development: Achieving the Millenium Development Goals. Winnipeg, Manitoba, Canada: International Institute for Environment and Development (IISD). Accessed at: http://www.iisd.org/pdf/2005/ investment_iias.pdf in March 2006

Dabo, Shuaibu & Jibril, Mohammed & Suberu, Habibat & Murjanatu, & Ayawa, Garba. (2024). The Role of Geographic Information Systems (GIS) as an Effective Tool in Environmental Planning and Management.

de Oliveira, M. C. C., Machado, M. C., Jabbour, C. J. C., & de Sousa Jabbour, A. B. L. (2019). Paving the way for the circular economy and more sustainable supply chains: Shedding light on formal and informal governance instruments used to induce green networks. *Management of Environmental Quality*, 30(5), 1095–1113. DOI: 10.1108/MEQ-01-2019-0005

Droj, G. (2012). GIS and remote sensing in environmental management. *Journal of Environmental Protection and Ecology*, ●●●, 2.

Earth Negotiations Bulletin, 2005., Volume 15 Number 129, 2 October. On final decision UNEP/FAO/RC/COP.2/CRP.5/Rev.1

Elliot, L. (2005). Expanding the Mandate of the United Nations Security Council. In Chambers, W. B., & Green, J. F. (Eds.), *Reforming International Environmental Governance: From Institutional Limits to Innovative Reforms* (pp. 204–226). United Nations University Press.

Forss, K. 2004. Strengthening the Governance of UNEP: A Discussion of Reform Issues. Commissioned by the Ministry for Foreign Affairs, Sweden. http://www.sweden.gov.se/content/1/c6/03/24/92/ f60748c8.pdf accessed March 2006

Frondel, M., Horbach, J., & Rennings, K. (2007). End-of-pipe or cleaner production? An empirical comparison of environmental innovation decisions across OECD countries. *Business Strategy and the Environment*, 16(8), 571–584. DOI: 10.1002/bse.496

Glenn, J. C. and Gordon, T., 1999. 2050 Global Normative Scenarios. American Council for the United Nations University in cooperation with The Foundation for the Future

Goguichvili S., Linenberger A., Gillette A., The Global Legal Landscape of Space: Who Writes the Rules on the Final Frontier? Wilson Centre, 1 October 2021

Guha, S., Jana, R. K., & Sanyal, M. K. (2022). Artificial neural network approaches for disaster management: A literature review (2010–2021). *International Journal of Disaster Risk Reduction*, 81, 103276. DOI: 10.1016/j.ijdrr.2022.103276

Gupta, J., & Ringius, L. (2001). The EU's Climate Leadership: Reconciling Ambition and Reality. [Springer Science Business Media B.V]. *International Environmental Agreement: Politics, Law and Economics*, 1(2), 281–299. DOI: 10.1023/A:1010185407521

Hao, L.-N., Umar, M., Khan, Z., & Ali, W. (2021). Green growth and low carbon emission in G7 countries: How critical the network of environmental taxes, renewable energy and human capital is? *The Science of the Total Environment*, 752, 141853. DOI: 10.1016/j.scitotenv.2020.141853 PMID: 32889278

Khalid, R. M., & Maidin, A. J. (2022). *Introduction: Governance issues towards achieving SDGs in Southeast Asia. Good Governance and the Sustainable Development Goals in Southeast Asia.* Routledge. DOI: 10.4324/9781003230724-1

Khan, R., Chaudhry, I. S., & Farooq, F. (2019). Impact of human capital on employment and economic growth in developing countries. *Review of Economics and Development Studies*, 5(3), 487–496. DOI: 10.26710/reads.v5i3.701

Kloppenburg, S., Gupta, A., Kruk, S., Makris, S., Bergsvik, R., Korenhof, P., Solman, H., & Toonen, H. (2022). Scrutinizing environmental governance in a digital age: New ways of seeing, participating, and intervening. *One Earth*, 5(3), 232–241. DOI: 10.1016/j.oneear.2022.02.004

Lanjouw, J. O., & Mody, A. (1996). Innovation and the international diffusion of environmentally responsive technology. *Research Policy*, 25(4), 549–571. DOI: 10.1016/0048-7333(95)00853-5

Lanoie, P., Patry, M., & Lajeunesse, R. (2008). Environmental regulation and productivity: Testing the porter hypothesis. *Journal of Productivity Analysis*, 30(2), 121–128. DOI: 10.1007/s11123-008-0108-4

Le, T., Ha, P., Tram, V., Thuy, N., Phuong, D., Tran, T. N., & Nguyen, L. (2023). GIS Application in Environmental Management: A Review. *VNU Journal of Science: Earth and Environmental Sciences.*, 39(2). Advance online publication. DOI: 10.25073/2588-1094/vnuees.4957

Leese, L. AI for Earth: How NASA's Artificial Intelligence and Open Science Efforts Combat Climate Change, The National Aeronautics and Space Administration, updated on 23 May 2024.

Lens M. GDPR & Convention 108: Adequate Protection in a Big Data Era? Tilburg University, 8 June 2018

Mori, S. (2004). Institutionalization of NGO Involvement in Policy Functions for Global Environmental Governance. In Kanie, N., & Haas, P. M. (Eds.), *Emerging Forces in Environmental Governance* (pp. 157–175). United Nations University Press.

Okusimba, George. (2019). Role of GIS as a Tool for Environmental Planning and Management. 2454-9444. .DOI: 10.20431/2454-9444.0501002

Osman, O., & Yassin, E. E. H. (2024). Fostering environmental and resources management in Sudan through geo-information systems: A prospective approach for sustainability. *Journal of Degraded and Mining Lands Management*, 11(3), 5647–5657. DOI: 10.15243/jdmlm.2024.113.5647

Paramati, S. R., Mo, D., & Huang, R. (2021). The role of financial deepening and green technology on carbon emissions: Evidence from major OECD economies. *Finance Research Letters*, 41, 101794. DOI: 10.1016/j.frl.2020.101794

Parhamfar, M., Sadeghkhani, I., & Adeli, A. M. (2024). Towards the net zero carbon future: A review of blockchain-enabled peer-to-peer carbon trading. *Energy Science & Engineering*, 12(3), 1242–1264. DOI: 10.1002/ese3.1697

Patel, R., Taneja, S., Singh, J., & Sharma, S. (2024). Modelling of surface run-off using SWMM and GIS for efficient stormwater management. *Current Science*, 126, 463–469. DOI: 10.18520/cs/v126/i4/463-469

Pedrycz, W., Deveci, M., & Chen, Z.-S. (2024). BIM-based building performance assessment of green buildings -A case study from China. *Applied Energy*, 373, 123977. DOI: 10.1016/j.apenergy.2024.123977

Pierre, J. (2005). Comparative urban governance. Uncovering complex causalities. *Urban Affairs Review*, 40(4), 446–462. DOI: 10.1177/1078087404273442

Principles Relating to Remote Sensing of the Earth from Outer Space, United Nations Office for Outer Space Affairs, www.unoosa.org/oosa/en/ourwork/spacelaw/principles/remote-sensing-principles.html

Round, C., & Visseren-Hamakers, I. (2022). Blocked chains of governance: Using blockchain technology for carbon offset markets? *Frontiers in Blockchain.*, 5, 957316. DOI: 10.3389/fbloc.2022.957316

Rutherford, J. (2011). Rethinking the relational socio-technical materialities of cities and ICTs. *Journal of Urban Technology*, 18(1), 21–33. DOI: 10.1080/10630732.2011.578407

Sira, M. (2024). Potential of Advanced Technologies for Environmental Management Systems. *Management Systems in Production Engineering.*, 32(1), 33–44. DOI: 10.2478/mspe-2024-0004

Smith, A., & Raven, R. (2012). What is a protective space? Reconsidering niches in transitions to sustainability. *Research Policy*, 41(6), 1025–1036. DOI: 10.1016/j.respol.2011.12.012

SPACE LAW & POLICY. CRITICAL STUDY OF INTERNATIONAL LEGAL REGULATION, https://legalvidhiya.com/space-law-policy-critical-study-of-international-legal-regulation/

Spiezia, V., & Tscheke, J. International agreements on cross-border data flows and international trade: A statistical analysis, OECD Science, Technology, and Industry Working Papers, 2020

Sustainable Development Goals. United Nations, www.un.org/sustainabledevelopmentTong, W.W.; Zhang, J.M. Can Environmental Regulation Promote Technological Innovation. Financ. Sci. 2012, 11, 66–74

Taleb, M., & Kadhum, H. (2024). *The Role of Artificial Intelligence in Promoting the Environmental, Social and Governance (ESG)*. Practices., DOI: 10.1007/978-3-031-63717-9_17

Testa, F., Iraldo, F., & Frey, M. (2022). The effect of environmental regulation on firms' competitive performance: The case of the building & construction sector in some EU regions. *Journal of Environmental Management*, 92(9), 2136–2144. DOI: 10.1016/j.jenvman.2011.03.039 PMID: 21524840

The Rio Conventions | UNFCCC; United Nations Climate Change, www.unfccc.int

Tironi, M., & Lisboa, D. (2023). Artificial intelligence in the new forms of environmental governance in the Chilean State: Towards an eco-algorithmic governance. *Technology in Society*, 74, 102264. DOI: 10.1016/j.techsoc.2023.102264

Wang, S., Li, J., & Razzaq, A. (2023). Do environmental governance, technology innovation and institutions lead to lower resource footprints: An imperative trajectory for sustainability. *Resources Policy*, 80, 103142. DOI: 10.1016/j.resourpol.2022.103142

Wang, W., Li, Y., Lu, N., Wang, D., Jiang, H., & Zhang, C. (2020). Does increasing carbon emissions lead to accelerated eco-innovation? Empirical evidence from China. *Journal of Cleaner Production*, 251, 119690. DOI: 10.1016/j.jclepro.2019.119690

Woo, Junghoon & Asutosh, Ashish & Li, Jiaxuan & Ryor, Wolfgang & Charles, & Kibert, Charles & Shojaei, Alireza. (2020). Blockchain: A Theoretical Framework for Better Application of Carbon Credit Acquisition to the Building Sector. .DOI: 10.1061/9780784482858.095

Xu, C., Chen, X., & Dai, W. (2022). Effects of Digital Transformation on Environmental Governance of Mining Enterprises: Evidence from China. *International Journal of Environmental Research and Public Health*, 19(24), 16474. DOI: 10.3390/ijerph192416474 PMID: 36554353

Zhao, H. (2008). Empirical Analysis on the Impact of Environmental Regulation on Technological Innovation of Chinese Enterprises. *Manag. Mod.*, 3, 3–5.

Chapter 12
Integrating Climate Resilience Into Educational Curricula:
Frameworks, Challenges, and Opportunities

Md Ikhtiar Uddin Bhuiyan
Jahangirnagar University, Bangladesh

Sawmeem Sajia
https://orcid.org/0009-0001-3386-6412
Jahangirnagar University, Bangladesh

ABSTRACT

Climate change is one of the most vital challenges of the 21st century which pertains, directly or indirectly, to virtually every sphere of human life and natural systems. As education determines the values, behavior, and responses of society and individuals towards ecological issues, it is important to address the existing gap between the traditional curriculum and incorporation of climate change-focused knowledge in the existing framework. There are several defining reasons for this study in the area of incorporation of climate resilience into the visionary curricula. First, a growing necessity to adopt and implement educational frameworks which could assist in the process of coping with or adapting to climate change effects is driven by a specific social need. The key focus of this research is to generate a framework for the integration of climate resilience into educational curricula at multiple levels. The study follows a mixed-methods pathway in order to create an inclusive structure for climate resilience integration in the curricula of education.

DOI: 10.4018/979-8-3693-7001-8.ch012

1. INTRODUCTION

Climate change is one of the key challenges facing humanity in the 21st century and affects almost all aspects, both natural ecosystems as human life (Reis & Ballinger, 2020). Climate change is driving more frequent and intense climate events like hurricanes, droughts floods wildfires and heatwaves leading to stresses for communities globally (Islam, 2020). As these predation challenges increase, societies look more than ever to adapt with greater resiliency. In recent years, climate resilience the capacity of communities, systems and individuals to anticipate prepare for respond to recover from adverse climatic events has dominated discussions in mainstream discourse (Hügel & Davies, 2024).

According to (Leal Filho et al., 2021) education is the centerpiece for influencing values, behavior approaches and responses of individuals and societies to environmental issues. Traditional educational systems have centered on delivering what are considered to be the essential body of knowledge and proficiency in standard academic subjects while neglecting environmental education. Nonetheless, as impacts of climate change become ever clearer, there is a shift towards the recognition that everything from primary education upwards should also include resilience (Seritan et al., 2022). This is where education can be an essential contribution to the creation of a resilient community that strives toward sustainability: By empowering students with the ability to process and respond effectively, even solve climate change difficult challenges.

The inclusion of climate resilience in syllabi has therefore been a giant leap towards training youth to wade through the murky waters presented by climate warming. It aims not only to make sure students know the science of climate change, but also that they understand in which social, economic and environmental ways will be these changes (Lotz-Sisitka et al., 2021). It gives students a role in helping to fix climate change while they prepare for their future as the next generation of world citizens. Environments matter where integration is key especially in the light of achieving The United Nations Sustainable Development Goals (SDGs) and even more so, SDG 4(Quality Education), SDG13(Climate Action) and SGD17(Partnerships for the goals) (Salinas et al., 2022).

While the need to embed climate resilience in educational curricula is apparent, there remains a void between what defines traditional curriculum and inclusion of knowledge on management from a changing climate. Curriculums are crowded, teachers get little training, resources fail and there is great resistance from policymakers & educational establishments (Rabin et al., 2020). However, the task ahead is in crafting and enacting educational models which are capable of integrating climate resilience without replicating within existing systems.

This study was conducted to look into frameworks, challenges and opportunities in the integration of climate resilience issues with educational curriculum. Through the examination of current curricula, key stakeholder interviews and surveying educators, this study will work to establish a strong foundation climate resilience education framework. These research findings will make a valuable contribution to educators, policymakers and curriculum developers to support the development of practical climate resilience education strategies that contribute towards building ongoing capacity in sustainability and strengthening sustainable development efforts.

2. LITERATURE REVIEW

The purpose of the literature review was to conduct a broad analysis on available research, and theoretical frameworks in relation to climate resilience integration within educational curricula. Meanwhile, this chapter considers the meaning of climate resilience and establishes a foundation for understanding existing educational approaches to teaching about climate change and aligning with what can be learned from theoretical frameworks as applied within education. This review also notes limitations in the literature and considers what these findings could mean for future research and practice.

Climate resilience is the ability to anticipate, prepare for, respond and recover from climate change. Climate resilience "is linked to the reduction of vulnerability to climate-related hazards and stresses, as well as increased adaptive capacity which ensures contemporary infrastructure will be able adjust effectively over time in response changing conditions" (Seritan et al., 2022)

Climate Resilient has been defined in many ways across multiple sectors. In environmental science, it refers to the resistance and resilience of ecosystems under climatic perturbations, while in social sciences adaptation "relates primarily to the capacity of communities and individuals manage different temperatures due or exacerbated by global warming" (Ruiz-Mallén et al., 2022). According to (Hurlimann et al., 2021) climate resilience is "the capacity of a system and its component parts to anticipate, absorb, accommodate or recover from the effects of a hazardous event in as timely and efficient manner as possible

The way we prepare and educate the next generation is changing paradigmatically with the advent of climate resilient education. It is a complete solution to the difficulties of education and the climate, not merely an instructional strategy. It integrates climate change information, skills, and resilience-building into official and informal education systems. To improve local communities' resilience, particularly their ability to withstand climate change, the Framework for Resilient Development in the Pacific (FRDP) was created (Takinana & Baars, 2023) 2023).

Adaptation and mitigation methods for the effects of climate change are included in the framework of climate-resilient education. Its complex nature is highlighted by recent research, which also stresses the value of interdisciplinary approaches and community involvement.

Researchers have long recognized that climate resilience has three components: reducing vulnerability, increasing adaptive capacity and supporting system transformation. The adaptation component will help partner countries to reduce vulnerability by reducing their exposure and ongoing sensitivity to climate risks, as well as helping them mitigate "projected changes in the frequency or intensity of extreme weather events," (Eilam, 2022) said, Perhaps unsurprisingly, various definitions of adaptive capacity are presented in the literature but what it tends to be is essentially the ability to change practices, processes or structures that might cause environmental harm and prevent exploitation of opportunities potentially created by climate-induced changes (Habiba et al., 2013). Finally, system transformation involves the changes that modify key properties of a system in ways that build its ability to adapt and continue serving human needs (Yu & Chiang, 2017)

There is growing acknowledgment of the importance of preparing climate resilient educates among educational communities; adaptively meeting challenges in addressing climate and environmental changes for future generations. Education can be a cornerstone in providing knowledge, skills and understanding of how local communities could view impending climate risks (Apollo & Mbah, 2021).

The extent to which climate resilience is incorporated in a manner consistent with other subjects offered within the educational systems of countries varies widely. Climate resilience is introduced in both primary and secondary schooling by means of subjects such as science, geography, and social studies. In Finland, climate resilience is an interdisciplinary concept addressed through the integrated science-geography-social studied learning themes to help students comprehend the complexity of climate change (Apollo & Mbah, 2021). The US takes a more piecemeal approach, that would likely result in very different kinds of climate change education across states (Kumar et al., 2023). In addition, also the universities and tertiary institutions are starting to include climate resilience in their curriculums (mainly degrees related with environmental science economy or engineering). Specialized courses – and modules within existing ones (often in participation with governmental and non-governmental organizations) on climate resilience are being developed by universities of countries such as the United Kingdom or Australia, for instance (Chowdhury et al., 2021). These programs provide students with a comprehensive overview of climate risk and response by sectors so that they are familiar enough to be able to initiate these discussions.

Notwithstanding these efforts, several challenges from the survey analysis and submissions received that relate to education is having a standard curriculum in place with climate resilience being central (Yu & Chiang, 2017). Also, they are often underprepared and not equipped with the appropriate knowledge to effectively teach for climate resilience (Harper, 2023). In addition, climate science moves so quickly in terms of new discoveries that creating educational content ends up being also very time bound.

2.1 Theoretical Frameworks Supporting Climate Resilience Education

The integration of climate resilience into educational curricula is strongly supported by environmental education theories. David Orr's concept of ecological literacy underscores the need for students to develop complex systems thinking and to understand human-environment interactions, fostering environmental responsibility essential for climate resilience (Blair & Hitchcock, 2001). Additionally, Albert Bandura's social learning theory emphasizes learning through observation and collaboration, which is critical in climate resilience education as students work together on local environmental issues, developing both knowledge and skills to build resilient communities (Jena & Nayak, 2020). Constructivist theory further supports this approach, suggesting that students best understand climate resilience when they connect prior knowledge with new information through hands-on, reflective activities, such as project-based learning focused on real-world environmental improvements (Quayson, 2018; Eilam, 2022).

2.2 Gaps in the Literature

Although the research on climate resilience education is rapidly growing, there are several gaps in the existing literature.

2.2.1 Limited Focus on Interdisciplinary Approaches

The lack of attention to interdisciplinary approaches to climate resilience education is one of the main gaps in the literature. Given that climate resilience education may be applied not only within science and geography educational frameworks but also in history, literature, economy, and other subjects, it is essential to evaluate its implementation interdisciplinary. However, such studies are currently rather scarce.

2.2.2 Need for Longitudinal Studies

The existing literature lacks longitudinal studies related to the development of climate resilience education; most of the existing studies are cross-sectional and provide a clear image of the current situation without tracking any changes. Since cross-sectional studies lack a broader perspective, there is a need for longitudinal research to identify the dynamics in students' knowledge, attitudes, and practices.

2.2.3 Insufficient Research on Educator Training

Another gap in the literature is the lack of studies focused on the training and retraining of educators. Although it is widely assumed that teacher training is crucial, few models of climate resilience education incorporate strategies to improve educators' competencies. Studies are needed to compare various training programs existing in the world, evaluate their impacts, and develop more efficient strategies to create climate resilience learning environments.

2.2.4 Lack of Research on Student Perspectives

The literature also lacks an understanding of students' perceptions that may influence their learning. Most research studies and approaches to developing a curriculum are based on educators' or sometimes policymakers' perspectives. However, it is essential to understand how students perceive climate resilience education, what they learn in schools, and its impact on their attitudes and practices. Such information is vital to designing more engaging, relevant, and successful curricula.

2.3 Implications for Practice

Some of the most evident are that all regions need to develop across-the-board climate resilience curricula combined with real-world adaptation examples and practices in teaching at different levels (Molthan-Hill et al., 2019). Such curricula should be designed to ensure that students gain comprehensive knowledge of all the dimensions scientific, social and economic of climate resilience. It can also include clear guides and resources for teaching about climate resiliency.

With the obstacles identified in the literature, it is important to bolster educator education and training processes for a stronger Curriculum connection of climate resilience (Reis & Ballinger, 2020). So, professional development programs should be developed that can provide educators with the necessary knowledge and skills to promote climate resilience. Programs will also offer professional development in climate science and teach strategies to help students tackle tough challenges related to our changing planet (Tasnim et al., 2023).

Tl on promoting interdisciplinary and project-based learning in climate resilient education. These methods provide multiple facets with which to explore climate resilience and apply such knowledge into real-world challenges for students. Schools must enable students to engage in projects interdisciplinary, local or global that address climate challenges and make them feel both agency as well as a sense of care. Academics that work around this space should identify organizations leading in climate resilience and looking to partner with them (Khan et al., 2014). These partnerships can offer students real-world, on-the-ground learning opportunities and help educators create and teach climate resilience curricula.

2.4 Assessment Strategy

Various assessment strategies in climate resilience education, such as project-based learning (PBL), reflective journals, community engagement projects, and digital tools like AI and IoT, provide unique benefits in enhancing student engagement and understanding. Project-based learning, highlighted by Walumbwa et al. (2017) and Harper (2023b), encourages students to tackle real-world climate problems, fostering higher-order thinking and a proactive approach toward climate action. Reflective journal writing, as discussed by Burck et al. (2019) and Kisinger & Matsui (2021), allows students to internalize complex climate topics by connecting theoretical knowledge with personal experience, aligning with constructivist learning principles. Community engagement projects, supported by Kisinger & Matsui's (2021) research, cultivate civic responsibility as students collaborate with communities to address climate challenges, emphasizing the social dimensions of climate resilience. Digital tools such as AI and IoT further enhance education, with Yu & Chiang (2017b) demonstrating AI's role in personalized learning, and Nazrul Islam et al. (2013) showcasing IoT's ability to offer real-time climate data for experiential learning. Additionally, simulations and VR, as studied by Sarker et al. (2022) and the Ministry of Environment, Forest, and Climate Change (2022), immerse students in realistic climate scenarios, enhancing empathy and urgency to address climate issues. Together, these strategies collectively enrich the learning experience and prepare students with both knowledge and practical skills for climate resilience.

3. RESEARCH METHODOLOGY

3.1 Research Design

This study presents the research methodology used to investigate the process of integrating climate resilience into the educational curriculum. The used research methodology is a mixed-method one, involving both qualitative and quantitative research methods. The research approach allowed the researcher to gather a comprehensive understanding of the ways of integrating climate resilience in the educational field. The quantitative research method utilized an online survey administered to the study participants (Dubey & Kothari, 2022). The qualitative methods included in-depth interviews with the educators and policymakers involved. Specifically, questions aimed to collect data on the past and current curriculum, climate-related issues, and integration in climate resilience into a curriculum in general.

3.2 Sampling Strategy

The purposive sampling strategy was used to include all possible subgroups active in the educational field on a daily basis and see the curriculum changes happening. The study participants involved five main groups of the educational sector: students, teachers, curriculum developers, school administrators, and policymakers. The sample for the quantitative survey included 100 respondents in Bangladesh, distributed as follows:

- Students: 20 respondents (20%)
- Teachers: 30 respondents (30%)
- Curriculum Developers: 20 respondents (20%)
- School Administrators: 15 respondents (15%)
- Policymakers: 15 respondents (15%)

In the qualitative component of the study, 15 in-depth interviews were conducted with educators and policymakers in Bangladesh. This data collection method was chosen because the target group was expected to be small and diverse, providing insights from different perspectives. Indeed, the inclusion of both educators and policymakers in the study enriched the results, as the experience of each interviewee was discussed in the context of their role. Therefore, the primary data source allowed accessing the most valuable information in this case and creating a complete picture of the impact and prospects of climate resilience integration.

3.4 Data Collection Methods

3.4.1 Survey

The primary method of quantitative data collection was a survey conducted with the 100 selected respondents. The structured questionnaire, provided online, included five sections on demographic information, familiarity with climate resilience, integration of climate resilience, challenges and barriers and asked for respondent's suggestions on possible ways to improve integration. The survey included both open and closed question types.

3.5 Data Analysis

The analysis utilized both quantitative and qualitative approaches to examine data on climate resilience integration. Quantitative analysis included descriptive statistics, which provided insights into demographic characteristics and familiarity levels with climate resilience, alongside staffing changes and average percentages. Inferential statistics were applied to investigate correlations between respondents' perceptions and factors related to protection viability, with cross-tabulations examining the relationship between study level and trade aspects. For qualitative analysis, thematic analysis was used, involving steps such as familiarization, coding, theme development, and interpretation. The themes identified were then discussed in relation to the research questions, providing a structured interpretation of the results.

4. FINDINGS AND DISCUSSION

4.1 Introduction

Chapter 4 of the study presents both the qualitative and quantitative findings of the study aimed at determining the adoption of the education curricula in terms of climate resilience. The chapter was driven exclusively by primary data. From this standpoint, the chapter represents the result analysis of the data obtained as a result of the mixed-methodology approach. The latter allowed considering the subject under study from various perspectives and enriched the statistical quantitative data with the context received through qualitative interviews. The chapter is divided into four sections where the data analysis, familiarization with climate resilience, challenges of implementation, opportunities for growth in the area, and the perspectives for the study in these areas are discussed.

4.2 Demographic Overview

The survey was conducted following the purposive sampling method. Overall, 100 primary stakeholders took part in the study. The total number of five groups was equally distributed among the sample is as follows: 20% – students; 30% – teachers; 20% – curriculum developers; 15% – school administrators; 15% – policymakers. Figure 1 shows the distribution of roles among respondents.

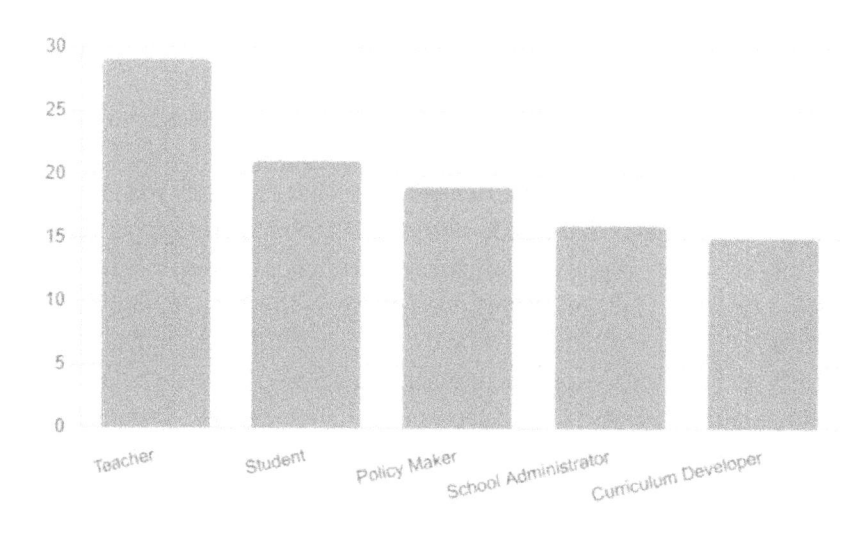

Figure 1. Distribution of roles among respondents

Figure 1 Distribution of Roles Among Respondents

The majority of interviewees and survey primary stakeholders belonged to the group of 35-45-year-olds, thus providing stable data for comparison.

4.3 Familiarity with Climate Resilience

4.3.1 Survey Findings

The survey results presented the mixed nature of responses aimed at determining how familiar the education curriculum stakeholders were with climate resilience. According to the survey data, 20% of the primary stakeholders considered themselves to be extremely familiar with climatic resilience. At the same time, 30% of the stakeholders were "moderately familiar" with the notion; 20% were "very familiar" with it, and 15% were "slightly familiar" with the term. The number of primary stakeholders "not at all familiar "with climate resilience totaled 10%.

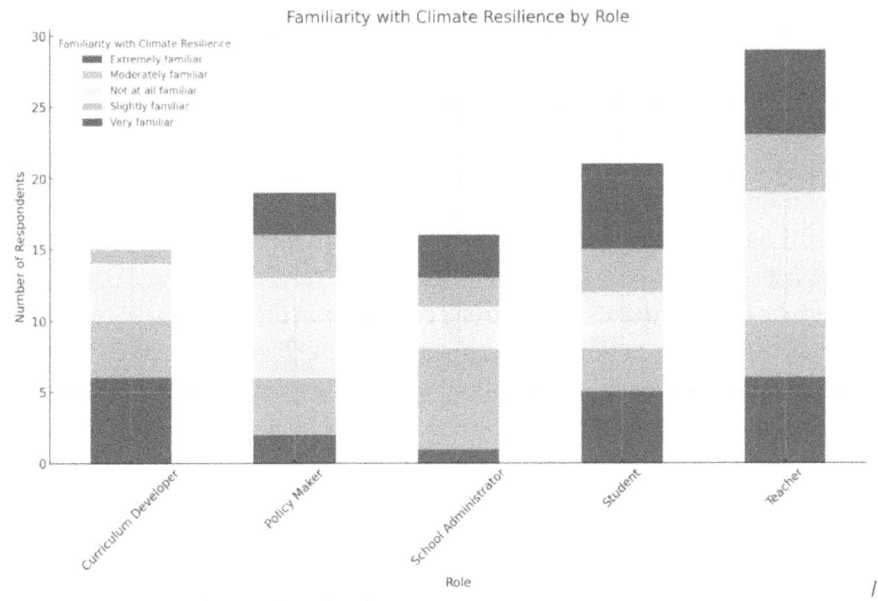

Figure 2. Familiarity with climate resilience by role

[Stacked bar chart titled "Familiarity with Climate Resilience by Role" showing number of respondents across five roles: Curriculum Developer, Policy Maker, School Administrator, Student, and Teacher, with categories Extremely familiar, Moderately familiar, Not at all familiar, Slightly familiar, and Very familiar.]

Figure 2 Familiarity with Climate Resilience by Role

This distribution implies that a significant proportion of the educational community recognizes and understands climate resilience. However, there is still a large gap on the other side. This gap is located among those in the roles that are less directly involved in curriculum development – primarily students and most administrators. If these groups are not familiar with climate resilience, it can hamper future efforts to integrate the subject across the whole span of formal and informal education.

4.3.2 Interview Insights

Educators and curriculum developers generally understood the subject better than other professionals, often mentioning it in the context of broad environmental education. Policymakers, however, were less clear about the meaning of climate resilience and tended to confuse it with general sustainability.

As one curriculum developer commented, *"Climate resilience is still a somewhat abstract concept for many in the education sector. We talk a lot about sustainability, but resilience – how we adapt and respond to climate impacts – isn't always clearly understood.*" This statement once again indicates that better communication and education exposing specific elements of climate resilience and distinguishing it from other environmental issues are needed.

4.4 Importance of Integrating Climate Resilience into the Curriculum

4.4.1 Survey Findings

The great majority of respondents, 60%, said that they "Strongly agree" that it is important to integrate the subject into the curriculum, and 25% somewhat agreed. Merely 10% have selected 'Neutral', and 5% of the participants did not agree with the statement.

The results indicate that there is a strong consensus in the educational community that students should be taught to be climate resilient. The high percentage of the positive answers suggests that a large proportion of respondents felt that it is indisputable that students must be taught the necessary skills, tools, and way of thinking necessary to cope with the challenges of climate change. However, the remaining 15% of the respondents show that there still are people in the educational sector that resist the inclusion of climate resilience or do not consider it a high priority.

4.4.2 Interview Insights

Interviews had shown that while the importance of climate resilience is acknowledged, the reasons for claiming it differ. For the educators, the primary concern was preparing students for the future challenges. Teachers and lecturers believed that education should focus on the ability to become resilient individuals and specialists. One teacher said, "Our students are the ones who will have to face the consequences of climate change. It's important that they know not just the science but how to adapt and respond." For the policymakers, their answers focused on the implications for society. In particular, they had stated that their decision to include the topic in education was based on the belief that it is necessary to raise a generation of citizens who both understand the climate problem and can contribute to increasing the level of resilience in their communities. One of the policymakers said, *"Education is central to our national strategy on climate resilience. We start the work straight with children to plant an idea in them and in the future in the curriculum that will live for years."*

4.5 Current Integration of Climate Resilience in the Curriculum

4.5.1 Survey Findings

Although climate resilience is gaining recognition as an important aspect for the curriculum, the level of integration across institutions is considerably uneven. Figure 3 shows the current levels of climate resilience integration into the curriculum.

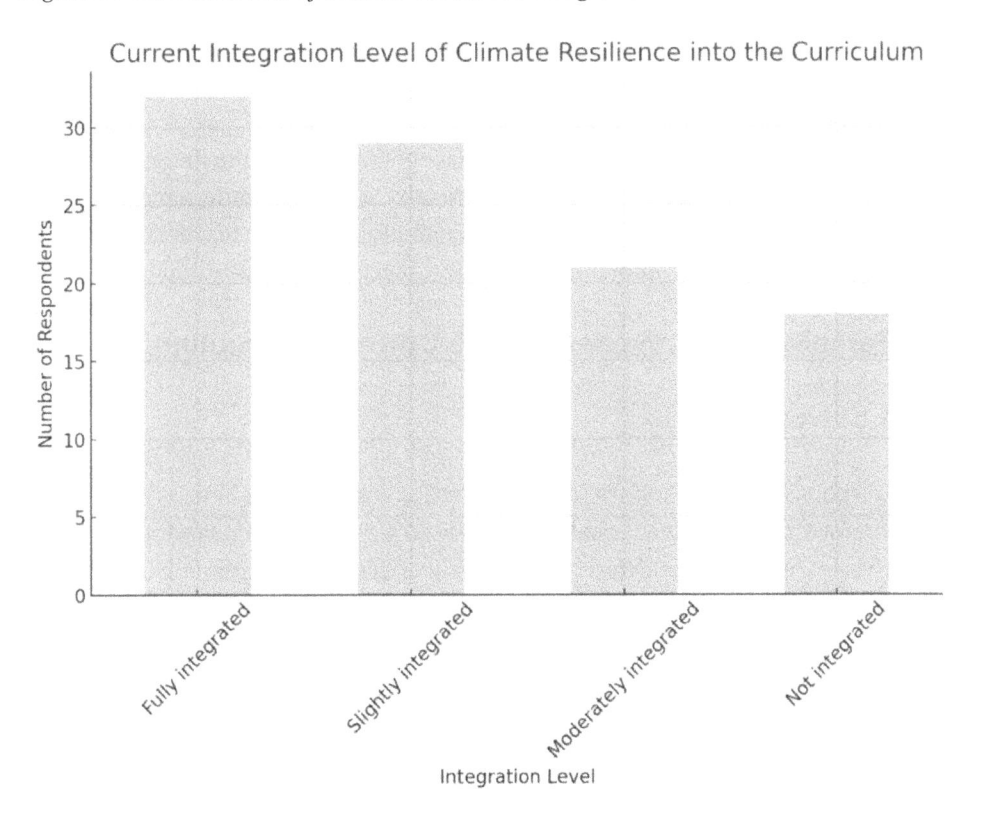

Figure 3. Current levels of climate resilience integration into the curriculum

Figure 3 Current levels of climate resilience integration into the curriculum

As per the survey data, 15% of the respondents reported that climate resilience is Fully integrated into their curriculum, while 25% reported that it is 'Moderately integrated.' However, the largest group of 35% reported that it is Slightly integrated, and 25% said it is Not integrated at all.

4.5.2 Interview Insights

The findings from the interviews indicate that the survey results do reflect the current situation regarding unequal integration across institutions and educational levels. For instance, although a few curriculum developers reported successful integration in subjects such as geography and environmental science, they felt that it was lacking in other subject areas, such as social studies or economics. Regarding this, one school administrator indicated, *"While we've managed to incorporate climate resilience into our science curriculum, it's been much harder to get it into something like history or literature it's just given a passing glance. I think that it should be integrated everywhere. I see social studies teachers looking at us science teachers and giving us some feedback that it's our business to teach our subject, it's their own duty to teach social studies."*

Thus, from the interviews, it becomes evident that if at all, the integration is not systematic and is not a matter of common practice for curriculum implementation. At the moment, it is rather an issue of good intentions of some teachers rather than a concern systematically implemented in all subjects at all educational levels.

4.6 Perceived Effectiveness of the Current Curriculum

4.6.1 Survey Findings

In response to the question of how effective the current curriculum is in preparing students for climate resilience challenges, only 20% of the respondents found it "Very effective," while 5% considered it "Extremely effective". Additionally, the current approach is described as "Moderately effective" by 30% and "Slightly effective" by 25% of the sample, while 20% found the curriculum to be "Not effective" at all. Figure 4 shows the perceived effectiveness of the current curriculum.

Figure 4. Perceived effectiveness of the current curriculum

Figure 4 Perceived Effectiveness of The Current Curriculum

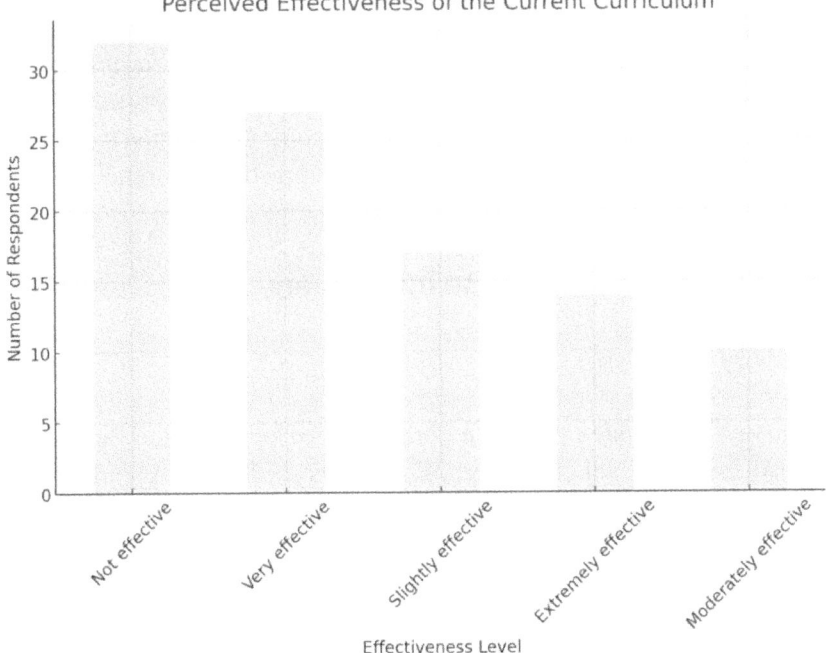

Some of the institutions have taken some actions in this context, but the dissatisfaction with the present level of climate resilience education is evident. It can be noted that these results closely correspond to the findings regarding the partial integration and issues connected with climate resilience education.

4.6.2 Interview Insights

When discussing the survey results, it would be useful to present the insights gained from the interviews to provide some additional context. From the interviews, it becomes evident that many educators are unsatisfied with the current curriculum. Some of them note that while the topic of climate resilience is raised in their programs, it is frequently mentioned in passing, while it should be central to education. For example, one of the teachers stated that, *"We touch on climate resilience, but it's usually just a brief mention. There's no depth or continuity, so students don't really grasp the importance."* As far as policymakers are concerned, some of them admitted the lack of depth of the current curriculum and the difficulties that are related to it. For instance, one of them said, "climate science is advancing so quickly that our curricula can't keep pace. By the time we develop new materials, they're already outdated".

Beyond these significant challenges, there were also examples of promising or effective practices. For example, some educators discussed their successes in using project-based learning with students on real climate resilience projects. Not only did students' understanding grow, but they also became empowered to take action. While overall effectiveness was limited, positive practices do remain and offer opportunities for wider adoption.

4.7 Challenges and Barriers to Integration

4.7.1 Survey Findings

The survey investigated the main barriers faced by educators in integrating climate resilience into their curriculum. The results showed that the most frequent barrier was "Insufficient teacher training" with a 40% respondent rate. Moreover, 30% mentioned "Lack of resources" as a common obstacle. Meanwhile, "Competing priorities" and "Institutional resistance" had a response rate of 25% and 20% respectively. Finally, "Lack of time in the curriculum" had a 15% rate. Figure 5 shows some challenges and barriers to integration.

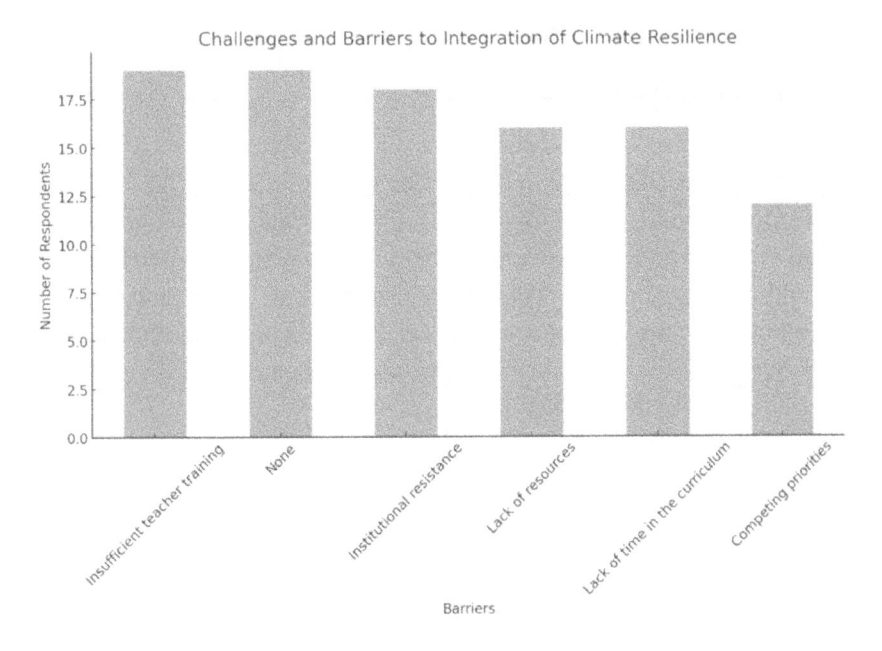

Figure 5. Challenges and barriers to integration

Figure 5 Challenges and Barriers to Integration

The findings reflect some of the practical limitations that educators and their institutions face in the effort to incorporate climate resilience into its teaching. The lack of training and resources indicates that there is a demand for more targeted or specific programs of professional development and educational investments. In the same manner, competing priorities within the curriculum imply the importance of the systemic approach that ensures that the subject matter will not be sidelined.

4.7.2 Interview Insights

The interviews provided some additional insights into these barriers. It becomes apparent that the lack of specialized training is a significant source of impediment. A curriculum developer added: *"Many teachers feel keenly about the topic, but they have no idea about what resources to use or how best to teach it. So, the training is essential, and far too little is invested in it"*. *"When it gets to school level and you have a headteacher that's a bit skeptical about climate resilience, then you have to really fight to do things"*. A school administrator argued further: *"There's sometimes a feeling that pushing climate resilience too much is certain to become controversial, or political, so people sometimes shy away."*

4.8 Opportunities for Enhancing Integration

4.8.1 Survey Findings

Respondents perceive many opportunities that can improve the integration of climate resilience into the curriculum. The highest percentage of respondents, 50%, mentioned "Teacher professional development," while "Project-based learning" and "Interdisciplinary approaches" were second and third, with 45% and 40%, respectively. In addition, 35% of respondents indicated "Collaboration with NGOs" as one of the opportunities. Figure 6 shows opportunities for enhancing the integration of climate resilience into the curriculum.

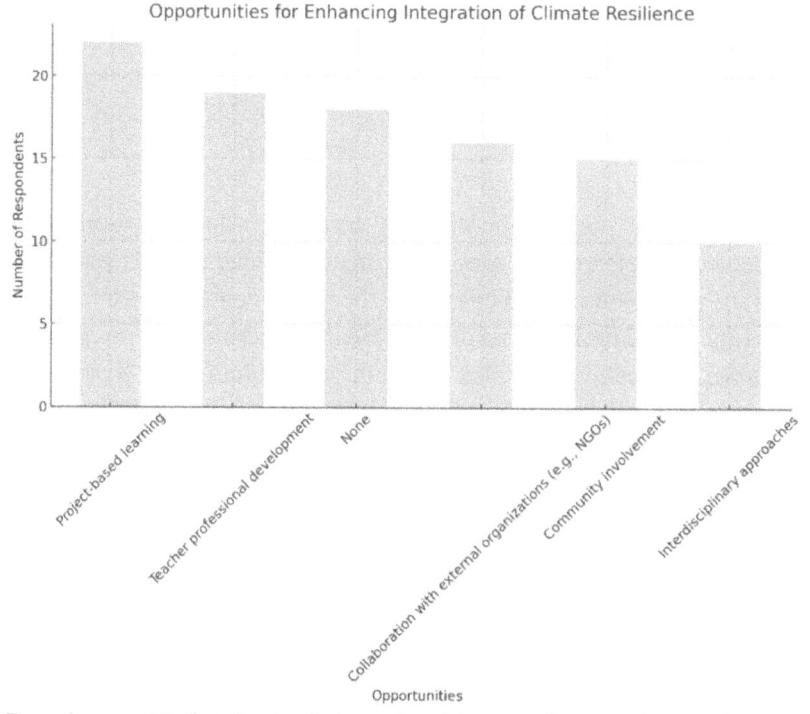

Figure 6. Opportunities for enhancing the integration of climate resilience into the curriculum

Figure 6 opportunities for enhancing the integration of climate resilience into the curriculum

The major opportunities identified in the survey resonate with the respondents' views about major challenges. In particular, a high percentage of respondents suggested the importance of professional development, and it can be reasonably expected due to the complexity of this area. The facts that teachers emphasized the importance of making learners engage in practical action and that they were eager to discuss success stories suggest this opportunity can be promising.

4.8.2 Interview Insights

Interviews helped to elaborate on survey findings because respondents provided examples of the opportunities and their perceptions. The interview mainly supported the survey results because teachers described approaches to realize these opportunities. Most teachers emphasized the importance of PBL, which can help make climate resilience a part of students' life and become less theoretical. For instance, one of the teachers talked about a project in which students had created a program to reduce school waste, and *"they were serious about it and took action. Now our school wastes less cardboard."*

Additionally, many professionals noted that collaboration with external organizations was seen as another opportunity. In a few cases, educators mentioned partnerships with environmental NGOs, which provided resources, guest speakers, and hands-on learning. The majority of professionals described these collaboration types as beneficial for both students and the organization, as the former obtained actual knowledge while the latter was able to proceed with the educational part.

4.9 Impact and Effectiveness

4.9.1 Survey Findings

The survey elicited mixed responses regarding the effectiveness of the current curriculum in preparing students for the challenges of climate resilience. The majority of 30% of the respondents felt that the curriculum was moderately effective, while another 25% reported that it was slightly effective. On the other hand, only 20% of the respondents ranked the curriculum as very effective, while 5% felt that it was extremely so. Finally, 20% of the participants believed that the curriculum was not effective at all. Figure 7 shows a comparison of familiarity with climate resilience in students.

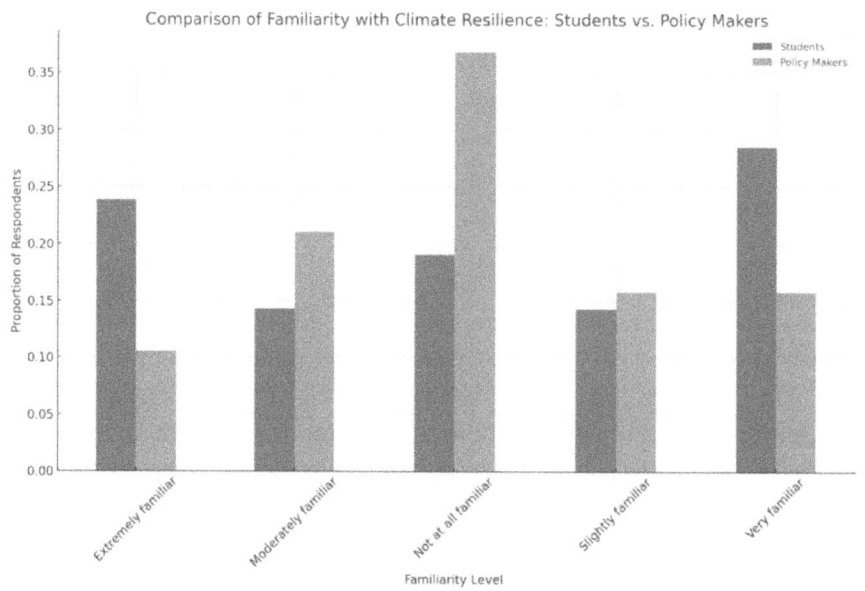

Figure 7. Comparison of familiarity with climate resilience: student

Figure 7 Comparison of Familiarity with Climate Resilience: Student

Thus, while the responses vary, it appears that there is some recognition of the curriculum's potential to prepare students. With that said, it appears that it consistently falls short of equipping students properly for the challenges of climate resilience. Figure 8 shows a familiarity with climate resilience in policymakers.

Figure 8. Familiarity with climate resilience among policy makers

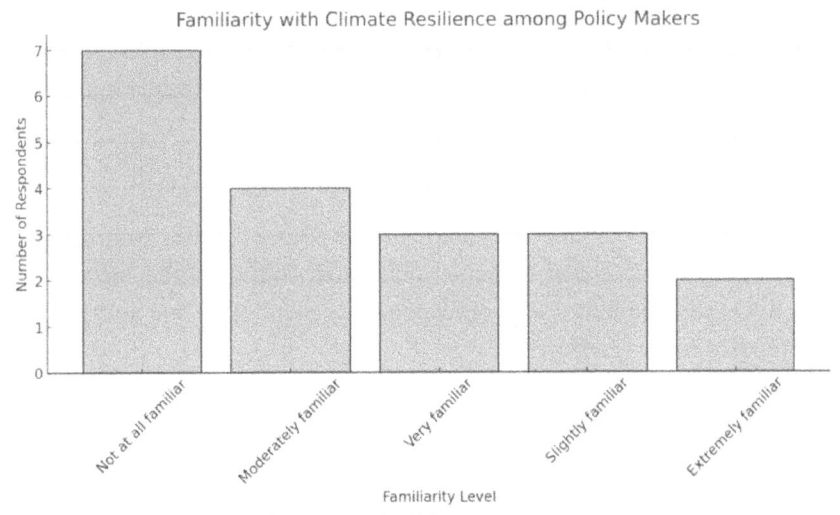

Figure 8 Familiarity with Climate Resilience among Policy Makers

The variance in its effectiveness appears to be related to the extent to which it integrates the focus on climate resilience across subjects and levels of education. Figure 9 shows a comparison of years of experience and perceived curriculum.

Figure 9. Comparison of years of experience and perceived curriculum

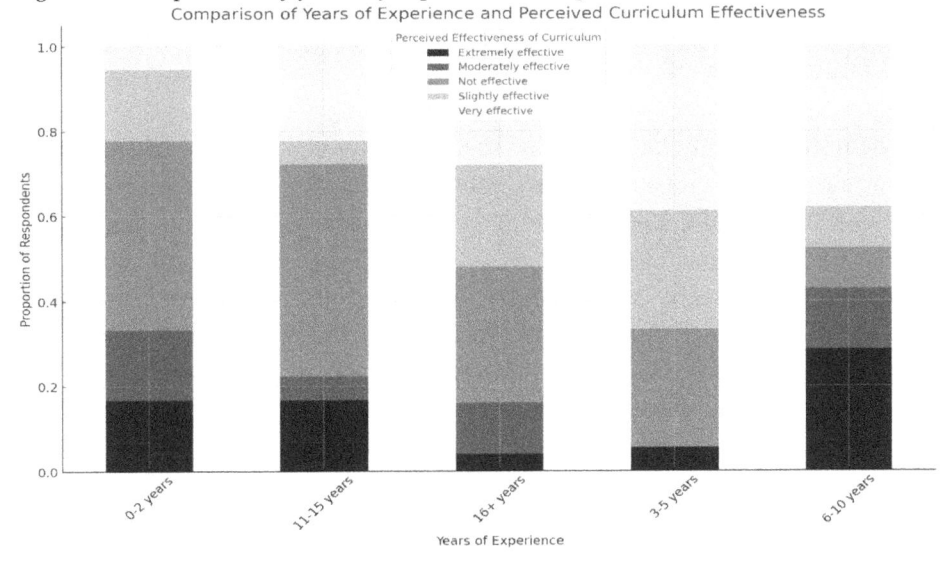

The first group, the less experienced educators falling into the category of between 0 and 2 years, often take a more critical stance towards the curriculum's effectiveness. The second group, teachers with moderate extents of teaching experience from 3 to 5 as well as from 6 to 10 years, are somewhat more positive. currently known to them.

4.9.2 Interview Insights

The interview participants provided further context to this information, with educators discussing both the successes and shortcomings of the current model of education. On the one hand, several respondents mentioned that some of the education efforts were very effective.

For example, one of the educators described an initiative in which students conducted a climate effect on community assessment that was highly successful in teaching students about the consequences of climate change. Further, it had the added benefit of engaging students with local stakeholders. At the same time, the educator also mentioned that it was entirely reliant on external donor funding and the enthusiasm of a small group of individual teachers, creating doubts about its sustainability.

Some of the policymakers also talked about the broader effects of climate resilience education and agreed that the entire system needs to change in order for climate resilience to be effective. For example, one of the policymakers said the following: *"We keep having these little projects and side-initiatives. Pilot-programs. But if we want to achieve anything, the only way to address it is to have climate resilience a part of the curriculum, at scale, supported by national policies and funding."*

4.10 Discussion

The results of both the survey and the interviews contribute to understanding the current status of climate resilience education. The key conclusion is that while there is a substantial consensus among educators and policymakers that climate resilience should be included in the curriculum, the degree of actual implementation is variable. Additionally, the perceptions of current efforts are mixed, with many respondents noting that while some progress has been made, it is not nearly sufficient for adequately preparing students for the challenges of climate change. The general conclusion that one can make is that, while the field is moving forward, it is still in somewhat of an experimental stage for many institutions as the approaches and strategies differ significantly.

Considering the opportunities that have been highlighted in the field, such as professional development, project-based learning, and interdisciplinary approaches, it should be noted that the effort would be required from educators and policymakers as well as external organizations for them to receive broader use. As a result of examining the current status of the issue and identifying the needs and opportunities that exist therein, the researchers will be able to devise some of the recommendations for addressing the most pressing challenges and using the opportunities to improve education in climate resilience.

4.11 Assessment Strategies for Evaluating Students' Understanding of Climate Resilience

Project-Based Learning

Many educators who talked with me mentioned the effectiveness of using a Project-Based Learning approach to assert whether students understand the concept of climate resilience. "*PBL is very effective because it's learning while doing. In the context of climate resilience, students must actually work on real world problems, such as creating sustainable gardens or designing practicable flood defenses. It's these projects that really show whether they have any idea what they're talking about or not,*" said a teacher who participated in the interview.

One of the examples that were shared with me was that students would work on activities where they would try to assess the vulnerability of their own community to floods or extreme heat. "*In one PBL activities, students worked on a new rainwater harvesting system for the school. They really had to think of risks and come up with very sensible solutions, and so their learning was also much more meaningful,*" said one of the respondents. Thus, the students are no longer assessed only in terms of acquiring climate resilience knowledge but also in their capacity to apply it in ways that matter.

It was also emphasized by the respondents that PBL allows interdisciplinary learning. One of the teachers said, "*Climate resilience is not just a science issue; it crosses into geography, economics, and social studies. Students need to pull from all these areas to create comprehensive solutions.*" The complexity of PBL complements the complexities of climate resilience in real life, as improving climate health often takes varied viewpoints from multiple disciplines.

Moreover, one of the most popular ENA assessment tools was reflective journals, and they requested to describe the relevant projects. The interviews stated that it could be useful as students can track the progress of their personal climate understanding increase. One of the educators added, "*When students keep journals, they can see how their understanding of climate resilience evolves. It's not just about facts but how they connect those facts to their own experiences and quite often for the first time with the issue on a more personal level,*". Moreover, according to several educators, reflective writing can help students better understand complex topics such as the social or economic impacts of climate change. One of the respondents gave such an example, "*At the start of the course, many students don't fully grasp the depth of climate resilience. But as they participate in projects and discussions, their journal reflections start showing a deeper appreciation of how interconnected these issues are*". This type of assessment allows teachers to see not only the knowledge gained by students in one subject but also how they can reflect on the topic and become emotionally involved in it.

Several of the responding educators also emphasize the importance of reflective journals for developing students' understanding of, and engagement with, the emotional and psychological aspects of the climate resilience education. Accordingly, "*Climate change is overwhelming for students. Journaling helps students process their feelings, whether it's fear or frustration and work out how they can engage to possibly resolve them.*" That is, as described by the respondents, reflective journals provide students with a channel to engage with and reflect on their emotional responses to climate-related problems, develop their own sense of agency in response to environmental issues, and, by extension, capacity to engage with them. In conclusion, the interviews demonstrate a strong consensus among educators and curriculum specialists that traditional modes of assessment must be integrated with more interactive and reflective measures such as PBL and reflective journals to effectively gauge students' knowledge of and engagement with climate resilience issues. Moreover, such measurements not only assess knowledge but also promote the development of critical thinking and problem-solving skills in addition to emotional engagement with individual students' potential for action on the issue.

Talking to educators and people who design curricula, "*I realized that community engagement is one of the most frequently used innovations or exercises that help evaluate climate resilience understanding among students.*" One of the teachers said, "When students work with a community, they are learning and at the same time, they see the real-world consequences of their 'projects.' It is incredibly motivating for them". In most cases, these projects are organized in collaboration with local environmental organizations. Almost all my interviewees noted that community engagement projects allowed determining students' leadership, ability to communicate and work in teams, and skills to solve problems and apply knowledge. One of the curriculum developers emphasized that community engagement projects are crucial because "*one the one hand, students learn how to preserve the environment, and on the other one, they learn to think about the critical areas in the context of real community problems and identify ways to solve them. It helps them learn to lead others and work in teams with others.*" In addition, community engagement gives students an understanding of a social and cultural aspect of climate resilience because they see that people view climate issues from different perspectives.

However, the greatest benefit, according to some teachers, is the role community engagement plays in closing the gap between theory and practice. According to one of the schoolteachers, *"These projects underscore what was taught in the classroom. However, we can use them to give students the learning experiences that are useful for their future work related to climate"*. Another teacher stated, *"Students continue working with organizations they have engaged with during the project, long after the experience is over."*

Digital Portfolios

The interviews also touched upon the topic of digital portfolios as an effective method of assessment of climate resilience education. As one respondent put it, *"Digital portfolios help the students record what they learn. They can keep everything from the research papers to the videos, and it will show their progression."* The use of digital media to assess students' learning allows for the demonstration of their work overtime, presenting a profile of what a student has learned and can now do. Of the teachers interviewed, some mentioned the possibility for digital portfolios to provide more "holistic" assessment compared to the traditional testing methods. One of the teachers stated, *"We can check just one test, or we can look at a range of things they have done, which shows different types of skills […] One piece can be research around the best project they can do on climate, the next can be video learning about sustainable farming, a third project might be volunteering at a shelter, recording and reflecting on it."* The teachers agreed, therefore, that digital portfolios provide a broader picture of students' learning and achievements.

Another important aspect of the use of digital portfolios for the described purpose is that they allow students to keep track of their learning and reflect on their progress. As another teacher put it, *"The students love it because they can see the huge improvement they have got […] they won't just be learning facts; they will be engaging with the different agenda. They link them and they offer a forum to present."* Summarizing the above, it can be stated that the interviews revealed that digital portfolios can be used as an effective moderate mode of assessment, checking students' understanding of a climate resilient topic from a variety of perspectives, while simultaneously promoting continuous learning and self-reflection.

4.11 Leveraging Digital Tools, AI, and IoT to Enhance Climate Resilience Education

In interviews with educators and technology specialists, respondents highlighted the significant role of AI-powered adaptive learning platforms in enhancing climate resilience education. One respondent explained, *"AI can personalize learning in ways that traditional methods can't. If a student is struggling to understand something like carbon footprints, the AI system can recommend additional materials that are tailored to their learning style.*" These platforms, as several educators mentioned, allow for real-time analysis of student performance, offering targeted content and resources based on individual needs.

Additionally, many respondents discussed the importance of AI in providing formative assessments. "*The AI system gives immediate feedback,*" noted one teacher. "*If a student finishes a lesson on climate adaptation strategies, they can take a quiz, and the AI analyzes the results to identify areas that need more focus.*" This process ensures that students fully understand the material before progressing to more complex topics, keeping them on track and enhancing overall comprehension. Some educators mentioned the development of AI virtual tutors that guide students through complex climate-related scenarios.

IoT for Real-Time Climate Monitoring

Respondents also discussed the potential of Internet of Things (IoT) devices to bring real-world data into the classroom. One educator remarked, "*We've installed weather sensors and air quality monitors at the school. The data collected allows students to track local environmental conditions, and they can analyze this information to understand how climate change is affecting their community.*" This hands-on approach gives students practical experience in using data to assess climate risks and propose solutions.

Another teacher shared, "*Students love seeing how their proposed solutions work in real-time. For example, we installed sensors on a green roof to monitor temperature changes and stormwater runoff. The students analyze the data to see if their design is effective and make adjustments based on the results.*" These IoT applications provide a tangible way for students to engage with climate science and contribute to climate resilience projects in their local environment.

Digital Simulations and Virtual Reality (VR)

Digital simulations and virtual reality (VR) were also mentioned frequently by respondents as valuable tools for immersing students in climate resilience learning. "*Using simulations, students can explore how rising sea levels affect coastal cities. They can experiment with different adaptation strategies and see the consequences of their choices,*" said one respondent. These simulations allow students to critically evaluate various resilience strategies and understand the long-term impacts of their decisions.

Several educators noted the advantages of VR in creating immersive learning experiences. *"In a VR simulation, students might respond to a natural disaster, like a hurricane. They have to make real-time decisions how to evacuate people, where to allocate resources and they experience the impact of those decisions firsthand,"* explained one respondent. This type of experiential learning helps students develop practical skills and decision-making abilities, which are critical for real-world climate resilience efforts.

Online Collaboration Tools

The interviews also highlighted the value of online collaboration tools in fostering global teamwork on climate projects. As one educator stated, *"Students from different countries are working together on projects using platforms like Google Classroom and Microsoft Teams. They compare how climate change is affecting their communities and share solutions."* These tools enable students to communicate and collaborate across geographic boundaries, broadening their perspective and helping them develop the teamwork and communication skills necessary to tackle global climate challenges.

One respondent remarked, *"The best part of these tools is that students aren't limited by location. They can work with peers from around the world, share data, and discuss strategies for resilience. This kind of global collaboration is essential for addressing climate change."* Many educators agreed that online collaboration tools enhance students' global awareness and prepare them for working in international settings to solve environmental problems.

5. CONCLUSION AND RECOMMENDATIONS

5.1. Implications for Education Policy and Practice

The results of this study have several important implications for education policy and practice, especially in the domain of preparing students for climate change challenges. The extent of integration of resilience into curricula appears to be uneven, which suggests that a more systemic and coordinated effort is required to ensure that all students receive a comprehensive education on this subject. However, the lack of suitable training for educators and scarcity of resources may complicate the achievement of this objective.

5.2 Enhancing Teacher Training and Professional Development

The most significant barrier identified in the study is the lack of appropriate teacher training on resilience. Educators are often unequipped to engage with this highly complex and rapidly changing subject, which undermines the overall performance of the curricula. Hence, it seems that targeted proposals for the professional development of the staff tasked with the education of climate resilience are needed. These programs should focus on the individual specifics of resilience education and feature teachings on student engagement for this purpose.

5.3. Making the Issue Part of Multiple Subjects

One of the reasons why the implementation of a new program is limited is the fact that respondent schools overwhelmingly included the new matter into a couple of subjects, primarily science and geography. However, as is the case with other subjects, and in this regard, climate resilience is interdisciplinary. For example, we know the economics of climate change and how it can be impacted by climate resilience. There also exist history and literature on the topic since stories about our past and what people had to do to avoid total destruction in the face of climate change often serve as inspiration. One way to improve the current state is to encourage use across the curriculum. The issue might be analyzed from different perspectives using multiple subjects. For instance, in social studies, students can look at the effect of climate change on different communities, both in social and economic contexts. In literature, they can evaluate works that touch upon environmental justice and resilience. Hence, embedding climate resilience in different subjects can help make the issue more understandable and more relevant to students.

Based on the findings of the survey, several recommendations can be made for enhancing the education of the climate resilience concept and its integration into educational curricula. A more detailed description of each of the initiatives is provided below.

5.4. Recommendations for Enhancing Climate Resilience Education

Based on the results of the study being performed, the following set of recommendations can be proposed for enhancing existing and promoting new initiatives of integrating climate resilience into educational curricula:

5.4.1. Develop and Implement Fair Climate Resilience Education Standards

One of the critical factors revealed by this research is the absence of a unified set of standards prescribing which topics must be covered in the teaching area of climate resilience. To eliminate this factor, an appropriate recommendation can be made to authorities in charge of the education, to elaborate a fair provision of standards indicating which concepts, topics, and competences can be stated as a component of climate resilience. Moreover, these standards shall be done under the guidance of the context of national and international frameworks in this regard, like the United Nations Sustainable Development Goals to be applicable at various educational levels.

The next step for implementing the results of this activity is related to the necessity to apply relevant frameworks to all educational institutions, including guidelines on their integration into non-kindred curricula. The recommendation given will allow conducting education in a manner eliminating disparities between students of different regions and educational levels.

5.4.2. Implement Targeted Development Programs for Educators

As was mentioned earlier, another factor influencing the quality of education in the area of climate resilience of students is the level of professional development of educators. To provide stakeholders with necessary knowledge and skills, relevant authorities, and institutions of education are highly encouraged to adopt targeted development programs in this regard. It shall be performed in the matter of including all levels of educators – from primary and secondary to tertiary. At the same time, the given programs have to be conducted in the form of fieldwork, using essential means of hands-on learning, implementing such tools as workshops, forums, etc. in order to ensure the participation of environmental organizations. Finally, educators shall be provided with a specific type of ongoing support, namely providing them with sufficient resources available including online databases, newsletters, and teaching materials.

5.4.3 Promote Interdisciplinary and Project-Based Learning

Educational institutions should facilitate the adoption of interdisciplinary and project-based learning to improve the integration of climate resilience across the curriculum. These approaches allow students to explore climate resilience from different perspectives and apply their knowledge to real-world problems. For instance, schools could promote project-based learning where students design different projects that enhance community resilience. They could explore different areas such as sustainable energy or enhancing disaster preparedness. The projects could be integrated into subjects such as science, social studies, and technology. Additionally, the projects could be implemented in collaboration with external organizations such as non-governmental and local government authorities.

5.4.4 Foster Collaboration with External Organizations

Collaboration with external organizations such as NGOs, research institutions, and local government authorities could enhance resources for promoting climate resilience across the curriculum. Greenberg et al. state that the external organizations could provide partnerships and resources such as research, teachers' workshops, field trips, and projects. It is recommended that education providers cultivate strong ties with these organizations and use them to bolster climate resilience education. For instance, external organizations could assist in the implementation of projects or workshops aimed at providing learners with practical and hands-on experiences. Furthermore, they could offer support such as teaching materials, funding, and professional development.

5.4.5 Advocate for Policy Support and Funding

The integration of climate resilience across the educational curricula requires policy support and funding at the national and local levels. Education providers should work with policymakers to ensure that climate resilience education is central in the overall education agenda. For example, they could develop national strategies and action plans to guide climate resilience education while advocating for the necessary resources for their implementation. Additionally, the education providers and stakeholders should push for the allocation of resources such as funding for professional development, curriculum development, and instructional tools for teachers.

5.5 Reflection on the Study's Contributions

This study has added to the literature on climate resilience education by providing a detailed overview of the existing context for integration, as well as the barriers and challenges that education professionals and policymakers are facing in this regard. Additionally, the use of a mixed-methods approach allowed for a more nuanced and detailed perspective on the issue. In general, this study has a number of important implications for both education policy and practice, and particularly in the context of preparing students to meet the challenges of climate change. Finally, by outlining the most significant barriers to integration and suggesting ways to address them, this study became an important resource for curriculum developers, policymakers, educators, and other stakeholders participating in the climate resilience education and attempting to improve its quality.

5.6 Suggestions for Future Research

There are multiple areas for future research that can contribute to the existing body of knowledge and address some of the gaps related to this study. These include:

5.6.1 Longitudinal Studies on Climate Resilience Education

One limitation of this study is that it is primarily cross-sectional, meaning that it provides a snapshot of the current state of climate resilience education but does not analyze or follow the ongoing trends over time. Therefore, another possible application for future research could be to create a longitudinal study that will stretch over the course of several years, monitoring changes in implementation and results of climate resilience education. This approach can allow for a better understanding of how this intervention works in the long term and how it affects students' knowledge, attitudes, and behaviors over time.

5.6.2 Comparative Studies Across Different Educational Contexts

Another limitation of this study is that it is limited to a specific set of educational contexts. With this in mind, one possible application for future research could be to conduct similar study or studies in different systems, regions, or cultural contexts, enabling a more comprehensive understanding of effective practices and strategies while trying to apply to other contexts.

5.6.3. Exploring Student Perspectives on Climate Resilience Education

This study deals primarily with the perceptions of educators and policymakers but, in the future, researchers may use surveys, interviews, or focus groups to explore the student perspective of climate resiliency education. It would be advisable to examine the perceptions of students from different educational levels to understand how exactly they perceive this term, how they are taught resiliency at school, and how the concept impacts their behavior. The inclusion of students' perspective may help to adopt new, more effective teaching methods which will make the teaching of resiliency more proper and practical.

5.7 CONCLUSION

This study has been aimed at the analysis of the effects of efforts to integrate climate resiliency into the education curriculum and includes opportunities and challenges that exist in this field. As such, the study points to the need for the broader, systemic approach to resiliency education which, in turn, should include improvements in teachers' training, more efficient integration of climate resiliency in the curricula, recognition of the institutional barriers to acquiring proficiency in resiliency education and, finally, establishing better communication and collaboration with various outside organizations. The recommendations, which are offered in this chapter, may help educational institutions to join the common endeavor of creating knowledgably, skillful, and motivated students who are able to face the challenges of a continually changing climate. The present study also suggests directions for further research which can contribute to our knowledge of how to integrate resiliency into education and raise the new generation which will be able to tackle the complexities of climate change successfully.

REFERENCES

Apollo, A., & Mbah, M. F. (2021). Challenges and opportunities for climate change education (Cce) in East Africa: A critical review. *Climate (Basel)*, 9(6), 93. DOI: 10.3390/cli9060093

Blair, A. M., & Hitchcock, D. H. (2001). *Environment and business*. Routledge.

Burck, J., & Hagen, U. Hãúhne, N., Nascimento, L., & Bals, C. (2019). Climate change performance index. *Bonn: Germanwatch, NewClimate Institute and Climate Action Network*.

Chowdhury, M. T. A., Ahmed, K. J., Ahmed, M. N. Q., & Haq, S. M. A. (2021). How do teachers' perceptions of climate change vary in terms of importance, causes, impacts and mitigation? A comparative study in Bangladesh. *SN Social Sciences*, 1(7), 1–35. DOI: 10.1007/s43545-021-00194-7 PMID: 34693329

Dubey, U. K. B., & Kothari, D. P. (2022). Research Methodology. In *Research Methodology*. DOI: 10.1201/9781315167138

Habiba, U., Abedin, M. A., & Shaw, R. (2013). Disaster education in Bangladesh: opportunities and challenges. *Disaster Risk Reduction Approaches in Bangladesh*, 307–330.

Harper, R. (2023). *Toward Climate-Smart Education Systems: A 7-Dimension Framework for Action*. Global Partnership for Education.

Hügel, S., & Davies, A. R. (2024). Expanding adaptive capacity: Innovations in education for place-based climate change adaptation planning. *Geoforum*, 150, 103978. DOI: 10.1016/j.geoforum.2024.103978

Hurlimann, A., Cobbinah, P. B., Bush, J., & March, A. (2021). Is climate change in the curriculum? An analysis of Australian urban planning degrees. *Environmental Education Research*, 27(7), 970–991. DOI: 10.1080/13504622.2020.1836132

Islam, M. R. (2020). Environment and disaster education in the secondary school curriculum in Bangladesh. *SN Social Sciences*, 1(1), 23. DOI: 10.1007/s43545-020-00025-1

Jena, L., & Nayak, U. (2020). Theories of Career Development: An analysis. *Indian Journal of Natural Sciences*, 10(60).

Khan, A. U., Hassan, M., & Atkins, P. (2014). International curriculum transfer in geography in higher education: An example. *Journal of Geography in Higher Education*, 38(3), 348–360. DOI: 10.1080/03098265.2014.912617

Kisinger, C., & Matsui, K. (2021). Responding to climate-induced displacement in bangladesh: A governance perspective. *Sustainability (Basel)*, 13(14), 7788. DOI: 10.3390/su13147788

Kumar, P., Sahani, J., Rawat, N., Debele, S., Tiwari, A., Emygdio, A. P. M., Abhijith, K. V., Kukadia, V., Holmes, K., & Pfautsch, S. (2023). Using empirical science education in schools to improve climate change literacy. *Renewable & Sustainable Energy Reviews*, 178, 113232. DOI: 10.1016/j.rser.2023.113232

Leal Filho, W., Sima, M., Sharifi, A., Luetz, J. M., Salvia, A. L., Mifsud, M., Olooto, F. M., Djekic, I., Anholon, R., Rampasso, I., Kwabena Donkor, F., Dinis, M. A. P., Klavins, M., Finnveden, G., Chari, M. M., Molthan-Hill, P., Mifsud, A., Sen, S. K., & Lokupitiya, E. (2021). Handling climate change education at universities: An overview. *Environmental Sciences Europe*, 33(1), 1–19. DOI: 10.1186/s12302-021-00552-5 PMID: 34603904

Lotz-Sisitka, H., Mandikonza, C., Misser, S., & Thomas, K. (2021). Making sense of climate change in a national curriculum. *Teaching and Learning for Change*, 92.

Ministry of Environment Forest and Climate Change. M. (2022). *Climate Change Initiatives of Bangladesh Achieving Climate Resilience*. www.moef.gov.bd, Molthan-Hill, P., Worsfold, N., Nagy, G. J., Leal Filho, W., & Mifsud, M. (2019). Climate change education for universities: A conceptual framework from an international study. *Journal of Cleaner Production, 226*, 1092–1101.

Nazrul Islam, A. K. M., Deb, U. K., Al Amin, M., Jahan, N., Ahmed, I., Tabassum, S., Ahamad, M. G., Nabi, A., Singh, N. P., Byjesh, K., & others. (2013). *Vulnerability to Climate Change: Adaptation Strategies and Layers of Resilience–Quantifying Vulnerability to Climate Change in Bangladesh*.

Quayson, F. O. (2018). *The Feasibility of Establishing A Private International Virtual High School In Ghanathe-feasibility-of-establishing-a-private-international-virtual-high-school-in-ghana. 1*(1).

Rabin, B. M., Laney, E. B., & Philipsborn, R. P. (2020). The unique role of medical students in catalyzing climate change education. *Journal of Medical Education and Curricular Development*, 7, 2382120520957653. DOI: 10.1177/2382120520957653 PMID: 33134547

Reis, J., & Ballinger, R. C. (2020). Creating a climate for learning-experiences of educating existing and future decision-makers about climate change. *Marine Policy*, 111, 111. DOI: 10.1016/j.marpol.2018.07.007

Ruiz-Mallén, I., Satorras, M., March, H., & Baró, F. (2022). Community climate resilience and environmental education: Opportunities and challenges for transformative learning. *Environmental Education Research*, 28(7), 1088–1107. DOI: 10.1080/13504622.2022.2070602

Salinas, I., Guerrero, G., Satlov, M., & Hidalgo, P. (2022). Climate change in chile's school science curriculum. *Sustainability (Basel)*, 14(22), 15212. DOI: 10.3390/su142215212

Sarker, M. N. I., Peng, Y., Khatun, M. N., Alam, G. M. M., Shouse, R. C., & Amin, M. R. (2022). Climate finance governance in hazard prone riverine islands in Bangladesh: Pathway for promoting climate resilience. *Natural Hazards*, 110(2), 1115–1132. DOI: 10.1007/s11069-021-04983-4

Seritan, A. L., Coverdale, J., & Brenner, A. M. (2022). Climate change and mental health curricula: Addressing barriers to teaching. *Academic Psychiatry*, 46(5), 551–555. DOI: 10.1007/s40596-022-01625-0 PMID: 35314961

Takinana, A., & Baars, R. C. (2023). *Climate change education in the South Pacific: Resilience for whom? Asia Pacific Viewpoint*. Business Source Index.

Tasnim, T., Annum, F., Curtis, A. E., Braun, M., Shuvo, S. R. S., Anjum, B., Amjath-Babu, T. S., & Khanam, F. (2023). *Bridging the climate information gap for adaptation: Mainstreaming climate services into higher education in Bangladesh*.

Walumbwa, F. O., Hartnell, C. A., & Misati, E. (2017). Does ethical leadership enhance group learning behavior? Examining the mediating influence of group ethical conduct, justice climate, and peer justice. *Journal of Business Research*, 72, 14–23. DOI: 10.1016/j.jbusres.2016.11.013

Yu, C.-Y., & Chiang, Y.-C. (2017). Designing a climate-resilient environmental curriculum—A transdisciplinary challenge. *Sustainability (Basel)*, 10(1), 77. DOI: 10.3390/su10010077

Chapter 13
Gamification in Climate Action:
Understanding the Role of Game Technologies and Participatory Engagement

Yigang Liu
https://orcid.org/0000-0003-2735-1882
Nanjing Forestry University, China

ABSTRACT

The imperative to address climate change has led to innovative approaches, one of which is the application of gamification. This paper explores the potential of gamification to mitigate climate change by enhancing eco-education, simulating scientific scenarios, and providing policy feedback. It argues that gamification, through its ability to engage users in a "magic circle" of play, can foster a participatory culture that encourages behavioral changes aligned with climate mitigation strategies. The objective is to shift the focus from merely improving gamification's effectiveness to leveraging the gaming industry's potential to directly contribute to reducing personal carbon footprints and shaping power consumption habits through gaming. It concludes that gamification and the gaming industry can play a significant role in climate change mitigation by transforming players' behaviors and by establishing a virtual carbon credit market that incentivizes sustainable practices.

DOI: 10.4018/979-8-3693-7001-8.ch013

INTRODUCTION

In 2022, a heat wave is roasting the whole Eurasische Platte. According to the land surface air temperatures indicated by NASA (NASA, 2022), the summer will probably be the hottest since the late 19th century. Instead of affecting the biosphere gradually, climate change is now a potential threat to all biont types. To address the climate crisis, humans set global agendas and contribute endless efforts, from Stockholm and Kyoto to Paris (UN, 2007). Can we reverse the dead end of an eco-catastrophe and develop a sustainable lifestyle? Or is this goal only an ectopia in a digital game simulation? In light of this question, applying a gaming approach to address climate change benefits eco-education, science simulation, and policy feedback. Relentless exploration by game scholars has reached a pivot point, resulting in an imaginative concept based on reviews that can not only impact the climate through pedagogy or simulation but also functions as an engine that can drive humanity toward a low-carbon society directly.

Dealing with a retaliation from nature, this COVID-19 pandemic, impacts humans through our disputes and discordances. This is also the case when we encounter global climate change. From a socioeconomic standpoint, climate change enlarges the gap between wealth and poverty, which results in great inequity regionally (Tol, R. S. J., 2009). Such inequity affects food production, causing a significant polarization that developing nations may experience with the risk of hunger (Parry M. L. and et.al., 2004). Through the means of the social cost of carbon (SCC) and climate change impacts, adaptation and vulnerability (IAV), analytical results are used to help enact policies managing economic risk and greenhouse gas emissions, namely, carbon dioxide (CO^2) (Diaz, D., & Moore, F., 2017). Instead of focusing on public organizations and private sectors of business (Martinich, J. & Crimmins, A., 2019), we must include individuals and households as well, as they are the basic cell of human society. Based on the carbon footprint Gini coefficient, Chinese households' carbon consumption reveals inequities in rural and urban areas and among wealth and impoverished population distributions (Wiedenhofer, D. et al., 2017). The inequity of economic risk does not only stretch spatially; if we use a more farsighted scope, it has a long-term effect on intergenerational equity (Mejean, A. et al., 2020). Hence, climate change, especially CO^2 emissions, is highly integrated into the interaction of humans and ecosystems and remains dynamic and multifactor. Thus, the innovation of green energy is on one side; on the other side, humankind requires another approach to sustainable collaboration to reduce its carbon footprint without causing social and economic inequity.

Backgrounds: Game Studies on Climate Change

One of the most accepted research projects on games and their application in fighting climate change is in education. Especially when participants engage in the gameplay on the topic of environmental issues. Taking KEEP KOOL for example, Jasper N. Meya and Klaus Eisenack (2018) found out that the sense of personal responsibility, as well as the confidence in politics for climate change mitigation will be affected through the gaming experience, which means responsibility and confidence could be improved. These games like KOOL simulate policy decision-making and cross-nation cooperation agendas to face the climate challenge, which provides the players with an immersive experience of the climate problem-solving process. Another case comes from the perspective of climate literacy. The study (Harker-Schuch et al., 2020) points out that the simulation of climate science and physical mechanisms helps students better understand climate change as a system that can be influenced by solar radiation, gravity, atmosphere, and greenhouse gases.

From social science and physical knowledge learning gaming objectives, it demonstrates empirical evidence to support the argument that climate simulation gaming does benefit the climate knowledge pursuit. But when we are facing the public, can games still function as the experiments conducted in the school or in a limited area? Robert Kwok (2019) gives us an elaboration, which indicates that players are more likely to believe afterwards that humans are realizing the catastrophic climate requires global cooperation to stop. So, Robert makes his point that it is also vital to express the gaming experience not only to the players but to the public of the new audiences. To reach the wild spreading of the gaming experience, Sebastian Seebauer (2013) provides us with a fresh idea that social media games like Climate Quiz can improve the layperson's knowledge on climate change, which might boost climate knowledge spreading.

Besides that, it is worth noticing that the studies on utilizing games to strengthen the ability of humans to adapt to the coming climate change. The current study suggests that players intend to reflect on their climate adaptation behaviour and decisions based on the 2030 goals (Neset T.S. et al., 2020). Studies reveal that playing games around climate change provides an approach to developing resilience (Buck, M. et al., 2023). So, the gaming experience from a virtual future perspective usually develops the players' resilience and adaptation, making them rethink and take action in this virtual future while facing various climate consequences. The ability and the mindset cultivated in the fictional scenes are not predictions to the coming climate disasters, but a warning sign to the players who live in nowadays.

Overall, this research frame the narrative that the behavioural changes required to address climate change. It highlights how interactive and immersive methods, such as serious games, can influence climate-related behaviours by enhancing understanding and engagement. A lot of evidence emphasises the importance of psychological insights in designing effective climate interventions through games (Douglas, B. D. and Brauer, M., 2021). It suggests that these tools can make the abstract concept of climate change more concrete and actionable for players. According to the scholars Gerber, A. and his teams (Gerber, A. et al., 2021), most of the games were found to be educational, and aimed at increasing awareness and understanding of climate issues among diverse audiences. Their study also identified gaps in the current landscape of climate games, particularly a lack of games targeting professional development and those that create a direct impact outside the game environment. The gap they mentioned is exactly the potential solution this article is going to offer, which might generate a direct impact on the climate via the gaming and the platforms of the game industry. To sum up, the current studies underscore the transformative potential of games in climate education and action. Games can simplify complex climate concepts, making them more accessible and engaging, thus fostering a deeper understanding and prompting behavioural changes necessary for addressing climate challenges. They also highlight the need for further development of games that not only educate but also lead to tangible environmental benefits and cater to professional audiences.

PLAY THEORY BEHIND CLIMATE CHANGE

Critical Definition of Magic Circle

In the modern academic era, the intellectual structure of game studies has connected multiple fields, including communications, education, humanities and the social sciences, and health (Martin, P., 2018). Before exploring specific discipline, it is more useful to understand what a game is in essence. Fundamental game research can be traced to John Huizinga's *Homo Ludens*, which emphasizes that a game is a cultural phenomenon even older than culture itself (Huizinga, J., 1949, pp.10-11). Another concept derived from Huizinga is the boundary of a game, the controversial magic circle, which separates the gaming world from our reality (Huizinga, J., 1949). His elaboration on a game's boundary allowed his successors to translate the effect from a game's metaverse to our universe. One significant contributor is Jesper Juul, who argues that the definition of a game involves a static object and activity, where the spectrum moves from non-games to ambiguous forms and eventually authentic games (Juul, J., 2005, pp.56-57) (Fig. 1). Since the magic circle that separates not-games from authentic games tends to be rather ambiguous, this is theoretical evidence that social issues can be approached with virtual forms of play. As an activity, play is the other side of the coin with a game. In contrast to games, play emerges in an even more primitive form. Pioneer theorist of play Roger Caillois (2001) points out that during the play, property is exchanged but no goods are produced. Only when players decide to accept freely will play affect and realize. Based on the theory, each basic play has socialized aspects, does it somehow evolve as a socially legitimate cultural form (Caillois, R., 2001, pp.5-10). The main characteristic of play, Caillois states, determines play's influence on reality and is constrained to individuals and limited in scope. However, play is also powerful in its sphere of influence. Another play theorist, Brain Sutton-Smith, elucidates why play is so powerful. He asserts that the application of force, skill, and leadership in actions and strategies within a game are intrinsic power concerns (Sutton-Smith, B., 1997, p.90). These two play theorists thus prove that play activity is not only limited to within the magic circle but can exert power over certain areas. Therefore, the boundary of the magic circle is not fixed. It is a dynamic relationship within a game's contents and a player's will that cultivates the culture from play to joy to the seriousness of performance. Within a review article of 52 climate change games published in 2013, in its analysis of target games, it reveals that rich climate change issues are considered in over 40% of the articles it reviews, including basic knowledge, such as socioeconomic, mitigation technologies, mitigation of climate change, climate change politics, and the impacts that climate change adaptation has (Reckien, D. & Eisenack, K., 2013). Even though this review categorizes such topics, there is still an urgent need to provide an overview of such game studies, which can facilitate the ongoing efforts to avert potential climate catastrophes.

Figure 1. Magic circle diagram

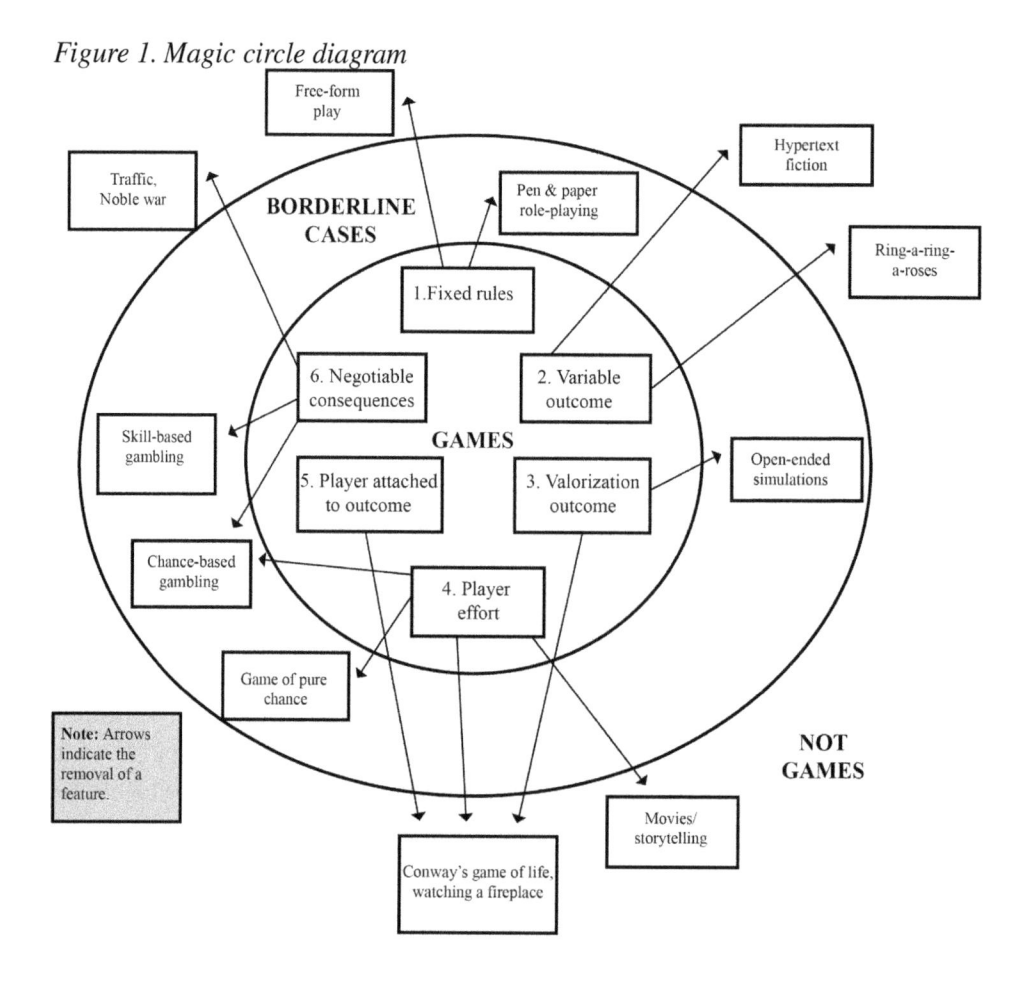

Jesper Juul's magic circle interpretation in Level Up: Digital Games Research Conference Proceedings, in Utrecht, November 4th – 6th 2003 (Juul, J., 2003).

Gamification for Climate Change Education

The research community has shown great interest in serious games, gamification and various forms of game-based learning, which can provide the foundation for further climate change knowledge learning or spreading. Despite the different definitions of the relevant game-like forms, such research emphasizes the concept that game forms and contents can be used to deploy gaming's influences on users. The first author who used the term gamification in academic research was Sebastian Deterding, who defined gamification as the use of game design elements in a non-game context (Deterding, S., Dixon, D., Khaled, R. & Nacke, L., 2011). Obviously, this definition is too general. Moreover, several gamification theories have been applied in mechanism design. For example, motivation theory, achievement goal theory, and flow theory drive extrinsic and intrinsic desire, set types of goals to enhance user engagement, or arouse positive emotions based on Csikzentmihalyi's psychology theory (Stieglitz, S. et al., 2017).

For climate change education purposes, there are a wide range of layers for games on climate change to explore. For example, the well-known board game KEEP COOL is designed with objectives that introduce climate change thermology, elucidate the relevant challenges and opportunities, and convey systemic knowledge to player participants, who can play as groups of countries (Reckien, D. & Eisenack, K., 2013; Eisenack, K., 2013). When the game feedback is being evaluated, the results indicate that the simplified climate change rules provide an overview of policy-relevant awareness. However, education is a slow and long-term process, and quantitative evidence that demonstrates whether players will select environmentally friendly policies when they conflict with their own living conditions is lacking. Therefore, players' postgame behavior concerning energy use remains unchecked, which means that such an examination should is urgently needed (Fernandez, G. D. and et.al., 2021). Compared with an abstract board game, Derek L. Kauneckis and Matthew R. Auer provide a more pedagogical focused simulation in which students are permitted to form their own policies based on the focal countries' information, and they negotiate for economic development in international relationships (Kauneckis, D. L. & Auer, M. R., 2013). To students, as realistic politics entail more pragmatistic cultures, such simulation and debate certainly cannot represent the intricacy of ecological politics completely. The real-world action game GREENIFY applies motivation theory to generate content that offers participants a playful and positive experience (Lee, J. J., Ceyhan, P., Jordan-Cooley, W. & Sung, W., 2013). In recent gamification studies, scholars have expressed the following concerns: First, gamification design should align with the objective of climate change education. Second, the strong motivation of gaming might threaten this objective. Finally, the over justification effect of tangible rewards has a negative effect on intrinsic interests (Stieglitz, S. et.al., 2017). Despite the internal design focus of gamification, research teams can contribute many valuable transformations from gameplay to practice; however, multiple methods are required to examine their effectiveness.

Verifying such effectiveness, many results show a positive correlation between a fun experience and knowledge spreading (Seebauer, S., 2014; Blanchard, O. & Buchs, A., 2015; Wu, J. S. & Lee, J. J., 2015; Rumore, D., Schenk, T. & Susskind, L., 2016; Skains, R. L., Rudd, J. A, Horry, R. & Ross, H., 2022). When climate change gameplay provides topics such as knowledge learning, civic engagement, and peer communication, its effectiveness can hardly be measured without bias and subjectivity. Game academics often use debriefing to evaluate effusiveness. In the case of a simulated model United Nations, a research team used after-game feedback as raw material to debrief its records (Matzner, N. & Herrenbrück, R., 2017). In addition, game scholars adopt measurement tools from social science and statistical science. Effectiveness is usually tested through before and after gameplay interviews, forming data clusters across players' cultural backgrounds to find subtle evidence of climate change learning (Ando, K. et.al., 2019). For instance, pregame, in-game, and postgame data have been acquired from KEEP COOL gameplay, and researchers have employed least square linear regression to measure the effectiveness of belief in climate change policies (Meya, J. N. & Eisenack, K., 2018). Additionally, the typical control experimental group is used to address learning outcomes (Waddington, D. I. & Fennewald, T., 2018). However, there is an opposite conclusion regarding behavior difference before and after gameplay, which has been deemed statistically meaningless for this particular pair of correlations (Ouariachi, T., Gutiérrez-Pérez, J. & Olvera-Lobo, M-D., 2018). Thus, gamers' behavior will generate a substantial amount of data. A research team from New Zealand has pointed out, in a review of serious games for climate change, how players' understanding and application of real-world climate change management opinions can be used as a proxy for social learning (Flood, S. and et.al., 2018). Conversely, depicting perceptions of climate change and real-world living conditions can also play a key role in affecting the data a game generates. Furthermore, when we recruit players, in some extreme cases, players even pretend to advocate a climate-friendly position that may not reflect their extravagant behavior in life. Even through a game that reduces the psychological distance from the climate tipping point (van Beek, L. et al., 2022), we cannot confirm a solid action for climate change. The various background conditions remain a massive challenge to quantifying effectiveness precisely, which makes debriefing and statistics the most reliable evaluating method over a short period at present.

Since serious games and gamification in either traditional or digital forms affect only a small proportion of the population, its target users are mainly focused on students, who are powerless to make decisions and policies. Commercial games, in contrast, have a spectacular number of players and fans. According to Newzoo's forecast, there will be 3.2 billion gamers worldwide in 2022, and this number will grow to 3.5 billion by 2025 (Wijman, T., 2022). The potential growing number of engagers drives the game market and industry to be continuously burgeoning, and this process's massive influence on climate change cannot be neglected. Therefore, by applying grounded theory, scholars have concluded that there is a preference for constructivism; specifically, players must actively engage in cognitive, emotional, and behavioral processes to construct their own climate change knowledge (Ouariachi, T. et al., 2019). When a commercial game targets entertainment, it drives players to mainly engage with the game's mechanics, narrative, and the aesthetics of its virtual environment. Therefore, Benjamin Abraham (2022) has argued that game mechanics can hardly persuade players solely through interactions. Instead, the compelling aesthetic vision of an ecological utopia is more persuasive than didactic pedagogy (Abraham, B., 2018; Abraham, B., 2022; Chang, A. Y., 2019; Crowley, E. J., Silk, M. J. & Crowley, S. L., 2021). More importantly, the game industry itself has not yet met the goal of net zero emissions in the phases of development, device production, distribution and play (Abraham, B., 2022). The consequence of the intense CO^2 emissions of the game industry has eclipsed climate change education. Therefore, it is urgent to reconsider games, gaming and the whole industry from a socioeconomic viewpoint rather than superseding the gamification effect with design.

THEORY OF LABOR VALUE BEHIND GAMING FOR DECARBONIZATION

Power of Play in Metabolism

For a long time, games have set the course for fun, pleasure, and entertainment. Gaming is, as Huizinga argues, a primitive human activity. Today, games are not only triple-A games on computers, PlayStations and Xboxes, but part of a casual revolution (Juul, J., 2010), with countless games running on mobile devices. It seems that we have overlooked the potential power of a game, the power of pursuing pleasure. However, how can we connect the power of a game with climate change? Here, I propose an imaginative assumption that exerts this power to mitigate the greenhouse effect.

Deriving from Karl Marx's materialist conception, the conceptual framework of "metabolism" defines labor as "a process between man and nature, a process by which man, through his own action, mediates, regulates and controls the metabolism between himself and nature." (Foster, J. B., 2000) Hence, human labor becomes a one-way medium, from society's reproduction to the alienation of nature. This means that human labor in our society inevitably affects the nature in which humans live; meanwhile, estranged labor estranges nature from ourselves—a manifestation of the immanent contradiction of capitalism impelling the ecosystem crisis (Saito, K., 2017). Indeed, renewable energy and sustainable materials are the key to unlocking the green planet, preventing the climate crisis from worsening. Nevertheless, is there another route to detour the one-way medium, human labor in the capitalist world? In contrast with the reproduction of human labor, there is another activity, human play, which is unproductive. As Roger Caillois (2001) states, play creates neither goods, nor wealth, nor new elements of any kind except for the exchange of property among players, ending in a situation identical to that prevailing at the beginning of game [12]. Meanwhile, its unproductive play does not necessarily prove that a game is meaningless. In contrast, games are conceived as having instrumental goals built into them, and they are viewed as essential instruments for the achievement of prolusory goals (Suits, B. H., 1978, p.147). Despite such production in digital games, play remains largely unproductive, except for electrical power consumption during play. Therefore, human play and human labor together construct the bidirectional relationship between humans and nature and complete Marx's metabolism framework under capitalism's ideology.

The power of human play is equivalent to the power of human labor. Since we have built and shaped this world tremendously, it is conceivable that the power of human play can affect this planet as well. Human play is currently represented in the billion-dollar digital game industry, which makes the power of human play an expectation and the effect on climate more achievable.

Reversed Magic Circle

Under the circumstances of global capitalism and the digital game industry, Nick Dyer-Witheford (2009, p.6) has applied immaterial labor theory to illustrate gaming behavior. As he argues, virtual play, in the commodity form, involves a constant infusion of energy from players' vitality. Therefore, while avoiding harvest by capital, how can autonomy's capacity redeem the "playbor force (Dyer-Witheford, N. & Peuter, Gd., 2009; Kücklich, J. R., 2005)" to mitigate climate change? He contends that games potentially and critically enlighten, although the slow uptake of climate change as a game theme reveals the continuing ambivalence toward this prospect (Dyer-Witheford, N. & de Peuter, G., 2021). As discussed above, serious games, gamification and commercial games implementing climate change themes all have varying degrees of uncertainty of effectiveness.

Simplifying Juul's interpretation of magic circles, we find that games, broadline cases and not-games are situated in axipetal order. Beneath this axipetal order, there is an incremental feeling of pull, which is a subjective experience from outside to the center that depends on whether you are willing to give a game the time it asks for (Juul, J., 2010). On the one hand, it appears that sought-after games, usually triple-A commodities and popular casual games, possess an intense feeling of pull for potential players. On the other hand, gamification, according to the general definition, should overlap the areas of games and broadline cases, which usually lack the strong feeling of pull when fulfilling an education purpose. As a result, the power of play gradually intensifies as it approaches the center of the magic circle, which usually contains commercial game design elements, such as fixed rules, variable outcomes, valorization of outcomes, player effort, player attachment to outcomes, and negotiable consequences. Certainly, gamification can also employ these elements for a better playful experience. Nonetheless, the education agenda embedded in the instrumental goals of a game creates an ambivalent lusory experience. This ambivalent lusory experience leads to a threatened objective in terms of both education and entertainment. If we turn this magic circle inside out, then the gamification type becomes located in the center of the circle, while a game with various design elements for pure pleasure is located outside the circle. Then, the power of play also reverses in axipetal order. Therefore, the reversed magic circle makes gaming for climate change possible theoretically. Gaming as a behavior, including games on Steam, mobile phone games, and console games, can be designed to serve the purpose of climate change. Gamification can become a designed organizational social initiative, facilitating behavior-shaping for decreasing CO^2 emissions (Fig. 2). To achieve decarbonization by gaming, there are some key points to note when designing the core of the reversed magic circle: a) the design of this organizational initiative should reflect the objective of CO^2 emissions control through both the game production and distribution phases, limiting power consumption during game development; b) the gamification of this organizational initiative should make players attached to the outcome in a long-term process, allowing either serious or casual players to feel an accomplishment when reducing their carbon footprint; c) the games on different platforms, including Steam, should connect to the same grid of gamification, set for power consumption calculation and carbon credit exchange, thereby establishing a vigorous carbon market for trading and monetization. In short, gamification should be set to organize gaming behavior, making gaming a potential source of carbon credit and thus controlling millions of gamers' power use habits through a virtual carbon credit market.

In many cases, these carbon credits are designed as a virtual currency to encourage consumption, which is often seen in the game design and gamification design. The carbon credits share similarities with the money in virtual economies. In the early days of Chinese Internet, Tencent started its virtual currency project as Q coin, which can be used to purchase virtual goods like avatars, outfits and other properties (Lehdonvirta, V. and Castronova, E., 2014, p.15). The Q coin even went into the market for exchanges, which provides the opportunity for Tencent instant messenger (QQ) users to purchase with real money. This monetizing strategy has been widely accepted nowadays, and it can be designed to monetized the carbon credits to purchase goods in reality. For instance, Alibaba Ant Forestry is trying to use a gamification design to encourage users to select green behavior among 5 categories and 54 options, including green and low-carbon transportation, online business, cyclic utilization, reducing the usage of paper and plastic products, and energy saving. Each category offers various behavior options to collect virtual currency. And the more currency you collect, users will meet the requirements of monetizing, which allows users to redeem the farm products that come from the western desert areas. Based on the macroeconomic metaphors of wheels and pipes, the monetizing gamification design, on the one hand, the wheels establish the internal part of the virtual wealthy, encompassing everything that happens from the point that virtual money and goods enter the system through faucets up to the point that virtual money and goods leave the system through sinks; on the other hand, the pipes model, virtual stuff flows into the economy via a faucet at the top, whirls around inside for a while, and then flows out of the economy via a sink at the bottom (Lehdonvirta, V. and Castronova, E., 2014, pp.223-225). The macroeconomics in the virtual system empowers the "managers", who design and control the virtual currency, making the carbon credits monetizing possible. Furthermore, the game company uses an action-reward cycle to keep the user active for as long as possible to stimulate them to spend money in the game's virtual space. This design includes rewards, punishments, loot boxes, social incentives, and altruism, all of which are optimized to increase daily and monthly activity and even spending. Through these methods and design strategies, game companies not only occupy the daily leisure time of players. But more importantly, it converts real-world money into game currency to buy virtual time and goods, so that virtual time and goods become a cultural symbol of the real world. As a result, these cultural symbols of the virtual world have influenced people's attitudes toward game consumption in the real world.

    The initiative of gaming exerts the power of play on decarbonization. It elaborates a process other than human labor, i.e., unproductive human play, from nature to human in the metabolism framework. In that sense, although playing a game does not generate goods, it exchanges credit for CO^2 emissions among players. If a player wants to play more games more times, he or she must recalibrate his or her power consumption habits, saving more energy for gaming. Casual players can also save energy and earn more credit for their CO^2 emissions, which balances the time of play among the degrees of gamers.

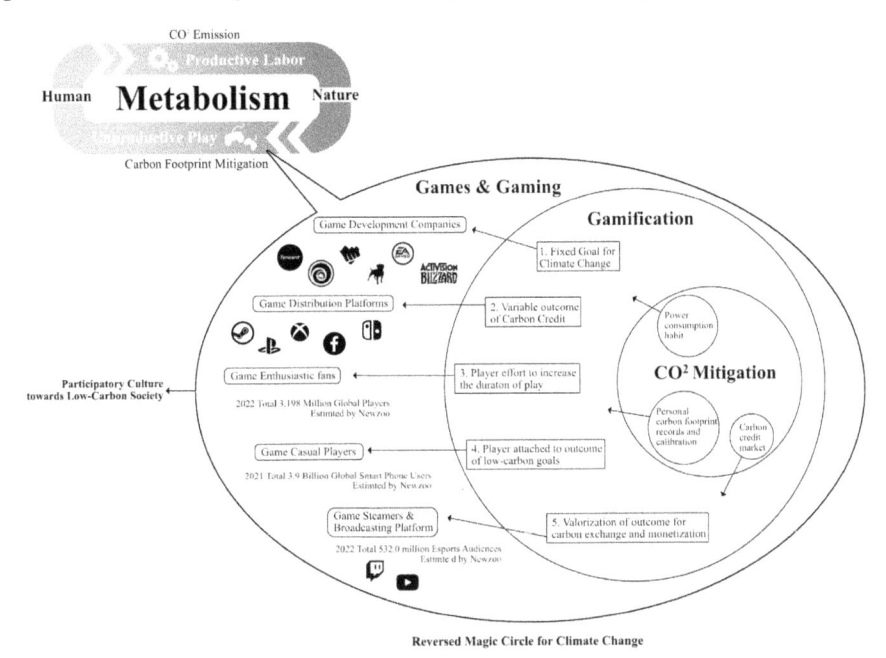

Figure 2. Reversed magic circle diagram for climate change

Under the metabolism framework, gamification of games and gaming mitigate the greenhouse effect.

DISCUSSION AND CONCLUSION

According to the labor theory of value, the process by which humans create value is fundamentally a process of labor production. Marx's framework of Metabolism establishes a cyclical relationship between nature and human labor. Historically, humans have extracted resources from nature to facilitate labor production, thereby obtaining the resources necessary for survival. This cyclical relationship, however, overlooks how nature reciprocates and provides feedback to humanity. Consequently, the unilateral extraction of natural resources to support labor production disrupts the balance between nature and humanity. This imbalance manifests directly as the rapid depletion and reduction of natural resources and indirectly as ecological imbalance, environmental degradation, and increasingly severe climate change. These feedback mechanisms from nature exacerbate the negative cycle within Marx's Metabolism framework. For example, in the context of climate change, humans extract resources from nature for production materials, industrial production and processing emit substantial amounts of CO_2. Not only does industrial production generate CO_2, but industrial products such as cars and airplanes continue to emit various greenhouse gases. These greenhouse gases, a byproduct of human labor production, negatively impact on the natural environment within the Metabolism framework. In contrast to these cyclical influences, nature's negative feedback can only be balanced through non-labor production means. But how can non-labor production positively impact nature? This paper posits that play, a non-labor activity, neither participates in production activities nor produces goods. Play, through its non-productive actions, can reduce CO_2 emissions. According to the Magic Circle theory, games are a negotiated activity occurring in a virtual time and space. However, at the boundary between game and the real world, there exists a category of game forms that straddle fiction and reality. These forms can transcend the boundaries of fiction and directly influence real-world people and events. For instance, gambling is a game that directly impacts real-world finances and wealth. Activities that lie between games and non-games significantly involve play. While play occurs in fictional times and space, it produces real-world effects.

Guided by this theoretical framework, this paper argues that gamification can be viewed as a medium of non-labor, non-productive activities. Through this medium, player participation will help build a low-carbon society. Specifically, the engagement behaviors in gamification, facilitated by digital gaming platforms, will promote the consumption of low-carbon products, enhance low-carbon industries, and strengthen the momentum of green industries. Gamification uses game mechanics to further encourage the formation of a participatory culture. Simultaneously, players' participation behaviors and culture will standardize green, low-carbon economic consumption behaviors, establishing a sustainable cyclical system. Through the non-labor activity of play, using digital platforms provided by gaming companies as a medium, the consumption behaviors of low-carbon products and green production will stabilize within the relationship between nature and humanity. Ultimately, this non-productive play participatory culture will mitigate the vicious cycle of humans continuously extracting natural resources for labor production, thereby liberating the human-nature relationship from the imbalance caused by CO_2 emissions and climate change. Thus, Marx's Metabolism framework identifies the counterpart to labor production, namely non-labor play. The Magic Circle in game theory implies that games are not only fictional activities but can also have real-world impacts, influencing human society and the natural environment. Research on Gamification highlights the effectiveness of environmental education, suggesting that digital gaming platforms and collective player participation can guide low-carbon behaviors, low-carbon consumption, and the construction of a green economy. This achieves a rebalancing of the climate impacts caused by labor production through play.

Since almost all modern societies depend on the extraction of natural resources, similar to the original depiction of the metabolism process, controversies exist between environmentalists and large businesses (Diamond, J., 2005). Even the metaverse and the AI is heavily relied on the energy consumption, which is not the so-called "green" industry as we imagined (Liu, Y., 2023). Hence, the game industry is also an energy-consuming one, that does not fit the scope of low-carbon production. Meanwhile, gaming, if used as an inverted metabolism process, can release the primitive power of play and provide an alternative where environmentalism, capitalism, and society coincide in a common interest. Although games are widely accepted as amusement media, there are still limitations to burdening them with climate challenges. The most pressing one is that the initiative of gaming with massive numbers of participants requires the collaboration of game companies, distribution platforms, and players worldwide. Another challenge limiting the potential power of play is individual carbon footprints, which remain ambiguous for recordings.

Despite these limitations, through the medium of play, we are revealing an expected and ecological future for participatory culture (Jenkins, H., 2006), which has already appeared during media development in the form of fandom within disciplines such as film, blogs, and popular culture. Future research should rigorously investigate the potential of game participation culture to serve as an educational tool for consumers, specifically focusing on its ability to foster the consumption of low-carbon products and contribute to the construction of a sustainable green economy. From a social science perspective, researchers could examine the sociocultural dynamics and community engagements fostered by gaming environments that advocate for green consumption patterns. And the design studies and game studies can facilitate the empirical research to the argument that aesthetics, mechanics and the dynamics between players and games can influence green and low carbon products consumption. Additionally, future studies should delve into the role of digital game distribution platforms in guiding and educating consumers about low-carbon and green products. This involves evaluating the efficacy of various platform features, such as in-game advertisements, virtual rewards for sustainable choices, and community-driven initiatives that highlight green products. By integrating interdisciplinary approaches, researchers can comprehensively assess how these platforms not only disseminate information about eco-friendly products but also actively shape consumer behaviors and social norms towards sustainability.

The complete game industry, which consists of game players, game manufacturers and game distribution platforms, is not only striving to transform itself to low carbon, but also plays an important role in building a participatory economy, a participatory group and a participatory culture as an important carrier of game participation culture. Therefore, the game industry should play a role in the relationship between nature and humans to provide a play platform, build a play culture, and guide play behavior. This function will balance the labor and production relationship between human beings and nature, slow down and even reduce the climate impact caused by labor production, transform non-labor behaviors into green consumption behaviors and culture, and realize the recyclable and sustainable development of metabolism. All in all, we have already failed once, when encountering the pandemic; we cannot afford such a failure to prevent further and mitigate existing climate change. This demands that we act with great imagination, vision, ambition, and courage via participation, cooperation and collaboration. In conclusion, gaming possesses enormous power of play theoretically, and the urgency of climate change compels us to realize it practically, without any hesitation.

REFERENCES

Abraham, B. (2018). Video Game Visions of Climate Futures:ARMA 3 and Implications for Games and Persuasion. *Games and Culture*, 13(1), 71–91. DOI: 10.1177/1555412015603844

J. Abraham, B. (2022). *Digital games after climate change*. Palgrave Macmillan.

Ando, K., Sugiura, J., Ohnuma, S., Tam, K.-P., Hübner, G., & Adachi, N. (2019). Persuasion Game: Cross Cultural Comparison. *Simulation & Gaming*, 50(5), 532–555. DOI: 10.1177/1046878119880236

Blanchard, O., & Buchs, A. (2015). Clarifying Sustainable Development Concepts Through Role-Play. *Simulation & Gaming*, 46(6), 697–712. DOI: 10.1177/1046878114564508

Buck, M., Sturzaker, J., & Mell, I. (2023). Playing games around climate change - new ways of working to develop climate change resilience. *Journal of Planning Literature*, 38(4), 622–622.

Caillois, R. (2001). *Man, play and games*. University of Illinois Press.

Chang, A. Y. (2019). *Playing Nature: Ecology in Video Games*. University of Minnesota Press. DOI: 10.5749/j.ctvthhd94

Crowley, E. J., Silk, M. J., & Crowley, S. L. (2021). The educational value of virtual ecologies in Red Dead Redemption 2 [Article]. *People and Nature*, 3(6), 1229–1243. DOI: 10.1002/pan3.10242

Deterding, S., Dixon, D., Khaled, R., & Nacke, L. (2011). From game design elements to gamefulness: defining "gamification" *Proceedings of the 15th International Academic MindTrek Conference: Envisioning Future Media Environments*, Tampere, Finland. DOI: 10.1145/2181037.2181040

Diamond, J. (2005). *Collapse: How societies choose to fail or succeed*. Penguin Books.

Diaz, D., & Moore, F. (2017). 2017/11/01). Quantifying the economic risks of climate change. *Nature Climate Change*, 7(11), 774–782. DOI: 10.1038/nclimate3411

Douglas, B. D., & Brauer, M. (2021). Gamification to prevent climate change: A review of games and apps for sustainability. *Current Opinion in Psychology*, 42, 89–94. DOI: 10.1016/j.copsyc.2021.04.008 PMID: 34052619

Dyer-Witheford, N., & de Peuter, G. (2021, May). Postscript: Gaming While Empire Burns. *Games and Culture, 16*(3), 371-380. *Article*, 1555412020954998. Advance online publication. DOI: 10.1177/1555412020954998

Dyer-Witheford, N., & Peuter, G. d. (2009). *Games of empire: Global capitalism and video games*. University of Minnesota Press.

Eisenack, K. (2013). A Climate Change Board Game for Interdisciplinary Communication and Education. *Simulation & Gaming*, 44(2-3), 328–348. DOI: 10.1177/1046878112452639

Fernandez Galeote, D., Rajanen, M., Rajanen, D., Legaki, N.-Z., Langley, D. J., & Hamari, J. (2021, June). Gamification for climate change engagement: Review of corpus and future agenda. *Environmental Research Letters*, 16(6), 063004. Advance online publication. DOI: 10.1088/1748-9326/abec05

Flood, S., Cradock-Henry, N. A., Blackett, P., & Edwards, P. (2018). 2018/06/01). Adaptive and interactive climate futures: Systematic review of 'serious games' for engagement and decision-making. *Environmental Research Letters*, 13(6), 063005. DOI: 10.1088/1748-9326/aac1c6

Foster, J. B. (2000). *Marx's ecology: Materialism and Nature*. Monthly Review Press.

Gerber, A., Ulrich, M., Wäger, F. X., Roca-Puigròs, M., Gonçalves, J. S. V., & Wäger, P. (2021). Games on Climate Change: Identifying Development Potentials through Advanced Classification and Game Characteristics Mapping. *Sustainability (Basel)*, 13(4), 1997. DOI: 10.3390/su13041997

Harker-Schuch, I. E. P., Mills, F. P., Lade, S. J., & Colvin, R. M. (2020). CO2peration-Structuring a 3D interactive digital game to improve climate literacy in the 12-13-year-old age group. *Computers & Education*, 144, 103705. DOI: 10.1016/j.compedu.2019.103705

Huizinga, J. (1949). *Homo ludens: A study of the play-element in culture*. Routledge.

Jenkins, H. (2006). *Fans, bloggers, and games: Exploring participatory culture*. New York University Press.

Juul, J. (2003). *The Game, the Player, the World: Looking for a Heart of Gameness*. https://www.jesperjuul.net/text/gameplayerworld/

Juul, J. (2005). *Half real: Video games between real rules and fictional worlds*. The MIT Press.

Juul, J. (2010). *A casual revolution: Reinventing video games and their players*. The MIT Press.

Kauneckis, D. L., & Auer, M. R. (2013). A Simulation of International Climate Regime Formation. *Simulation & Gaming*, 44(2-3), 302–327. DOI: 10.1177/1046878112470542

Kücklich, J. R. (2005). Precarious Playbour: Modders and the Digital Games Industry. *The Fibreculture Journal*(5).

Kwok, R. (2019). Can climate change games boost public understanding? *Proceedings of the National Academy of Sciences of the United States of America*, 116(16), 7602–7604. DOI: 10.1073/pnas.1903508116 PMID: 30992396

Lee, J. J., Ceyhan, P., Jordan-Cooley, W., & Sung, W. (2013). GREENIFY: A Real-World Action Game for Climate Change Education. *Simulation & Gaming*, 44(2-3), 349–365. DOI: 10.1177/1046878112470539

Lehdonvirta, V., & Castronova, E. (2014). *Virtual economies: Design and analysis*. The MIT Press. DOI: 10.7551/mitpress/9525.001.0001

Liu, Y. (2023). Ecological Thinking about the Metaverse from a Posthumanist Perspective. *Critical Arts*, •••, 1–18. DOI: 10.1080/02560046.2023.2276423

Martin, P. (2018). The intellectual structure of game research. *Game Studies, 18*(1). https://gamestudies.org/1801/articles/paul_martin

Martinich, J., & Crimmins, A. (2019). 2019/05/01). Climate damages and adaptation potential across diverse sectors of the United States. *Nature Climate Change*, 9(5), 397–404. DOI: 10.1038/s41558-019-0444-6 PMID: 31031825

Matzner, N., & Herrenbrück, R. (2017). Simulating a Climate Engineering Crisis: Climate Politics Simulated by Students in Model United Nations. *Simulation & Gaming*, 48(2), 268–290. DOI: 10.1177/1046878116680513

Mejean, A., Pottier, A., Fleurbaey, M., & Zuber, S. (2020, November). Catastrophic climate change, population ethics and intergenerational equity [Article]. *Climatic Change*, 163(2), 873–890. DOI: 10.1007/s10584-020-02899-9

Meya, J. N., & Eisenack, K. (2018). Effectiveness of gaming for communicating and teaching climate change. *Climatic Change*, 149(3-4), 319–333. DOI: 10.1007/s10584-018-2254-7

NASA. (2022). *GISS Surface Temperature Analysis (v4)* https://data.giss.nasa.gov/gistemp/maps/

Nations, U. (2007). *From Stockholm to Kyoto: A Brief History of Climate Change*. https://www.un.org/en/chronicle/article/stockholm-kyoto-brief-history-climate-change

Neset, T. S., Andersson, L., Uhrqvist, O., & Navarra, C. (2020). Serious Gaming for Climate Adaptation-Assessing the Potential and Challenges of a Digital Serious Game for Urban Climate Adaptation. *Sustainability (Basel)*, 12(5), 1789. DOI: 10.3390/su12051789

Ouariachi, T., Gutiérrez-Pérez, J., & Olvera-Lobo, M.-D. (2018). 2018/09/02). Can serious games help to mitigate climate change? Exploring their influence on Spanish and American teenagers' attitudes / ¿Pueden los serious games ayudar a mitigar el cambio climático? Una exploración de su influencia sobre las actitudes de los adolescentes españoles y estadounidenses. *PsyEcology*, 9(3), 365–395. DOI: 10.1080/21711976.2018.1493774

Ouariachi, T., Olvera-Lobo, M. D., Gutiérrez-Pérez, J., & Maibach, E. (2019). 2019/05/04). A framework for climate change engagement through video games. *Environmental Education Research*, 25(5), 701–716. DOI: 10.1080/13504622.2018.1545156

Parry, M. L., Rosenzweig, C., Iglesias, A., Livermore, M., & Fischer, G. (2004, April). Effects of climate change on global food production under SRES emissions and socio-economic scenarios. *Global Environmental Change*, 14(1), 53–67. DOI: 10.1016/j.gloenvcha.2003.10.008

Reckien, D., & Eisenack, K. (2013). Climate Change Gaming on Board and Screen: A Review. *Simulation & Gaming*, 44(2-3), 253–271. DOI: 10.1177/1046878113480867

Rumore, D., Schenk, T., & Susskind, L. (2016). Role-play simulations for climate change adaptation education and engagement. *Nature Climate Change*, 6(8), 745–750. DOI: 10.1038/nclimate3084

Saito, K. (2017). *Karl Marx's ecosocialism: Captialism, nature, and the unfinished critique of political economy*. Monthly Review Press. DOI: 10.2307/j.ctt1gk099m

Seebauer, S., & Ieee. (2013). Measuring climate change knowledge in a social media game with a purpose. Paper presented at the *5th International Conference on Games and Virtual Worlds for Serious Applications (VS-GAMES)*, Bournemouth Univ, Bournemouth, ENGLAND. DOI: 10.1109/VS-GAMES.2013.6624236

Seebauer, S. (2014). Validation of a social media quiz game as a measurement instrument for climate change knowledge. *Entertainment Computing*, 5(4), 425–437. https://doi.org/https://doi.org/10.1016/j.entcom.2014.10.007. DOI: 10.1016/j.entcom.2014.10.007

Skains, R. L., Rudd, J. A., Horry, R., & Ross, H. (2022, January 27). Playing for Change: Teens' Attitudes Towards Climate Change Action as Expressed Through Interactive Digital Narrative Play. *Frontiers in Communication*, 6, 789824. Advance online publication. DOI: 10.3389/fcomm.2021.789824

Stieglitz, S., Lattemann, C., Robra-Bissantz, S., Zarnekow, R., & Brockmann, T. (Eds). (2017). *Gamification: Using Game Elements in Serious Contexts*. Springer.

Suits, B. H. (1978). *The grasshopper: Games, life and utopia*. University of Toronto Press. DOI: 10.3138/9781487574338

Sutton-Smith, B. (1997). *The ambiguity of play*. Harvard University Press.

Tol, R. S. J. (2009). Spr). The Economic Effects of Climate Change. *The Journal of Economic Perspectives*, 23(2), 29–51. DOI: 10.1257/jep.23.2.29

van Beek, L., Milkoreit, M., Prokopy, L., Reed, J. B., Vervoort, J., Wardekker, A., & Weiner, R. (2022). 2022/02/16). The effects of serious gaming on risk perceptions of climate tipping points. *Climatic Change*, 170(3), 31. DOI: 10.1007/s10584-022-03318-x

Waddington, D. I., & Fennewald, T. (2018). Grim FATE: Learning About Systems Thinking in an In-Depth Climate Change Simulation. *Simulation & Gaming*, 49(2), 168–194. DOI: 10.1177/1046878117753498

Wiedenhofer, D., Guan, D., Liu, Z., Meng, J., Zhang, N., & Wei, Y.-M. (2017). 2017/01/01). Unequal household carbon footprints in China. *Nature Climate Change*, 7(1), 75–80. DOI: 10.1038/nclimate3165 PMID: 29375673

Wijman, T. (2022). *The Games Market Will Show Strong Resilience in 2022, Growing by 2.1% to Reach $196.8 Billion* Newzoo. Retrieved Aug 6th from https://newzoo.com/insights/articles/the-games-market-will-show-strong-resilence-in-2022

Wu, J. S., & Lee, J. J. (2015). 2015/05/01). Climate change games as tools for education and engagement. *Nature Climate Change*, 5(5), 413–418. DOI: 10.1038/nclimate2566

Chapter 14
Future Trends and Innovations in Environmental Governance for Sustainable Development

K. Anitha
https://orcid.org/0000-0002-1940-2101
Meenakshi Academy of Higher Education and Research, India

Indrajit Ghosal
https://orcid.org/0000-0003-0744-2672
Brainware University, India

J. Amala
https://orcid.org/0009-0002-9448-7106
DDGD Vaishnav College, India

Imran Hossain
Varendra University, Bangladesh

ABSTRACT

In response to escalating environmental challenges, this research explores emerging trends and innovations poised to revolutionize environmental governance for sustainable development. Key areas of focus include digital technology integration, green finance, circular economy models, community-driven governance, and international cooperation. Digital innovations, particularly artificial intelligence (AI) and

DOI: 10.4018/979-8-3693-7001-8.ch014

blockchain, enhance environmental monitoring, ensure transparency, and foster trust in governance processes. Green finance channels investments towards sustainable projects, with tools like green bonds facilitating a transition to low-carbon economies. The circular economy, emphasizing resource efficiency and waste reduction, is vital for sustainable growth, while community-driven governance empowers local involvement in decision-making for more responsive and innovative solutions. The chapter emphasizes a multifaceted approach in creating resilient and adaptive governance frameworks capable of addressing today's environmental crises.

1 INTRODUCTION

The escalating environmental crises, including climate change, biodiversity loss, and pervasive pollution, have underscored the urgent need for transformative and effective environmental governance. Traditional governance frameworks, often linear and fragmented, have struggled to keep pace with global environmental degradation's multifaceted and interdependent challenges. The complexity of these issues necessitates innovative governance models that can bridge gaps, anticipate future challenges, and foster resilience within ecosystems and societies alike as the global community shifts towards sustainability, emerging trends, and innovations in environmental governance present promising avenues for enhancing environmental stewardship.

This study aims to explore the future directions and innovative mechanisms that can bolster sustainable development, focusing on five key areas: the integration of digital technologies, the rise of green finance, the adoption of circular economy models, the role of community-driven governance, and the importance of international cooperation. Each of these elements offers a strategic pathway to reimagining governance structures that are not only more responsive but also more adaptive to the evolving environmental landscape.

The application of digital technologies, including artificial intelligence (AI), big data analytics, and blockchain, has the potential to revolutionize how environmental data is collected, analyzed, and utilized in policymaking (Wang et al., 2020). Green finance, another pivotal area, is emerging as a critical tool in directing capital flows toward sustainable projects and reducing the carbon footprint of industries (Zhang & White, 2021). Furthermore, circular economy models, which emphasize the reuse, recycling, and regeneration of resources, offer an alternative to the traditional linear economic model and align closely with sustainability objectives (Geissdoerfer et al., 2017).

Community-driven governance is increasingly recognized for its ability to engage local stakeholders in decision-making processes, ensuring that environmental policies are more inclusive and equitable (Ostrom, 2009). Finally, international cooperation remains a cornerstone of environmental governance, as transboundary environmental issues require coordinated efforts at the global level (Biermann & Pattberg, 2012). By examining these emerging trends and innovations, this study seeks to contribute to the ongoing discourse on how governance systems can evolve to meet the challenges of sustainable development in the 21st century.

1.1 Background and Significance

The Anthropogenic era is marked by noteworthy human impact on the Globe's geology and ecosystems, demanding a revolution in how we oversee environmental resources. Traditional governance methods often lack the flexibility and receptiveness required to deal with rapidly changing environmental conditions. Subsequently, there is a mounting interest in leveraging new technologies, financial mechanisms, and governance models to augment environmental sustainability. Digitalization such as AI and blockchain is reforming environmental monitoring and administration. AI's prognostic capabilities enable more operative environmental monitoring, permitting for timely and hands-on measures against effluence and climate change (Uriarte-Gallastegi N et.al, 2024). Blockchain technology enhances transparency and liability in environmental data management, nurturing trust, and competence in governance processes (Kshetri, N. 2021). This alignment promotes businesses to use environmentally friendly practices while simultaneously promoting sustainable projects (Anitha, K., 2024). A sustainable replacement for the conventional linear economy is offered by the circular economy model, which emphasizes waste reduction and resource efficiency. The circular economy minimizes its impact on the environment and encourages sustainable growth by rethinking production and consumption patterns to maximize the efficiency of resources.

The capacity of community-driven governance models to promote sustainable development through the inclusion of local communities in environmental decision-making is becoming more widely acknowledged. By ensuring that governance structures are more sensitive to the unique requirements and circumstances of various regions, this bottom-up strategy improves the legitimacy and efficacy of environmental policies (Enayat A. Moallemi, et.al, 2020). Addressing global environmental issues that cut beyond national boundaries requires international cooperation. To encourage a coherent and well-coordinated worldwide response, stronger international agreements and creative financing methods are essential (Kyriakopoulos, G.L. 2022). Technology transfer and capacity building are two aspects of this cooperation that help sustainable global development. This study's main goal is to find and evaluate new developments in environmental governance that have the potential to promote sustainable development. Moreover, the study also aims to investigate how blockchain and artificial intelligence may improve data management, environmental monitoring, and governance procedures.

2. LITERATURE REVIEW

2.1 Environmental Governance

Environmental governance refers to the frameworks, policies, institutions, and processes through which societies manage environmental issues. Traditionally, it has involved a top-down, state-centered approach, where governments impose regulations to control the use of natural resources and mitigate environmental degradation (Xing G, Zhang Y, Guo J, 2023). However, the complexity of contemporary environmental challenges, such as climate change, biodiversity loss, and pollution, requires more integrative and adaptive governance models. Scholars increasingly advocate for multi-level governance systems that involve various stakeholders, including governments, the private sector, civil society, and international organizations, working collaboratively to manage environmental issues (Wang, H., & Ran, B. 2021). This participatory approach fosters the inclusion of diverse perspectives and enhances the legitimacy of environmental policies.

In recent years, there has been a growing emphasis on adaptive governance, which promotes flexibility in decision-making to address the dynamic and unpredictable nature of environmental systems. This approach draws on the principles of resilience theory, emphasizing the need for governance systems that can learn from past experiences and adjust in response to emerging challenges (Ungar, M.,2018). Additionally, the shift towards polycentric governance has been significant, where multiple centers of decision-making authority operate at different scales (Heikkila, T., & Weible, C. M., 2018). This decentralized structure enables more localized responses to environmental problems while maintaining coordination across levels of governance.

The rise of global environmental governance has also been critical in addressing transboundary environmental issues, such as climate change. International agreements, such as the Paris Agreement, have been pivotal in fostering global cooperation (Hernández Guzmán, D., & Hernández García de Velazco, J.,2023). However, challenges remain, particularly in ensuring compliance and addressing disparities between developed and developing countries in terms of capacity to implement environmental policies. The literature highlights the need for stronger enforcement mechanisms and more equitable frameworks to ensure that environmental governance systems contribute effectively to sustainable development.

2.2 Artificial Intelligence

Artificial Intelligence (AI) is transforming various sectors, including environmental governance, by offering advanced tools for data collection, analysis, and decision-making. AI's ability to process vast amounts of data and provide predictive insights has made it an essential tool for monitoring and managing environmental issues, from climate change to biodiversity loss (DB Olawade et. al., 2024). AI technologies, such as machine learning and big data analytics, enable real-time tracking of environmental parameters, enhancing the precision and timeliness of policy responses. For instance, AI is used to monitor deforestation through satellite imagery and to predict the impacts of climate change through sophisticated climate models (Wang et al., 2020).

The integration of AI into environmental governance has also facilitated the rise of smart environmental management systems, which allow for the automation of monitoring and reporting processes. This has been particularly beneficial in industries such as energy and agriculture, where AI can optimize resource use and reduce environmental footprints. For example, AI-driven precision agriculture helps farmers minimize water and pesticide use, thereby reducing environmental impacts (T Talaviya et al., 2020). Additionally, AI has been instrumental in the development of circular economy models, where AI-powered systems help optimize resource loops by predicting product lifecycles and identifying materials for reuse and recycling (MS Pathan et al., 2023).

However, the literature also highlights challenges in the application of AI to environmental governance. Ethical concerns, such as data privacy and the risk of exacerbating existing inequalities, must be addressed (R Rodrigues, 2020). Moreover, the development of AI technologies is energy-intensive, raising concerns about the carbon footprint of AI itself. As AI continues to evolve, there is a need for governance frameworks that ensure the responsible and sustainable deployment of these technologies, balancing innovation with the broader goals of environmental sustainability.

2.3 Sustainability

Sustainability has become a central focus in environmental governance, emphasizing the need to balance economic development, environmental protection, and social equity. The concept gained prominence with the 1987 Brundtland Report, which defined sustainable development as "development that meets the needs of the present without compromising the ability of future generations to meet their own needs" (WCED, 1987). Since then, sustainability has evolved into a multi-dimensional framework encompassing environmental, social, and economic pillars. In this context, environmental sustainability focuses on the responsible management of natural resources to avoid long-term ecological degradation (TC Lee et.al., 2021).

The literature on sustainability highlights the critical role of corporate sustainability practices in achieving broader sustainability goals. Businesses are increasingly adopting sustainable practices, such as reducing carbon emissions, transitioning to renewable energy, and embracing circular economy models, to mitigate their environmental impact (MV Barros et. al., 2021). These corporate initiatives are often driven by a combination of regulatory pressures, market demands, and reputational concerns. Additionally, the rise of green finance has facilitated investments in sustainable projects, further integrating sustainability into economic systems (Zhang & White, 2021).

At the policy level, sustainable development goals (SDGs) have become a global framework for addressing sustainability challenges. The United Nations' 2030 Agenda outlines 17 SDGs that emphasize the interconnectedness of environmental, social, and economic sustainability (UN, 2015). However, achieving these goals requires overcoming significant challenges, including the need for better coordination across governance levels and sectors. The literature also highlights the importance of behavioral change in driving sustainability efforts. Consumer demand for sustainable products and services has grown, but aligning consumption patterns with sustainability goals remains a key challenge (Mensah, J., & Ricart Casadevall, S. (2019).

Overall, the sustainability literature underscores the need for a holistic approach to governance that integrates environmental, economic, and social considerations. As global challenges such as climate change and resource depletion intensify, sustainability must remain at the forefront of both policy and practice.

3. EMERGING TRENDS IN ENVIRONMENTAL GOVERNANCE

3.1 Overview of Current Environmental Governance Models

Environmental governance refers to the frameworks, institutions, and policies that guide how societies manage natural resources and address environmental issues. Current models are diverse, ranging from top-down regulatory approaches to collaborative and multi-stakeholder governance. Traditional governance models typically involve state-led mechanisms where governments implement environmental policies through legislation and enforcement. These often focus on compliance with regulations such as emission controls, resource management, and biodiversity conservation (Lockwood et al., 2021).

In contrast, more recent models emphasize participatory approaches involving local communities, businesses, and civil society. Collaborative governance, for example, seeks to engage diverse stakeholders in decision-making processes, fostering transparency and accountability (Brockhaus et al., 2021). Market-based mechanisms, such as carbon pricing and green certifications, have also gained prominence, encouraging private sector engagement in environmental stewardship. Overall, governance models are evolving towards more inclusive and flexible systems that integrate sustainability goals while balancing economic and social development.

3.2 Drivers of Change in Environmental Governance

Several factors are driving changes in environmental governance, reshaping how institutions and societies respond to environmental challenges. One key driver is climate change, which has led to a global reevaluation of policies and governance frameworks. The increasing frequency and severity of climate-related disasters have pushed governments and organizations to prioritize adaptation and resilience in their environmental strategies (Meadowcroft, 2021).

Technological advancements, such as big data, remote sensing, and AI, are also transforming governance models by providing better tools for monitoring environmental conditions and enforcing regulations. Another significant drive is the growing demand for sustainability from both consumers and investors. Corporations are under pressure to adopt green practices, driven by both regulatory mandates and consumer preferences for eco-friendly products (Wright & Nyberg, 2021). Additionally, global environmental agreements like the Paris Agreement have fostered cooperation among nations, further pushing changes in governance structures to meet international environmental commitments.

4. THEORETICAL FRAMEWORK FOR UNDERSTANDING INNOVATIONS AND TRENDS IN ENVIRONMENTAL GOVERNANCE

This study explores future trends and innovations in environmental governance by employing an integrative theoretical framework that includes Social-Ecological Systems (SES) Theory, Sustainability Science, and Ecological Economics Theory. These theories collectively offer a holistic approach to understanding how governance innovations can address environmental, social, and economic challenges in a manner that promotes sustainable development.

1. **Social-Ecological Systems (SES) Theory**

Social-Ecological Systems (SES) Theory posits that human and ecological systems are deeply interconnected, forming complex and adaptive systems (Berkes et al., 2000). SES theory emphasizes the idea that environmental challenges cannot be separated from social dynamics, including governance, economic policies, and cultural practices. For example, the over-exploitation of resources such as deforestation or overfishing is not just an ecological problem but also a societal issue that requires governance interventions, policy shifts, and stakeholder collaboration (Folke et al., 2005).

This theory is essential for understanding how innovations in governance, such as decentralized decision-making, community-based resource management, and adaptive governance models, can offer solutions to the complexity of global environmental problems (Ostrom, 2009). The SES framework encourages resilience-building through participatory governance, enabling systems to adapt to environmental changes while maintaining ecosystem services crucial for human well-being.

2. **Sustainability Science**

Sustainability Science is an interdisciplinary field that combines ecological, social, and economic knowledge to solve sustainability challenges (Kates et al., 2001). It moves beyond traditional disciplinary boundaries by focusing on the interactions between human and environmental systems. Sustainability science is centered on how to achieve long-term ecological balance while ensuring human welfare and social equity. It integrates technological innovations, policy reforms, and stakeholder engagement to create pathways that promote sustainable development (Clark & Dickson, 2003).

This framework is particularly relevant when exploring innovations such as green finance, circular economy models, and AI-driven environmental management systems. These innovations are designed to align economic development with environmental sustainability, balancing the need for economic growth with ecological preservation and social well-being. Sustainability science provides the lens to assess how these governance trends can contribute to meeting global sustainability goals, such as the United Nations' Sustainable Development Goals (SDGs).

3. **Ecological Economics Theory**

Ecological Economics Theory integrates ecological and economic systems, advocating for a shift from traditional economic models that prioritize profit and growth to those that account for the planet's finite resources and ecological limits

(Costanza et al., 1997). This theory asserts that economic systems are embedded within ecological systems, and thus, economic activities must respect environmental constraints to avoid ecological collapse.

In the context of this study, ecological economics informs the development of governance innovations like payment for ecosystem services, carbon trading schemes, and resource efficiency strategies. It highlights the importance of internalizing environmental costs into economic decision-making, recognizing the value of ecosystem services, and promoting sustainable consumption and production patterns (Daly & Farley, 2011). By integrating the principles of ecological economics, innovations in environmental governance can help steer economies toward sustainable trajectories that respect planetary boundaries.

5. DIGITAL TECHNOLOGIES IN ENVIRONMENTAL GOVERNANCE

5.1 Role of Artificial Intelligence

Artificial Intelligence (AI) is playing an increasingly vital role in environmental governance by enhancing the capacity of organizations and governments to monitor, analyze, and manage environmental challenges. AI's ability to process vast amounts of data in real-time allows for improved environmental monitoring and forecasting, particularly in areas such as climate change modeling, wildlife tracking, and pollution detection (Rolnick et al., 2021). AI-powered tools like satellite imagery and machine learning algorithms enable more accurate predictions of natural disasters, such as floods and forest fires, facilitating better preparedness and disaster management.

AI is also being used to optimize resource management. For example, AI-driven smart grids help reduce energy consumption by predicting demand and improving efficiency in energy distribution. Moreover, AI has a key role in advancing precision agriculture by providing real-time data to farmers on soil health, weather conditions, and crop management, which promotes sustainable farming practices (Zeng et al., 2021). However, concerns about data privacy and AI's environmental footprint, particularly its high energy consumption, highlight the need for balanced approaches in its deployment within environmental governance.

5.2 Applications of Blockchain Technology

Blockchain technology is emerging as a powerful tool for enhancing transparency, accountability, and traceability in environmental governance. Its decentralized and immutable nature enables the creation of trustworthy systems for tracking environmental data, carbon credits, and sustainable supply chains. One of the key applications of blockchain in this area is the monitoring of carbon emissions and the facilitation of carbon trading markets. Blockchain allows for accurate and verifiable tracking of emissions, helping companies meet their sustainability targets and ensuring compliance with environmental regulations (Draetta et al., 2021).

In addition, blockchain is being applied in the management of natural resources, particularly in sectors like fisheries, forestry, and agriculture. By creating a transparent record of resource extraction and usage, blockchain helps prevent illegal activities, such as overfishing and deforestation, while promoting sustainable practices. Another significant application is the use of blockchain to authenticate the origins of eco-friendly products, providing consumers with confidence in the environmental integrity of their purchases (Howson et al., 2021). Overall, blockchain offers innovative solutions for addressing governance challenges in environmental protection and sustainability.

6. GREEN FINANCE

In this rapid growing economic world, industrialization cause many adverse effects on environment (Bhutta et al., 2022). It is necessary to integrate sustainability into financial practices. Green Finance the rapidly growing field that directs investment to into sustainable development project which helps to reduce carbon emission and promote environmental sustainability.

6.1 Definition and Importance

According to (Hohne et al., 2012), the board sense Green Finance can be defined as investment in sustainable development which includes investing in environmental product and service to support sustainability. It is not limited to climatic finance but has a wider prospective to other environmental objectives like preservation of biodiversity, water quality restoration and control of industrial pollution. According to Pricewaterhouse cooper consultant (PWC) (2013) can be defined as green finance in the bank sector is investment in financial product and service by considering environmental variables in process of decision making, monitoring the process and analysing the risk to promote environmental sustainability that support low carbon technologies, project and organisation.

Green finance is the critical component to achieve sustainable development and reduce adverse effect on climatic change. This chapter highlights the importance of green finance and its role in driving the innovation in fostering sustainability and resilience of the economy. Green financing facilities in transforming into low carbon economy by allocating funds in infrastructure, energy management and renewable energy sources (Sachs et.al 2019). It also provides funds to face negative consequences of climatic change like rise of sea level, extreme weather conditions and shift in planting zone. In order to encourage investment in climate friendly projects with the help of financial instruments like green bonds, climate bonds and carbon credit that ensure sufficient funds to create meaningful climate action.

6.2 Green Bonds and Sustainable Investment Funds

Green Bonds (Maltais, A., & Nykvist, B. 2020) are financial instruments that are explicitly designed to finance the project that have favourable effect on the environment. Back in 2007, European investment bank is the first bank to issue climatic awareness bond followed by world bank's green bond in 2008. Green bonds are very similar to other commercial bonds except it focus on green projects like energy efficiency, pollution control, water management and sustainable development.

The International Capital Market Association provides guidelines to have transparency and integrity in issuing green bonds. Some of the benefits in issuing green bonds. It helps to generate funding for environmental projects like reduction of carbon emission, energy management and water management etc. On the investor side it yields more profit because of lower volatility when it is compared to any other commercial bond.

Sustainable investment funds focused on combining conventional investing goals with a dedication to environmental sustainability which represents a growing and expanding sector of the financial industry. The growth of sustainable investment is anticipated that as legislative framework change and awareness of Environment social and government factors. Technological and data analytic innovations will raise openness and accountability by improving ESG factor reporting and evaluation. By integrating ESG factors, it not only provides affordable financial return but also contributes to the well-being of society by focusing on welfare of the environment.

6.3 Impact on Business Practices

Green finance drives invention and sustainable practices. Green Finance will provide funds for environmentally sustainable projects that support the companies that develop and implement innovative technologies that reduce impact on environment. As a result, many businesses started to invest in Research and Development to create a product with sustainable material that is energy efficient with less carbon footprint. Green finance influences corporate strategy and governance that integrates ESG performance to satisfy the expectation of green investors (Yeow, K.E. and Ng, S.-H. 2021).

Climatic changes pose great risk to business, so green finance encourages the adoption of environmentally friendly practices that help business to manage risks. Businesses that pose green finance must follow strict guidelines when it comes to documenting and disclosing their environmental effects. This includes reporting on consumption of energy, waste management, water management and other sustainability indicators. Companies may improve their brand reputation, attract more customers, and obtain a competitive edge by Integrating green financing and establishing good ESG performance. As the need for environmental sustainability, green finance plays a vital role in assisting companies in adjusting and prospering in a green economy. Businesses may achieve long term success by positively contributing to global effort to address climate change and protect the planet by coordinating financial goals with environmental objectives

6.4 Policy Frameworks and Incentives

Policy framework and incentives play a crucial in promoting green finance. Governments and international organisation can encourage to adopt the environmentally friendly practices across the business by establishing regulations, providing financial incentives (Zhou and Cui,2019) and creating a supportive environment for sustainable investments. Emission Trading System a market-based approach that provide financial incentives for organisation which controls pollution by reducing the carbon emission. Many countries have developed national strategies and action plans to utilize green finance in order to achieve environmental sustainability goals. In the Paris Agreement many countries agreed to align financial goals with climate goals and also set global framework to reduce greenhouse gas emission. UN Sustainable Development Goals have a set of 17goals to act as a worldwide guide for building a better and sustainable future. Mechanism like carbon tax and cap-and-trade system encourage organisation to reduce carbon footprint by providing incentives, subsidies and grants. Many countries provide tax credits, tax deduction and exemptions for investments in energy efficiency, renewable energy and green projects. Purchasing policies of Government give importance to goods and services are ecologically friendly. By establishing a supportive environment and providing right incentives, government can mobilize private sector to speed up in shifting to green economy.

7 CIRCULAR ECONOMY MODELS

The "take, make, dispose" mentality of the conventional linear economy is challenged by the novel circular economy concept, which promotes sustainability. Rather, it places more emphasis on maximizing resource efficiency, reducing waste, and developing closed-loop systems that recycle, remanufacture, and reuse goods and materials. This part addresses the fundamental ideas and principles of the circular economy, looks at new developments in recycling technology, looks at product-as-a-service models, and talks about the advantages and difficulties of putting circular economy activities into practice.

7.1 Principles and Concepts

Several fundamental ideas set the circular economy apart from the conventional linear economy. Fundamentally, the circular economy aims to prevent waste and pollution, maintain the useful life of goods and materials, and replenish natural systems (Kirchherr et.al.,2023)

Designing for Longevity and Reusability: One of the foundational principles is designing products with their entire lifecycle in mind. This means creating products that are durable, easy to repair, and capable of being disassembled for recycling or repurposing. The goal is to extend the lifespan of products and materials, thereby reducing waste (Anne et al., 2021).

Keeping Products and Materials in Circulation: Another key concept is maintaining the value of products, materials, and resources within the economy for as long as possible. This involves strategies such as reusing, repairing, refurbishing, remanufacturing, and recycling. By keeping resources in circulation, the circular economy reduces the need for virgin materials and minimizes environmental impact (Ellen MacArthur Foundation, 2021).

Regenerating Natural Systems: Restoring and renewing natural systems is another important point made by the circular economy. This can be accomplished by employing techniques including conserving biodiversity, using renewable energy sources, and composting organic waste to restore soil nutrients. The goal is to design systems that complement nature rather than diminish it. (Geissdoerfer et al., 2020).

Decoupling Economic Growth from Resource Use: Disentangling the consumption of finite resources from economic growth is a crucial objective of the circular economy. This entails generating value through processes other than the extraction and use of raw materials, supporting a more sustainable kind of economic growth. (Ellen MacArthur Foundation, 2021).

By implementing these ideas, the circular economy presents a viable substitute for the conventional linear model and offers a framework for cutting waste, preserving resources, and promoting creativity.

7.2 Innovations in Recycling Technologies

Recycling technologies are essential to the circular economy because they make it possible to collect and reuse resources that would otherwise be thrown away. The efficacy and efficiency of material recovery procedures have been greatly increased by recent developments in recycling technologies.

Advanced Sorting Technologies: The accuracy and speed of sorting recyclable materials have increased thanks to technological advancements in sorting, including machine learning, artificial intelligence (AI), and optical sorting. Robots with artificial intelligence (AI) and computer vision systems can accurately recognize and sort various plastics, metals, and papers, improving the quality and purity of recycled materials. (Fang, B., Yu, J., Chen, Z. et al., 2023).

Chemical Recycling: Plastic trash can be recycled chemically to create new plastics by breaking it down into its constituent parts through processes like depolymerization and pyrolysis. Chemical recycling makes it possible to recycle plastics indefinitely without losing quality, in contrast to mechanical recycling, which gradually deteriorates plastic quality (Zou, Liang, 2023).

Biodegradable and Compostable Materials: Another important innovation is the creation of materials that are compostable and biodegradable. These materials can be broken down by biological processes, lowering the environmental impact of trash, because they are manufactured from renewable resources like plant-based polymers. Compostable items can be processed into nutrient-rich compost, which improves the fertility and health of the soil. (Rosenboom, JG et al., 2022).

Closed-Loop Recycling Systems: Within the same product lifecycle, materials can be recovered and reused with closed-loop recycling systems. To recycle end-of-life automobiles and recover metals, polymers, and other materials for use in new vehicles, the automotive industry, for instance, has devised procedures. This method lessens the industrial industry's environmental impact while simultaneously conserving resources. (Wei Guo et al., 2022).

The objectives of the circular economy, effective material recovery and reuse, and a decrease in the environmental impact of trash depend on these advancements in recycling technologies.

7.3 Products-as-a-Service Models

The usage-based economy is replacing the conventional ownership-based economy with the product-as-a-service (PaaS) paradigm. Under a PaaS model, businesses maintain ownership of their products and offer them to clients as a service, emphasizing the utility of usage above ownership.

Leasing and Renting: One common form of PaaS is leasing or renting products instead of selling them. For example, companies like Rent the Runway offer clothing rental services, allowing customers to rent high-quality garments for a period of time instead of purchasing them. This model reduces waste by extending the lifecycle of products and promoting multiple uses (Patwa et. Al., 2021).

Performance-Based Contracts: Customers pay for a product's performance or result under performance-based contracts rather than the product itself. For instance, users can pay for the light they receive rather than the light fixtures under Philips Lighting's "light as a service" approach. The fixtures are still owned by Philips, which is responsible for their upkeep and eventually recycling. (van Strien, J., et.al. 2019).

Product Stewardship: PaaS models frequently place a high priority on product stewardship, in which businesses remain in charge of their goods for their whole lifecycle. This covers upkeep, fixing, and recycling or end-of-life disposal. Companies have a stake in creating items that are recyclable, long-lasting, and repairable because they are the ones who own them. This is in line with the circular economy concept (Nitin Patwa et. Al., 2021).

Sharing Economy Platforms: Another illustration of PaaS is the growth of sharing economy platforms, like tool and car-sharing services like Zipcar. By allowing numerous users to share access to things, these platforms maximize their use and lessen the requirement for individual ownership.

PaaS models have several advantages, such as fewer resource consumption, less of an influence on the environment, and longer product life. Additionally, it pushes businesses to innovate in service delivery and product design, balancing commercial objectives with environmental ones.

7.4 Benefits and Challenges

While the circular economy offers numerous benefits, its implementation also presents several challenges that need to be addressed.

Benefits

Resource Conservation: The circular economy helps to conserve natural resources by extending the lifecycle of products and materials. By prioritizing reuse, remanufacturing, and recycling, it significantly reduces the demand for virgin resource extraction, which in turn lessens environmental degradation. For instance, extending the life of products through remanufacturing can reduce raw material usage by up to 80% (Geissdoerfer et al., 2020). This not only helps preserve finite resources but also minimizes the energy and water inputs associated with extraction and processing.

Waste Reduction: Circular economy strategies such as recycling, remanufacturing, and reusing products directly contribute to a reduction in waste sent to landfills or incinerators. In the European Union, circular practices have already led to a 10% reduction in waste disposal over the past decade (Dominko et al., 2023). This also results in lower greenhouse gas emissions, as landfills and incineration are major contributors to methane and carbon emissions. Additionally, reducing waste limits environmental pollution and the accumulation of hazardous materials in ecosystems.

Economic Opportunities: The shift to a circular economy creates a range of economic benefits. Not only does it stimulate innovation in product design, remanufacturing, and recycling, but it also generates new job opportunities. According to the Ellen MacArthur Foundation (2021), transitioning to a circular economy could create up to 700,000 new jobs in Europe alone by 2030. These opportunities extend across sectors, from waste management and resource recovery to new business models focused on product longevity, such as leasing and sharing platforms. Businesses also benefit from reduced costs as they implement resource-efficient practices that minimize inputs and lower production expenses.

Environmental Benefits: Circular economy models contribute to reducing the ecological footprint of production and consumption. By decreasing the need for resource extraction and lowering waste production, the circular economy helps protect ecosystems and biodiversity. For example, reducing the demand for raw materials like timber or metals can slow deforestation and minimize the environmental impacts of mining (Geissdoerfer et al., 2020). Additionally, by fostering sustainable material cycles, the circular economy mitigates climate change impacts through reduced greenhouse gas emissions across the value chain.

Challenges

Technological Barriers: One of the key challenges in implementing circular economy practices is the need for advanced technologies in product design, recycling, and material recovery. Many products are currently designed with linear lifecycles in mind, making them difficult to disassemble, remanufacture, or recycle. Developing technologies that enable efficient material recovery and closed-loop processes can be both costly and technically complex (Anitha, 2024). Furthermore, industries need to invest in upgrading manufacturing systems to accommodate circular designs, which may require significant financial resources and time.

Economic and Market Barriers: Existing economic systems heavily favor linear models due to established infrastructure, market practices, and cost structures. Shifting to circular models often necessitates significant upfront investment in new infrastructure, business models, and supply chains (Dominko et al., 2023). Additionally, industries entrenched in the linear economy, such as mining and manufacturing, may resist adopting circular practices due to the perceived threat to their profitability. Overcoming these economic barriers requires creating incentives, subsidies, or financial mechanisms to encourage the shift towards circular systems.

Regulatory and Policy Barriers: Effective implementation of circular economy models depends on robust regulatory frameworks that promote circular practices. However, many countries lack coherent policies or face inconsistencies in regulations that hinder circular initiatives (Velenturf et al., 2021). For example, regulations on waste management and product standards often differ across regions, creating obstacles for companies trying to scale circular solutions (Hossain et al., 2024a; Hossain et al., 2024b). Furthermore, in the absence of strong incentives, businesses and consumers may not be sufficiently motivated to adopt circular practices.

Consumer Behavior: Shifting consumer behavior is crucial for the success of the circular economy. Consumers play a vital role in extending product lifecycles through reuse, repair, and recycling. However, current consumption patterns are heavily oriented toward disposability and convenience (Kirchherr et al., 2023). Changing these behaviors requires educating consumers about the environmental and economic benefits of circular products and services, as well as encouraging sustainable consumption habits. Moreover, developing trust in the quality and safety of remanufactured or recycled goods is essential for fostering widespread adoption.

The circular economy presents a novel and sustainable way to manage resources and reduce waste, but its successful implementation depends on resolving related issues. The circular economy can significantly improve the environment, the economy, and society by removing these obstacles, which will aid in sustainable development.

8 COMMUNITY-DRIVEN GOVERNANCE

Community-Driven governance refers to local communities actively participating in decision making process that impact on resources and environment. This approach focuses on the importance of local knowledge, empowerment, and ownership in order to achieve a sustainable outcome.

8.1 Importance of Local Participation

Participation of local communities in governance is more likely to enhance trust in the process and outcome which results in greater acceptance and compliance. Input from local communities helps in improved decision making which results in contextual appropriate decisions according to current situation. Involving the local communities (Mahzouni,2008) in decision making process will increase accountability and transparency that reduce the possibility of corruption and encouraging an open environment. Participation in decision making can empower local communities by giving them a voice that impacts on their lives and environment. Involvement in governance processes helps community members develop their skill and capability so that they can make meaning contributions in the future. Local communities that are actively involved can mobilize group efforts to implement environmental projects to protect their interests.

8.2 Mechanisms for Empowering Communities

Capacity building and providing training to communities will empower the people with information and the ability to participate in governance procedures and decision-making processes (Etkind, el.at 2023). Training programs like workshops and seminar conducted to provide information about environmental issues, governance procedures and sustainable practices. Technical training program conducted on resource management, conservation techniques and environmental monitoring to empower the local communities. In order to enhance participation, community meetings are organised to discuss governance and environmental issues. Local communities are involved in planning and implementation of green projects. In the realm of technology, digital platforms can be used for community forum to discuss or to conduct survey for collecting information. Mobile technologies are used to promote engagement and provide information, specifically in remote areas. Grants and funded programs offered to communities to support sustainable and environmental projects. Empowering communities are essential to achieve sustainable development and effective environmental governance. Communities can be empowered by providing capacity building through training programs, improving participation, access to resources to take ownership of local environmental problems and contributing to sustainable solutions.

8.3 Case Studies of Successful Community-Driven Initiatives

This section explores several case studies on successful community-driven initiatives emphasizing strategies used and highlights positive outcome.

Case Study 1: The Chipko Movement India

This movement initiated in 1970 in the Garhwal Himalayas of India is famous example for community driven initiatives in which local villagers including women protected against deforestation and commercial logging. This movement has raised awareness of deforestation and its impact on the environment through education and communication. As a result of this movement, the Uttarakhand government passed a ban for 15 years on green felling in the Himalayan forests. (Pathak,2020)

Case Study 2: The Solar Mamas, Barefoot College, India

As part of this movement, Barefoot college have trained women from rural areas, most of them are illiterate or semi-illiterate to become solar engineering i.e. training them on installation and maintenance of solar panel and system. The main objective of this program is to raise the standard of living by providing renewal energy and reducing dependency on fossil fuels. This movement have empowered women by providing skills, creating job opportunities and provides leadership position among communities. (Narasimhan,2019)

Case Study 3: The Green Belt Movement, Kenya

This Initiative started on 1977 focus on tree planting in Kenya for environmental conservation and community development. This campaign focuses on creating awareness on negative impact of deforestation, protecting landscape and improves biodiversity. As a result of this movement over 51 million trees are planted across Kenya to improve environment health by promoting sustainable development. This movement has empowered the local communities by providing economic independence and leadership role. (Rukuni,2021)

9 INTERNATIONAL COOPERATION

International Cooperation (Sharif,1992) is necessary to solve global environmental challenges and achieve sustainable development. International cooperation enables countries to share technological, financial, and intellectual resources to tackle environmental problems. Integrated policies reduce conflict and improve environment protection by developing consistent and comprehensive policies across the national boundaries. International Cooperation encourages group action on international problems that are too difficult to handle for individual nations, for example climate change. International cooperation can develop the capacity of developing countries to face and protect their environment. Despite of many benefits in international cooperation but there are some challenges like Political differences, Economic disparities, Enforcement and Compliance. Political and economic disparities can have significant impact on the effective implementation of international agreements across the globe. It might be difficult to make sure that all countries abide by international obligations and agreements

9.1 Synergies Between Digital Technologies, Green Finance, and Circular Economy

The integration of Digital technologies, green finance and circular economy can have significant synergies that enhance their individual impact and hasten the transition towards the development of sustainability. Digital Technologies like AI, Blockchain, IOT and Big data Analytics have important roles in promoting the circular economy and green finance by offering new tools for effectiveness, honesty and scalability. AI can predict environmental impact and make sure optimize usage of resources which helps to reduce waste and improves decision making in circular economic activities. Machine learning algorithms can be automated to increase productivity and reduce the operation cost by recycling and waste management. Blockchain Technology. Blockchain technology maintains transparency and accountability of records in supply chain by providing unchangeable records of transactions (Zhang. Et.al 2018)). With the help of Smart contracts, provide transparency and minimize administrative complexity by enforcing adherence to sustainability standards and green finance agreement. IOT devices like sensors and smart devices assist in optimizing energy consumption, water usage and waste management contributing to sustainable goals (Hossain et al., 2024a).

Green finance involves allocating financial resources to projects that support environmental sustainability. It gives vital financial support for the shift to a circular economy and promotes advancement in digital technology. Circular economy aims to reduce waste and create the close loop on product life cycle through recycling, reuse and refurbishing to maximize resource utilization.

9.2 Integrating Community-Driven and International Efforts

To accomplish sustainable development objectives, local, grassroots projects must be combined with global plans and policies through integration of community-Driven and international environmental governance efforts. Community based natural resource management formed among local communities involved in environmental decision for sustainability and helps to avoid exploitation of natural resources. Generating environmental education and awareness programs like grassroots campaigns to address environmental issues like waste management, water management and energy management etc., among the local communities.

Some of the international efforts like Sustainable Development goals develop a framework to address the comprehensive global effects and motivate countries to address various aspects of sustainable development. International and national organizations (Global Environment Facility) offer financial resources to local projects that address environmental issues that support community-driven initiatives. UNDP's Impact Assessment System analyses the project and identifies the best practices and opportunities for improvement that contribute to SDGs. (Clausen, el.at,2011). Community-based Forest monitoring programs supported by international organizations might track the changes in habitat and biodiversity which provide information to preserve nature. Integrating community-driven and international efforts helps to achieve comprehensive and governance of the system by framing goals, effective utilization of resources, improving capacity building, and promoting policy integration that increases the participation of both local and global communities which ensures resilient sustainable development.

9.3 Framework for Adaptive and Resilient Governance

Adaptive and resilient governance are considered more important in the current era of unpredictable environmental change, socioeconomic upheaval and growing uncertainty (Benson, et.al 2017) In foresee, address and recover from disruption. Integrated policy frameworks are designed to support long term sustainability by assisting local communities (Hossain et al., 2024c). Frameworks include flexibility in governance designed in such a way that capacity of institutions, rules and procedures to adapt to new information, situation and issues that arise. Adaptability is an important framework that involves evaluating results, adjusting framework according to which strategies are working and what are not. Information is gathered through continuous feedback loops which help in the implementation of policies about future changes. Participation from stakeholders in policy development, implementation and evaluation which results in building trust, enhance reliability and ensure that the policies are accepted and supported. Proactive frameworks help to identify the future problems or challenges and preparing the strategies to overcome the risk by taking preventive actions for example rise in sea level can damage the coastal areas in the future, it can be overcome by investing in green infrastructure like wetlands and mangroves helps to solve the problem. With the help of continuous learning and collaboration, adaptive and resilient governance frameworks that not only survive disruption but also achieve success in the face of change.

10 CHALLENGES AND BARRIERS

Environmental governance for sustainable development faces many challenges and barriers while integrating environmental protection, economic growth, and social justice (Hossen et al., 2024). Some of the challenges are financial constraints, technological and infrastructure limitations, political, social and cultural factors. By addressing these issues helps to provide solutions for sustainable development and environmental protection. (de Oliveira et al., 2024)

10.1 Technological and Financial Constraints

Access to cutting-edge environmental technologies is unevenly distributed, so developing countries may find it difficult in implementing creative strategies for environmental preservation and sustainable development due to lack of resources and capability to acquire those technologies. Digital gap is uneven, access to information and communication technologies among global countries may result in resistance of adopting innovative approach to environmental governance.

Many green projects require substantial investment which can be quite difficult, particularly for developing countries. These countries limited financial resources make it difficult in implementation of sustainable environmental projects. Access to finance is one of the major barriers because many financial institutions hesitate to invest on sustainable project because of perceived risks, lack of knowledge and security (Mishra et al., 2020). Environmental projects require longer time horizons to become profitable ones. As a result, many financial institutions are not interested in investing in green projects. Lack of government support for sustainable initiatives like subsidise for green project, tax benefits, grants which leads to difficult in sustainable development.

10.2 Political and Social Barriers

Environmental governance and sustainability development policies can be disrupted by political instability (Misleh, et.al, 2024) can cause frequent changes in government, political unrest and conflicts. Lack of politics will result in weak environmental policies and inadequate funds for green projects. International cooperation is very important in dealing with global environmental problems like climate change and degradation of biodiversity. Countries with political disputes may face difficulties in implementing international agreements and policies effectively. Lack of awareness and education among individuals and communities were unable to understand the importance of environmental protection or actions to contribute sustainability.

Sustainable development objectives may not always align with cultural values and traditions. It is difficult to adopt sustainability goals because of resistance to change (Singh, el.at 2024), so it is essential to promote goals while interacting with communities and preserving cultural values. Large scale of environmental initiatives can lead to communities' displacement can arise social unrest and opposition to sustainable development projects. Social and economic barriers can prevent youngsters from getting involved in environmental governance and sustainable development initiatives.

11 IMPLICATIONS AND FUTURE RESEARCH DIRECTIONS

Managerial and Ethical Implications

There are major administrative advantages to integrating digital technologies like blockchain and artificial intelligence (AI) into environmental regulation. AI improves environmental monitoring and prediction skills, allowing for pre-emptive actions against environmental deterioration. Blockchain technology promotes trust among stakeholders by guaranteeing accountability and transparency in data sharing and resource management (Kshetri, N., 2021). In order to increase environmental governance's efficacy and efficiency, managers must implement these technologies. When it comes to data privacy, security, and equity, the use of digital technology creates ethical questions. It is crucial to make sure that these technologies are applied fairly and ethically. Regulations and ethical standards need to be put in place to safeguard people's privacy and stop data misuse. To prevent escalating already-existing disparities, access to these technologies should also be made more widely available (Dhirani, L.L et al., 2023).

Financial flows are matched with environmental objectives through the use of green finance, which is essential in directing investments toward sustainable initiatives. Growing awareness of the necessity to finance the shift to a low-carbon economy is demonstrated by the emergence of green bonds and sustainable investment funds (Alamgir, M.; Cheng, M.-C, 2023). Financial sector managers need to devise plans for integrating environmental factors into investment choices and encouraging openness in disclosing the environmental effects of investments.

Future Research Directions

Subsequent investigations ought to concentrate on resolving the constraints mentioned in this analysis and investigating novel approaches to augment environmental governance. Additionally, research ought to look into how to lower the cost and increase accessibility to these technologies, especially for underdeveloped nations. Subsequent studies must investigate novel policy tools capable of encouraging environmentally friendly behaviors. analyzing methods to encourage stakeholders and customers to adopt new behaviors. This includes researching how societal norms, incentives, and educational efforts affect patterns of sustainable production and consumption. This entails assessing these practices' affordability as well as their capacity to generate new business ventures and employment possibilities. The main goal of research should be to improve the efficacy of international initiatives and multilateral agreements by removing geopolitical obstacles.

12 CONCLUSION

Considering the escalating environmental issues of pollution, climate change, and biodiversity loss, there is a growing need for strong and flexible environmental governance. With an emphasis on digital technologies, green finance, circular economy models, community-driven governance, and international cooperation, this study has investigated upcoming trends and developments in environmental governance. We can build governance frameworks that are more robust, adaptable, and equipped to handle the challenges of sustainable development by utilizing these new trends. A viable route forward for sustainable development is provided by the combination of digital technology, green financing, circular economy models, community-driven governance, and global collaboration. Even if there are obstacles to overcome and constraints to work within, continued research and innovation can offer the knowledge and answers needed to build robust and flexible environmental governance frameworks. We can clear the path for a sustainable future by addressing the ethical and managerial ramifications and encouraging cooperation across industries and geographical boundaries.

REFERENCES

Alamgir, M., & Cheng, M.-C. (2023). Do Green Bonds Play a Role in Achieving Sustainability? *Sustainability (Basel)*, 15(13), 10177. DOI: 10.3390/su151310177

Anitha, K. (2024). Emerging Trends in Sustainability: A Conceptual Exploration. In Kulkarni, S., & Haghi, A. K. (Eds.), *Global Sustainability. World Sustainability Series*. Springer., DOI: 10.1007/978-3-031-57456-6_2

Anne, P. M. Velenturf, Phil Purnell (2021) Principles for a sustainable circular economy, Sustainable Production and Consumption, Volume 27, 2021, Pages 1437-1457, ISSN 2352-5509, https://doi.org/DOI: 10.1016/j.spc.2021.02.018

Barros, M. V., Salvador, R., do Prado, G. F., Carlos de Francisco, A., & Piekarski, C. M. (2021). Circular economy as a driver to sustainable businesses, Cleaner Environmental Systems, Volume 2, 2021, 100006, ISSN 2666-7894, https://doi.org/ DOI: 10.1016/j.cesys.2020.100006

Benson, M. H., & Craig, R. K. (Eds.). (2020). *The End of Sustainability: Resilience and the Future of Environmental Governance in the Anthropocene*. University of Kansas Press.

Berkes, F., Colding, J., & Folke, C. (Eds.). (2000). *Navigating social-ecological systems: Building resilience for complexity and change*. Cambridge University Press.

Bhutta, U. S., Tariq, A., Farrukh, M., Raza, A., & Iqbal, M. K. (2022). Green bonds for sustainable development: Review of literature on development and impact of green bonds. *Technological Forecasting and Social Change*, 175, 121378. DOI: 10.1016/j.techfore.2021.121378

Biermann, F., & Pattberg, P. (2012). Global environmental governance revisited. *Environmental Politics*, 21(1), 115–134. DOI: 10.1080/09644016.2012.738749

Brockhaus, M., Di Gregorio, M., & Mardiah, S. (2021). Multi-level governance and its role in environmental sustainability. *Environmental Policy and Governance*, 31(1), 65–79.

Clark, W. C., & Dickson, N. M. (2003). Sustainability science: The emerging research program. *Proceedings of the National Academy of Sciences of the United States of America*, 100(14), 8059–8061. DOI: 10.1073/pnas.1231333100 PMID: 12794187

Clausen, A., Vu, H. H., & Pedrono, M. (2011). An evaluation of the environmental impact assessment system in Vietnam: The gap between theory and practice. *Environmental Impact Assessment Review*, 31(2), 136–143. DOI: 10.1016/j.eiar.2010.04.008

Costanza, R., Cumberland, J. H., Daly, H. E., Goodland, R., & Norgaard, R. B. (1997). *An Introduction to Ecological Economics*. CRC Press. DOI: 10.1201/9781003040842

Daly, H. E., & Farley, J. (2011). *Ecological economics: Principles and applications* (2nd ed.). Island Press.

David, B. Olawade, Ojima Z. Wada, Aanuoluwapo Clement David-Olawade, Oluwaseun Fapohunda, Abimbola O. Ige, Jonathan Ling (2024). Artificial intelligence potential for net zero sustainability: Current evidence and prospects,Next Sustainability,Volume 4,2024,100041,ISSN 2949-8236, https://doi.org/DOI: 10.1016/j.nxsust.2024.100041

de Oliveira, U. R., Menezes, R. P., & Fernandes, V. A. (2024). A systematic literature review on corporate sustainability: Contributions, barriers, innovations and future possibilities. *Environment, Development and Sustainability*, 26(2), 3045–3079. DOI: 10.1007/s10668-023-02933-7 PMID: 36687736

Dhirani, L. L., Mukhtiar, N., Chowdhry, B. S., & Newe, T. (2023). Ethical Dilemmas and Privacy Issues in Emerging Technologies: A Review. *Sensors (Basel)*, 23(3), 1151. DOI: 10.3390/s23031151 PMID: 36772190

Dominko, M., Primc, K., Slabe-Erker, R., & Kalar, B. (2023). A bibliometric analysis of circular economy in the fields of business and economics: Towards more action-oriented research. *Environment, Development and Sustainability*, 25(7), 5797–5830. DOI: 10.1007/s10668-022-02347-x PMID: 35530441

Draetta, L., Dabbicco, M., & Bisogno, M. (2021). Blockchain and carbon markets: A new horizon for environmental sustainability. *Technological Forecasting and Social Change*, 170, 120930.

Ellen MacArthur Foundation. (2021). The circular economy in detail. Retrieved from https://www.ellenmacarthurfoundation.org/explore/the-circular-economy-in-detail

Enayat, A. Moallemi, Shirin Malekpour, Michalis Hadjikakou, Rob Raven, Katrina Szetey, Dianty Ningrum, Ahmad Dhiaulhaq, Brett A. Bryan (2020). Achieving the sustainable Development Goals Requires Trans disciplinary Innovation at the Local Scale, One Earth, Volume 3, Issue 3, Pages 300-313, ISSN 2590-3322, https://doi.org/DOI: 10.1016/j.oneear.2020.08.006

Etkind, E., & de Vries, B. J. (2023). Co-creating a sustainability transition through community empowerment and education: A case study from Israel. *Journal of Cleaner Production*, 382, 135194. DOI: 10.1016/j.jclepro.2022.135194

Fang, B., Yu, J., Chen, Z., & Chen, Z. (2023). Artificial intelligence for waste management in smart cities: A review. *Environmental Chemistry Letters*, 21(4), 1959–1989. DOI: 10.1007/s10311-023-01604-3 PMID: 37362015

Folke, C., Hahn, T., Olsson, P., & Norberg, J. (2005). Adaptive governance of social-ecological systems. *Annual Review of Environment and Resources*, 30(1), 441–473. DOI: 10.1146/annurev.energy.30.050504.144511

Geissdoerfer, M., Pieroni, M. P. P., Pigosso, D. C. A., & Soufani, K. (2020).Circular business models: A review, Journal of Cleaner Production, Volume 277, 123741, ISSN 0959-6526, https://doi.org/DOI: 10.1016/j.jclepro.2020.123741

Geissdoerfer, M., Savaget, P., Bocken, N. M. P., & Hultink, E. J. (2017). The circular economy–A new sustainability paradigm? *Journal of Cleaner Production*, 143, 757–768. DOI: 10.1016/j.jclepro.2016.12.048

Guo, W., Li, K., Fang, Z., Feng, T., & Shi, T. (2022).A sustainable recycling process for end-of-life vehicle plastics: A case study on waste bumpers, Waste Management, Volume 154, Pages 187-198, ISSN 0956-053X, https://doi.org/DOI: 10.1016/j.wasman.2022.10.006

Heikkila, T., & Weible, C. M. (2018). A semiautomated approach to analyzing polycentricity. *Environmental Policy and Governance*, 28(4), 308–318. DOI: 10.1002/eet.1817

Hernández Guzmán, D., & Hernández García de Velazco, J. (2023). Global Citizenship: Towards a Concept for Participatory Environmental Protection. *Global Society*, 38(2), 269–296. DOI: 10.1080/13600826.2023.2284150

Hossain, I., Haque, A. M., Kılıç, Z., Ullah, S. A., Azrour, M., & Mabrouki, J. (2024b). Exploring Household Waste Management Practices and IoT Adoption in Barisal City Corporation. In *Smart Internet of Things for Environment and Healthcare* (pp. 27–45). Springer Nature Switzerland. DOI: 10.1007/978-3-031-70102-3_2

Hossain, I., Haque, A. M., & Ullah, S. A. (2024a). Assessing sustainable waste management practices in Rajshahi City Corporation: An analysis for local government enhancement using IoT, AI, and Android technology. *Environmental Science and Pollution Research International*, 1–19. DOI: 10.1007/s11356-024-33171-7 PMID: 38581631

Hossain, I., Haque, A. M., & Ullah, S. A. (2024c). Policy evaluation and performance assessment for sustainable urbanization: A study of selected city corporations in Bangladesh. *Frontiers in Sustainable Cities*, 6, 1377310. DOI: 10.3389/frsc.2024.1377310

Hossen, M. S., Haque, A. M., Hossain, I., Haque, M. N., & Hossain, M. K. (2024). Towards comprehensive urban sustainability: Navigating predominant urban challenges and assessing their severity differential in Bangladeshi city corporations. Urbanization. *Sustainability Science*, 1(1), 1–17.

Howson, P., Oakes, S., & Baynham-Hughes, J. (2021). Blockchain's role in sustainability: Transformative or problematic? *Environment, Development and Sustainability*, 23(1), 613–631.

Kates, R. W., Clark, W. C., Corell, R., Hall, J. M., Jaeger, C. C., Lowe, I., McCarthy, J. J., Schellnhuber, H. J., Bolin, B., Dickson, N. M., Faucheux, S., Gallopin, G. C., Grübler, A., Huntley, B., Jäger, J., Jodha, N. S., Kasperson, R. E., Mabogunje, A., Matson, P., & Svedin, U. (2001). Sustainability science. *Science*, 292(5517), 641–642. DOI: 10.1126/science.1059386 PMID: 11330321

Kirchherr, J., Yang, N.-H. N., Schulze-Spüntrup, F., Heerink, M. J., & Hartley, K. (2023). Conceptualizing the Circular Economy (Revisited): An Analysis of 221 Definitions, Resources, Conservation and Recycling, Volume 194, 107001, ISSN 0921-3449, https://doi.org/DOI: 10.1016/j.resconrec.2023.107001

Kshetri, N. (2021). Blockchain Technology for Improving Transparency and Citizen's Trust. In Arai, K. (Ed.), *Advances in Information and Communication. FICC 2021. Advances in Intelligent Systems and Computing* (Vol. 1363). Springer., DOI: 10.1007/978-3-030-73100-7_52

Kyriakopoulos, G. L. (2022). Energy Communities Overview: Managerial Policies, Economic Aspects, Technologies, and Models. *Journal of Risk and Financial Management*, 15(11), 521. DOI: 10.3390/jrfm15110521

Lee, T.-C., Anser, M. K., Nassani, A. A., Haffar, M., Zaman, K., & Muhammad, M. Q. A. (2021). Managing Natural Resources through Sustainable Environmental Actions: A Cross-Sectional Study of 138 Countries. *Sustainability (Basel)*, 13(22), 12475. DOI: 10.3390/su132212475

Lockwood, M., Davidson, J., & Curtis, A. (2021). Governance principles for natural resource management. *Environmental Management*, 57(2), 317–332.

. Mahzouni, A. (2008). Participatory local governance for sustainable community-driven development.

Maltais, A., & Nykvist, B. (2020). Understanding the role of green bonds in advancing sustainability. *Journal of Sustainable Finance & Investment*, ●●●, 1–20. DOI: 10.1080/20430795.2020.1724864

Meadowcroft, J. (2021). Climate change governance: The challenge of steering social-ecological systems towards sustainability. *Environmental Politics*, 30(1), 1–20.

Mensah, J., & Ricart Casadevall, S. (2019). Sustainable development: Meaning, history, principles, pillars, and implications for human action: Literature review. *Cogent Social Sciences*, 5(1), 1653531. Advance online publication. DOI: 10.1080/23311886.2019.1653531

Mishra, P. S., Kumar, A., & Das, N. (2020). Corporate sustainability practices in polluting industries: Evidence from India China and USA. *Problemy Ekorozwoju*, 15(1), 161–168. DOI: 10.35784/pe.2020.1.17

Misleh, D., Dziumla, J., De La Garza, M., & Guenther, E. (2024). Sustainability against the logics of the state: Political and institutional barriers in the Chilean infrastructure sector. *Environmental Innovation and Societal Transitions*, 51, 100842. DOI: 10.1016/j.eist.2024.100842

Narasimhan, H., & Biswas, P. (2019). The Barefoot Approach to Empowerment: Solar Mamas' Contributions to Sustainable Development. *International Journal of Gender and Entrepreneurship*, 11(4), 377–391. DOI: 10.1108/IJGE-06-2019-0089

Obaideen, K., Albasha, L., Iqbal, U., & Mir, H. (2024). Wireless power transfer: Applications, challenges, barriers, and the role of AI in achieving sustainable development goals-A bibliometric analysis. *Energy Strategy Reviews*, 53, 101376. DOI: 10.1016/j.esr.2024.101376

Ostrom, E. (2009). A general framework for analyzing sustainability of social-ecological systems. *Science*, 325(5939), 419–422. DOI: 10.1126/science.1172133 PMID: 19628857

Ostrom, E. (2009). A general framework for analyzing sustainability of social-ecological systems. *Science*, 325(5939), 419–422. DOI: 10.1126/science.1172133 PMID: 19628857

Pathak, N. (2020). Chipko Movement: Lessons Learned for Community-based Conservation. *Environmental Development*, 34, 100516. DOI: 10.1016/j.envdev.2020.100516

Pathan, M. S., Richardson, E., Galvan, E., & Mooney, P. (2023). The Role of Artificial Intelligence within Circular Economy Activities—A View from Ireland. *Sustainability (Basel)*, 15(12), 9451. DOI: 10.3390/su15129451

Patwa, N., Sivarajah, U., Seetharaman, A., Sarkar, S., Maiti, K., & Hingorani, K. (2021). Towards a circular economy: An emerging economies context,Journal of Business Research,Volume 122, Pages 725-735, ISSN 0148-2963, https://doi.org/ DOI: 10.1016/j.jbusres.2020.05.015

Rodrigues, R. Legal and human rights issues of AI: Gaps, challenges and vulnerabilities, Journal of Responsible Technology, Volume 4, 2020, 100005, ISSN 2666-6596, https://doi.org/DOI: 10.1016/j.jrt.2020.100005

Rolnick, D., Donti, P. L., & Kaack, L. H. (2021). Tackling climate change with machine learning. *Nature Communications*, 12(1), 1–14. PMID: 33397941

Rosenboom, J. G., Langer, R., & Traverso, G. (2022). Bioplastics for a circular economy. *Nature Reviews. Materials*, 7(2), 117–137. DOI: 10.1038/s41578-021-00407-8 PMID: 35075395

Rukuni, R. (2021). The Legacy of the Green Belt Movement in Kenya: Lessons for Sustainable Development. *Journal of Environmental Policy and Planning*, 23(5), 652–670. DOI: 10.1080/1523908X.2021.1927870

. Sachs, J. D., Woo, W. T., Yoshino, N., & Taghizadeh-Hesary, F. (2019). Why is green finance important?

Sharif, N. (1992). Technological dimensions of international cooperation and sustainable development. *Technological Forecasting and Social Change*, 42(4), 367–383. DOI: 10.1016/0040-1625(92)90080-D

Sharma, M., & Dikshit, O. (2021). Air Pollution and Health Management in Delhi: Current Challenges and Prospects. In *Air Pollution Sources, Statistics and Health Effects*. Nova Science Publishers.

Singh, P. K., & Maheswaran, R. (2024). Analysis of social barriers to sustainable innovation and digitisation in supply chain. *Environment, Development and Sustainability*, 26(2), 5223–5248. DOI: 10.1007/s10668-023-02931-9 PMID: 36687733

Talaviya, T., Shah, D., Patel, N., Yagnik, H., & Shah, M. (2020). Implementation of artificial intelligence in agriculture for optimisation of irrigation and application of pesticides and herbicides, Artificial Intelligence in Agriculture, Volume 4, 2020, Pages 58-73, ISSN 2589-7217, https://doi.org/DOI: 10.1016/j.aiia.2020.04.002

Ungar, M. (2018). Systemic resilience: Principles and processes for a science of change in contexts of adversity. *Ecology and Society*, 23(4), art34. https://www.jstor.org/stable/26796886. DOI: 10.5751/ES-10385-230434

Uriarte-Gallastegi, N., Arana-Landín, G., Landeta-Manzano, B., & Laskurain-Iturbe, I. (2024). The Role of AI in Improving Environmental Sustainability: A Focus on Energy Management. *Energies*, 17(3), 649. DOI: 10.3390/en17030649

van Strien, J., Gelderman, C. J., & Semeijn, J. (2019). Performance-based contracting in military supply chains and the willingness to bear risks. *Journal of Defense Analytics and Logistics*, 3(1), 83–107. DOI: 10.1108/JDAL-10-2017-0021

Walker, W. S., Baccini, A., Schwartzman, S., Ríos, S., Oliveira-Miranda, M. A., Augusto, C., & Smith, P. (2020). The role of forest conversion, degradation, and disturbance in the carbon dynamics of Amazon indigenous territories and protected areas. *Proceedings of the National Academy of Sciences of the United States of America*, 117(6), 3015–3025. DOI: 10.1073/pnas.1913321117 PMID: 31988116

Wang, H., & Ran, B. (2021). Network governance and collaborative governance: A thematic analysis on their similarities, differences, and entanglements. *Public Management Review*, 25(6), 1187–1211. DOI: 10.1080/14719037.2021.2011389

Wang, Q., Su, M., & Li, R. (2020). The impact of the digital economy on the efficiency of environmental governance in China. *Journal of Cleaner Production*, 276, 123716. DOI: 10.1016/j.jclepro.2020.123716

Wright, C., & Nyberg, D. (2021). Climate change, capitalism, and corporations: Processes of creative self-destruction. *Organization Studies*, 42(5), 747–766.

Xing, G., Zhang, Y., & Guo, J. (2023, March 10). Environmental Regulation in Evolution and Governance Strategies. *International Journal of Environmental Research and Public Health*, 20(6), 4906. DOI: 10.3390/ijerph20064906 PMID: 36981813

Yeow, K. E., & Ng, S.-H. (2021). The impact of green bonds on corporate environmental and financial performance. *Managerial Finance*, 47(10), 1486–1510. DOI: 10.1108/MF-09-2020-0481

Zeng, Y., Zhang, Y., & Yang, J. (2021). The role of artificial intelligence in sustainable agriculture: A review. *Journal of Cleaner Production*, 307, 127164.

Zhang, L., & White, M. (2021). Green finance: A pathway to sustainable development? *Journal of Sustainable Finance & Investment*, 11(3), 205–217. DOI: 10.1080/20430795.2020.1868752

Zhang, X., Aranguiz, M., Xu, D., Zhang, X., & Xu, X. (2018). Utilizing blockchain for better enforcement of green finance law and regulations. In *Transforming climate finance and green investment with blockchains* (pp. 289–301). Academic Press. DOI: 10.1016/B978-0-12-814447-3.00021-5

Zhou, X., & Cui, Y. (2019). Green bonds, corporate performance, and corporate social responsibility. *Sustainability (Basel)*, 11(23), 6881. DOI: 10.3390/su11236881

Zou, L., Xu, R., Wang, H., Wang, Z., Sun, Y., & Li, M. (2023). Chemical recycling of polyolefins: a closed-loop cycle of waste to olefins. China Science Publishing & Media. 10. DOI: 10.1093/nsr/nwad207

Chapter 15
Environmental Governance for Promoting and Application of Integrated Water Resources Management

Shahriar Shams
https://orcid.org/0000-0001-6233-5365
Universiti Teknologi Brunei, Brunei & INTI International University, Malaysia

ABSTRACT

Water crisis has been ranked amongst the top 10 global challenges with estimates predicting up to a 40 percent fall in freshwater resources by 2030. On top of that, water demand has far outpaced current population growth and experiencing water scarcity due to rising average global temperatures exacerbated by climate change. Hence, the environmental and sustainable management of water resources is a global imperative, and Integrated Water Resource Management (IWRM) has emerged as a key paradigm to address the multifaceted challenges associated with water management. This chapter discusses the dynamic interplay between IWRM and environmental governance through capacity developments emphasising their pivotal roles in promoting sustainable water resource management. In terms of the governance process, it is only possible to promote and apply IWRM through a well-defined policy for capacity development, ensuring transparency and accountability in water management and addressing inequality, gender biasness supported by available tools under Global Water Partnership (GWP).

DOI: 10.4018/979-8-3693-7001-8.ch015

1 INTRODUCTION

Water is a finite and indispensable resource, essential for life, livelihoods, and the well-being of our planet. Every human has the right to have access to clean water. According to the UN, water crisis has been ranked amongst the top 10 global challenges with estimates predicting up to a 40 percent fall in freshwater resources by 2030 (United Nations, 2018). On top of that, water demand has far outpaced current population growth with half of the global population experiencing severe water scarcity. It is only expected to increase due to rising average global temperatures exacerbated by climate change (United Nations, 2015). As a result, the availability of water globally is becoming more disproportionate and less predictable with some countries experiencing drought such as Ethiopia, Sudan, Eritrea, Afghanistan, China (provinces along the Yangtze river basin), Pakistan (Southwest Baluchistan, South Punjab, and Southeast Sindh), Iran and Somalia (Anderson et al., 2021; Ashraf et al., 2021; Liu et al., 2023; Rafiq et al., 2023) and others with intense storms such as Bangladesh, Philippine, Vietnam, Pakistan, Brazil (north-east) (Das et al., 2020; Yuen et al., 2024; Takagi 2019; Cotrim et al., 2022). The scarcity as well as abundance of rainfall, both negatively impacting and threatening the health and well-being of humans for sustainable socio-economic and environmental development. The escalating need for water in countries with limited water resources, combined with the rising occurrence of declining water quality in numerous areas, has placed further strain. Hence, the environmental and sustainable management of water resources is a global imperative, and Integrated Water Resource Management (IWRM) has emerged as a key paradigm to address the multifaceted challenges associated with water management. The term 'water governance' has gained momentum all over the world due to inception of integrated water resources management (IWRM) which focuses on effective, efficient and sustainable water management. Absence of water governance, laws and policy lead to conflicts in water abstraction, water sharing and allocation, management of water pollution and mitigation measures for floods. The foremost obstacles to water governance are poorly defined roles and responsibilities, improper coordination among ministries and agencies related to water supply and distribution, lack of accurate information between central and sub-national governments. Lack of awareness about water policy and poor involvement of water users' associations, and participation in water related projects are major obstacles to water governance. Good water governance is the driving force behind successful water management.

Water resource management agencies should be urged to embrace this comprehensive approach to water resource management. It suggests utilising contemporary technology such as geographic information systems (GIS), remote sensing, and data analytics to facilitate the adoption of IWRM. India uses GIS and remote sensing to discover aquifers and build up bore wells, rainfall collecting, and groundwater recharge facilities to sustain drinking water supplies (Sharma, 2019). For IWRM capacity building, open-source desktop GIS technologies are proposed to give developing country users a financial and technologically viable solution (Chen et al., 2010).

Technology is essential in water governance, enhancing management, distribution, and conservation of water resources. Satellite photography and remote sensing technologies furnish data on aquatic systems, land use, and climatic trends. This facilitates the oversight of water supply, quality, and use. Internet of Things (IoT) devices and sensors are utilised in aquatic environments, pipelines, and reservoirs to gather real-time data on water levels, flow rates, and quality (Suhaili et al, 2023). Advanced software tools employ data analytics to enhance the distribution and utilisation of water, minimising waste and assuring sustainable supply. Machine learning and AI models forecast water demand, droughts, and floods, facilitating proactive management and emergency intervention.

Real-time water quality monitoring sensors and data platforms provide the continuous assessment of water quality, identifying contaminants and assuring the safety of water for consumption and agricultural use (Silva et al., 2021; Manjakkal et al., 2021). Biosensors in biotechnology may identify toxins at minimal quantities, enhancing the safety of potable water. Mobile applications and platforms furnish real-time information on water consumption, quality, and availability, therefore empowering individuals and enhancing transparency in water administration. Real-time data and GIS mapping facilitate disaster response by pinpointing impacted regions and orchestrating relief operations.

1.1 What is IWRM?

Integrated Water Resources Management (IWRM) is defined as the process for promoting the coordination, development and management of water, land, and other related resources as a means to maximizing economic and social welfare in an equitable manner without compromising the sustainability of vital ecosystems (GWP, 2007). It is designed to illustrate the inherent interconnectedness of hydrological resources by highlighting the interdependency between various uses of water resources. It serves as an alternative to the fragmented and hierarchical management approach that has historically resulted in inefficient and unsustainable use of water resources. The acceptance of the principles of Integrated Water Resources Management (IWRM) is motivated by the acknowledgement of two crucial factors: firstly, the need to manage all aspects of the water cycle as a unified entity rather than individual components, and secondly, the importance of involving all stakeholders more extensively in decision-making processes to enhance the acceptance and legitimacy of management outcomes (Turton et al., 2007). The IWRM approach is to be considered as a long-term process rather than an exclusive approach as its implementation differs in each country due to differing history, socio-economic, cultural, political, and environmental context. The approach relies on the four pillars as its framework (i) a conductive atmosphere created by appropriate policies, tactics, and legislation to support sustainable development and management (ii) Implementation of institutional frameworks to enable the policies, strategies and legislations to be put into practice (iii) Setting up management instruments for institutions to perform their roles (iv) Creation of financial tools required by the developed management instruments.

In a global effort to overcome water challenges, IWRM can play a significant role in contributing to the UN Sustainable Development Goals, specifically goal number 6, which aims to ensure sustainable availability, safe and easy access to clean water for all. Goal 6 has six targets to be met by the year 2030 which target both water quantity and quality (United Nations, 2015). There are however two other goals which indirectly meet Goal 6:

- Goal 6A which aims to expand international cooperation and capacity building
- Goal 6B which aims to support and strengthen the participation of local communities to improve water management.

The goals are aligned with increasing sector-wide investment and capacity building and so the following key strategies are recommended:

- Enhancing investment and capacity-building across the entire sector.
- Encouraging the development and implementation of new ideas and practices based on
    solid evidence.
- Improving collaboration and coordination among all sectors and all parties involved.
- Implementing a comprehensive and interconnected strategy for water management.

These strategies are crucial for stimulating economic growth and enhancing productivity, while ensuring the sustainable utilisation of water resources and promoting access to safe water and sanitation. These factors are vital for the well-being and survival of both individuals and the environment.

1.2 Role of Environmental Governance for Promoting IWRM

The prevailing viewpoint of the Second World Water Forum was that the water crisis is not due to a scarcity of the resource, but rather a crisis in the management of existing water resources (World Water Forum 2000). The changes in management philosophy have led to a gradual shift from traditional engineering approaches to a more integrated planning approach. This new approach considers both conventional and nonconventional options to reconcile water supply and demand, including water conservation and demand management measures. The complete range of potential advantages of Integrated Water Resources Management (IWRM) has not yet been fully achieved. There is a growing agreement that this failure may be attributed to insufficient focus on establishing suitable governance structures. This is likely due to differing interpretations of what constitutes effective governance. Governance is a procedural framework, where the successful management of an ecosystem (or Integrated Water Resources Management) is the outcome of "good" governance. The likelihood of good governance increases in environments where there is a dominant political culture of transparency, accountability. Consequently, the shift in management approaches towards embracing the principles of Integrated Water Resources Management (IWRM) has led to a process of institutional decentralisation and democratisation. The pillars of IWRM are fundamental ideas that direct the formulation and execution of water management plans. The implementation of these elements guarantees the integrated, fair, and sustainable management of water resources. The three fundamental components of IWRM are the enabling environment, institutional framework, and management instrument (GWP, 2002; Garcia, 2008).

An enabling environment refers to the establishment of suitable policies, strategies, and regulations to facilitate the sustainable development and management of water resources. The implementation of new policies, initiatives, and laws can foster the establishment of an institutional framework.

The institutional framework refers to the practices and principles that promote social fairness, economic efficiency, and ecological sustainability in water management. It facilitates the implementation of policies, strategies, and regulations pertaining to water management. The institutional design also facilitates capacity building through the organisation of IWRM training, workshops, seminars, and other related activities. The concept of capacity building also encompasses the involvement of entities and platforms that might contribute to the improvement of institutions. The institutional structure can enhance awareness of water usage and support water conservation efforts. Water consumers, non-governmental organisations (NGOs), municipal authorities, water supply authorities, and local government authorities can discuss and coordinate their water usage practices within the established institutional structure and rules. The creation of Catchment Councils, Water User Associations, and Catchment Management Agencies in various regions of the world is a clear indication of these institutional changes (Turton and Meissner 2000; Turton 2002). The government actor-cluster has the responsibility of creating an environment and institutions that promote socio-economic growth by incorporating relevant scientific and social factors (Hattingh et al. 2005). At the core of environmental governance lies an implicit agreement between society and the government, referred to as a hydro-social contract (Warner and Turton 2000; Turton and Meissner 2002). This hydro-social contract encompasses the societal norms and values that govern the interactions between important stakeholders.

Management instrument focuses on establishing the necessary management instruments for these institutions enables the implementing authorities to carry out their duties with greater efficiency. For instance, implementing higher tariffs for water use as a component of economic management might incentivise customers to save water and utilise it more effectively.

The capacity development for stakeholders to implement good water governance is essential for the successful execution of IWRM as it establishes the legal, institutional, and regulatory structure required for the management of water resources. Formulating a comprehensive policy for capacity building in IWRM requires pinpointing the deficiencies in water management capabilities at different levels (national, regional, local), comprehending the requirements of diverse stakeholders such as government agencies, water consumers, civil society groups, and the private sector.

2. CAPACITY DEVELOPMENT FOR ENVIRONMENTAL GOVERNANCE AND IWRM

Capacity development refers to the process of developing and strengthening of soft and hard skills, explicit and tacit knowledge, abilities, processes, and resources that individuals, communities or organizations need to adapt to survive in a fast-changing environment and achieve their own development objectives over time. Capacity development in the context of IWRM is an important tool to raise the awareness of the IWRM pillars discussed previously and to enhance the overall quality of water governance structures under the boarder aspect of environmental governance. It requires an understanding of the obstacles that prevent a people, organization, or other relevant components within the institutional framework from fully envisaging their development goals. Through such initiatives, applicable mechanisms can then be found and discovered to overcome the challenges of environmental governance and implementation of IWRM. This may be taken through tangible or intangible forms with tangible capacity development examples including the provision of training manuals, holding workshops and implementation of new digital technologies. Intangible capacity development however may take shape in the form information sharing networks or initiatives to engage in self-reflection. Whereas the tangible aspects have been traditionally favored, tangible aspects are explicit, visible outcomes that can be readily quantified and assessed, such as the number of people trained for fixing leaks of pipe network system or the treatment of water by chemical dosing. These results are typically easier to communicate and rationalise to stakeholders, donors, and funding agencies.

2.1 Key Attributes of Capacity Development

The conceptual shift with capacity development into environmental governance and IWRM is made in line with the overall development policy changes from focusing on lending financial and human resource aid through foreign expertise to developing countries to focusing on nurturing and developing local capacities (UNDP, 2009). This shift in focus emphasises the importance of acquiring and utilizing knowledge in developing a stronger local capacity, ensuring effective long-term development in a more sustainable manner (Wehn de Montalvo and Alaerts, 2013). The main key attributes for capacity building are discussed below.

2.1.1 Knowledge and Capacity Development

Capacity with regard to environmental governance and IWRM can be defined as the capability of a society or organisation to identify and understand issues related to water and environment, act to address them, and to learn from experience and accumulate knowledge for the future. Capacity development should constitute an essential mindset for all parties involved. Capacity development itself is not a one-way process of transferring knowledge or skills from one party to another. It demands collaboration and participation that involves active engagement of all the stakeholders who have a role or interest in the capacity development outcome. This includes all the beneficiaries, partners, donors, authorities, and others who are able to influence or be influenced by the intervention for capacity gains. Engaging the stakeholders will ultimately help build trust, ownership, and commitment, as well as ensure relevant, effective, and sustainable efforts within pursuing capacity development.

Knowledge transfer can be categorized in three different levels of the individual level, the institutional level, and the societal level through an enabling environment.

- **Individual:** Initiated with improving skills and gaining new knowledge via training and experiences driven by self-motivation, values, and favourable incentives such as financial allowances. Importance should be on the type

of knowledge that is gained, categorized as explicit and tacit knowledge. Explicit includes spoken knowledge and can be well transferred through education and training. Tacit knowledge, however, is more difficult as it requires a deeper cultural understanding of the content to be understood. This can be done through peer learning or learning by doing.
- **Organizational:** Involves the improvement of performance through strategies, plans, partnerships, and departmental roles. Within the water sector, knowledge management techniques that could be adopted range from human resources management to internal knowledge dissemination through modes via peer-to-peer exchanges, encouraging networking or through communities of practice.
- **Enabling Environment:** Involves the improvement in the economic, socio-political environment, laws, policies and social norms within which people and organizations function. This can be realized through awareness raising, press and mass communication and by peer learning also.

A summary of this actions that are required to gain capacity through knowledge sharing can be found in Figure 1 providing the development tools and indicators of such actions.

Figure 1 Capacity and knowledge sharing through individual, organisation and enabling environment.

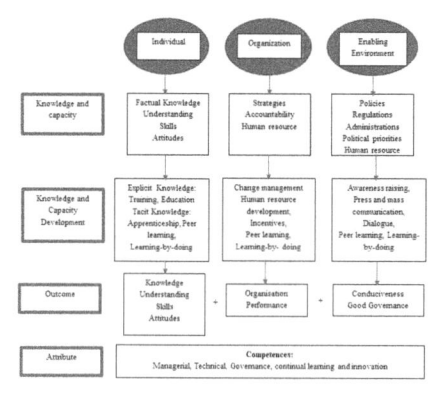

Capacity building of policymakers is essential for the effective execution of IWRM. When policymakers possess the requisite information, skills, and instruments, they can formulate, execute, and supervise policies that efficiently tackle water management constraints. A more comprehensive knowledge of water resources, ecosystems, and socio-economic aspects enables policymakers to make well-informed decisions that consider the interdependencies within the water cycle and across different sectors. The process of capacity building empowers policymakers to engage in strategic thinking and formulate plans for the enduring sustainability of water resources, rather than just concentrating on urgent or short-term concerns. It empowers policymakers to structure and execute inclusive procedures that engage all pertinent stakeholders, such as communities, non-governmental organisations (NGOs), the commercial sector, and other governmental entities. Engaging in capacity development for policymakers can facilitate the acquisition of skills in effective negotiation and mediation, therefore enhancing their ability to handle disputes related to water resources. Highly competent policymakers have the ability to enhance openness and accountability in water management, hence fostering increased public confidence in water policies and institutions.

2.2.2 Training Environment and Water Professionals

Developing the capacity of environment and water professionals allows for the human resource gaps to close in, allowing professionals to rapidly adapt to changing scenarios, as well as provide knowledge themselves, ultimately contributing to the learning culture needed in improving environmental and water governance. Techno-pedagogical content knowledge has emerged as a fundamental strategy in contemporary education and training (Shohel et al., 2021).

In building a multi-level capacity development programme, training is needed to equip individuals, organizations or communities with the right skillset and competencies to solve a variety of water-related problems. It is highly necessary due to insufficient availability of water professionals as many countries suffer certain water crises due to this. In a rapidly changing environment, flexibility and adaptability is also needed for professionals to acquire new skills or knowledge. The three main modes for training activities (Figure 2) to be carried out include:

- **Face to Face Training:** where a target group gains knowledge and skills in person in real time.
- **Online:** With the increasing popularity of online webinars, training opportunities have been widened with training programmes being offered online.
- **Blended Training:** Combining the in-person and online training allows for a sense of freedom whilst still grounded in reality as online training may be perceived as a virtual educational experience which can seem detached from reality.

Along with this, the method of which these training can be exchanged could be through individual or a group setting such as peer-to-peer exchange refers to a training format that may be used to stimulate knowledge sharing between multi-disciplinary professionals. Training of trainers is used for knowledge sharing with the individuals involved becoming trainers and are expected to teach their knowledge to others. It is considered a more sustainable method for training as it is both a dissemination process and an iterative process for the trainer themselves to learn and understand deeply of their own knowledge.

Figure 2 Modes of training for capacity building.

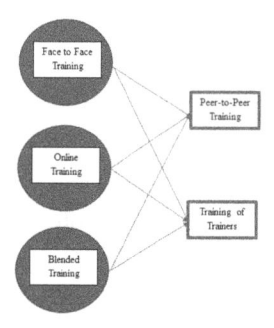

Digital courses, webinars, and virtual reality simulations may educate water managers, policymakers, and community leaders on the ideas, tools, and practices of IWRM. These platforms provide education accessible, irrespective of geographical location. Digital platforms for knowledge exchange and engagement among stakeholders (government agencies, NGOs, communities) can improve collective learning and promote partnerships for enhanced water management.

2.2.3 Gender Perspectives

Although women make up more than 50% of the worldwide population, they continue to be underrepresented and marginalised in the fields of policy formation and decision making at the local, national, and global levels. Although there has been progress in increasing women's presence in parliamentary representation in around 90 percent of nations in the past twenty years, women only make up 20 percent of members of parliament (UN, 2015). This lack of adequate representation is reflected in the decision-making and policy-making processes related to environmental challenges. The current social and institutional systems that result in women being more prevalent in precarious social and economic sectors often restrict their ability to shape the course of environmental progress (UNEP, 2015). The absence of women's involvement not only denies them the opportunity to contribute to decision-making processes, but also prevents society from benefiting from the unique perspectives, ideas, and experiences of half of the global population. Consequently, the needs of women, particularly those who are economically disadvantaged, are often disregarded. Gender disparity significantly affects the Sustainable Development Goals (SDGs) and good governance. A study (Women, 2020) has found that women are disproportionately impacted by poverty and a lack of access to financial, medical, and educational opportunities compared to males. Based on the data, the current rate of progress suggests that it will take an additional 25.7 decades for the gender wage gap to be completely closed worldwide (World Economic Forum, 2020). The involvement of women in policy decision-making is essential for attaining sustainable water management. Women frequently possess distinct ideas and experiences about water utilisation and management, especially in areas where they serve as the principal collectors and users of water. Incorporating women into decision-making processes guarantees that policies are more inclusive, just, and efficacious. Gender parity enhances economic growth and eliminating prejudice can potentially increase Gross Domestic Product (GDP) by as much as 34%. Addressing the gaps in water and sanitation, specifically in relation to SDG 6, has the effect of empowering women and girls (Jeevanasai et al., 2023).

2.2.4 Local Ownership and Leadership

Local can be defined through many levels such as national, subnational or on a community level. For simplicity, the term local is defined as the set of people or community of a specific geographical location affected. The capacity development implemented must be owned and managed by the individuals whose capacity is to be developed. External partners should only be able to support the process and create the right incentives. Local ownership and leadership are essential elements of capacity building for sustainable water management. When local communities assume responsibility for water management efforts and exhibit leadership, the results are more likely to be sustainable, culturally relevant, and successful over the long run. Local leaders and communities possess a profound comprehension of their unique water issues, cultural traditions, and environmental circumstances. This knowledge facilitates the creation of water management plans customised to the specific requirements of the region. Community-owned programs are more likely to be sustained and modified over time, as the community has a direct investment in their success.

2.2.5 Sustainable Change

Sustainable change is central for capacity development in terms of knowledge, skills, attitudes and formal rules that influence people's behaviours and relationships with one another. Change can only be sustained over a period of time as through a change in behaviour as social structures are one of the main factors in providing sustained change. To achieve change, attention needs to be given to the necessary changes needed in attitudes of individuals, professionals, and organizations as they are deeply rooted and embedded into them. Additionally, to create and sustain an enabling environment for capacity development, the main actions tackling the structure and framework of government that need to take place include:

- Analysis of the role of government policies and institutions in supporting capacity building for IWRM.
- Discussions of any relevant legislative or regulatory changes that have facilitated capacity-building initiatives.
- Exploring the sustainability of capacity building efforts in IWRM.
- Discuss the long-term impacts of capacity building on water resource management.
- Provide case studies showcasing enduring success.

Social change refers to the persistent transformations within the societal framework, marked by changes in cultural and behavioural norms, social structures, and value systems. The influence of positive changes in socio-cultural institutions, regulations, conduct, and value systems can significantly impact the management of water resources. There is a necessity to enhance the environmental awareness of water users by fostering a sense of shared responsibility. Changing and transforming people's mindset is ultimately a gradual process that cannot be hastily achieved; nevertheless, it has proven successful (Wals, 2015). Enhanced water governance is not solely reliant on tangible infrastructure and technological interventions but should also consider the attitudes in society and whether they align with fostering sustainability in the water sector. Therefore, it becomes crucial to alter the mindsets of stakeholders and individuals in positions of influence positively, fostering an appreciation for water as a tangible resource and recognizing the significance of water-related institutions, policies, laws, and plans.

It is extremely important that social change is targeted, as it is the programming of the internal mindset towards tackling greater water issues of the future. This will require a paradigm shift from current norms, attitudes, beliefs, and modes of operation in an attempt to build a more future proof and resilient water sector. This can be empowered through several initiatives:

- **Social Involvement:** Individuals should work to ensure effective participation of marginalized groups that have previously been excluded from the benefits of water i.e., women, indigenous groups, disabled individuals as they also make up a significant percentage of a population. With effective mentorship, educational exchanges and learning can foster social change.
- **Engagement of Private Sectors:** The imbalanced dispersion of wealth, societal turmoil, climate change, deteriorating biodiversity, and water scarcities have influenced business executives to grow increasingly mindful of the planet's pressing issues. This shift in awareness has persuaded them to reconsider the very foundations of their corporate operations, recognizing the imperative to adopt sustainable practices. Consequently, multinational corporations need to depart from conventional business models and embrace novel and fundamentally distinct approaches that prioritize profitability alongside ethical, social, and environmental responsibility.
- **Communication:** Using communication in a specific and thoughtful way can make people think differently, leading to new ways of dealing with challenges in managing water resources. When a problem is found, it is important to share information in a way that encourages individuals and communities to adopt better and more sustainable practices. Moving from random communications to carefully planned messages that speak directly to the audience and convince them to take action is crucial for bringing about meaningful change i.e., through meetings.

2.3 Challenges to Capacity Development for Environmental Governance and IWRM

Barriers to social change in IWRM can be complex and multifaceted. Some common barriers that can hinder social change. Communities characterised by entrenched norms, traditions, and beliefs owing to cultural rigidity can hinder societal advancement. The public image and societal acceptance of wastewater reuse can lead to the failure of several reuse initiatives (Azhoni, Jude and Holman, 2018). Perception frequently establishes obstacles to behavioural change, impeding total societal advancement. In the absence of education, social transformation may be significantly obstructed. To effect societal change, it is important to expand an individual's knowledge base. In the context of Integrated Water Resources Management (IWRM), water professionals must enable communities globally to get specialised knowledge regarding water-related concerns to achieve comprehensive literacy on the subject (Pouladi et al., 2021). The presence of laws, policies, and regulations at local, regional, or national levels may obstruct or inhibit the adoption of sustainable behaviour, constraining social development. Corruption, inadequate institutional frameworks, limited competence, and insufficient investment funding are all factors that influence water governance (Shunglu et al., 2022). Physical and constructed infrastructure can impede social transformation.

3 CASE STUDY

3.1 Background

India is considered to be one of the world's major irrigated countries with major shares in its water resource used for irrigation. However, irrigated agriculture was faced with water distribution and management problems, as the demand for the non-agricultural industries was also increasing rapidly, resulting in increasing pressure for available water resources. One of the major states heavily involved in agricultural production is Andhra Pradesh, with concentrated activity within its southern region in the densely populated catchments of Krishna and Godavari rivers. About 60 percent of Andhra Pradesh is in agricultural production and around 70 percent of its population relies on the agriculture for their livelihood. At the state level, irrigation developments in the upstream of the river catchments in the state of Karnataka and Maharashtra are increasing the overall uncertainty of water availability for irrigation use. This has resulted in tail end farmers receiving their water late or not at all. Because of such uncertainty, people have turned to groundwater supply, exploiting it to alleviate from the shortage where in rainfed areas. This could present a significant environmental hazard as the lower groundwater levels, along with increasing canal irrigation demands, have resulted in waterlogging, particularly at the southern end. Furthermore, the excessive use of groundwater has also caused the deterioration of wetlands, leading to salinisation through the infiltration of saltwater along the coastal areas and providing temporary assurance of an adequate drinking water supply for Hyderabad, which may have negative consequences in the future.

The problems and challenges experienced within agricultural irrigation was termed to be in a vicious cycle whereby the poor irrigation services resulted in low yields and income. This in turn contributed to low-cost recovery which resulted in poor operation and maintenance (O&M) due to inadequate funding. Cumulatively, it has resulted in poor irrigation systems.

3.2 Actions Taken

Funded by the Government of Andhra Pradesh, the Irrigation and Command Area Development Department (ICADD) proposed a reform programme to break this cycle through a whole-target approach, transforming it into the termed 'Virtuous

Cycle'. This involved the introduction of several simultaneous and rapid sequenced actions including a threefold increase in water tariffs, passing the Andhra Pradesh Farmers Management of Irrigation Systems Act and Rules, c creation of over 10,000 Water User Associations (WUAs) covering the entire irrigated landscape of Andhra Pradesh, introduction of the maintenance and rehabilitation programme and Irrigated Agriculture Intensification Programme (IAIP). Within the reform programme, capacity development relating to promoting IWRM involved central themes of public participation, grassroot movements, government cultural changes, and empowerment of people focusing on self-help and self-reliance. Knowledge and experience within the problems and challenges faced by the water sector was regulated through the creation of a network project of operational research known as the Andhra Pradesh Water Management project (APWAM).

3.3 GWP Toolbox

The APWAM project utilized the three main sub toolboxes under capacity development of (i) information gathering and sharing networks (ii) training water professionals (iii) communities of practice. With the main stakeholders involved in the APWAM project along with their roles included The Agricultural Department, the Irrigation and Command Area Development Department (ICADD).

The APWAM project's main aim was to contribute to increasing knowledge and experience in the IWRM by enhancing research participation and developing and implementing enhance IWRM skills and techniques to be adopted by farmers as well as long term monitoring of the IWRM skills. Additionally, the project aims to generate and disseminate knowledge and build research capacity for the improvement of agricultural water resources within canal and tank irrigations. This is realized through the facilitation of information exchange and data was further up scaled to allow all departments to use by dividing between three working groups: Remote Sensing (RS) and Geographical Information System (GIS) for the evaluation of the Canal Irrigation System performance, Hydrological Modelling and Irrigation Modernisation and Management Improvement. The project objectives were also to strengthen the partnership between researchers, policy makers, departments, NGOs and Water Users Association (WUAs). This was done on an individual level where training was done amongst scientists, farmers and government staff through activities relating to educational tours, seminars and conferences, awareness campaigns (On the State and farmer's level) and advisory service. The project was designed so that the network operates within the structure of the university, simultaneously providing a level of freedom to test with changes in the focus and implementation of the research. Stakeholders involved in the project include the engineers from agricultural officers from the Agriculture Department, to farmers from WUAs and scientists from research institutions, ensuring that sustainable use of water is made by the agricultural sector. Through this, each stakeholder's capacity is gained towards IWRM.

3.4 Outcomes

The APWAM initiative has achieved its objectives under several dimensions within the irrigated agricultural sector. Enhanced agricultural crop and water management tools implemented in the experimental regions led to a general increase in water productivity. Achieving up-scaling of pilot area findings using remote sensing enables the realisation of canal water savings of up to 40% in the Krishna Western Delta. Engaging in participatory irrigation water management led to water conservation of up to 30% and eliminated water conflicts among farmers, ensuring that even the most economically disadvantaged farmers have access to enough tank water. The strategic reserve status of groundwater has been established, with a proposed increase of cultivated area up to 30%. Implementation of sub-surface drainage systems to address deltaic environmental deterioration and salinization, achieving a 50% increase in water productivity.

3.5 Lessons Learned

The pilot demonstrations, together with the capacity development process, are essential for advancing enhanced IWRM practices across all stakeholders. Extensive time is necessary to establish mutual confidence between stakeholders and project personnel. The project accomplishments represented a paradigm for enhancing tank irrigation in areas including technical, managerial, and participatory aspects. The application of a capacity development plan may not be suitable for a distinct Integrated Water Resources Management IWRM specific problems. Enhanced involvement may be attained by collaborative efforts focused on local community leaders exchanging effective strategies and collaborating to jointly develop solutions for comparable water challenges.

4 CONCLUSION

IWRM offers a never-ending sequence of challenges that are always unique to a local region or community. This can be caused by various problems within the engineering side of delivering and managing water resources to the right people but can also be caused by the capacity of the people and organizations behind the delivery of such contents that can play a bigger role in the efficiency of the water resource management. One of the many ways this can be improved is through good water governance as part of environmental governance whereby the working capabilities and capacity of individuals can be developed even further to allow better connections between different stakeholders. In terms of the governance process, it is only possible to promote and apply IWRM through a well-defined policy for capacity development, ensuring transparency and accountability in water management and addressing inequality, gender biasness supported by GWP toolbox. One solution may not fit with a different management structure. Therefore, it is an iterative process which should have in mind a continuously evolving framework rather than a one-off end goal. The use of technology into IWRM is essential for capacity building and environmental governance, as it enhances water management, improves decision-making, and promotes sustainable development. Satellite images, drones, and Geographic Information Systems (GIS) can assess aquatic environments, land use, and ecological transformations. These instruments deliver real-time data, allowing the precise evaluation of water quality, volume, and ecosystem vitality. Sensors installed in aquatic environments, pipelines, and watersheds may incessantly assess characteristics such as water flow, quality, and pollution levels. This data facilitates the early identification of problems and underpins proactive management.

5 RECOMMENDATIONS

In promoting IWRM through capacity development within developing countries, the case study is a good example which can be used but not implemented without considering social, economic and environmental contexts. Relevant stakeholders would include the water works department, academic researchers, and engineers whereby meetings could be held in identifying significant challenges faced with streamlining cooperation amongst one another. Through this evaluation, the capacity development can be programmed in more depth. Challenges in integrating such programmes as previously pointed out is to deal with the social changes and lack of capacity itself. Initiation of such changes need to be inflicted internally by those trying to gain capacity. This must be coordinated by the respective ministries such as local government, water resources, through policies that provide an enabling environment for the individual and organizational levels to enhance their capabilities.

6 REFERENCES

Anderson, W., Taylor, C., McDermid, S., Ilboudo-Nébié, E., Seager, R., Schlenker, W., Cottier, F., De Sherbinin, A., Mendeloff, D., & Markey, K. (2021). Violent conflict exacerbated drought-related food insecurity between 2009 and 2019 in sub-Saharan Africa. *Nature Food*, 2(8), 603–615. DOI: 10.1038/s43016-021-00327-4 PMID: 37118167

Ashraf, S., Nazemi, A., & AghaKouchak, A. (2021). Anthropogenic drought dominates groundwater depletion in Iran. *Scientific Reports*, 11(1), 9135. DOI: 10.1038/s41598-021-88522-y PMID: 33911120

Azhoni, A., Jude, S., & Holman, I. (2018). Adapting to climate change by water management organisations: Enablers and barriers. *Journal of Hydrology (Amsterdam)*, 559, 736–748. DOI: 10.1016/j.jhydrol.2018.02.047

Chen, D., Shams, S., Carmona-Moreno, C., & Leone, A. (2010). Assessment of open source GIS software for water resources management in developing countries. *Journal of Hydro-environment Research*, 4(3), 253–264. DOI: 10.1016/j.jher.2010.04.017

Cotrim, C. D. S., Semedo, A., & Lemos, G. (2022). Brazil wave climate from a high-resolution wave hindcast. *Climate (Basel)*, 10(4), 53. DOI: 10.3390/cli10040053

Das, M. K., Islam, A. S., Karmakar, S., Khan, M. J. U., Mohammed, K., Islam, G. T., Bala, S. K., & Hopson, T. M. (2020). Synoptic flow patterns and large-scale characteristics of flash flood-producing rainstorms over northeast Bangladesh. *Meteorology and Atmospheric Physics*, 132(5), 613–629. DOI: 10.1007/s00703-019-00709-1

Garcia, L. E. (2008). 'Integrated water resources management: A 'small'step for conceptualists, a giant step for practitioners'. *International Journal of Water Resources Development*, 24(1), 23–36. DOI: 10.1080/07900620701723141

GWP. (2002) Integrated Water Resources Management, TAC Background Paper #4, Global Water Partnership Technical Advisory Committee, Stockholm: 2000.

GWP. (2007) *IWRM Explained*. Available at: http://www.gwptoolbox.org/learn/iwrm-explained#:~:text=of%20GWP's%20strategy-What%20is%20IWRM%3F,vital%20ecosystems%20and%20the%20environment.

Hattingh, J., Claassen, M., Leaner, J., Maree, G., Strydom, W., Turton, A. R., Van Wyk, E., & Moolman, M. (2005a) Towards innovative ways to ensure successful implementation of government tools. Report ENV-P-I 2004-083. Pretoria, CSIR.

Jeevanasai, S. A., Saole, P., Rath, A. G., Singh, S., Rai, S., & Kumar, M. (2023). Shades & shines of gender equality with respect to sustainable development goals (SDGs): The environmental performance perspectives. *Total Environment Research Themes*, 8, 100082. DOI: 10.1016/j.totert.2023.100082

Liu, Y., Yuan, S., Zhu, Y., Ren, L., Chen, R., Zhu, X., & Xia, R. (2023). The patterns, magnitude, and drivers of unprecedented 2022 mega-drought in the Yangtze River Basin, China. *Environmental Research Letters*, 18(11), 114006. DOI: 10.1088/1748-9326/acfe21

Manjakkal, L., Mitra, S., Petillot, Y. R., Shutler, J., Scott, E. M., Willander, M., & Dahiya, R. (2021). Connected sensors, innovative sensor deployment, and intelligent data analysis for online water quality monitoring. *IEEE Internet of Things Journal*, 8(18), 13805–13824. DOI: 10.1109/JIOT.2021.3081772

OECD. (2008). The Challenge of Capacity Development. *OECD Journal on Development*, 8(3), 233–276. DOI: 10.1787/journal_dev-v8-art40-en

Pouladi, P., Badiezadeh, S., Pouladi, M., Yousefi, P., Farahmand, H., Kalantari, Z., Yu, D. J., & Sivapalan, M. (2021). Interconnected governance and social barriers impeding the restoration process of Lake Urmia. *Journal of Hydrology (Amsterdam)*, 598, 126489. DOI: 10.1016/j.jhydrol.2021.126489

Rafiq, M., Cong Li, Y., Cheng, Y., Rahman, G., Zhao, Y., & Khan, H. U. (2023). Estimation of regional meteorological aridity and drought characteristics in Baluchistan province, Pakistan. *PLoS One*, 18(11), e0293073. DOI: 10.1371/journal.pone.0293073 PMID: 38033048

Sharma, S. K. (2019). Role of Remote Sensing and GIS in Integrated Water Resources Management (IWRM). In Ray, S. (Ed.), *Ground Water Development - Issues and Sustainable Solutions*. Springer., DOI: 10.1007/978-981-13-1771-2_13

Shohel, M. M. C., Ashrafuzzaman, M., Islam, M. T., Shams, S., & Mahmud, A. (2021). 'Blended teaching and learning in higher education: Challenges and opportunities', *Handbook of research on developing a post-pandemic paradigm for virtual technologies in higher education*, pp.27-50. DOI: 10.4018/978-1-7998-6963-4.ch002

Shunglu, R., Köpke, S., Kanoi, L., Nissanka, T. S., Withanachchi, C. R., Gamage, D. U., Dissanayake, H. R., Kibaroglu, A., Ünver, O., & Withanachchi, S. S. (2022). Barriers in Participative Water Governance: A Critical Analysis of Community Development Approaches. *Water (Basel)*, 14(5), 762. DOI: 10.3390/w14050762

Silva, G. M. E., Campos, D. F., Brasil, J. A. T., Tremblay, M., Mendiondo, E. M., & Ghiglieno, F. (2022). Advances in technological research for online and in situ water quality monitoring—A review. *Sustainability (Basel)*, 14(9), 5059. DOI: 10.3390/su14095059

Suhaili, W. H., Patchmuthu, R. K., & Shams, S. (2023). October. The Internet of Things in the Rearing of Giant Freshwater Prawn: A Pilot Study. In *2023 6th International Conference on Applied Computational Intelligence in Information Systems (ACIIS)* (pp. 1-6). IEEE.

Takagi, H. (2019). Statistics on typhoon landfalls in Vietnam: Can recent increases in economic damage be attributed to storm trends? *Urban Climate*, 30, 100506. DOI: 10.1016/j.uclim.2019.100506

Turton, A. R., Hattingh, J., Maree, G. A., Roux, D. J., Claassen, M., & Strydom, W. F. (2007). *'Governance as a Trialogue: Government-Society-Science in Transition', Water Resources Development and Management Series, ISBN-10 3-540-46265-1*. Springer-Verlag Berlin Heidelberg. DOI: 10.1007/978-3-540-46266-8

Turton, A. R., & Meissner, R. (2000) Analysis of the influence from the international hydropolitical environment, driving the process of change in the water sector (Appendix F). In: Report of the institutional support task team – shared rivers initiative of the Incomati river. Scottsville and Harare, Institute of Natural Resources (INR) and Swedish International Development Agency (SIDA). Available online at https://www.up.ac.za/academic/libarts/polsci/awiru

UNDP (2009) Capacity Development: A UNDP Primer, Available at https://iwrmactionhub.org/resource/capacity-development-undp-primer.

UNEP, 'Policy and Strategy for Gender Equality and the Environment 2014-2017', February 2015, para. 21.

United Nations. (2015) *Water and Sanitation - United Nations Sustainable Development*. Available at: https://www.un.org/sustainabledevelopment/water-and-sanitation/

United Nations. 'The Millennium Development Goals Report 2015', 2015, p.5. Available at: https://www.undp.org/content/dam/undp/library/MDG/english/UNDP_MDG_Report_2015.pdf p. 5

United Nations. (2018) *Water Action Decade - United Nations Sustainable Development, United Nations*. Available at: https://www.un.org/sustainabledevelopment/water-action-decade/

Wals, A. E. J. (2015). Social Learning-Oriented Capacity-Building for Critical Transitions Towards Sustainability. In *Schooling for Sustainable Development in Europe* (pp. 87–107). Springer International Publishing., DOI: 10.1007/978-3-319-09549-3_6

Wehn de Montalvo, U., & Alaerts, G. (2013). Leadership in knowledge and capacity development in the water sector: A status review. *Water Policy*, 15(S2), 1–14. DOI: 10.2166/wp.2013.109

World Economic Forum. (2020) Global gender gap report 2020, Retrieved from https://www.weforum.org/reports/gender-gap-2020-report-100-years-pay-equality

World Water Forum. (2000) Final report: second world water forum and ministerial conference. The Hague, Ministry of Foreign Affairs.

Yuen, K. W., Switzer, A. D., Teng, P. P., & Lee, J. S. H. (2024). Statistics on Typhoon Intensity and Rice Damage in Vietnam and the Philippines. *GeoHazards*, 5(1), 22–37. DOI: 10.3390/geohazards5010002

Chapter 16
Green Socio-Ecological-Technological Innovation Policymaking and Strategies

José G. Vargas-Hernandez
https://orcid.org/0000-0003-0938-4197
Tecnològico Nacional de Mèxico, ITS Fresnillo, Mexico

Francisco Javier J. González
Tecnológico Nacional de México, ITSF, Mexico

Selene Castañeda-Burciaga
https://orcid.org/0000-0002-2436-308X
Universidad Politécnica de Zacatecas, Mexico

Omar Guirette
https://orcid.org/0000-0003-1336-9475
Universidad Politécnica de Zacatecas, Mexico

Omar C. Vargas-González
https://orcid.org/0000-0002-6089-956X
Tecnológico Nacional de México, Ciudad Guzmán, Mexico

ABSTRACT

This study aims to analyze the relationships between policymaking and strategies for socio-ecological, environmental, and green socio-technological innovation. The analysis departs from the assumption that the environment of a socio-ecological and socio-technological innovation can be fostered by policymaking and strategies

DOI: 10.4018/979-8-3693-7001-8.ch016

aimed to provide the transformation of pure research on scientific knowledge into innovative commercial products and services. The method employed is the analytic-reflective based on the theoretical and empirical literature review. It is concluded that policymaking and strategies accelerates the creation and development processes of socio-ecological, environmental, and green socio-technological innovation.

INTRODUCTION

Currently, the relevance of generating innovation processes is recognized as a way to achieve economic development, as well as the sustainability of organizations. To this end, it is necessary to promote green technological innovation, in addition to public intervention to promote various political strategies, which, in turn, have an impact on society. In other words, we are talking about the collegiate participation of the public, social and business sectors.

The technological innovation trajectories are inseparable from the scale and scope economies, mass production, consumption, and the expansion of global markets. The global commitment to economic growth and social development is linked to technological and socio technological innovations unmeshed in complex systems in scale and scope, involving supranational institutions, national and local governments, transnational and multinational corporations, trading blocks, etc.

Global economic development, political revolution, technological innovation change and resource scarcity are some factors that have an impact on socio-ecological interactions at regional and local levels (Haber et al., 2006). Global economic growth has become an unsustainable pattern of socio-ecological and technological innovation development for solving some operational challenges such as investments and incentives, global supply chains and delivery service models, etc. Innovative service models are perceived as turning innovation into solutions for problems.

In the logic of mass production and technological innovation systems characteristic of market spaces make possible to conceive uncoupling high technology and global scale innovations increasingly used in medicine, biotechnology, electronics, etc. (Oladapo et al., 2021). The economic and socio-ecological subsystems lead to an integrated renewable resources innovation model based on quantitative methods and information technologies to forecast processes and objects suitable for systemic crises. Technological innovation is stimulated by direct access information. Integrated models of socio-ecological economy and socio-humanitarian systems are based on quantitative methods and innovative technologies to forecast and control environmental conditions of uncertainty, stability, and risks.

Theoretical models of systemic sustainability promote the emergence of a global living technological laboratory network innovative initiatives and the emergence of an auto-catalytic socio-eco-technical system focusing on individual projects of systemic sustainability that contribute to the collective creation for a thrivable planet (Laszlo, et al., 2012). Integrated models based on quantitative methodologies and innovative technologies are used to forecast and control the dynamics of socio ecological-economic and humanitarian systems under conditions of complexity, uncertainty, and instability (Ramazanov, 2019).

Scientific knowledge on the innovation of socio-ecological economic development using methods for science and technology policies in aggregated economic models have been developed by Heets & Seminozhenko (2006), Klebanova & Kizima (2010), Kleiner (2013), Ayvazyan et al. (2012) and others. Socio-ecological innovation for economic development supports individual and group creativity to create and develop technologies to market more effectively (Marxt & Brunner, 2012).

Research and development have integrated mathematical models, methods, and innovation technologies to make decisions, forecast and manage the dynamics of socio-ecological, economic, and socio-humanitarian systems under conditions of uncertainty, complexity, and risk. The integrated models used to approach socio-ecological-economic and humanitarian systems are supported by the emerging concept of sustainable development as the combination of the economic, social, and environmental dimensions framed by knowledge creation, technology innovation and other factors influencing the environment and the structure of economic growth (Heets & Seminozhenko, 2006; Ramazanov, 2018; Babenko et al., 2019; Solou, 2002).

Research and development of an integrated stochastic model of the dynamics of socio-ecological and economic, innovative, and humanitarian systems and processes for optimal forecast and control in innovation economy is necessary to function at the macro and micro levels in complicated conditions of crises, uncertainty, and instability. Integrated stochastic models of the dynamics of socio-ecological economic growth and development and socio-humanitarian systems, including other sectors such as production, R&D and education considering socio-ecological innovation based on mathematical methods and innovation technology in conditions of uncertainty, instability, and risk, must be created and developed.

Innovation studies can identify three sub-bodies of literature towards responsible research and innovation, mission-oriented policy of innovation and capacious innovation. Responsible research on innovation is a theoretical construct and a policy concept is linked to tradition of assessment technology (Schot & Rip 1997). Responsible research innovation is the transparent and interactive processes where the societal actors and innovators are mutually responsible with ethical acceptability, sustainability and social desirability of the innovation process, products and services with control over the research direction, development of technology and innovation to stakeholders (von Schomberg, 2012). Responsible research innovation advocates democratic control intervening in innovation processes.

This study is created based on the interrelationships of three main topics. It begins with an analysis on technological innovation followed by the socio-ecological, environmental, and green socio-technological innovation and finally it analyzes the policymaking and strategic implications to provide some conclusions.

TECHNOLOGICAL INNOVATION

Economic globalization brings changes in world-wide challenges to improve the economic growth at the cost of competition and degradation of natural resources and socio-ecosystems, and labor division supported by technological innovation and post-industrialization innovative technologies aimed to create new opportunities in changing production modes and consumption patterns (Dimitrova, 2013). The techno-economic is the dominant technological paradigm resulting from technological and organizational innovations that cause economic, social, organizational, and institutional impact (González Álvarez, 1997). The technological paradigm shift affects the traditional economic sectors requiring the modernization of production processes and the adoption of management techniques. The techno-economic standard is formed by a set of principles applied in the generation of technological and organizational innovations producing economic growth.

Modernity has increased the per capita access to energy, resources, and materials associated to complexity overhead available by a consistent innovation pattern of innovative technologies and gadgets affluence of consumer durables. Modernization refers to the development of specific applications of technological innovation that have an impact on efficiency and quality. Revolutionary innovations and scientific shamanism, remains as had pertained since the mid-18th century, the same as state peasants in their relationship to land. The technological innovations and management structures and practices after 1960s made the herders to change techniques and practices (Pelto, 1973). Orthodox innovation institutions are facilitative or constraining, static and inert with capacious innovation conceptualizations leading to institutional entrepreneurship operating with technological innovation and change.

The component of technological innovation versus traditional approaches as for example the Porter's Five Forces Model, organizational business themselves for competitive advantage. Technology innovation is a change factor responding to creative management action in the strategic role within the innovation process and not in a static system such as the 5 Forces model of Porter (Henderson, 1994). The contributions of Porter (1980) to the analysis of innovation related to processes and products in corporate strategy although he underestimates the ability of management to implement innovation strategies the capacity for change and technological discontinuities of the industrial structure.

The innovation modes are the science and technology model, the science, technology and innovation and the doing, using, and interacting models, they have differentiated knowledge bases of innovation. These models aimed to predict the state and the economy innovation and development result from the dynamics of capital, labor knowledge and innovative technology resources, consumption, investments, and natural resources.

Evolution provides the analytical instrument to encompass the organizational network with the conditions to assimilate technological innovations (Trist, 1976; OECD, 2001c). Change framework considers that social transformation and evolution benefits from the utilization of localized visions leading to local diversification of social and technological innovation for attracting the postcapitalist world through the real time and responsive networks to support the nods for the needed learning the innovations developed. The evolutive societal innovation interactions from a macro-perspective links the global economic, demographic, and environmental developments with energy metabolism and technological innovation changes (Krausmann & Fisher-Kowalski, 2013)

Open innovation diplomacy is related to the different ecosystems and organizations able to collaborate and bridges the separated domains of social, organizational, technological, and cultural (Carayannis & Campbell, 2011). Incumbent organizations can dominate the technological frontier, increasing their preeminence with their tenure. Sustainable society is built with knowledge and experience development in transformation and innovation processes becoming strategic variables implying economic, social, technical, technological, political, and cultural innovation.

Management of complex organizational systems develop and function considering the conditions of instability to ensure increasing economic growth supported by achievements of innovation technology and science. Management innovation strategies lead to processes that combine information technologies, quality, and innovation processes, etc. to achieve significant market positions. The managerial component of the strategic advantage of well-used technology is crucial because managing innovation technology effectively requires a clear understanding of the evolution, maturation, and diffusion of the technology itself throughout the global economy (Price, 1996).

Organizational technological change is paradoxically associated with organizational stability. Defenses imposed by the management inabilities to face situations of strategic nature decisions affect organizations in case of technological innovation discontinuities and rejection of change. The balance of an organization can be seriously affected many times by social impacts produced by the introduction of a technological innovation. Models of change management contribute to balance evolution and the process of adaptation become more sensitive internal and externally to management experience, financing, and innovative technological resources, making the organizational technological and social aspects in an innovation system.

The organizational size has positive effects in the technological industry is because the market tends to be more independent. Recent cites of new patents are from firms elaborating in current developments of leading age technological innovation fields.

Setting the focus to provide a platform for contextual technologically innovative designs framed by a meta-level objective for a thrivable planet to catalyze the emerging networking initiatives and activities through the system level.

Urban living labs are defined as the physical region where stakeholders form public-private-people partnerships among public agencies, businesses, universities, and users can collaborate to create, develop, validate and test new and innovative technologies, products, services, and systems in contexts (Juujärvi & Pesso, 2013). The strategy Research and Innovation for Smart Specialisation (RIS3) is an ex-ante condition for receiving support to achieve a shift in the structure to solve the problems of standardized regional innovation policies (Landabaso, 2014; Tödtling & Trippl, 2005).

Socio-Ecological, Environmental, and Green Socio-Technological Innovation

Innovation is creating knowledge and technologies to be used in socio-ecological sustainability and the use of natural resources. Natural resources and commitments develop innovative socioeconomic, ecological, and technological contexts, leading to diversification and transitioning to alternative production forms.

The inclusion of natural environment in the quintuple helix for knowledge and innovation model is the component for knowledge innovation and production serving for the survival, preservation and vitalization of humankind, green technologies, sustainable development and socio-ecology become the societal knowledge and innovation production (Carayannis & Campbell, 2010). Environmental pressures result in technological innovations such as demands of market, competition, government regulations and internal pressures such as improvement of working quality of life and active planning that enable organizations to cope with changes. Active planning starts from the normative level including the organizational, group and members values with its actors, and the community in which they work (de Melo, 1997b).

The dynamism of innovative environments based on the sociological perspective treats it as and organizational ecology leading to the technological innovation component compared to traditional approaches to small businesses seeking competitive advantages.

Every day, human future may exceed technoscientific innovation engaged on socioecological futures as cultural inheritances are being within particular localities engaged with future scenarios (Gibson et al., 2015).

The socio-ecological model is more implicit in the transformation towards sustainable society, deserving more attention in smart specialization policies which has different ends that the science and technology model (Coenen & Morgan, 2019). The socio-ecological model of innovation is the research and innovation strategies discerned in RIS3 (Jensen et al. 2007; Asheim et al. 2017; European Commission 2012). The smart specialization promoted by research and innovation strategies is predicated on an innovation model of conventional science and technology (European Commission 2012).

Smart urban innovation technology is a means to improve urban life leading to the risk of surveillance and loss of privacy are prevalent (Hollands, 2008; Kitchin, 2014). The increased automation and roboticization of services and other domains using artificial intelligence and smart innovation technologies increases the risks of innovation (Karvonen et al. 2019).

Ecosystems services of high nature added value delivery through public interventions aimed to empower communities aimed to foster technological innovation in multifunctional landscapes. This emerging situation attracts the systemic sustainability to demonstrate the socio-ecological innovations spanning the economic, social, technological, infrastructural, and agricultural domains (Laszlo, et al., 2012). Socio-ecological innovation for economic development is supported by creativity of individuals and groups to create and develop knowledge and technology (Marxt & Brunner, 2012).

The effectiveness of socio-ecological and technological innovation in transition to sustainable development implementation must consider parameters such as innovation regulatory policies, the rational use of production resources, the political situations, functioning of the economy and market performance approach to analysis of renewable industrial purposes resources, goods, and services, investments, demographics (The European Sustainable Development Strategy, 2001).

The socio-ecological sustainable transitions approaches bring changes in the societal system in which the socio-political actors share a set of rules while the subordinated social groups share different sets of rules leading to distinguishing the regimes among the technological innovation, socio-ecological, socio-cultural, socio-technical and policy regimes. The socio-technological system is a set of heterogenous and interlinked material and technical elements such as structures and artefacts that encompasses societal production, diffusion, and use of technology with societal functions such as communications, transports, nutrition, etc., represented through networking of actors and social groups with formal, cognitive, and normative rules guiding activities (Papachristos, Sofianos, & Adamides, 2013).

The sociotechnical concepts from both social and technological innovation perspectives provide the basis for formulating and implementing innovations making possible to design flexible organizational structures to facilitate communication more compatible with innovative management methods and content (de Melo, 1997). Socio-economic and socio-cultural factors are linked between them and affect socio-technological innovation and development, and therefore all these factors have a relevant significance in the determination of the socio-ecological niche including group values, norms, attitudes, labor, and land tenure allocations to community and household tasks, livelihood strategies, labor organization and marketing, food supply and demand, household food preferences and habits, etc.

The socio-ecological development poses biophysical limits to economic growth on a planet of finite resources unless technological innovations and systems overcome the biophysical costs at the expenses of socio-ecological spaces and enhance health and quality of life of people in local communities (Daly & Farley 2004). Large-scale transition to socio-ecological and technological innovation has unpredictable and inevitable uncertain consequences in economic growth, social justice, equality, equity, and environmental sustainability of communities, improvement of quality of life in a horizon moral enhancement (Beck & Beck-Gernsheim, 2002; Bauman, 2000). From the socio-ecological perspective, technology innovation has no power but with human agency fulfills functions associated with social and organizational structures (Geels, 2002) and other combinations of transition forms are required to achieve sustainability transitions.

Environmental sustainability strategies are the efficiency, consistency in eco-innovation, and sufficiency which can be measured to address the environmental sustainability crisis. The consistency strategy of environmental sustainability aims to the restructuring of economies though socio-ecological and technological innovations supported with resource productivity and enhanced efficiency measures for decoupling economic growth and environmental damage (Gouverneur & Netzer, 2014). All types beyond technological innovation involving actors and agents in the strategy setting and development

Greening economy and investments in eco-innovation and clean innovation technologies are aimed to reduce the destructive levels of temperature increase (Rockström et al. 2016). And the demand for green innovation and knowledge society solutions utilizing socio-ecological and economic resources, technology, and entrepreneurship as the drivers. Environmental regulations stimulate the green technological innovation activities. Environmental regulation affects green technological innovation (Luo et al. 2021). More intense and strict environmental regulations strengthen the effectiveness and efficiency of green finance on green technological innovation. As environmental regulations become more stringent, it motivates the firm to achieve green technological innovation with green finance investment (Luo et al. 2021).

Environmental regulations positively moderate the relationship between green finance and green technological innovation. Firms adapt their green technological innovation strategy to become more outcome oriented, increasing the effect of green finance. Green finance has a positive effect in driving green technological innovation. An adaptive approach addresses the issues of managing and planning the technological innovation processes, technology transfer and training in innovative environments. Social and technological innovation in organizations leads to competitive advantage able to mobilize technological knowledge, skills, and experience to develop the innovation strategy to create and develop processes innovation and market leading products and services that allow to adapt and respond to the threats of the changing environment and become successful (Tidd et al, 1997).

Green technology innovation comprises the stages of research and development, start-up, patenting, upscaling, dealing with risk, commercialization, and adoption. The factors influencing these stages are the organizational setting, the technology-specific characteristics, the funding available, and the commercial market conditions (Pizzol & Andersen, 2022). Adoption of new green technology innovation requires support from training and preparation to avoid the harmful effects on organizational development in less developed and peripheral regions (González Álvares, 1997), ranging from the waste of technology underuse to alterations of organizational social fabric.

Technological innovation autonomy through adequate technical training of efficiency in human resources and infrastructure must be available to organizational and institutional actors (Dias, Rosenthal & de Melo, 2000). Technological infrastructure must be compatible with the local capacities for innovation development of efforts by national and local development agencies which results in less risks and lower costs (OECD, 1999a). Building the infrastructure and capital investment for scientific research and technological innovation lead to consulting and training people to operate successfully (OECD, 1999a, 2001b, 2001c).

The local socio-ecological factors contribute to the design of the socio-ecological niche concept in the context of biophysical variables at the local level that co-determine and constrain productivity. These factors in combination with interactions delineate the socio-ecological niche for innovation technology. The concept of a sociological niche and the concept of innovation share similarities to accept the existence of technology if there is an appropriate balance of mix between technical development and socio-organizational arrangements (Leeuwis & van den Ban, 2004).

Socio-ecological niche is a concept like ecological niche of an organism in classical ecological theory. The concept of socio-ecological niche is defined and used in research and development as a framework to integrate technologies in farming systems, to analyze the factors operating at various levels and delineate a procedure to identify compatible technology development. A niche is defined as the region n-dimensional and multidimensional space and hyper-volume of environmental factors affecting the welfare of species (Hutchinson, 1957). The concept of the socio-ecological niche in research and technology development can be tailored according to the specific needs of farming systems.

The concepts of socio-ecological niche and innovation or complete technology share similarities. The niche defined by a legume innovation technology combine several factors to fulfill the needs of farmers to find a suitable socio-ecological niches (Leeuwis & van den Ban 2004). The definition of socio-ecological niche is dependent on the notions of farmers and cannot receive blanket recommendations for technology. Institutional support services are cross-cutting and relevant to technological innovation integrated into the socio-ecological niche and not shown as separate layer.

POLICYMAKING AND STRATEGIES

The societal ecosystems and environmental services recognition delivered by high nature value systems lead to public policies interventions aimed at empowering rural communities, foster technological innovation, and promote multifunctional landscapes. Innovations production for sustainability in the nation state leading to circulation of knowledge aimed to promote sustainability through innovations obtained from research in science and technology. Innovation in policymaking is tight to research in geography innovation leading to place-based innovation approach (Tödtling & Trippl 2005; Barca et al. 2012). The geography of innovation literature is limited to the conditions, and skewed to kinds of innovations such as technology driven and market-based (Shearmur et al. 2016)

Policymaking must enact more diverse well-designed environmental policies regulations to stimulate firms to develop green technological innovation, considering the use of the latest technological innovations without implementing huge research and development funds, green subsidies, and green innovation protection policies (Percival et al. 1992). Policymaking can formulate and implement market-based environmental regulations and policies to leverage market forces and encourage the development of green innovation technologies (Zameer and Yasmeen 2022).

There are diverse types of designed innovation institutions including formal organizations and informal networks to facilitate information flow, resources, and knowledge between firms aimed to add economic value to the scientific knowledge and technological innovation generated in the research institutions and universities and transferred in terms of research results to market institutions. Institutional set-ups of successful regional economies support by programs of regional innovation policies such as innovation support agencies, technological funding schemes, science parks, cluster policies, smart specialization industries, etc. (McCann & Ortega-Argilés, 2013).

Regional innovation policy consists of horizontal measures in financial schemes, technology transfer, etc. Smart strategic specialization is a vertical non-neutral measure. Regional innovation policies are aimed at supporting regional endogenous innovation potential by the diffusion of innovative technologies supported by strategic alliances and partnerships between academic institutions, public research centers, small, medium-sized, and large enterprises in vertical co-operation and between enterprises themselves in horizontal cooperation.

Science and technology models of innovation focusing on entrepreneurial activities may be supported by the implementation of smart strategic specialization to achieve sustainable transition considering that policymaking pursuing economic competitiveness may contradicts with social well-being, ecological integrity, and environmental cleanness. Smart strategies, specialization policies and cluster initiatives are part of the regional innovation policy. The place-based approach of regional development policy argue in favor of endogenous potential regionalization (Barca et al., 2012; OECD, 2011). Smart strategy specialization is a policy instrument to foster regional innovative capability predicated on conventional science and technology innovation models of policies 1.0 and 2.0 and the socio-ecological model of innovation has been less relevant and mentioned (Foray et al., 2009).

Innovation policymaking is framed by innovation policies 1.0, 2.0 and 3.0. Innovation policy 1.0 becomes part of science and innovation technology policy based on research and development (R&D) based innovation with an emphasis on commercialization of scientific and technological knowledge. Innovation policy 2.0 is the system with the objectives of economic growth, competitiveness, and job creation linked between discovery and commercial use of applied knowledge.

Innovation policy 3.0 is the explicit mobilization of science, technology and innovation knowledge for meeting sustainable societal needs addressing the Sustainable Development Goals, sustainable and inclusive societies (Schot & Steinmueller, 2018). Innovation policy 3.0 is under debate including various scalar and spatial variations (Fagerberg 2018). Innovation policy 3.0 mobilizes science, technology, and innovation for meeting the societal needs and the sustainable development goals addressed by the United Nations. Previous generations, innovation 1.0 and innovation 2.0 only motivated to resolve structural failures Coenen et al., 2015; Schot & Steinmueller, 2018).

Green technology development distinguishes the shared framework conditions from technology specific ones leading to supporting the green innovation policies. Diffusion of findings and innovations in scientific and technological research support policymaking processes facilitated by communication (Klabbers et al. 1996).

Sustainable Development Goals are required to implement along with major transformations in policies and technologies, modes of transformative innovations involving stakeholders in inclusive and multi-scale.

The historical context had shaped science and technology policies to incorporate innovation as the strategic component for scientific and technological development. Business incubators, technological parks, and other institutions of support of technological innovation environments aimed to increase the creation and development of innovative technologies for innovation in products and services and the promotion of innovative organizational performance.

An environment of innovation systems has the objective to transform scientific knowledge and innovation technology into organizational business, products and services with social value, economic growth and environmental sustainability facilitated through business incubators, technology parks, technology transfer offices and marketing. Private or public incubators located in universities and research centers, house traditional, technology or mixed firms linked to technological sectors in need of modernization and with resources limitations.

Innovation environment systems have objectives such as to promote technological knowledge innovation; encourage science-based knowledge innovations, establish relationships of cooperation and trust, provide return on capital investments, generate an organizational culture, etc. Innovation environments apply a broad set of objectives, some of which may be related to universities linking with firms for resources exploitation, creation and support of new technological firms' regional development policies and economic revitalization of depressed regions.

There are various similar terms to name a technological innovation park such as science park, research park, innovation park, innovation center, business park, etc. (González Álvarez, 1996; Monck et al., 1988; Löfsten & Lindelöf, 2001). The environment of a technological innovation park can provide the transformation of pure research into commercial products and services under the assumption that scientific research is the origin of technological innovation (Löfsten & Lindelöf, 2001). Innovation parks need to significantly increase the specialized knowledge and technological assets centered in the universities to attract firms and institutions to provide research facilities and critical mass of technicians and researchers helping to develop industrial clusters in areas such as information technology, biotechnology, telecommunications, biomedical equipment, pharmaceuticals, chemicals, plastics, fibers, education, etc. (Porter, 2001).

Technological innovation parks serve private and public interests promoting a favorable innovation environment through making risky decisions, exchanging technical and market information, investing and cooperation in infrastructure, and making efficient gains meeting the needs and expectations of regional and local communities (OECD, 2001a). Technological innovation parks on the public interest are a strategic alternative to industrial policy aimed to accelerate the regional development taking advantage of local characteristics with low level intervention and public resources which can be leveraged by a private interests motivated by a tangible cost/benefit proportion (Pimienta-Bueno, 1999).

Firms involved in technological innovation have as objectives to access to the agendas, laboratories and equipment of research institutes and universities through contracts, involve students in industrial practices and projects, recruit graduates, promote consulting and growth of technology-based parks and incubators, encourage synergy in scientific research, development, and technological innovation, etc. Identifying and attracting firms with innovation projects for joint innovation action are process required to provide and support with resources and capabilities, scientific knowledge and technological innovation, infrastructure and facilities, financing, equipment, technology transfer and dissemination to expand the potential of firms (Lalkaka, 1999).

Technological-based firms need resources and capabilities which can made available by research centers and university settings in an innovation environment with technological base such as incubators and technology parks facilitates the attraction of other actors and agencies such as government institutions promoting innovation, consulting agencies, financing and investing, etc. University-based local innovation systems as a model are a source of inspiration acting as mechanisms for the technology generation, dissemination and transfer as their raison d'être innovation and entrepreneurial activity in an environment of technology-based companies. Firms in local innovation systems change their profile since a university based local innovation system is surrounded by emerging technology-based firms with the research centers and institutes as the sources of technology innovation and access to qualified resources and capabilities.

Generation of an innovation dynamic environment for firms can enable regional economic growth and social value based on a scientific and technological agenda for research institutions and universities based on research and development to attend and meet the demands of firms, communities, and society. These changes of the innovation dynamics contribute positively to the transformation of the research university as an institution capable of creating and disseminating scientific knowledge and technological innovation to benefit the society linking the specificities of the regional vocation.

Governments and organizations are focusing on establishing socio-ecological economy system through the intensification of economic and management institutional transformations supported by science and technology innovation policies aimed to improve the quality of the local community's development. Government policies support local innovation systems and innovation removing barriers and promoting the transformation of scientific and technological advances into improvement of economic performance and enhancement of living conditions (Wiig & Wood, 1995). The different government instances are expected to contribute to enable the policymaking processes to support innovation, the creation and formation of local innovation systems, the development of technological innovation capabilities.

The local innovation system has as a central element the research and higher education center as the technological resources provider and systems innovation. Research institutes and universities are active actors in creation, dissemination and application of scientific knowledge and technological innovation that may have the objectives to encourage and facilitate links with firms and industries, access to the state-of the art R&D activities, transfer of technological innovation commercialization of academic research and technological innovation, encourage the emergence of spin-off and start-up industries, create jobs and consulting opportunities, generate return on investments, etc.

The concept of local innovation models and systems demonstrates the advantages in relation to the conceptualization of university-based innovation systems in local contexts. Local innovation systems are community-driven in line with local demands of the community and better understanding of local environments, requiring more planning, technological sophistication, and financial support (Schlapfer & Marinova, 2001; de Melo, 2001). Local Innovation System builds local economic and social institutional structures aimed at the scientific and technological knowledge and skills in the local communities. The local innovations systems are a form in which the university is the larger provider of technological resources for innovation.

Local university-based innovation systems are characterized by their research and development institutes and centers in a research and development university with an infrastructure, laboratories, and the interactions of students with scientists and technologists with converged objectives, access to resources and communication which become more efficient under an institutional framework. Local innovation systems research universities based have a prominent function in research and development of scientific knowledge and innovation technology acting as elements of an innovation structure activities of Research and Development institutions, firms, anchor and collaborating companies aimed to generate new ventures, university start-up and emerging research university.

Under the local innovation systems approach, local and community development strategies seek the creation and formation of innovation skills for economic growth, social development and environmental sustainability supported by research centers, universities and training institutions linked to the relationships with the productive sector. These innovation linkages are aimed to stimulate the creation, construction and formation of organizational capabilities and competencies leading to the formation and development of innovation networks and cooperation in the production, dissemination, transfer, and use of scientific and technological knowledge (D'Avila

Garcez, 2000; de Melo, 2001). Local innovation systems may organize technological innovation events aimed to attract potential users and financial sources to entrepreneurs in technological innovation

Research centers and universities are engines of socio-technological innovation, knowledge creation and transfer to the regional economic growth (Saxenian, 1994; Kenney, 2000; Druilhe and Garnsey, 2000). Research institutes and universities can unite research projects around local community common goals, creating scientific and technological innovation knowledge, disseminating, transferring, and using it, training qualified work, and becoming a decisive component of socio-economic performance in some regions effecting societal changes in an economy of knowledge.

Local industrial systems are the result of an evolution process centered around the university as the genesis. Stanford University was instrumental in the establishment of the Silicon Valley becoming as the model for regional development from the agricultural sector to the leader in emerging companies and start-ups in microelectronic and information technology with leading roles in dictating the technological innovation trends markets for performance. This region is home of highly specialized firms of technological innovation organized to recombine the different subsystems formed by specialized suppliers.

Local innovation system by The Massachusetts Institute of Technology – (MIT) as the spatial concentrations of companies, including specialized suppliers of equipment and services and customers, and associated non-commercial institutions, universities, research institutes, training institutions, standard setting agencies, local trade associations, regulatory agencies, technology transfer agencies, trade associations, relevant government agencies, and departments that combine to create new products and/or services in lines of business (Lester, 2005).

In England, University of Cambridge has become the sources of technological innovation developed, abilities and capabilities to create technology intensive firms in their environment and interacting with the market. The policy of the university free professors to exploit their inventions protecting their intellectual property encouraging to create new firms in business incubators and technological parks aimed to launch start-ups.

Local innovation systems give support to multiple goals of agents and actors giving evidence that there is a localized physical proximity for relationships of trust transactions to share scientific and technological knowledge perspectives in local environments that are attractive to produced results in socio-technical innovation and entrepreneurial activities. The local innovation system must be able to converge multiple institutional objectives, manage resources to preserve integrity and provide feedback to raise the identification, identity, coordination, and governance aimed to service the creation, diffusion, spread and use the scientific and technological knowledge among the market, non-market organizations and anchor companies with the greatest influence.

The feedback loop of the local innovation system produced is explicit on the environmental factors produced through the analysis of performance. The proximity of its physical and interest agents facilitates to install processes of creation, dissemination, transfer, and use of scientific and technological innovation knowledge in the innovative activities of organizations. Local innovation Systems provide their communities with scientific and technological knowledge that suit their needs.

Local innovation systems have similar characteristics the national systems of innovation aimed to transform scientific and technological advances into goods and services with potential economic use and social value. Local innovation systems focus more rationally on regional and communities' interests (Schlapfer & Marinova 2001). National and local governments, authorities and development agencies have as objects technological innovation to stimulate the creation and formation of technology-based companies, generate employment, improve performance, reduce decline and imbalances in research and development activities in depressed regions, reproduce international experiences and technological innovation transfers to local reality.

The concept of National Innovation Systems builds on national institutions is constituted by a set of agencies and organizations of a society that promote the transformation of scientific and technological advances of goods and services, facilitate learning and transfer processes. These national institutions develop to become the lobby of powerful groups of interest to meet the emerging economic, social, cultural, and ecological needs and to promote the development and implementation of innovative technologies with long-term benefits.

Innovative technology transfer is an imperative to invest considering the local context in less developed countries (Schlapfer & Marinova, 2001) to be compatible with local indigenous sources of knowledge.

Transition management is a process of variation and selection leading to the gradual transformation of societies from existing innovative technologies and practices by coevolutionary steering (Lachman, 2013; Kemp, Loorbach, & Rotmans, 2007). The transition management of a societal system is a cyclical process that includes strategic levels and networks to initiate visions and paths upon transition agenda implemented at tactical level focusing on structural institutional regulatory barriers, economic conditions, specific innovative technologies, physical infrastructures, consumer routines, etc. leading to long-term goals.

The cooperation perspective (Melo, 1997) provided through collaboration agreements is a strategic skill to develop technologies and complementary skills as it has been considered in technological research (Price, 1996) and technological training of the technological innovation processes through the innovation networks (González

Álvarez, 1997; OECD, 2001c; Porter, 2001; de Melo, 2002). Organizational, processes and product arrangements conducted in cooperatives and modernization poles of traditional technology sectors incorporate international advances in development efforts in innovation technologies through joint and shared efforts among institutions, government agents and other actors to generate technological knowledge. These efforts of cooperation are essential for small enterprises (González Alvarez, 1997).

Comparative Analysis: Innovation Policies and Strategies

The formulation of environmental regulation policies has been approached from different contexts, either through a governmental approach or through policies implemented within companies. It should be noted that policies and strategies for socio-ecological innovation are closely related, since environmental regulations can encourage innovation in companies, either through the application of incentives or through legislative regulations that contemplate certain types of sanctions.

With regard to the above, the research conducted by Sheng et al. (2020), whose study focuses on analysing stakeholders in environmental regulation, is noteworthy, the research was carried out considering environmental policies in China. The interactions between the national government, local governments and companies were analysed. The study was able to identify the factors that influence stakeholder strategies and enable the development of policies that are compatible with various incentives. The relevance of the study by Sheng et al. (2020) is that it highlights the role of governments and the sanctions they impose to reduce stakeholder conflicts. It also details the need for governments to carry out monitoring activities to ensure the effectiveness of environmental regulatory policies and to increase penalties and incentives.

Another study that looks at environmental regulations is Song et al. (2020), who explored the impact of environmental regulations and R&D tax incentives in the Chinese context. The authors analysed thirty regions over nine years, which allowed them to identify the factors that impact on the development of sustainable practices. The results of the study showed that there is a complex relationship between business innovation, environmental sustainability and government intervention in the context of a rapidly changing economy. It highlighted that greater intensity of environmental regulation is reflected in inhibition to promotion.

However, within the local context of companies with respect to the development of internal policies, the study by Carrasco et al. (2023), who determined the profitability of thirty-four Ecuadorian companies with environmentally friendly policies accredited by certifiers, stands out. They used panel data models, which describe the significance level of the capital structure indicators and observe how the data relate to the profitability of the firms. Their results show that companies with environmental policies do not have problems in meeting their financial obligations; however, there are no differences in capital structure and profitability with respect to companies that do not use environmental policies.

On the other hand, eco-innovation is also influenced by environmental policies. For example, Tu and Wu (2020) analysed how eco-innovation has a direct impact on firms' competitive advantage. The study looked at a number of Chinese firms and aimed to discover the role of the mediator of organisational learning in the relationship between eco-innovation and competitive advantage, considering policies and stakeholders as moderating factors. Its results led to the development of a theoretical model with various proposals for companies that want to become more sustainable and competitive.

Another study that revisits eco-innovation strategies is Ahmed et al. (2023), who investigated the impact of eco-innovation on environmental performance that has a direct influence on organisational performance. The study was conducted in textile companies in Pakistan, the results reflected the importance of eco-innovation in driving sustainable development and thereby improving organisational performance.

It is important to mention that there is still a theoretical gap between environmental regulation and so-called eco-innovation. However, these concepts are now beginning to be analysed, together with that of sustainable performance. Such is the case of the study by Maldonado et al. (2024), who investigated these concepts in 460 manufacturing firms in Mexico through confirmatory factor analysis and covariance-based structural equation modelling. They concluded that environmental regulations have a positive effect on the sustainable performance of firms, as they also enhance both eco-innovation activities.

Case Studies: The Debate on the Role of SMES

Socio-ecological, environmental and green socio-technological innovation is an issue that is becoming more relevant for the development of any organization, where it is essential for its processes to be efficient, supported by new technologies, in addition to promoting sustainability and a friendly treatment of the environment.

In this regard, there are several success stories that can be taken up, especially within small and medium-sized companies, whose operation differs in terms of equipment and availability of resources, compared to larger companies. In accordance with what was previously pointed out, the study conducted by Vargas et al. (2017), who conducted research focused on hotel SMEs in Mexico and Colombia, stands out. Their main results show different priority issues, such as energy recovery from waste, collection and use of rainwater, design of new services or products, as well as bioclimatic construction.

Another study that has addressed environmental issues in small and medium-sized companies is Bermeo et al (2024), who analyzed the case of an Ecuadorian company in this sector. According to the authors, companies face challenges in complying with environmental standards, which impacts their competitiveness and innovation. The company analyzed stands out for its use of various sustainable strategies, such as international alliances, as well as the use of drones for environmentally friendly spraying.

Similarly, the study carried out by Cañas (2023), which analyzes the strategies for the sustainability of small and medium-sized companies, based on the experience of triple impact companies in Colombia, stands out; according to the author, micro, small and medium-sized companies play a very important role in the business fabric since they implement social and environmental actions that have an impact on society.

In this same sense, the research by López (2021), carried out in the Peruvian context, stands out with respect to technological innovations in the export market of small and medium-sized companies. His main results showed that the increase in technological innovation in the value chain is associated with an increase in the levels of internationalization of companies, because it improves the participation of companies in foreign markets by increasing their advantage.

It is possible to recognize how the relationship between technological innovation and environmental sustainability, which together can contribute to a greener and more resilient future (Plúas-Arias et al., 2024), becomes more relevant in small and medium-sized companies, which must compete in a globalized environment and with larger companies with greater capacity for action.

CONCLUSION

The aims of this study is to analyze the relationships between policymaking and strategies for socio-ecological, environmental, and green socio-technological innovation assuming that the environment of a socio-ecological and socio-technological innovation are fostered by policymaking and strategies to provide the transformation of pure research on scientific knowledge into innovative commercial products and services, leading to the conclusions that policymaking and strategies accelerates the creation and development processes of socio-ecological, environmental, and green socio-technological innovation.

Social and technological innovation models in high natural value systems, environmental and ecosystems services must be stimulated linked by rewarding the stakeholders by delivery to society. Technological innovation must adapt the capabilities and develop where there are information resources, institutions, organizations and individuals with knowledge and competences to make available to society.

The historical context involving the legal frameworks, government and financing agencies and business and academic communities shaped by science and technology incorporates innovation as the strategic component for science and technological development. Creation and development of technological knowledge and innovation policy in resource management is a context dependent variable and institutional process.

In intensive competitive environments, technological innovation in organizational innovation is an imperative for the stagnant survival. Industrial ecology combines technological innovation changes in processes and products and organizational innovation. Technological innovation as a source of competitive advantage for organizations has advantages on the components of social capital leading to facilitate socio structural change in organizations and society. Applying the socio-ecological niche concept, technological innovation products can be rationalized based on biophysical performance, socio-cultural and economic issues fitting in the socio-ecological systems.

The socio-ecological transition needs further exploration to contribute to the different sustainability dimensions to increase the capabilities for flourishing and reconciling with economic growth, employment, productivity, and technological innovations. Transformation is a long-term perspective that integrates social and technological innovations with forms of cooperation, innovations in socio-ecological sustainability and the interests of the action's groups with knowledge, social and human resources.

Depending on the societal paradigm with impacts on productive organizations requires the cooperation of necessary skills to overcome the arising difficulties from innovative technologies to obtain better opportunities and advantages offered by technological innovation. Institutions supporting socio-ecological innovation play a relevant role in raising cooperation for sustainable development and green growth in organizations facilitating the flow and transfer of technological innovation knowledge and ideas, information and human resources and nature to enhance the environment and maintain their vitality.

From the conception to the large-scale adoption, environmental innovation using green technologies remains a challenge. Upscaling of green technology innovations is a complicated process. Green financing process of small-, medium- and large-sized enterprises need to be supported by environmental government policies and regulations to increase their capabilities of green technological innovation.

The public policies and objectives of technological innovation should be known, clearly analyzed, and interpreted to be adapted organizational activities subject to coordination structures of the established environment in specific locations in the context of local innovation systems

Further research needs to be conducted in technological innovation and change with institutional entrepreneurship. Future research must analyze the contingency of environmental regulations as the moderators of green finance on green technological innovation.

REFERENCES

Ahmed, R. R., Akbar, W., Aijaz, M., Channar, Z. A., Ahmed, F., & Parmar, V. (2023). The role of green innovation on environmental and organizational performance: Moderation of human resource practices and management commitment. *Heliyon*, 9(1), e12679. DOI: 10.1016/j.heliyon.2022.e12679 PMID: 36660461

Asheim, B., Grillitsch, M., & Trippl, M. (2017). Introduction: Combinatorial knowledge bases, regional innovation, and development dynamics. *Economic Geography*, 93(5), 429–435. DOI: 10.1080/00130095.2017.1380775

Ayvazyan, S. A., Afanase, M. Y., & Rudenko, V. A. (2012). Some questions of specification of three-factor models of the company's production potential, taking into account intellectual capital. *Applied Economics*, 3(27), 56–66.

Babenko, V.; Perevozova, I.; Mandych, O.; Kvyatko, T.; Maliy, O. & Mykolenko, I. (2019). World informatization in conditions of international globalization: factors of influence. *Global. J. Environ. Sci. Manage.*, 5(SI): 172-179, 2019. DOI: DOI: 10.22034/gjesm.2019.SI.19

Barca, F., McCann, P., & Rodríguez-Pose, A. (2012). The case for regional development intervention: Place-based versus place-neutral approaches. *Journal of Regional Science*, 52(1), 134–152. DOI: 10.1111/j.1467-9787.2011.00756.x

Bauman, Z. (2000). *Liquid Modernity*. Polity.

Beck, U., & Beck-Gernsheim, E. (2002). *Individualization: Institutionalized Individualism and Its Social and Political Consequences* (1st ed.). SAGE Publications Ltd.

Bermeo Tinoco, E. O., & Ortiz Cañares, S. N. (2024) *Estrategias ambientales y prácticas sostenibles en pymes: caso Aeroagripac S.A.* (trabajo de titulación). UTMACH, Facultad de Ciencias Empresariales, Machala, Ecuador. http://repositorio.utmachala.edu.ec/handle/48000/22631

Cañas, D. (2023). *Estrategias para la sostenibilidad de las Pequeñas y Medianas Empresas – MiPyMES – en Santander, basado en la experiencia de las empresas de Triple Impacto*. Universidades Tecnológicas de Santander. Pp. 1-69. http://repositorio.uts.edu.co:8080/xmlui/handle/123456789/12730

Carayannis, E. G., & Campbell, D. F. J. (2010). Triple Helix, Quadruple Helix and Quintuple Helix and how do knowledge, innovation and the environment relate to each other? A proposed framework for a trans-disciplinary analysis of sustainable development and social ecology. *International Journal of Social Ecology and Sustainable Development*, 1(1), 41–69. https://www.igi-global.com/bookstore/article.aspx?titleid=41959. DOI: 10.4018/jsesd.2010010105

Carayannis, E. G., & Campbell, D. F. J. (2011). Open Innovation Diplomacy and a 21st Century Fractal Research, Education, and Innovation (FREIE) Ecosystem: Building on the Quadruple and Quintuple Helix Innovation Concepts and the "Mode 3" Knowledge Production System. *Journal of the Knowledge Economy*, 2(3), 327–372. DOI: 10.1007/s13132-011-0058-3

Carrasco, L., Valdivieso, P., & Herrera, A. (2023). Estructura de capital y rentabilidad de las empresas ecuatorianas amigables con el medio ambiente en el período 2015 – 2021. *Iberian Journal of Information Systems and Technologies,* (60). 240-252. https://media.proquest.com/media/hms/PFT/1/6a1EV?_s=v3kFLwTSzfZnTOK49nTKNry4nzI%3D

Coenen, L., Hansen, T., & Rekers, J. V. (2015). Innovation policy for grand challenges: An economic geography perspective. *Geography Compass*, 9(9), 483–496. DOI: 10.1111/gec3.12231

Coenen, L., & Morgan, K. (2019). Evolving geographies of innovation: Existing paradigms, critiques, and possible alternatives. *Norsk Geografisk Tidsskrift*.

D'avila Garcez, C. M. (2000). Sistemas Locais de Inovação na Economia do Aprendizado: Uma abordagem conceitual. [Dezembro]. *Revista do BNDES*, 14(7), 351–366.

Daly, H. E., & Farley, J. (2004). *Ecological Economics: Principles and Applications*. Island Press.

de Melo, A. (2001). *The Innovation Systems of Latin America and the Caribbean*. Inter-American Development Bank. Agosto.

de Melo, M.A.C. (1977b). Articulated incrementalism: a strategy for planning (with special reference to the design of an information system as an articulative task). Tese de Doutorado, University of Pennsylvania, Filadélfia.

de Melo, M. A. C. (1997). *Processo de Planejamento e as Inovações Tecnológicas e Sociais: uma perspectiva sócio-ecológica*. Anais do 5o. Seminário de Modernização Tecnológica.

de Melo, M.A.C. (2002a). *Enriquecendo a Atuação de Incubadora de Emrpesas. Tecnologia e Inovação: experiências de gestão nas micro e pequenas empresas.* PGT/USP. de Melo, M.A.C. (2002b). Inovação e Modernização Tecnológica e Organizacional MPMEs: o domínio interorganizacional. *Seminário Internacional: Políticas para Sistemas Produtivos Locais de MPME*: 2002

Dias, A. B., Rosentha, L. D., & de Melo, M. A. C. (2000) Enriquecendo a Atuação de Incubadoras de Empresas. *XXI Simpósio de Gestão da Inovação Tecnológica. Co-autores: Adriano Batista Dias e David Rosenthal.* São Paulo: 2000.

Dimitrova, A., ... (2013). Literature review on fundamental concepts and definitions, objectives, and policy goals as well as instruments relevant for socio-ecological transition, *WWWforEurope Working Paper*, No. 40, WWWforEurope, Vienna.

Druilhe, C., & Garnsey, E. (2000), 'The incubation of Academic Spin-off Companies', Proceedings of the High Technology Small Firm Conference 2000, 22-23 May, University of Twente, Enschede, The Netherlands (published as ' Academic spin-off ventures: a resource-opportunity approach' in During, W., Oakey, R. and Kauser, S. Eds. 2001, New Technology Based Firms in the New Millennium, Oxford: Pergamon).

European Commission. (2012). *Guide to Research and Innovation Strategies for Smart Specialisations (RIS 3)*. Publications Office of the European Union.

Fagerberg, J. (2018). Mobilizing innovation for sustainability transitions: A comment on transformative innovation policy. *Research Policy*, 47(9), 1568–1576. DOI: 10.1016/j.respol.2018.08.012

Geels, F. (2002). Technological transitions as evolutionary reconfiguration processes: A multilevel perspective and a case-study. *Research Policy*, 31(8-9), 1257–1274. DOI: 10.1016/S0048-7333(02)00062-8

Gibson, C., Head, L., & Carr, C. (2015). From incremental change to radical disjuncture: Rethinking everyday household sustainability practices as survival skills. *Annals of the Association of American Geographers*, 105(2), 416–424. DOI: 10.1080/00045608.2014.973008

González Álvarez, M. D. (1997). *Processos de Planejamento nos Pólos Tecnológicos: um enfoque adaptativo. Tese de Doutorado, Departamento de Engenharia Industrial*. PUC-Rio.

Gouverneur, J., & Netzer, N. (2014). Take the Wheel and Steer! Trade Unions and the Just Transition. In *State of the World 2014. State of the World*. Island Press., DOI: 10.5822/978-1-61091-542-7_21

Haberl, H., Winiwarter, V., Andersson, K., Ayres, R. U., Boone, C., Castillo, A., Cunfer, G., Fischer-Kowalski, M., Freudenburg, W. R., Furman, E., Kaufmann, R., Krausmann, F., Langthaler, E., Lotze-Campen, H., Mirtl, M., Redman, C. L., Reenberg, A., Wardell, A., Warr, B., & Zechmeister, H. (2006). From LTER to LTSER: Conceptualizing the socio-economic dimension of long-term socio-ecological research. *Ecology and Society*, 11(2), 13. Retrieved September 3, 2010, from www.ecologyandsociety.org/vol11/iss2/art13/. DOI: 10.5751/ES-01786-110213

Heets, V. M., & Seminozhenko, V. P. (2006). Innovative Prospects of Ukraine, Kh.: Constant.

Henderson, R. (1994). Managing Innovation in the Information Age. *Harvard Business Review*, 72(01), 100–107.

Hollands, R. G. (2008). Will the real smart city please stand up? Intelligent, progressive, or entrepreneurial? *City (London, England)*, 12(3), 303–320. DOI: 10.1080/13604810802479126

Hutchinson, G. E. (1957). *Concluding remarks*. On WWW at http://www.gypsymoth.ento.vt.edu/sharov/PopEcol/lec11/niche.html. Accessed 11.11.2005

Jensen, M. B., Johnson, B., Lorenz, E., & Lundvall, B. Å. (2007). Forms of knowledge and modes of innovation. *Research Policy*, 36(5), 680–693. DOI: 10.1016/j.respol.2007.01.006

Juujärvi, S., & Pesso, K. (2013). Actor Roles in an Urban Living Lab: What Can We Learn from Suurpelto, Finland? *Technology Innovation Management Review*, 11(3), 22–27. http://timreview.ca/article/742. DOI: 10.22215/timreview/742

Karvonen, A., Cugurullo, F., & Caprotti, F. (Eds.). (2019). *Inside Smart Cities: Place, Politics and Urban Innovation*. Routledge.

Kemp, R., Loorbach, D., & Rotmans, J. (2007). Transition management as a model for managing processes of co-evolution towards sustainable development. *International Journal of Sustainable Development and World Ecology*, 14(1), 78–91. DOI: 10.1080/13504500709469709

Kenney, M. (Ed.). (2000). *Understanding Silicon Valley: the anatomy of an entrepreneurial region*. Stanford University Press. DOI: 10.1515/9781503618381

Kitchin, R. (2014). The real-time city? Big data and smart urbanism. *GeoJournal*, 79(1), 1–14. DOI: 10.1007/s10708-013-9516-8

Klabbers, J. H. G., Swart, R. J., Janssen, R., Van Vellinga, P., & van Ulden, A. P. (1996). Climate science and climate policy: Improving the science/policy interface. *Mitigation and Adaptation Strategies for Global Change*, 1(1), 73–93. DOI: 10.1007/BF00625616

Klebanova, T.S. & Kizima, N.A. (2010). Models of estimation, analysis and forecasting of social and economic systems, *Monogr. Kh.:* FOP Liborkina L.M., VD "INZHEK".

Kleiner, G. B. (2013). System Economics and System-Oriented Modeling. *Economics and Mathematical Methods*, 3, 71–93.

Krausmann, F., Fischer-Kowalski, M., Grunbuhel, C., & Krausmann, F.. (2008). Socio-ecological regime transitions in Austria and the United Kingdom. *Ecological Economics*, 65(1), 187–201. DOI: 10.1016/j.ecolecon.2007.06.009

Lachman, D. (2013). A survey and review of approaches to study transitions. *Energy Policy*, 58, 269–276. DOI: 10.1016/j.enpol.2013.03.013

Lalkaka, R., & Abetti, P. (1999). Business Incubation and Entreprise Support Systems in Restructuring Countries. *Creativity and Innovation Management*, 8(3), 197–209. DOI: 10.1111/1467-8691.00137

Laszlo, A., Blachfellner, S., Bosch, O., Nguyen, N., Bulc, V., Edson, M., Wilby, J., & Pór, G. (2012). *Proceedings of the International Federation for Systems Research Conversations,* 14-19 April 2012, Linz, Austria / G. Chroust, G. Metcalf (eds.); pp.41-65 https://hdl.handle.net/2440/77156

Leeuwis, C., & van den Ban, A. (2004). Communication for Innovation in Agricultural and Rural Resource Management. *Building on the Tradition of Agricultural Extension.* Oxford: Blackwell Science (in cooperation with CTA).

Lester, R. K. (2005). Universities, Innovation, and the Competitiveness of Local Economies A Summary Report from the Local Innovation Systems Project – Phase I Industrial Performance Center Massachusetts Institute of Technology 13 December 2005 MIT Industrial Performance Center Working Paper 05-010

Löfsten, H., & Lindelof, P. (2001). Science parks in Sweden — Industrial renewal and development? *R & D Management*, 5(3), 309–322. DOI: 10.1111/1467-9310.00219

López, C. (2020). *Relación entre la innovación tecnológica y el proceso de internacionalización de las pymes exportadoras de café verde de la selva central* [Tesis de Licenciatura]. Universidad Continental, Huancayo, Perú. https://hdl.handle.net/20.500.12394/8468

Luo, Y., Salman, M., & Lu, Z. (2021). Heterogeneous impacts of environmental regulations and foreign direct investment on green innovation across different regions in China. *The Science of the Total Environment*, 759, 143744. DOI: 10.1016/j.scitotenv.2020.143744 PMID: 33341514

Maldonado-Guzmán, G., Molina-Morejón, V., & Juárez-del Toro, R. (2024). Efectos de las regulaciones medioambientales en la eco-innovación y el rendimiento sustentable en la industria automotriz mexicana. La granja. *Revista de Ciências da Vida*, 39(1), 78–91. DOI: 10.17163/lgr.n39.2024.05

Marxt, C., & Brunner, C. (2012). *Analyzing and improving the national innovation system of highly developed countries: the case of Switzerland Technol*. Forecast. Soc. Chang., DOI: 10.1016/j.techfore.2012.07.008

McCann, P., & Ortega-Argilés, R. (2013). Modern regional innovation policy. *Cambridge Journal of Regions, Economy and Society*, 6(2), 187–216. DOI: 10.1093/cjres/rst007

Monck, C. S. P., Porter, R. B., Quintas, P. R., & Storey, D. J. (1988). *Science Parks and the Growth of High Technology Firms*. Routledge.

OECD. (1999a). *Managing National Innovation Systems*. OECD.

OECD. (2001a). *Innovative Clusters: drivers of national innovation systems*. Paris: OECD, 2001a

OECD. (2001b). *Drivers of Growth: information technology, innovation, and entrepreneurship. Science and Industry Outlook*. OECD.

OECD. (2001c). *Innovative Networks: co-operation in national innovation systems*. OECD.

OECD. (2011). *Regions and innovation policy, OECD Reviews of regional innovation*. Author.

Oladapo, B. I., Ismail, S. O., Afolalu, T. D., Olawade, D. B., & Zahedi, M. (2021). Review on 3D printing: Fight against COVID-19. *Materials Chemistry and Physics*, 2021(258), 123943. DOI: 10.1016/j.matchemphys.2020.123943 PMID: 33106717

Papachristos, G., Sofianos, A., & Adamides, E. (2013). System Interactions in Socio-technical Transitions: Extending the Multi-level Perspective. *Environmental Innovation and Societal Transitions*, 7, 53–69. DOI: 10.1016/j.eist.2013.03.002

Pelto, P. J. (1973) The Snowmobile revolution: technology and social change in the arctic. Cummings Publishing Company, Menlo Park, CA (revised Waveland Press, Prospects Heights, IL 1987 Percival RV, Miller AS, Schroeder CH, Leape JP (1992) *Environ-mental regulation: law, science, and policy*. Little, Brown, and Company, Boston, USA.

Pimenta-Bueno, J.A. (1999). Parque de Inovação Tecnológica e Cultural da Gávea: A Visão da PUC-Rio. Documento Interno, 23 págs, 1999.

Pizzol, M., & Andersen, M. S. (2022). Green tech for green growth? Insights from Nordic environmental innovation. In *Circular Economy Oriented Business Model Innovations: A European Perspective*. Springer. DOI: 10.1007/978-3-031-08313-6_8

Plúas-Arias, N., & Mejía-Caguana, D. (2024). Relación entre innovación tecnológica y sostenibilidad ambiental: Hacia un futuro resiliente y verde. *CIENCIAMATRIA*, 10(2), 482–500. DOI: 10.35381/cm.v10i2.1377

Porter, M. (1980). *Competitive Strategy: techniques for analyzing industries and competitors*. Free Press.

Porter, M. (2001). *Clusters of Innovation: regional foundations of U.S. competitiveness. Council on Competitiveness*. Monitor Group.

Price, R. M. (1996). Technology and Strategic Advantage. *California Management Review*, 38(3), 38–56. DOI: 10.2307/41165842

Ramazanov, S., Antoshkina, L., Babenko, V., & Akhmedov, R. (2019). Integrated model of stochastic dynamics for control of a socio-ecological-oriented innovation economy. *Periodicals of Engineering and Natural Sciences*, 7 (2), August 2019, pp.763-773 Available online at: http://pen.ius.edu.ba

Ramazanov, S. K. (2018). *Prediction and management of innovative economics based on the integral static model in phase space.*

Rockström, J., Schellnhuber, H. J., Hoskins, B., Ramanathan, V., Schlosser, P., Brasseur, G. P., Gaffney, O., Nobre, C., Meinshausen, M., Rogelj, J., & Lucht, W. (2016). The world's biggest gamble. *Earth's Future*, 10(10), 465–470. DOI: 10.1002/2016EF000392

Saxenian, A. (1994). *Regional Advantage: culture and competition in Silicon Valley and Route 128*. Harvard Business Press.

Schlapfer, A., & Marinova, D. (2001). Local Innovation Systems: nature, importance and role. *Conference Proceedings: International Summer Academy on Technological Studies:* user involvement in technological innovation. Deutschlandsberg. Austria.

Schot, J., & Rip, A. (1997). The past and future of constructive technology assessment. *Technological Forecasting and Social Change*, 54(2–3), 251–268. DOI: 10.1016/S0040-1625(96)00180-1

Schot, J., & Steinmueller, W. E. (2018). Three frames for innovation policy: R&D, systems of innovation and transformative change. *Research Policy*, 47(9), 1554–1567. DOI: 10.1016/j.respol.2018.08.011

Shearmur, R., Carrincazeaux, C., & Doloreux, D. (Eds.). (2016). *Handbook on the Geographies of Innovation*. Edward Elgar. DOI: 10.4337/9781784710774

Sheng, J., Zhou, W., & Zhu, B. (2020). La coordinación de los intereses de las partes interesadas en la regulación ambiental: Lessons from China's environmental regulation policies from the perspective of the evolutionary game theory. *Journal of Cleaner Production*, 249, 119385. DOI: 10.1016/j.jclepro.2019.119385

Solou, R. (2002). Theory of growth. Panorama of the economic thought of the end of the twentieth century, Ed. D. Greenway, M. Blini, I. Stewart, *SPb.: Econ. Shk.*, vol. 1. pp. 479-506, 2002.

Song, M., Wang, S., & Zhang, H. (2020). Could environmental regulation and R&D tax incentives affect green product innovation? *Journal of Cleaner Production*, 258, 120849. DOI: 10.1016/j.jclepro.2020.120849

The European Sustainable Development Strategy (2001). *The potential for sustainable development of 2001*. A sustainable Europe for a better World // enterprise of chemical industry. M.: Publishing house Europe – Sustainable Development. SDS 2005-2010.

Tidd, J.. (1997). *Managing Innovation: integrating technological, market and organizational change*. John Wiley & Sons.

Tödtling, F., & Trippl, M. (2005). One size fits all? Towards a differentiated regional innovation policy. *Research Policy*, 34(8), 203–1219. DOI: 10.1016/j.respol.2005.01.018

Trist, E. L. (1976). *A Concept of Organizational Ecology: an invited address to the three Melbourne universities*.

Tu, Y., & Wu, W. (2020). ¿Cómo mejora la innovación ecológica la ventaja competitiva de las empresas? The Role of Organizational Learning. *Sustainable Production and Consumption*. Advance online publication. DOI: 10.1016/j.spc.2020.12.031

Vargas, E., Montes, J., & Zartha, J. (2017). *Prospectiva sobre innovación ambiental. Un estudio para mipymes hoteleras de Colombia y México*. Pp. 167-187. Instituto de Ciencias Económico Administrativas. https://repository.uaeh.edu.mx/books/163/ti.pdf#page=169

von Schomberg, R. (2012). Prospects for technology assessment in a framework of responsible research and innovation. Dusseldorp, M. & Beecroft, R. (eds.) *Technikfolgen abschätzen lehren*, 39–61. Wiesbaden: VS Verlag für Sozialwissenschaften. DOI: 10.1007/978-3-531-93468-6_2

Wiig, H., & Wood, M. (1995). *What Comprises a Regional Innovation System? An empirical study*. Studies in Technology, Innovation and Economic Policy Group.

Zameer H, Yasmeen H (2022) *Green innovation and environmental awareness driven green purchase intentions.* Mark Intell Plan (ahead-of-print). Rodrigo Ramos Hospodar Felippe Valverde https://doi.org/DOI: 10.11606/issn.2179-0892.geousp.2023.194172.es

Chapter 17
Local Government Challenges in Implementing the 3R Strategy for Sustainable Waste Management

Imran Hossain
Varendra University, Bangladesh

A.K.M. Mahmudul Haque
University of Rajshahi, Bangladesh

S.M. Akram Ullah
University of Rajshahi, Bangladesh

Abdul Kadir
Dhaka University of Engineering and Technology, Bangladesh

Kabir Hossain
Varendra University, Bangladesh

ABSTRACT

This research delves into the multifaceted landscape of sustainable waste management in the context of Barisal City Corporation, emphasizing the challenges of implementing the 3R strategy. The city's waste management practices and perceptions among its residents are examined through a comprehensive primary data collection process, offering critical insights into the current state of waste management. The study

DOI: 10.4018/979-8-3693-7001-8.ch017

reveals formidable challenges such as insufficient local government initiatives and supervision, irregular waste transportation and segregation, insufficient manpower, unconventional waste storage and disposal practices, unaddressed environmental impact, and the lack of a secondary transfer station. At the same time, the research uncovers promising prospects, including source segregation, energy-efficient waste transportation, and an emerging recycling culture. By addressing these challenges, the city can significantly advance its waste management initiatives, aligning them with the 3R strategy and striving for a cleaner, more sustainable urban environment.

1.0 INTRODUCTION

Waste management is a fundamental component of urban development, affecting every aspect of public health, environmental sustainability, and the quality of life for urban inhabitants (Mesjasz-Lech, 2014). The problem of waste management in metropolitan areas is closely connected with the ongoing urbanization process (Hossain et al., 2024a; Hossen et al., 2024). The increasing number of people living in cities around the world has put more pressure than ever on cities to manage waste effectively (Hossain et al., 2024d; Oteng-Ababio et al., 2013). The percentage of people who live in urban areas worldwide is predicted to increase from 55% in 2021 to 68% in 2050 (Klohe et al., 2021; United Nations, 2019). Cities and governments face a challenge in managing the increasing volumes of waste as a result of ongoing urban growth and the resulting increase in waste generation (Mesjasz-Lech, 2014).

The management of waste in urban areas has become a critical task in the fast-paced urbanization period, requiring innovative thinking, responsible leadership, and comprehensive plans (Marshall & Farahbakhsh, 2013). According to Zhang et al. (2021), policymakers, urban planners, and researchers share a critical concern regarding the sustainable management of waste materials and water, particularly in metropolitan areas (Hossain et al., 2024b). The amount of solid waste produced worldwide has increased by 50% in just ten years, from 0.68 billion tons in 2000 to an unbelievable 1.3 billion tons in 2010 (Hoornweg & Bhada-Tata, 2012). Alarmingly, estimates indicate that this tendency will continue, with predictions of 2.2 billion tons annually by 2025 and an alarming 4.2 billion tons annually by 2050 (Hassan et al., 2019). These numbers highlight the challenges growing urban populations face when trying to dispose of waste in an eco-friendly manner. Bangladesh's cities are facing a growing waste problem as a result of increased urbanization. Bangladesh had a growth in urbanization from 28.97% to 36.63% between 2008 and 2018, with a peak in waste creation of 0.41 kilos per person per day in 2012 (Mostakim et al., 2021). The current growing trend indicates that the daily generation of waste might reach an estimated 47,064 tons, or 0.602 kg per person (Mostakim et al., 2021). The expanding urban population and the increase in the amount of waste produced per person draw attention to the critical state of waste management and its consequences for both society and the environment (Riyad et al., 2014).

Ineffective waste management has a variety of effects. Ineffective waste management not only immediately endangers the public's health and the quality of the air and water (Hossain et al., 2023a; Hossain et al., 2023b), but it also decreases limited resources and increases climate change (Hussain & Reza, 2023; Hossain et al., 2024c). Inefficient waste collection, disposal, and recycling practices are especially common in urban areas due to their high population density, which frequently coexists with economic inequality, inadequate infrastructure, and administrative problems (Moh, 2017). Consequently, municipalities find themselves in a challenging and pressing situation. The 3R strategy, which promotes "Reduce, Reuse, and Recycle," is a sustainable and comprehensive approach to waste management that prioritizes resource optimization, waste minimization, and minimizing environmental impacts (Visvanathan et al., 2007; Pariatamby & Fauziah, 2014). The 3R strategy has earned popularity in the global effort to manage waste sustainably because of its ability to lessen the negative effects of waste generation and disposal. Due to the high population density and peak waste generation found in metropolitan areas, this method is very important (Azevedo et al., 2021).

Waste management requires active local government participation because it is the foundation of urban governance. According to Fernando (2019), local governments are expected to play a significant role in the creation and execution of waste management policies and procedures due to their proximity to the citizens under their jurisdiction. The safety, health, and sanitation of their respective jurisdictions are directly impacted by their policies and actions. Nonetheless, there are regional differences in the efficiency of local administration when it comes to waste management, and there are numerous obstacles and difficulties that prevent the complete implementation of sustainable waste management approaches (Hoornweg & Bhada-Tata, 2012).

This research contributes to the growing amount of knowledge on sustainable waste management and local governance. It was carried out in the context of Bangladesh's Barisal City Corporation (BCC). The objectives of this study include examining potential solutions to enhance waste management in cities and examining the challenges local governments and citizens experience when implementing the 3R strategy. The study aims to shed light on the current state of waste management and governance procedures by examining the case of BCC, in addition to providing insights that could inform policy decisions for sustainable urban development.

2.0 LITERATURE REVIEW

2.1 Sustainable Waste Management and the 3R Strategy

Sustainable waste management is an important concept in the context of urban growth and environmental management. According to Sharma (2023) and Seadon (2010), it is a comprehensive strategy aimed at reducing the negative effects of waste generation and disposal on the environment and optimizing the use of resources. According to Seadon (2009), sustainable waste management is based on three main pillars: reducing waste at the source, encouraging recycling and material reuse, and implementing environmentally friendly disposal techniques. As previously indicated, the 3R strategy promotes "Reduce, Reuse, and Recycle," and these ideas are strongly aligned with it. Reduced sources are the foundation of the 3R strategy. The technique is designed to reduce the amount of waste generated overall by stopping waste generation at its source. In order to accomplish this reduction, Pariatamby & Fauziah (2014) and Visvanathan et al. (2007) address ways to discourage the use of single-use items, encourage sustainable consumption practices, and reduce the quantity of packaging used. The 3R method encourages a more sustainable approach to patterns of consumption and production by fostering a sense of accountability and awareness regarding waste generation.

The 3R strategy heavily emphasizes the "reuse" principle in addition to source reduction. Reuse means to use materials or goods again instead of throwing them away after just one use (Hussain et al., 2012). According to Mohammed et al. (2020) and Wichai-Utcha & Chavalparit (2019), this strategy increases the lifespan of products, conserving resources and lowering the need for new ones. Reusing materials and goods reduces the need for natural resources and waste production, which is crucial in this day and age as resource scarcity is becoming a greater concern.

Sustainable waste management relies heavily on recycling, which is the third pillar of the 3R strategy (Azevedo et al., 2021). Recycling involves the systematic collection and processing of recyclable materials to reintroduce them into the production (Cycle-Kofoworola, 2007). Yasuhara (2006) states that this procedure decreases the environmental impact of disposing of waste while also conserving precious resources and reducing waste. Energy consumption, which is frequently high during the manufacturing of new materials, can be reduced through recycling.

A key component of sustainable urban development is the application of these 3R principles to waste management. Using the 3R strategy correctly helps to move toward a circular economy and lessens the negative environmental effects of garbage (Kurniawan et al., 2021). A circular economy seeks to optimize the use of resources by minimizing waste and maximizing resource recovery (Ghisellini et al., 2016). The Sustainable Development Goals (SDGs) of the United Nations (especially Goal

12, which addresses responsible consumption and production) are closely aligned with this initiative. According to the United Nations (2017), adopting sustainable waste management techniques is essential to reaching these international goals. By implementing sustainable waste management practices, such as recycling, composting, and waste-to-energy technologies, countries can significantly contribute to achieving the SDGs and move towards a more sustainable future.

2.2 Local Governance in Waste Management

The efficient management of waste in urban areas is largely dependent on effective local administration. "Local governance" refers to the processes and procedures used by municipal bodies, such as city governments, to monitor and manage the goods and services provided to their communities (Chandler, 2001; Andrew & Goldsmith, 1998). In the context of waste management, local governance encompasses a wide array of responsibilities, including the formulation and implementation of waste collection, disposal, and recycling programs, as well as the provision of necessary infrastructure and services (Benito et al., 2021; Gross, 2018). Additionally, it is the responsibility of local governments to promote community involvement and public understanding of sustainable waste management approaches (Shandas & Messer, 2008).

There are large regional differences in the efficiency of local governance in waste management. The performance of waste management initiatives is heavily influenced by a variety of elements, such as institutional capability, financial resources, and the dedication of local leadership (Domingo & Manejar, 2021). For instance, it can be easier for local governments with strong administrative frameworks and adequate resources to establish and enforce waste management regulations. In contrast, those facing resource constraints may struggle to provide adequate services, which can lead to haphazard waste disposal practices (Wright & Reames, 2020).

Another essential component of local governance in waste management is public participation. It is imperative to involve the community and ensure their active participation in waste reduction and recycling programs. Engaging the public promotes environmental knowledge and a sense of shared responsibility in addition to improving compliance with waste management policies (Silva & Morais, 2021).

Furthermore, in order to guarantee the effective collection, transportation, and disposal of waste products, local governments must coordinate with a variety of stakeholders, including for-profit waste management organizations (Hondo et al., 2020). The overall performance of waste management initiatives can be strongly impacted by the quality and efficacy of these collaborations (Gross, 2018).

Waste management-related municipal governance practices can also be significantly influenced by the sociodemographic characteristics of metropolitan populations (Olukanni et al., 2020). Waste management strategies must be customized since different urban populations may have different waste generation patterns and disposal practices (Wilson, 2019). Urban households, for example, can differ greatly in terms of income, education, and occupation, which may have an impact on how they generate and dispose of waste.

Numerous criteria are necessary for local governance to be successful in waste management. Important components include strong leadership, adequate financial resources, infrastructure, and effective policies. The commitment and awareness of local leaders and the greater community, however, cannot be disregarded. Sustainable waste management approaches improve the environment and public health when they are developed in balanced environments (Gross, 2018).

In BCC, the effectiveness and quality of waste management techniques are significantly influenced by local governance. BCC has its own possibilities and challenges, just like any other city. The local government of BCC needs to address issues with inefficient waste collection, uneven disposal practices, and low public awareness. This study aims to investigate how local governments might support sustainable waste management practices, identify areas for improvement, and assess the role of local governance in these particular difficulties.

3.0 METHODOLOGY OF THE STUDY

This study uses a mixed-methods research approach, which utilizes methods for gathering both quantitative and qualitative data. In order to obtain data on demographics, waste generation patterns, and opinions of local governance, a survey of BCC citizens is conducted as part of the quantitative component. In-depth interviews with important waste management stakeholders and representatives of the local government are conducted as part of the qualitative component to obtain a better knowledge of governance practices, obstacles, and opportunities.

3.1 Data Collection

A stratified random sampling technique is used to ensure the representativeness of the survey sample. The population of BCC is divided into categories based on geographical locations within the city, and random samples are drawn from each category. This approach ensures that diverse areas of the city are covered, accounting for variations in waste management practices and governance perceptions. In this aspect, a structured questionnaire is employed as the primary instrument for quantitative data collection. The survey is designed to gather information from a diverse sample of BCC's residents. The survey includes questions related to demographics, waste generation practices, waste disposal habits, and perceptions of the local government's role in waste management. A pilot survey is conducted to test the questionnaire's effectiveness and gather feedback for refinement.

In-depth interviews are conducted with key informants, including local government officials, waste management experts, and representatives from non-governmental organizations (NGOs) involved in waste management activities in the BCC. These interviews are semi-structured, allowing for open-ended questions to explore the specifics of local governance practices, policy implementation, challenges faced, and prospects for improvement. Interviews are recorded and transcribed for later analysis.

3.2 Data Analysis

Quantitative data from the survey is analyzed using statistical software like SPSS V.22, employing descriptive and inferential statistical techniques. Descriptive statistics are used to summarize demographic information and respondents' perceptions of local governance. Qualitative data from in-depth interviews is subjected to thematic analysis. This involves identifying common themes, patterns, and connections in the interview transcripts. The data are coded to categorize responses and opinions related to local governance practices, challenges, and prospects. Thematic analysis provides a comprehensive understanding of the qualitative data and supports the development of recommendations and conclusions.

4.0 RESULTS AND DISCUSSION

4.1 Demographics of the Respondents

To understand BCC's waste generation procedures and opinions regarding local governance, one must acquire knowledge about the demographics of survey respondents. The study found that the gender distribution of the respondents was almost equal. In this study, 54.3% of respondents identified as women, and 45.7% of respondents identified as men. As seen by the equally distributed gender representation, the study collected a broad range of perspectives and experiences regarding waste management practices and local government (Table 1). Other studies indicate that women tend to be the primary sources of household waste; therefore, their perspectives and involvement are crucial in developing efficient waste reduction and recycling policies (UNEP, 2018). This supports our findings and highlights how crucial it is to implement gender-sensitive, inclusive waste management programs.

The vast range of professions among the respondents reflects the diversity of occupations found among BCC's inhabitants. Interestingly, a sizable proportion of respondents worked in the private sector (10.3%) and conducted business (19.3%). The public sector employed 16.3% of the respondents who were government employees. A significant proportion of the participants (34.7%) were households, indicating the variety of households found in urban environments. Moreover, the educational backgrounds of the respondents varied, ranging from those with little to no formal education to those with post-graduate degrees. The study found that 21.7% of respondents had only completed elementary school, while about 7.0% of respondents were illiterate. The respondents' varied educational backgrounds influence the comprehension of their knowledge and awareness levels about waste management and local governance processes (Table 1).

Table 1. Socio-demographic characteristics of the respondents

Characteristics		F	%
Gender	Male	137	45.7%
	Female	163	54.3%
Age	16-25	46	15.3%
	26-35	100	33.3%
	36-45	86	28.7%
	>46	68	22.7%
Profession	Gov jobs	49	16.3%
	Private jobs	31	10.3%
	Business	58	19.3%
	Household	104	34.7%
	Farmer	0	0.0%
	Day Labor	19	6.3%
	Students	30	10.0%
	House maid	7	2.3%
	unemployed	0	0.0%
	abroad	0	0.0%
	Taylor	0	0.0%
	Retired	2	0.7%
	Teacher	0	0.0%

continued on following page

Table 1. Continued

Characteristics		F	%
Education qualification	Illiterate	21	7.0%
	Primary	65	21.7%
	Secondary	34	11.3%
	Higher Secondary	65	21.7%
	Graduate	65	21.7%
	Post Graduate	50	16.7%
Monthly Income	<35	8	2.7%
	35-115	57	19.0%
	116-200	52	17.3%
	201-300	74	24.7%
	>300	109	36.3%

[Field Survey, 2022]

However, our study identified a significant proportion of respondents with primary, secondary, and upper-secondary education, followed by graduates and post-graduates, in terms of educational attainment. The general trend in Bangladesh's metropolitan areas is reflected in this educational profile (World Bank, 2018). The higher levels of education among our respondents, however, point to a reasonably well-educated urban population in BCC. This finding is in contrast to certain literature that contends that a sizeable population may lack formal education in some metropolitan regions (Sujauddin et al., 2008). The range of educational backgrounds emphasizes the necessity for specialized waste management awareness programs that consider residents' diverse degrees of literacy and environmental concern.

4.2 Household Solid Waste Management Practices

Effective waste management at the household level is essential to maintaining clean and sustainable cities (Hossain et al., 2024c). In the study, participants were asked how they usually dispose of solid waste. The study revealed that a significant percentage of BCC's inhabitants (46.7%) depend on city corporation employees for the collection of waste (Table 2). One alarming finding is that 11.7% of respondents acknowledged disposing of their household solid waste in open spaces or beside roads (Table 2). This behavior not only makes urban areas look unpleasant, but it also poses major risks to public health and the environment.

Table 2. Household solid waste disposal practices

Variables	F	%
Giving to City worker	140	46.7%
Throwing in drain	21	7.0%
Open place/road	35	11.7%
Specified places	87	29.0%
Dumps in water holes	17	5.7%

[Field Survey, 2022]

Table 2 shows that around 29.0% of respondents said they disposed of their solid waste in designated locations. This activity still needs to be monitored, even though it is more responsible than throwing waste outside. The study revealed a worrying and harmful practice for the environment. Table 2 indicates that 7.0% of participants acknowledged dumping their home solid waste down the drain. This technique is risky for blocking the drainage system, which can result in flooding and damage to the urban infrastructure, in addition to being filthy. According to previous research showing that informal waste collectors play a critical role in waste collection in many developing cities (Wilson et al., 2015), households in BCC frequently dispose of solid waste by giving it to City Corporation (CC) workers (Hossain et al., 2024d). However, despite the availability of designated collection places, our studies also show that a sizable fraction of waste still ends up in open spaces and sewers. This is consistent with research suggesting that modifying disposal practices might be difficult, even when alternatives are offered (Sujauddin et al., 2008). It highlights the necessity of behavior modification initiatives that not only encourage appropriate garbage disposal but also address the root causes of open dumping behaviors.

4.3 Role of BCC in Waste Management

In order to maintain urban sustainability and cleanliness, the local government plays a critical role in waste management (Hossain et al., 2024c). When asked about the local government's waste management initiatives, 99.0% of respondents said they believed the government was failing in this area (Table 3). The disparity between the expected and actual initiatives shows that there is a clear conflict between the public's expectations and the local government's actions. Similarly, in response to the question of whether the local government is in charge of waste management, 100.0% of respondents claimed that it is not (Table 3). The lack of noticeable supervision raises the possibility that BCC's waste management practices are not properly regulated and observed. A sizable percentage of respondents (80.0%) stated that they received regular waste collection services, despite the apparent lack of initiative and oversight (Table 3). According to this, residents benefit from waste collection services even though they may not believe that their government is keeping an eye on waste management.

Table 3. Role of BCC for waste management

Variables		F	%
Proper initiatives	yes	3	1.0%
	no	297	99.0%
Proper supervision	yes	0	0.0%
	no	300	100.0%
Regular drain cleaning	yes	83	27.7%
	no	217	72.3%
Regular waste collection	yes	240	80.0%
	no	60	20.0%
Regular waste disposal	yes	15	5.0%
	no	285	95.0%
Preventing connecting drains and toilet	yes	0	0.0%
	no	300	100.0%
Punishing the violators	yes	0	0.0%
	no	300	100.0%
City corporation plays no role	yes	0	0.0%
	no	300	100.0%

[Field Survey, 2022]

On the other hand, a significant 95.0% of respondents said they did not receive regular waste disposal services at all (Table 3). This implies that waste may not always be disposed of properly, even after it has been collected. This raises concerns about what happens to the waste after it is collected and how it will impact the environment. Our findings, which show a glaring absence of appropriate actions and oversight by the local government in BCC, are consistent with previous research highlighting governance issues in waste management (Guerrero et al., 2013; Aleluia & Ferro, 2016). According to the small percentage of respondents who acknowledged proper initiatives and monitoring, municipal waste management frequently suffers from governance deficiencies in underdeveloped countries (Hoornweg & Bhada-Tata, 2012). This finding is consistent with the widespread perception. It highlights the urgent need for institutional capacity building and governance reforms to boost waste management systems.

4.4 Frequency of Waste Collection

The foundation of effective waste management in urban areas is regular and efficient waste collection. This study delves into the experiences of residents with waste collection, the obstacles they encounter, and the effects of collection frequency on waste management and public health. The residents gave insightful answers when asked how often waste was picked up. Because daily collection services are offered in certain locations of BCC, the majority of respondents (50.7%) stated that waste collection happens every day. Waste was picked up every two days, according to a sizable 36.7% of respondents. However, a shocking 9.3% of respondents claimed that there was never any trash collection in their communities (Figure 4). The fact that a small but significant percentage of residents reported never having any waste collection raises concerns about the overall effectiveness of waste management practices in these communities.

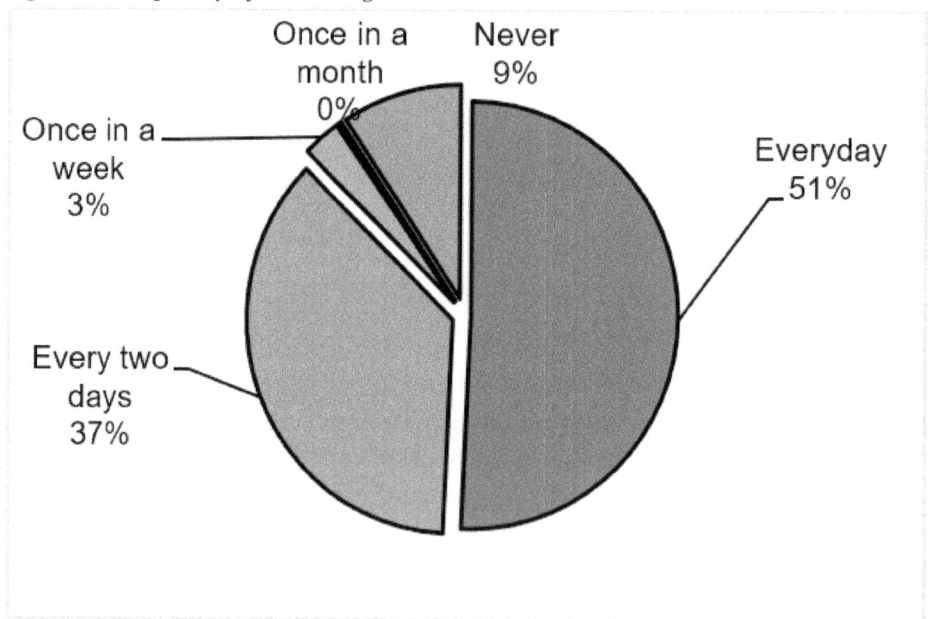

Figure 1. Frequency of collecting wastes

Our research on the frequency of waste disposal and collection in BCC shows that there are differences between the two, with regular waste collection being far more often than waste disposal. This discrepancy is consistent with literature that highlights the difficulties municipalities experience in guaranteeing regular garbage disposal, frequently as a result of insufficient resources and infrastructure (Scarlat et al., 2015). Studies emphasizing the varied quality of urban services and infrastructure in developing cities are similarly evocative of the differing assessments of the drainage and waste management systems in BCC.

4.5 Challenges in Sustainable Waste Management

4.5.1 Improper Solid Waste Disposal

The improper disposal of solid waste is a significant issue in BCC that has far-reaching effects. According to the study, approximately 11.7% of participants said they disposed of solid waste by dumping it in open spaces or on public highways. In addition to seriously threatening the environment and public health, this behavior negatively impacts the city's appearance. Ineffective methods for disposing of solid waste have a lasting impact on urban surroundings because they cause waste to build up in open areas, which attracts insects and disease-carrying organisms. Additionally, it maintains pollution in the environment and eventually affects public health. It is concerning that these kinds of acts are so common and that they require immediate action. In this aspect, a respondent argues:

"In our neighborhood, it's disheartening to witness improper solid waste disposal practices. Many residents simply throw their household waste into open spaces or the nearby drain without any consideration for the consequences. It not only makes our locality look untidy but also poses serious health and environmental risks. The lack of proper waste bins and collection services aggravates the situation," (KIIs, BCC, Bangladesh).

These results highlight how urgent it is to enhance BCC's waste disposal procedures. The incorrect disposal of solid waste requires a diverse approach. Public awareness initiatives are essential, first and foremost. Ensuring residents are informed about proper waste disposal methods, the need for waste separation, and the negative consequences of improper disposal is essential. The study highlights the different educational backgrounds of BCC citizens, which can be taken into consideration when customizing these messages to the town's specific demographics. Additionally, infrastructure investments related to waste collection can also help to lessen this problem. Establishing regular collection schedules and offering improved waste collection services that cover the entire city can help lessen the tendency to dispose of waste in open spaces or on highways.

Municipal authorities also need to think about how to enforce laws prohibiting inappropriate dumping of waste. This means that anyone engaged in illegal dumping operations needs to be under constant observation and face consequences. The results of the study, which showed that people's opinions of local government efforts differed from the services actually provided, require greater enforcement and oversight, as Davis & Wilson (2020) point out. Inappropriate solid waste disposal in the BCC is a serious issue with wide-ranging effects on the environment, public health, and the general standard of living in urban areas.

4.5.2 Insufficient Local Government Initiatives and Supervision

In order to ensure effective waste management in metropolitan settings, effective municipal government is essential. Regarding the local government's involvement in waste management in the BCC, residents' expectations and perceptions of reality diverge significantly. About 80% of the residents said they received regular waste collection services from the local government, compared to only one percent who thought it took appropriate action. However, the perception of a total lack of appropriate monitoring (100.0%) indicates the inhabitants' mistrust of the local government's regulatory and supervision role in waste management. A respondent demonstrates:

"We often find our streets filled with waste, and it seems there is no one overseeing this garbage. There's an urgent need for the local government to take the lead in waste management. We need them to set up proper collection and disposal systems, regularly monitor these activities, and enforce regulations to ensure that waste is managed in an environmentally responsible manner. It's not just about cleanliness; it's about our overall well-being and the health of our city," (KIIs, BCC, Bangladesh).

The study highlights the discrepancy between the perceived quality of services provided by the local government and the actual offerings of said services. The fact that most respondents said they regularly receive waste collection services suggests that the local government is actively involved in delivering this essential municipal function. However, the small percentage of respondents (1.0%) who said the local government had acted appropriately suggests that residents may not be aware of the entire extent of the government's waste management initiatives. According to this study, local government supervision and regulation of waste management are inadequate in the eyes of the residents. It brings up issues with how waste management laws are applied and how garbage collection and disposal companies are managed. The disparity between the expected monitoring and the supply of services highlights the necessity for enhanced communication and transparency between the local administration and the inhabitants.

4.5.3 Challenges with Waste Transportation and Segregation

The BCC's solid waste management system is essential to the implementation of the 3R waste management policy. To create effective plans and deal with problems as they arise, it is essential to understand current standards and the challenges the municipality faces in managing trash. At present, the BCC is largely in charge of managing the transportation and collection of solid waste in Barisal City, with a significant amount of community participation. Notably, the city demonstrates its dedication to good waste management techniques by treating and disposing of more than 90% of its solid waste in an environmentally responsible manner. To improve the effectiveness of waste management, there are still several areas in which the current protocols should be strengthened.

The BCC is confronted with a significant issue when residents fail to separate their sources, resulting in an increased amount of mixed waste. Ineffective source segregation hinders material recycling and reuse, which increases the difficulty of putting the 3R method into practice. Ineffective waste segregation raises the percentage of garbage that cannot be recycled, which in turn leads to increased landfill usage and lost chances for resource recovery. In order to maximize waste management in Barisal City, raising public awareness and promoting source segregation techniques should be top priorities.

The impact of garbage collection and treatment on the environment is another matter that must be taken into account. It is necessary to evaluate the energy usage and greenhouse gas emissions related to waste management operations. Developing waste management techniques that are environmentally sustainable requires careful consideration of these variables. A crucial component of putting the 3R approach into practice should be finding ways to cut down on energy use and greenhouse gas emissions.

The BCC uses motorized and non-motorized vehicles as its two main types of vehicles for the transportation of trash. Hand trolleys and rickshaw vans are examples of non-motorized vehicles that help move trash around the city efficiently. They are very helpful for distribution and short-distance transit. On the other hand, motorized vehicles are essential for moving larger trash loads and for collecting and distributing solid waste over greater distances. By ensuring regular garbage collection and transfer, motorized vehicle use lowers the risk of waste gathering and related health hazards. Furthermore, in keeping with the 3R strategy's principles, these cars help reduce carbon emissions and foster a greener urban environment. However, to improve waste management and move the 3R plan closer to implementation, it is imperative that residents address these issues and encourage source segregation. Utilizing motorized and non-motorized vehicles in equal measure for trash transportation offers an additional advantage in terms of resource optimization and environmental impact reduction.

4.5.4 Challenges of Sufficient Manpower

The availability of a sufficiently staffed and well-organized municipal waste management system is essential for the successful implementation of a 3R waste management plan. However, the BCC authority faces a major challenge in this area due to a severe staffing shortage, which makes it impossible to carry out municipal solid waste (MSW) management operations effectively. The BCC does not only lack the appropriate number of employees, but it also does not maintain the logical and technical ratios between various staff categories with varying technical specializations. The administrative and technical aspects of MSW management are greatly impacted by this lack of both technical and non-technical people. During KIIs, a responsible government official says:

"The issue of having sufficient manpower for waste management is a real concern for us. Our existing staff is working tirelessly, but they are becoming weak given the enormity of the task. We understand that a shortage of staff impacts the efficiency of our waste management efforts. To address this challenge, we are actively looking into strategies to recruit, train, and retain skilled personnel. We are committed to finding a solution to this challenge and providing a more effective waste management system for our citizens," (KIIs, BCC, Bangladesh).

Making sure the right procedures are taken in relation to MSW becomes a difficult task in the absence of the necessary administrative and technical staff. In order to effectively tackle this challenge, the BCC needs to give top priority to the recruitment of appropriately positioned and skilled employees to guarantee that technical and administrative requirements are fulfilled and to support well-informed decision-making. Moreover, the uncertainty linked to the temporary employment statuses of laborers and sweepers negatively impacts their dedication to their jobs, which potentially hinders the effective implementation of waste management programs.

The unique ways that residents of BCC store and dispose of their waste add to the problem of waste management. Contrary to recommended guidelines for appropriate waste management, many residents choose to store their waste in polythene bags, plastic containers, and metal containers. Furthermore, there are serious health and environmental risks associated with some regional habits, such as throwing trash on the main streets of busy neighborhood markets and immediately disposing of waste into surrounding canals. These non-traditional disposal techniques endanger the health of nearby ecosystems as well as human populations, in addition to causing pollution. It is essential to start educational programs aimed at residents and company owners in light of these behaviors. These programs ought to stress the need for implementing appropriate waste management procedures as well as the possible consequences of using inadequate disposal methods.

4.5.5 Lack of Secondary Transfer Stations (STSs)

One major obstacle to effectively implementing the 3R plan for waste management in the BCC is the lack of a secondary transfer station. An important part of MSW is secondary transfer stations, or SDS. They act as temporary phases, allowing MSW to be moved from primary sources to designated processing or treatment facilities with ease. SDS's ability to maximize waste transportation to more expansive disposal sites play a vital role. Both powered and non-powered vehicles are used in this operation, and they are all essential to the collection and movement of waste. These cars all share responsibilities for the efficient and safe delivery of waste. In this aspect, it becomes essential to create a secondary transfer station in the BCC in order to address this difficulty. A secondary transfer station is an essential link in the waste management chain that facilitates the efficient transfer of waste to larger disposal locations while allowing the 3R method to be applied. Enhancing the city's waste management infrastructure would be possible with the establishment of a centralized facility for temporary storage and garbage transfers. This tactical intermediate step would offer opportunities for proper waste segregation, recycling, and resource recovery in addition to expediting the waste transportation process. Additionally, it would lessen the negative effects of waste management operations on the environment.

Apart from the actual construction of a backup transfer station, maintenance and training should also be prioritized. Ensuring the safe and effective management of MSW requires ensuring the competence of both powered and non-powered vehicles in waste transportation. To fully realize their potential in supporting an environmentally responsible and sustainable approach to garbage management, the waste management framework should include regular maintenance schedules and adequate training for vehicle operators.

4.5.6 Challenges of Recycling Waste

The implementation of the 3R strategy for waste management is still filled with difficulties, despite the fact that the BCC has established recycling stores throughout the city, including well-known spots like the Sadarghat, Rupatoli, and Nottulabad bus stops, Chanmari market, Port Road, and Hat Khola hackers' market. By actively recycling and reusing a range of materials, the city's recycling stores contribute significantly to the promotion of environmental sustainability values. To fully realize the benefits of recycling in the waste management process, a few obstacles must be overcome. Above all, there is still work to be done in terms of increasing public knowledge and engagement in recycling initiatives. Comprehensive teaching efforts that emphasize the value of recycling, its advantages for the environment, and each person's involvement in the process are necessary, even as the city works to encourage citizens to participate in recycling initiatives. The effectiveness of recycling programs and the degree to which the 3R technique may be applied successfully can both be hampered by a lack of awareness.

Making sure recyclables are efficiently collected and transported to these recycling centers is another difficulty. A well-organized system is needed for the collection and transportation of recyclables from homes and businesses to the appropriate recycling centers. In the absence of an efficient system for collection and transportation, recyclables may not be processed for recycling and instead end up in landfills. The infrastructure and logistics needed for this operation need to be robust and well-thought-out in order to minimize any potential delays in the recycling process.

In addition, market demand and resource availability provide challenges for the recycling sector as a whole. The market's desire for recycled materials is a major factor in recycling's economic feasibility. It can be difficult to guarantee that there is a steady demand for certain resources. Furthermore, the smooth operation of recycling enterprises depends on a stable supply of recyclables. Enough recyclables must be collected in order to sustain the recycling sector. A shortage in supply or demand may cause the recycling process to be postponed.

4.5.7 Insufficient Waste Disposal Method

All types of MSW, including specific medical waste, are finally disposed of at the BCC's ultimate disposal site (UDS) in Kawnia. It is important to note that the open trash that is dispersed around the UDS area generates aesthetic and environmental issues because there isn't a suitable method for filling empty spaces. The absence of adequate space-filling systems in the UDS area not only hinders the overall aesthetics but also may result in environmental degradation. A more effective waste management system could help allay these worries and guarantee a cleaner and safer disposal procedure.

Gawashar (BCC Ward No. 3), the Kirtonkhola River, and Rupatoli are the main trash disposal locations used by Barisal City Corporation. The Gawashar site, which is located in Kaunia, receives the most waste out of all of these, making up over 62% of the entire waste volume. When the Gawashar site becomes overloaded as a result of things like severe rain, repairs, and upkeep of unloading platforms, other dumping sites come into play. Notably, the Gawashar dump site has been in use since 1999 and currently receives an average of 15 to 20 tons of trash every day. The management of trash disposal in the area is greatly aided by the dumping areas along the Kirtonkhola and Rupatoli rivers. These locations are necessary to sustain cleanliness and stop environmental contamination brought on by poor waste management. The Gawashar site has been operating continuously since 1999, underscoring its importance in managing a sizeable amount of the waste produced in the area. The presence and operation of these dumps have had a significant positive impact on the neighborhood's cleanliness and environmental health. Regular waste disposal at these locations guarantees that the waste is appropriately managed and does not endanger the local ecosystem. The Gawashar facility has also been operating for a long time, demonstrating its efficiency in managing the significant amount of trash produced in the area, making it an essential part of the waste management system.

5.0 CONCLUSION

The research's conclusions clarify the complexities of the BCC's waste management procedures while highlighting the opportunities and problems associated with building a greener, more sustainable urban environment. Urban governance necessitates waste management since it affects public health, environmental sustainability, and the general standard of living in cities. To sum up, this study has produced important insights that might guide urban waste management systems that are sustainable.

The difficulties are evident and complex. Effective waste management is significantly hampered by improper solid waste disposal, inadequate liquid waste management, inadequate local government oversight and initiatives, an inadequate drainage system, and ineffective waste collection services. To address each of these issues, particular investments, interventions, and community engagement initiatives are needed. Nevertheless, there is hope for sustainable waste management in the BCC despite these obstacles. Education and community involvement provide the means to cultivate environmentally conscious citizens who take an active role in waste management. Waste collection services can be optimized through technology-driven solutions, increasing their equity and efficiency. Initiatives like waste-to-energy and recycling can produce renewable energy while lessening the load on landfills. While including waste management in urban design might result in cleaner, more orderly urban places, government commitment and transparency can also promote confidence and cooperation. In addition, gender, occupation, level of education, and family status are important demographic variables that influence waste management procedures. An urban environment that is more environmentally sensitive and sustainable can be achieved by designing interventions according to these demographic characteristics.

There are significant consequences for local government. To improve waste management, local governments must consider a number of crucial factors, including community involvement, infrastructure development, open communication, regulations, and public awareness. As administrators of the public good, local government must take the initiative to influence and assist the shift to a more sustainable and cleaner urban environment. However, this study lays the groundwork for further efforts to achieve sustainable waste management and emphasizes the need for everyone to work together to create a more sustainable, healthy, and clean urban environment for the BCC's citizens.

ACKNOWLEDGEMENTS

The research was funded by the Bangladesh Bureau of Educational Information & Statistics (BANBEIS), Ministry of Education of the People's Republic of Bangladesh under Grants for Advanced Research in Education (GARE) program. The CP no. is SD20191014 and the grant number is: 37.20.0000.004.033.020.2016.1053, dated: October 13, 2019. The authors would like to give special thanks to BANBEIS for providing support in research grants. They are also thankful to the study participants, the research assistants and the staff of the selected City Corporations of this research.

REFERENCES

Aleluia, J., & Ferrão, P. (2016). Characterization of Urban Waste Management Practices in Developing Asian Countries: A New Analytical Framework Based on Waste Characteristics and Urban Dimension. *Waste Management (New York, N.Y.)*, 58, 415–429. DOI: 10.1016/j.wasman.2016.05.008 PMID: 27220609

Andrew, C., & Goldsmith, M. (1998). From local government to local governance—And beyond? *International Political Science Review*, 19(2), 101–117. DOI: 10.1177/019251298019002002

Azevedo, B. D., Scavarda, L. F., Caiado, R. G. G., & Fuss, M. (2021). Improving urban household solid waste management in developing countries based on the German experience. *Waste Management (New York, N.Y.)*, 120, 772–783. DOI: 10.1016/j.wasman.2020.11.001 PMID: 33223248

Benito, B., Guillamón, M. D., Martínez-Córdoba, P. J., & Ríos, A. M. (2021). Influence of selected aspects of local governance on the efficiency of waste collection and street cleaning services. *Waste Management (New York, N.Y.)*, 126, 800–809. DOI: 10.1016/j.wasman.2021.04.019 PMID: 33895563

Carley, M., Smith, H., & Jenkins, P. (2013). *Urban development and civil society: The role of communities in sustainable cities*. Routledge. DOI: 10.4324/9781315071725

Chandler, J. A. (2001). *Local government today*. Manchester University Press.

Domingo, S. N., & Manejar, A. J. A. (2021). *An analysis of regulatory policies on solid waste management in the Philippines: Ways forward* (No. 2021-02). PIDS Discussion Paper Series.

Fernando, R. L. S. (2019). Solid waste management of local governments in the Western Province of Sri Lanka: An implementation analysis. *Waste Management (New York, N.Y.)*, 84, 194–203. DOI: 10.1016/j.wasman.2018.11.030 PMID: 30691892

Ghisellini, P., Cialani, C., & Ulgiati, S. (2016). A review on circular economy: The expected transition to a balanced interplay of environmental and economic systems. *Journal of Cleaner Production*, 114, 11–32. DOI: 10.1016/j.jclepro.2015.09.007

Guerrero, L. A., Maas, G., & Hogland, W. (2013). Solid Waste Management Challenges for Cities in Developing Countries. *Waste Management (New York, N.Y.)*, 33(1), 220–232. DOI: 10.1016/j.wasman.2012.09.008 PMID: 23098815

Hassan, S. S., Williams, G. A., & Jaiswal, A. K. (2019). Moving towards the second generation of lignocellulosic biorefineries in the EU: Drivers, challenges, and opportunities. *Renewable & Sustainable Energy Reviews*, 101, 590–599. DOI: 10.1016/j.rser.2018.11.041

Hondo, D., Arthur, L., & Gamaralalage, P. J. D. (2020). Solid Waste Management in Developing Asia: Prioritizing Waste Separation.

Hoornweg, D., & Bhada-Tata, P. (2012). What a Waste: A Global Review of Solid Waste

Hossain, I., Haque, A. M., Kılıç, Z., Ullah, S. A., Azrour, M., & Mabrouki, J. (2024b). Navigating the Challenges of Urban Water Sustainability in Bangladesh: Towards a Sustainable Approach for Urban Governance. In *Smart Internet of Things for Environment and Healthcare* (pp. 1–25). Springer Nature Switzerland. DOI: 10.1007/978-3-031-70102-3_1

Hossain, I., Haque, A. M., Kılıç, Z., Ullah, S. A., Azrour, M., & Mabrouki, J. (2024d). Exploring Household Waste Management Practices and IoT Adoption in Barisal City Corporation. In *Smart Internet of Things for Environment and Healthcare* (pp. 27–45). Springer Nature Switzerland. DOI: 10.1007/978-3-031-70102-3_2

Hossain, I., Haque, A. M., & Ullah, S. A. (2023a). Assessment of domestic water usage and wastage in urban Bangladesh: A study of Rajshahi City corporation. *The Journal of Indonesia Sustainable Development Planning*, 4(2), 109–121. DOI: 10.46456/jisdep.v4i2.462

Hossain, I., Haque, A. M., & Ullah, S. A. (2024a). Policy evaluation and performance assessment for sustainable urbanization: A study of selected city corporations in Bangladesh. *Frontiers in Sustainable Cities*, 6, 1377310. DOI: 10.3389/frsc.2024.1377310

Hossain, I., Haque, A. M., & Ullah, S. A. (2024c). Assessing sustainable waste management practices in Rajshahi City Corporation: An analysis for local government enhancement using IoT, AI, and Android technology. *Environmental Science and Pollution Research International*, •••, 1–19. DOI: 10.1007/s11356-024-33171-7 PMID: 38581631

Hossain, I., Ullah, S. A., & Haque, A. M. (2023b). Water and sanitation services at the local government level in Bangladesh: An analysis of SDG 6 implementation status and way forward. *Asia Social Issues*, 17(3), e265358. DOI: 10.48048/asi.2024.265358

Hossen, M. S., Haque, A. M., Hossain, I., Haque, M. N., & Hossain, M. K. (2024). Towards comprehensive urban sustainability: Navigating predominant urban challenges and assessing their severity differential in Bangladeshi city corporations. *Urbanization. Sustainability Science*, 1(1), 1–17.

Hussain, M., Balsara, K. P., & Nagral, S. (2012). Reuse of single-use devices: Looking back, looking forward. *The National Medical Journal of India*, 25(3), 151. PMID: 22963293

Hussain, S., & Reza, M. (2023). Environmental Damage and Global Health: Understanding the Impacts and Proposing Mitigation Strategies. *Journal of Big-Data Analytics and Cloud Computing*, 8(2), 1–21.

Klohe, K., Koudou, B. G., Fenwick, A., Fleming, F., Garba, A., Gouvras, A., Harding-Esch, E. M., Knopp, S., Molyneux, D., D'Souza, S., Utzinger, J., Vounatsou, P., Waltz, J., Zhang, Y., & Rollinson, D. (2021). A systematic literature review of schistosomiasis in urban and peri-urban settings. *PLoS Neglected Tropical Diseases*, 15(2), e0008995. DOI: 10.1371/journal.pntd.0008995 PMID: 33630833

Kofoworola, O. F. (2007). Recovery and recycling practices in municipal solid waste management in Lagos, Nigeria. *Waste Management (New York, N.Y.)*, 27(9), 1139–1143. DOI: 10.1016/j.wasman.2006.05.006 PMID: 16904308

Kurniawan, T. A., Lo, W., Singh, D., Othman, M. H. D., Avtar, R., Hwang, G. H., Albadarin, A. B., Kern, A. O., & Shirazian, S. (2021). A societal transition of MSW management in Xiamen (China) toward a circular economy through integrated waste recycling and technological digitization. *Environmental Pollution*, 277, 116741. DOI: 10.1016/j.envpol.2021.116741 PMID: 33652179

Marshall, R. E., & Farahbakhsh, K. (2013). Systems approaches to integrated solid waste management in developing countries. *Waste Management (New York, N.Y.)*, 33(4), 988–1003. DOI: 10.1016/j.wasman.2012.12.023 PMID: 23360772

Mesjasz-Lech, A. (2014). Municipal waste management in context of sustainable urban development. *Procedia: Social and Behavioral Sciences*, 151, 244–256. DOI: 10.1016/j.sbspro.2014.10.023

Moh, Y. (2017). Solid waste management transformation and future challenges of source separation and recycling practice in Malaysia. *Resources, Conservation and Recycling*, 116, 1–14. DOI: 10.1016/j.resconrec.2016.09.012

Mohammed, M., Shafiq, N., Abdallah, N. A. W., Ayoub, M., & Haruna, A. (2020, April). A review on achieving sustainable construction waste management through application of 3R (reduction, reuse, recycling): A lifecycle approach. [). IOP Publishing.]. *IOP Conference Series. Earth and Environmental Science*, 476(1), 012010. DOI: 10.1088/1755-1315/476/1/012010

Mostakim, K., Arefin, M. A., Islam, M. T., Shifullah, K. M., & Islam, M. A. (2021). Harnessing energy from the waste produced in Bangladesh: Evaluating potential technologies. *Heliyon*, 7(10), e08221. DOI: 10.1016/j.heliyon.2021.e08221 PMID: 34729441

Olukanni, D. O., Pius-Imue, F. B., & Joseph, S. O. (2020). Public perception of solid waste management practices in Nigeria: Ogun State experience. *Recycling*, 5(2), 8. DOI: 10.3390/recycling5020008

Oteng-Ababio, M., Arguello, J. E. M., & Gabbay, O. (2013). Solid waste management in African cities: Sorting the facts from the fads in Accra, Ghana. *Habitat International*, 39, 96–104. DOI: 10.1016/j.habitatint.2012.10.010

Pariatamby, A., & Fauziah, S. H. (2014). Sustainable 3R practice in the Asia and Pacific Regions: the challenges and issues. *Municipal solid waste management in Asia and the Pacific Islands: Challenges and strategic solutions*, 15-40.

Riyad, A. S. M., Hassan, M., Rahman, M. A., Alam, M., & Akid, A. S. M. (2014). Characteristics and Management of Commercial Solid Waste in Khulna City of Bangladesh. *International Journal of Renewable Energy and Environmental Engineering*, 2.

Scarlat, N., Motola, V., Dallemand, J. F., Monforti, F., & Mofor, L. (2015). Evaluation of Energy Potential of Municipal Solid Waste from African Urban Areas. *Renewable & Sustainable Energy Reviews*, 50, 1269–1286. DOI: 10.1016/j.rser.2015.05.067

Seadon, J. K. (2010). Sustainable waste management systems. *Journal of Cleaner Production*, 18(16-17), 1639–1651. DOI: 10.1016/j.jclepro.2010.07.009

Shandas, V., & Messer, W. B. (2008). Fostering green communities through civic engagement: Community-based environmental stewardship in the Portland area. *Journal of the American Planning Association*, 74(4), 408–418. DOI: 10.1080/01944360802291265

Sharma, R. (2023). Plastic Waste Management: Challenges and Potential Solution With the application of AI. *International Journal of Sustainable Development Through AI. ML and IoT*, 2(1), 1–27.

Silva, W. D. O., & Morais, D. C. (2021). Transitioning to a circular economy in developing countries: A collaborative approach for sharing responsibilities in solid waste management of a Brazilian craft brewery. *Journal of Cleaner Production*, 319, 128703. DOI: 10.1016/j.jclepro.2021.128703

Sujauddin, M., Huda, S. M. S., & Hoque, A. T. M. R. (2008). Household Solid Waste Characteristics and Management in Chittagong, Bangladesh. *Waste Management (New York, N.Y.)*, 28(9), 1688–1695. DOI: 10.1016/j.wasman.2007.06.013 PMID: 17845843

UNEP (2018). Municipal solid waste: Is it garbage or gold?" *UNEP Global Environmental Alert Service (GEAS)*, October

Visvanathan, C., Adhikari, R., & Ananth, A. P. (2007). 3R practices for municipal solid waste management in Asia. *Linnaeus Eco-Tech*, 11-22.

Wichai-Utcha, N., & Chavalparit, O. (2019). 3Rs Policy and plastic waste management in Thailand. *Journal of Material Cycles and Waste Management*, 21(1), 10–22. DOI: 10.1007/s10163-018-0781-y

Wilson, D. C., Velis, C. A., & Rodic, L. (2013). Integrated Sustainable Waste Management in Developing Countries. *Proceedings of the Institution of Civil Engineers: Waste Management Resource*, 166, 52–68.

World Bank. (2018). *Enhancing opportunities for clean and resilient growth in urban Bangladesh: Country environmental analysis 2018*. The World Bank Group.

Wright, N. S., & Reames, T. G. (2020). Unraveling the links between organizational factors and perceptions of community sustainability performance: An empirical investigation of community-based nongovernmental organizations. *Sustainability (Basel)*, 12(12), 4986. DOI: 10.3390/su12124986

Zhang, J., Qin, Q., Li, G., & Tseng, C. H. (2021). Sustainable municipal waste management strategies through life cycle assessment method: A review. *Journal of Environmental Management*, 287, 112238. DOI: 10.1016/j.jenvman.2021.112238 PMID: 33714044

Chapter 18
Environmental Governance for Promoting Dental Public Health

Sadia Chowdhury
 https://orcid.org/0009-0004-9229-2668
Independent Researcher, Brunei

ABSTRACT

Environmental change has a profound effect on the physical environment as well as all elements of natural and human systems. This includes social and economic conditions, as well as the functioning of health systems. Climate change poses an additional risk to the environment, since it leads to more frequent and severe storms, floods, high temperatures, droughts, and wildfires. This chapter examines the importance and role of environmental governance in promoting oral public health. The discussion will centre around the multiple determinants that affect dental public health, encompassing the importance of fluoridation, the consequences of air and water pollution, the accessibility of uncontaminated water, sustainable approaches in dentistry, minimising sugar intake and advocating for nutritious diets, the availability of dental care (including healthcare facilities and workforce training), and the role of education and awareness. It is essential to include sustainability education into the curriculum of both undergraduate and postgraduate students.

DOI: 10.4018/979-8-3693-7001-8.ch018

1. INTRODUCTION

The widespread deterioration of the environment presents a basic peril to human well-being. Environmental change impacts the physical environment and all aspects of natural and human systems, encompassing social and economic situations and the operation of health systems. The state of the environment is further threatened by climate change, with more frequent and intensifying storms, floods, extreme temperatures, droughts, and wildfires. From 2030 to 2050, it is projected that climate change-induced environmental deterioration will lead to an estimated annual increase of 250,000 deaths due to under nutrition, malaria, diarrhea, and heat stress alone. The estimated direct health damage costs, excluding costs in areas such as agriculture and water and sanitation, are projected to range between US$ 2 and 4 billion years by 2030. The Global Burden of Disease (GBD) research has demonstrated that the occurrence of common oral disorders continues to be a substantial global health concern (Bernabe et al., 2020), impacting around 3.5 billion individuals globally (WHO, 2022).

Dental public health is an essential aspect of general public health, with a primary emphasis on preventing and managing dental problems and advocating for oral health. The primary objective of oral health professionals is to advance universal oral health by addressing preventable and/or treatable diseases, as outlined in the FDI Vision 2030 statement, which emphasises the need for immediate action in this area. The delivery of oral health care, whether through preventive measures, therapeutic interventions, or ongoing maintenance, generates pollution and contributes to a substantial carbon footprint. For instance, the National Health Service (NHS) in England generates 22.8 million tonnes of carbon equivalent emissions, which accounts for 3% of the total carbon footprint of the entire country (846 million tonnes). In comparison, the similar health care services in the United States and Australia contribute 10% (Booth, 2022) and 7% (Malik, et al., 2021) to their respective countries' carbon footprints. The health of individuals is being impacted by air pollution, with 10% of air pollution emissions being attributed to the health care sector and the utilisation of substances such as nitrogen oxide and sulphur dioxide (Duane and Fennell, 2023). Furthermore, it should be noted that there is a substantial amount of energy released as a result of patients travelling to get dental services. In the field of dentistry, it was discovered that staff travel for work and commuting to work accounted for the highest amount of carbon emissions (33.4%) in dental services throughout England in the period from 2013 to 2014 (Duane and Fennell, 2023). This was followed by patient travel to dental practices (31.1%), procurement (19.0%), electricity and gas consumption (15.3%), release of nitrous oxide (0.9%), waste management (0.2%), and water usage (0.1%) (Duane and Fennell, 2023) (Steinbach et al., 2018).

Both the creation of a sustainable health care system necessitates accelerated efforts and system transformation. Ensuring appropriate service delivery methodologies, workforce, health information systems, funding, preventative care, and leadership in dentistry are essential for maintaining long-term advantages for both environmental and human health, with a focus on environmental sustainability. The response of the World Health Organisation (WHO) to these difficulties revolves around three primary objectives. (i) Encourage measures that simultaneously decrease carbon emissions and enhance health outcomes. (ii) Construct improved health systems that are more resilient to climate change and environmentally sustainable. (iii) Safeguard health from the diverse array of climate change impacts. This chapter aims to offer a practical method, with a focus on public health, for creating guidelines that promote ecologically sustainable dental treatment in line with broader sustainability objectives. The goals are as follows:

- Enhance knowledge and comprehension of the ecological impact of oral health care within the dental field and among the general population.
- Offer guidance to dental professionals on adapting their practices to deliver environmentally sustainable oral health care.
- Minimise carbon emissions and waste by promoting good oral health practices.

There are many health professionals in the dental field who have adopted sustainable practices and thus improving the patient care and environment. Such as Australia's O'meara dental has reduced 70% waste by using ecofriendly products. They have also reduced energy use and have become the first carbon neutral dental office in Australia as well as within the first five globally (case study, cool planet). In the USA, Artisan dentals have become first carbon neutral practice by 2020. From the opening of the office in 2014, they started using energy star appliances, compact fluorescent lighting, natural day lighting, recycled, renewable electricity and began working with carbon credit capital to measure total carbon footprint, which enable them to gradually reduce the carbon footprints (World Dental Federation). These examples can help health professionals to implement these practices and become a part of the solution, not part of the problem.

2. ENVIRONMENTAL GOVERNANCE AND DENTAL PUBLIC HEALTH

Dental public health is commonly known as "the science and art of preventing and controlling dental diseases and promoting dental health through organized community efforts". The role of oral health professionals is not only related to the treatment of oral diseases but to conserve the dentition and to prevent the preventable diseases from occurring in the first place. On one hand the revolutionary change in dental treatment has given the oral health professionals knowledge and ease in providing their patients the best and safe treatment but it has also left its negative mark on environment, whether an atmospheric pollution through increased greenhouse gas emissions, or waste disposal based on its various activities.

An environmental impact refers to any alteration in the environment, whether positive or negative, that arises from the actions, goods, or services of an organisation. The field of dentistry is highly dependable on supplies, energy, fuel and water for daily clinical activities. This consumption of resources has an unavoidable adverse impact on environmental pollution and climate change (de Leon, 2020). Hence, dental health care has a duty to mitigate adverse environmental effects. This is where the connection between dental public health and environmental governance becomes crucial. It is a key determinant of the success or failure of conservation and environmental management efforts. The goals of boosting governance, advocating for oral disease prevention, strengthening the health workforce, integrating oral care into primary health care, developing information systems, and expanding research include the complex and multifaceted character of oral health (Cabrera et al, 2024).

The impact on environment from the dental settings occur in several ways, from CO_2 emission to improper management of medical waste, all contribute to environmental pollution. According to the world dental federation (FDI) and other authors (Martin and Mulligan, 2022) (Martin, et al., 2024) (Mulligan et al., 2021) four domains Preventive care (i) operative care (ii) integrated care (iii) ownership of care (iv) incapsulates the delivery of oral health care.

And while preventing and protecting oral health, therapeutic procedures or long-term maintenance, a dental team creates pollution and carbon footprint (Martin and Mulligan, 2022). Majority of the CO_2 emission by the dental clinic occurs from three sources (Martin and Mulligan, 2022) such as (i) travel by dentist, other members of the team and patient (ii) procurement of materials and sun dries from the supplier along with the manufacture and distribution and (iii) management of waste generated by the dental office as well as single use plastics (SUPs). A large dental office which is equipped with more sophisticated technologies which serve a considerable number of patients daily, produces a significant amount of waste and carbon footprints than a small dental office (European Federation of Periodontology).

A healthy mouth is also responsible for less environmental impact. An example of a case study of two 50 years old patients done by (Martin and Mulligan, 2022) can be used here for better understanding. Patient A who has good oral health with no active disease and some tooth surface loss does not need any restorative intervention. As the patient has low disease risk, regular hygiene maintenance is enough. On the other hand, patient B who has a failing dentition with new and recurrent active diseases, tooth loss and extensive restorative treatment is under high risk and is needed repeat intervention and management of active diseases.

In these two cases, the environmental impact for each of them is different. Patient A has less impact on the environment than patient B who will cause greater carbon footprint and waste production due to more travel, professional services, laboratory services and use of materials, sundries and personal protection equipment. So, an effective public health approach of prevention and quality care management is needed for lesser environmental impact.

3. FACTOR AFFECTING DENTAL PUBLIC HEALTH

Public health is influenced by a complex interplay of various factors, which can be broadly categorized as water fluoridation, pollution, access to clean water, access to dental care, dietary practices, education and awareness as shown in Figure 1.

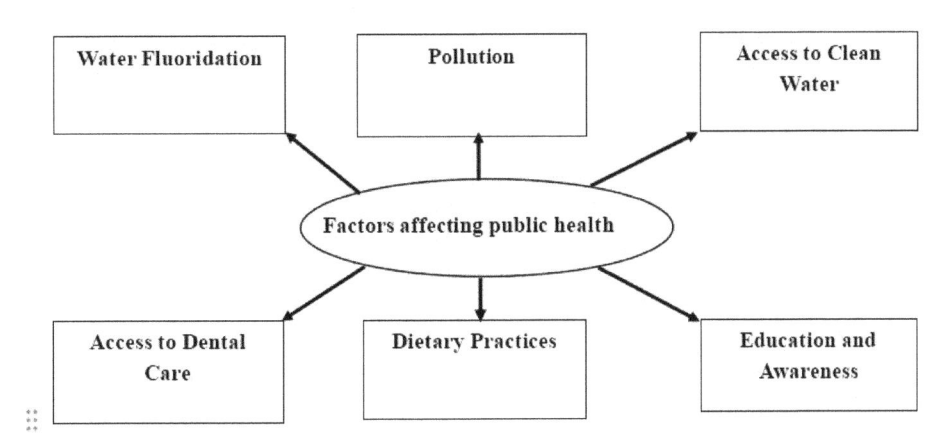

Figure 1. Factors affecting Dental Public Health

3.1 Water Fluoridation

Water fluoridation is a very impactful measure in public health that occurred in the twentieth century. However, as the climate continues to evolve, it is crucial to assess the impact of health care and illness prevention. Globally, 35 percent of the population has access to water fluoridation, which has been proven to result in substantial decreases in dental cavities. A study conducted by researchers from Trinity College Dublin and University College London compared the environmental impact of water fluoridation to other preventive measures for tooth decay. The study concluded that water fluoridation has a low carbon footprint and is an effective tool for preventing tooth decay. Fluoride consumption has both advantageous effects, such as lowering the occurrence of dental cavities, and detrimental consequences, such as producing tooth enamel and skeletal fluorosis at extended high levels of exposure (Iheozor, et al., 2015). The prevalence of caries in permanent teeth is projected to be the highest among all studied conditions, affecting around 2.4 billion individuals worldwide. Additionally, around 486 million children experience caries in their primary teeth. Public health interventions are necessary to ensure enough fluoride consumption in regions where it is deficient, in order to reduce the occurrence of dental caries. One way to do this is by implementing drinking water fluoridation. Alternatively, if this is not feasible, fluoridation can be done through the addition of fluoride to salt or milk, or by using dental care products that include fluoride toothpaste and fluoride varnish, pits and fissure sealant (Walsh, et al., 2019) (Yeung, et al., 2015) (Kashbour, et al., 2020). Additionally, promoting a low-sugar diet can also contribute to this goal (Sheiham and James, 2015).

Excessive fluoride intake typically happens when people consume groundwater that naturally has high levels of fluoride. This is more common in warm climates where people tend to drink more water, or in areas where water with high fluoride content is utilised for cooking or irrigating crops. Exposure to fluoride can result in dental fluorosis or severe skeletal fluorosis, characterised by osteoporosis, calcification of tendons and ligaments, and bone abnormalities.

Trinity College Dublin (TCD) and University College London (UCL) did a study on environmental impact of water fluoridation for an individual five-year-old child over a one-year period. They carried out a life cycle assessment (LCA) by measuring the travel, weight and quantity of all products and processes involved in tooth brushing, fluoride varnish programmes and water fluoridation, entered the data into a specific environmental programme (Open LCA) and used the *Ecoinvent* database, calculate environmental outputs, including the carbon footprint, the amount of land and water used for each product. The researchers concluded that water fluoridation had the lowest environmental impact of the three categories studied (Duane, et al., 2022). Also the EU commission's scientific committee on health and environmental risk (SCHER) in a report in 2011 on the basis of the evidences found, concluded that "Exposure of environmental organisms to the levels of fluoride used for fluoridation of drinking water is not expected to lead to unacceptable risks to the environment".

3.2 Pollution

The largest share of carbon emissions from NHS dental services in England is attributed to the travel of patients and dental personnel, accounting for 31.1% and 30.3% of the total emissions, respectively (Public Health, England). The determining factors are primarily the distance and the mode of transportation employed. The commuting behaviour might vary as individuals may opt to travel by means of walking, cycling, utilising public transportation, or engaging in carpooling. However, dentists are willing to travel large distances in order to obtain a good work for many reasons, which can influence their choice of transportation. Most dental treatments mostly involve brief procedures with minimal material expenses. Specifically, procedures such as scale and polish, fluoride varnish application, radiographs, and examinations constitute 73% of all dental treatments for both adults and children. Approximately 41.5% of procedures conducted on both adults and children consist of examinations, which typically last for approximately 15 minutes. For most dental treatments, patient travel contributes to a very modest but nonetheless significant amount of the carbon footprint. This is because these treatments often involve 2 to 4 round trips to complete.

Procurement accounts for the second largest proportion of greenhouse gas emissions in NHS dentistry. However, the carbon footprint associated with procurement, which stands at 19%, is quite modest compared to the overall carbon footprint of procurement for the entire NHS in England, which is 58%. The costs associated with dental procurement can be categorised into three distinct types: administrative costs, material costs, and laboratory service costs. The carbon content in these groups accounts for 31.2%, 33.8%, and 35.1% correspondingly.

The yearly carbon footprint resulting from the consumption of electricity and gas in NHS dental facilities accounts for 7.7% and 7.6% respectively of the overall carbon footprint of NHS dental services in England. These data are derived from the mean power and gas consumption per surgical procedure. Therefore, if the dental surgery has a greater number of surgical rooms and education suites, their consumption of electricity and gas should be correspondingly higher. Empirical data collected from a dental practice in England demonstrates reduced energy consumption in terms of real measurements of electricity and gas usage.

Dental hand devices, including the dental mirror, dental probe, and dental tweezers, are utilised in every dental consultation. Additionally, certain operations like tooth extraction require the usage of different types of forceps. Dental instruments are commonly fabricated from stainless steel, although they can also be composed of ceramics or brass. Stainless steel has the biggest carbon footprint, measuring 6.15 kg CO_2e/kg, when compared to ceramics, which have a carbon footprint of 1.14 kg CO_2e/kg, and brass, which has a carbon footprint of 2.42 kg CO_2e/kg.

3.3 Access to Clean Water

The relationship between water and climate change is inseparable. Climate change has intricate impacts on the Earth's water. According to UN Water, the majority of the effects of climate change, such as irregular rainfall, reduction in ice sheets, increased sea levels, and occurrences of floods and droughts, may be attributed to water. Currently, around two billion individuals globally lack access to potable water (according to the SDG Report 2022). Additionally, around half of the global population faces significant water scarcity at some periods of the year (as reported by the IPCC). The numbers are projected to rise, intensified by climate change and population expansion (WMO). Merely 0.5 percent of the water present on Earth is accessible and suitable for use as freshwater, and the availability of this vital resource is being gravely impacted by climate change. The food supply will face significant challenges due to climate change, population expansion, and escalating water scarcity, as highlighted by the Intergovernmental Panel on Climate Change (IPCC). This is because the majority of freshwater resources, approximately 70 percent on average, are utilised for agricultural purposes. Oral disorders such as dental caries and fluorosis are affected by the quality of food and water in a certain geographical area.

3.4 Access to Dental Care

The single most important factor to improve sustainability in dentistry is prevention of oral and dental disease, which would reduce the need for travel, production and use of dental materials and waste production - ultimately reducing the clinical burden currently placed on dental care provision. With travel accounting for over 60% of the carbon footprint for NHS dental services, it would be expensive and impractical to move established premises - travel for patients and staff in some form is also unavoidable. Lack of appropriate oral health care interventions is prevalent in underdeveloped nations (Nakre and Harikiran, 2013). Consequently, underprivileged groups continue to experience a higher impact from oral disorders and are more likely to encounter obstacles in obtaining and using oral health care services (Bayne et al., 2013).

However, when commissioning new contracts and services, consideration should be given to patient travel patterns and accessibility via public transport. Providers should also be encouraged to implement a travel plan, promoting walking, cycling and car sharing by dental practice team members whenever possible. Installing electric charging points for both staff and patient use would reduce emissions and promote use of electric vehicles. Reducing the number of patient appointments and use of appropriate recall intervals will play a key part in a more sustainable model of dental care for NHS and private dental practice. Increased training of carers and availability of domiciliary care in dentistry would prevent travel required by elderly, vulnerable and medically compromised patients. The use of technology - as seen over the course of the COVID-19 pandemic - can also be used, in certain instances, to reduce the need for travel. Some professional and work-related activities such as meetings, conferences and educational courses have also been shown to work quite successfully online, improving access and increasing participation. Providing remote clinical services such as dental triage, consultations and retainer reviews may, in certain circumstances, be indicated and desirable. The use of digital technology could also reduce waste in dentistry; for example, by using digital study models rather than traditional plaster casts.

3.5 Dietary Practices

Common dental conditions encompass dental caries, enamel developmental abnormalities, tooth erosion, and periodontal disease. Dental cavities, primarily caused by food, is the main factor contributing to tooth loss. Factors affecting dental caries include excessive sugar consumption, poor dietary habits, inadequate teeth cleaning, and insufficient or excessive fluoride exposure (Kabil and Eltawil, 2017). These factors can all contribute to the early development of caries in children or teenagers. Diet is a major factor in causing tooth erosion, which appears to be becoming more common. Certain components of the diet can also contribute to the development of enamel abnormalities, such as enamel hypoplasia and fluorosis. It is specifically linked to a diet that lacks fruits, vegetables, and non-starch polysaccharides (NSP), as well as having a low level of vitamin C in the blood. Additional factors include tobacco usage and alcohol consumption. In areas with limited access to fluoride and where populations have greater exposure to free sugars and other fermentable carbohydrates, the incidence of dental decay is on the rise. It is advisable for individuals to limit their consumption of meals containing free sugars to four times a day in order to lower the overall amount of free sugars consumed. Individuals residing in countries where fluoride toothpaste is accessible and reasonably priced should be strongly urged to practice twice-daily tooth brushing using fluoride toothpaste. Food makers should persist in creating low-sugar or sugar-free alternatives to items that are high in free sugars, such as beverages, and decrease the erosive capacity of soft drinks. Furthermore, it is imperative to have transparent and impartial labelling of food products in terms of their sugar content to provide accurate information to consumers. Incorporating oral health education into nutrition education programmes at schools and prenatal classes is essential. It is advisable to promote health education initiatives and health promotion websites. The World Health Organisation (WHO) should actively support initiatives aimed at identifying effective methods for supplying optimal levels of fluoride to countries with insufficient exposure. International institutions, including the WHO, should acknowledge nutrition as a crucial component of training for oral health practitioners and addressing dental health concerns. It is imperative for governments to guarantee that teachers, learners, and health professionals obtain comprehensive instruction regarding diet, nutrition, and oral health matters. Governments should establish explicit directives about the use and substance of educational resources. Health education and awareness can aid in diminishing sugar consumption and acidic beverage intake, as well as decreasing the frequency of sugar consumption. Health promotion aims to encourage dental brushing, the use of fluoride, moderation in alcohol consumption, and the cessation of tobacco use. An optimal oral health condition results in reduced treatment requirements, decreased utilisation of resources, and lower expenses, ultimately leading to enhanced quality of life and contentment for both the patient and the health care provider. Maintaining good oral health results in less visits to the dentist and the consumption of fewer dental supplies. This will result in a decrease in the production and transportation of resources, leading to a reduction in packaging, clinical waste, as well as a decrease in CO_2 emissions and pollution.

3.6 Education and Awareness

An examination of the lives of children and adolescents is essential in order to identify the most effective strategies that can enhance present educational programmes for promoting and preventing oral health. These programmes should encompass all the constituents of their socio-educational community, including primary school educators, students, and dental school instructors. This approach aims to promote appropriate oral hygiene and dietary practices in order to sustain optimal oral health and holistic growth among elementary pupils (Tenelanda-López et al., 2020). By adopting community-based programmes for oral health promotion, there may be an enhancement in community participation, resulting in the advancement and enhancement of information, attitudes, and behaviours about oral health (Nghayo et al., 2024).

Incorporating sustainability education into the curriculum of undergraduate and postgraduate students is crucial (Dixon et al., 2024) (Tancu, et al., 2023). Additionally, it is possible to create guidelines for the dental sector to manufacture environmentally-friendly products. It is necessary to prioritise research. It is crucial to enhance research capabilities by providing training to students and establishing virtual or physical centres dedicated to sustainability (Bauer, et al., 2020). Funding is required for critical research fields. Nevertheless, the successful execution of guidelines necessitates backing from several stakeholders, such as government institutions, public health experts, dentistry teams, and members of the general public (Hackley, 2021). It is feasible to incorporate a climate perspective into dental education programmes, specifically in relation to practice readiness, waste management, and service delivery models (Hackley, 2021). Within organisations, sustainability committees consisting of several stakeholders should strive to decrease the use of fossil fuels, optimise energy efficiency, and enhance waste management (Hackley, 2021).

4. ENVIRONMENTAL HEALTH POLICIES

Environmental health policies have a substantial impact on dental public health since they tackle the elements that affect oral health at a community level. These policies cross in numerous crucial areas. The process of adding fluoride to water fluoridation is a very successful public health intervention for reducing dental caries, often known as tooth decay. The objective of policies advocating for the addition of fluoride to public water supplies is to decrease the occurrence of dental cavities, especially in communities that have limited availability to dental services. Dental Amalgam Regulation, a frequently utilised substance for dental fillings, contains mercury, which might present environmental and health hazards if not appropriately handled (Fairbanks et al., 2021). Policies govern the utilisation, disposal, and reutilisation of dental amalgams in order to minimise the contamination of mercury and safeguard the well-being of both patients and the ecosystem (Fairbanks et al., 2021).

Insufficient air quality, particularly elevated concentrations of pollutants such as particulate matter, can impact general well-being, including oral health. Efforts to mitigate air pollution have a positive impact on respiratory and dental health. Provision of Potable Water

Having access to uncontaminated and secure drinking water is essential for sustaining optimal dental hygiene and overall health. Environmental laws that guarantee the availability and upkeep of uncontaminated water supplies aid in the prevention of dental problems caused by polluted water sources. Dental clinics produce hazardous waste, which encompasses chemicals, biological substances, and sharp tools. Regulatory policies governing the management, processing, and elimination of hazardous waste in dental clinics serve to safeguard environmental and public health. Environmental regulations that facilitate the availability of nutritious meals and restrict the accessibility of sugary beverages and snacks have a positive impact on oral health outcomes. For instance, implementing policies that promote urban agriculture and address the issue of food deserts can contribute to the maintenance of healthy diets within communities, which in turn can have a positive impact on dental health. Through the implementation of comprehensive policies, public health authorities have the ability to greatly enhance oral health outcomes and general well-being by addressing these environmental issues.

5. SUSTAINABLE DENTAL PRACTICES

Sustainable dental practices strive to provide the best possible care while reducing their negative effects on the environment. These procedures are meant to strike a balance between the requirements of patient care and the need to preserve resources and the environment. Here are some important strategies for sustainability in dental practices.

5.1 Waste Management

Efficient waste management in dental public health is essential for reducing environmental harm and safeguarding the health of patients, staff, and the community. The process entails the appropriate management, categorization, and elimination of different forms of waste produced in dental environments. An all-encompassing strategy for waste management in dental public health involves the crucial practice of effectively segregating dental waste to guarantee that various types of waste are processed and disposed of in the appropriate manner (Saxena, et al., 2024). The primary classifications of dental waste comprise non-hazardous waste, encompassing materials like paper, packaging, and routine office waste; infectious waste, encompassing items contaminated with blood or bodily fluids, gloves, masks, and gauze; hazardous waste, encompassing items such as amalgam used for a restoration, nitrous oxide used for conscious sedation, sharp objects like needles and blades; and dental waste ((Saxena, et al., 2024), encompassing specific materials like extracted teeth, orthodontic wires, and dental impressions, sterilisable tools, e.g., dental probe and tweezers. An important issue regarding dental waste is the management and containment of items such as Personal Protective Equipment

(PPE). The purpose of this is to guarantee that all personnel utilise suitable personal protective equipment (PPE) while managing garbage in order to avoid any contact with hazardous substances (Saxena, et al., 2024). Employ distinctly marked and chromatically organised containers for various categories of refuse to expedite accurate disposal. For instance, red containers are designated for sharps, yellow containers are designated for infectious trash, and black containers are designated for general rubbish. Following the processes of handling and storage, the subsequent step is the appropriate disposal. Needles, knives, and other sharp materials must be discarded in a specifically designated container solely intended for sharp objects. These containers must be resistant to punctures and should be disposed of in accordance with local legislation. Incineration is the appropriate method for disposing of infectious trash. Properly discard hazardous trash, such as dental amalgam and chemicals, by utilising specialised disposal services that adhere to regulatory rules for handling dangerous substances. Nevertheless, it is crucial to prioritise recyclable items such as paper, plastics, and metals and ensure that recycling procedures conform to local requirements.

The carbon emissions associated with the disposal of all waste streams amount to roughly 1,493 metric tonnes of CO_2 equivalent (CO_2e), representing 0.22% of the total carbon footprint of NHS dental services in England (Steinbach et al., 2018). Domestic trash accounts for 90.3% of the greenhouse gas emissions from dental waste disposal, making it the largest contributor to the waste stream. The reason for this is that the volume of residential garbage is larger compared to other waste streams, and the carbon conversion factor for domestic waste disposed of in landfills is likewise higher. Clinical waste accounts for 6.4% of the carbon footprint of garbage disposal, while amalgam waste contributes 0.8% and recycling cardboard contributes 2.5% (Steinbach et al., 2018).

5.2 Sustainable Materials

The objective of incorporating sustainable materials in dental public health is to minimise the ecological footprint while ensuring the efficacy and safety of dental care. The use of many restorative materials impacts the environment. The placement and removal of resin-based composite restorations lead to the release of microparticle pollutants into the nearby environment through human excretion. These pollutants have the potential to pollute the waste system (Batsford et al., 2022). Dentists ought to employ environmentally-friendly dental bibs and drapes that are biodegradable or recyclable, crafted from sustainable materials such as bamboo or recycled paper. It is recommended to use disposable suction tips made from biodegradable plastics or recyclable materials. In addition, dentists are encouraged to utilise stainless steel tools that can be sterilised and reused, thereby minimising the necessity for disposable alternatives. Furthermore, it is advisable to use hand-pieces that are designed for durability and efficiency, as this can help reduce the frequency of replacements. Autoclavable plastic or metal can be used as a substitute for single-use disposables including suction tips, cups, examination kits, and impression trays (Sitterson, 2017). Choose dental restoratives composed of environmentally-friendly and biocompatible materials, such as ceramics and composites. Certain types of cement utilise recycled glass as a primary component, thereby minimising both trash generation and environmental consequences. In addition, it is advisable to use dental goods that are packaged in materials that can be recycled or composted in order to reduce waste. Develop and establish mechanisms for replenishing supplies such as disinfectants or cleaning solutions, with the aim of minimising the usage of disposable packaging. Endorse producers that employ sustainable packaging practices by utilising minimum packaging or eco-friendly materials for their product packaging. Choosing cleaning products that are non-toxic, biodegradable, and eco-friendly. Employ digital technology such as CAD/CAM systems to minimise material waste and enhance efficiency. Whenever feasible, obtain source materials from nearby providers in order to minimise transportation emissions and bolster local economies. Through the implementation of sustainable materials and methods, dental public health can actively contribute to the reduction of environmental impact, preservation of resources, and promotion of a healthy lifestyle, all while upholding high standards of care and efficacy.

5.3 Reducing Travel

Dental offices can introduce and encourage active travel (Harford and Duane, 2018) such as cycling and walking where possible to reduce carbon emission through travel (Hurley, et al., 2018). Certain localities possess mobile dental units that deliver services directly at local schools, community centres, or workplaces which can cut down emission. Besides, emphasizing preventive care can diminish the necessity for more comprehensive treatments that frequently necessitate travel. During purchasing of the materials, if the products come in the same delivery, same logistic centre then it can reduce carbon emission and air pollution (Duane, et al., 2020). Furthermore, for the laboratory services, the closest laboratory and a bulk delivery of products can reduce impact of carbon emission and air pollution (Duan et al., 2019).

5.4 Innovative Products

Dental laboratories are progressively adopting 3D printing technology to fabricate crowns, bridges, and aligners with enhanced speed and precision, hence minimising turnaround time and expenses. Handheld intraoral scanners supplant conventional imprints, facilitating expedited, more pleasant, and accurate digital scans of patients' oral cavities. Smart toothbrushes are advanced devices that offer real-time feedback on brushing behaviours via applications, assisting individuals in enhancing their dental hygiene (Chen, et al, 2021) (Jeon, et al, 2021). Furthermore, laser dentistry instruments are employed for various dental operations, such as cavity preparation, excision and gum contouring, frequently yielding reduced discomfort and expedited recovery periods. There are many alternatives for dental practitioners in the market now to choose from. Biodegradable toothbrushes, nontoxic cleaning agent, etc. that can reduce environmental impact as well as enable the dentist to maintain a high standard of care (Beşiroğlu, et al., 2021) (Top green dental initiatives, 2024).

5.5 Digital Dentistry

Advances in digital technology also have helped revolutionize dental offices as well as reducing waste and travel. The CAD/CAM process in dentistry refers to an indirect restoration created by computer-aided design and fabricated using a computer-assisted milling machine. The process can be categorised into three distinct phases: data acquisition, indirect restoration design, and the manufacture of the prosthesis (Samra et al., 2016). For example, Digital radiography have reduced chemical based film processing while computer aided design and manufacturing (CAD/CAM) systems have helped to deliver high standard of dental procedures as well as reduced material waste (Marchesi et al., 2021). Digital patient records, online appointment scheduling, electronic billing have reduced waste and administrative burden. Telemedicine, electronic patient forms, SMS for communication with patient for reminder, follow up, confirmation, cloud based practice management system, virtual staff training and meeting, mobile app to access health information can contribute to sustainable practice and healthy environment (Top green dental initiatives, 2024).

6. CONCLUSION

The occurrence of climate change is undeniable, and the field of dentistry should actively contribute to the mitigation of carbon emissions and waste. It is imperative for clinicians, manufacturers, and key stakeholders to comprehend and be in agreement while addressing the environmental disaster that our planet is currently facing. The paramount component in enhancing sustainability in dentistry is the prevention of oral and dental diseases, which would diminish the necessity for travel, the manufacturing and utilisation of dental materials, and waste generation—ultimately alleviating the clinical strain on dental care services. Innovative technologies, like 3D printing, handheld intraoral scanners, smart toothbrushes, and CAD/CAM technologies, along with sustainable practices and educational programs, can reduce turnaround time and costs. Emphasising the importance of preventing and treating oral diseases to be given a top priority, along with conducting additional research to determine the specific measures that dental teams may take at a community level. Health care professionals should be motivated to contemplate and implement modifications to their therapeutic approach in order to enhance their environmental consciousness.

REFERENCES

Batsford, H., Shah, S., & Wilson, G. J. (2022). A changing climate and the dental profession. *British Dental Journal*, 232(9), 603–606. DOI: 10.1038/s41415-022-4202-1 PMID: 35562450

Bauer, N., Megyesi, B., Halbac-Cotoara-Zamfir, R., & Halbac-Cotoara-Zamfir, C. (2020). Attitudes and environmental citizenship. *Conceptualizing environmental citizenship for 21st century education, 4*, 97-111.

Bayne, A., Knudson, A., Garg, A., & Kassahun, M. (2013). Promising practices to improve access to oral health care in rural communities. *Rural Evaluation Brief*, 7, 1–6.

Bernabe, E., Marcenes, W., Hernandez, C. R., Bailey, J., Abreu, L., & Arora, A. (2020). GBD 2017 oral disorders collaborators. global, regional, and national levels and trends in burden of oral conditions from 1990 to 2017: A systematic analysis for the global burden of disease 2017 study. *Journal of Dental Research*, 99(4), 362–373. DOI: 10.1177/0022034520908533 PMID: 32122215

Beşiroğlu, S., Tağtekin, D., & Beşiroğlu, Ş. (2021). Sustainability In Dentistry. *European Journal of Research in Dentistry*, 5(2), 91–98.

Booth, A. (2022). Carbon footprint modelling of national health systems: Opportunities, challenges and recommendations. *The International Journal of Health Planning and Management*, 37(4), 1885–1893. DOI: 10.1002/hpm.3447 PMID: 35212060

Cabrera, M., Bedi, R., & Lomazzi, M. (2024). The Public Health Approach to Oral Health: A Literature Review. *Oral*, 4(2), 231–242. DOI: 10.3390/oral4020019

Carbon modelling within dentistry: towards a sustainable future. (2018). *Public Health England & The Centre for Sustainable Healthcare.* Available at: https://assets.publishing.service.gov.uk/government/uploads/system/uploads/attachment_data/file/724777/Carbon_modelling_within_dentistry.pdf

Chen, C. H., Wang, C. C., & Chen, Y. Z. (2021). Intelligent brushing monitoring using a smart toothbrush with recurrent probabilistic neural network. *Sensors (Basel)*, 21(4), 1238. DOI: 10.3390/s21041238 PMID: 33578684

de Leon, M. L. (2020). Barriers to environmentally sustainable initiatives in oral health care clinical settings. *Canadian Journal of Dental Hygiene : CJDH = Journal Canadien de l'Hygiene Dentaire : JCHD*, 54(3), 156. PMID: 33240375

Delivering better oral health: an evidence-based toolkit for prevention. (2021). *UK Government*. Available at https://www.gov.uk/government/publications/delivering-better-oral-health-an-evidence-based-toolkit-for-prevention

Department of Health and Public Health England. (2017). Delivering better oral health: an evidence-based toolkit for prevention. *PHE Gateway Number*: 2016224. https://assets.publishing.service.gov.uk/government/uploads/system/uploa

Dixon, J., Field, J., Gibson, E., & Martin, N. (2024). Curriculum content for environmental sustainability in dentistry. *Journal of Dentistry*, 147, 105021. DOI: 10.1016/j.jdent.2024.105021 PMID: 38679135

Duane, B., & Fennell-Wells, A. (2023). *Clinical guidelines for environmental sustainability in dentistry*. Trinity College Dublin.

Duane, B., Lyne, A., Parle, R., & Ashley, P. (2022). The environmental impact of community caries prevention-part 3: Water fluoridation. *British Dental Journal*, 233(4), 303–307. DOI: 10.1038/s41415-022-4251-5 PMID: 36028695

Duane, B., Stancliffe, R., Miller, F. A., Sherman, J., & Pasdeki-Clewer, E. (2020). Sustainability in dentistry: A multifaceted approach needed. *Journal of Dental Research*, 99(9), 998–1003. DOI: 10.1177/0022034520919391 PMID: 32392435

Duane, B., Steinbach, I., Ramasubbu, D., Stancliffe, R., Croasdale, K., Harford, S., & Lomax, R. (2019). Environmental sustainability and travel within the dental practice. *British Dental Journal*, 226(7), 525–530. DOI: 10.1038/s41415-019-0115-z PMID: 30980009

Fairbanks, S. D., Pramanik, S. K., Thomas, J. A., Das, A., & Martin, N. (2021). The management of mercury from dental amalgam in wastewater effluent. *Environmental Technology Reviews*, 10(1), 213–223. DOI: 10.1080/21622515.2021.1960642

Green dentistry: a way forward for oral-health. *European Federation of Periodontology*, accessed at https://www.efp.org/publications/perio-insight/green-dentistry-a-way-forward-for-oral-health-professionals/

Hackley, D. M. (2021). Climate change and oral health. *International Dental Journal*, 71(3), 173–177. DOI: 10.1111/idj.12628 PMID: 34024327

Harford, S., & Duane, B. (2018). Sustainable dentistry: how-to guide for dental practice s sustainable dentistry how to guide for dental practices sustainable dentistry: howto guide for dental practices. *Cent. Sustain. Healthc*.

Hurley, S., & White, S. (2018). *Carbon Modelling within Dentistry: Towards a Sustainable Future*. Public Health England.

Iheozor-Ejiofor, Z., Worthington, H. V., Walsh, T., O'Malley, L., Clarkson, J. E., Macey, R., Alam, R., Tugwell, P., Welch, V., & Glenny, A. M. (2015). Water fluoridation for the prevention of dental caries. *Cochrane Database of Systematic Reviews*, 2015(9), 6. DOI: 10.1002/14651858.CD010856.pub2 PMID: 26092033

Jeon, B., Oh, J., & Son, S. (2021, March). Effects of tooth brushing training, based on augmented reality using a smart toothbrush, on oral hygiene care among people with intellectual disability in Korea. []. MDPI.]. *Health Care*, 9(3), 348. PMID: 33803836

Kabil, N. S., & Eltawil, S. (2017). Prioritizing the risk factors of severe early childhood caries. *Dentistry Journal*, 5(1), 4. DOI: 10.3390/dj5010004 PMID: 29563410

Kashbour, W., Gupta, P., Worthington, H. V., & Boyers, D. (2020). Pit and fissure sealants versus fluoride varnishes for preventing dental decay in the permanent teeth of children and adolescents. *Cochrane Database of Systematic Reviews*, 11. PMID: 33142363

Malik, A., Padget, M., Carter, S., Wakiyama, T., Maitland-Scott, I., Vyas, A., Boylan, S., Mulcahy, G., Li, M., Lenzen, M., Charlesworth, K., & Geschke, A. (2021). Environmental impacts of Australia's largest health system. *Resources, Conservation and Recycling*, 169, 105556. DOI: 10.1016/j.resconrec.2021.105556

Marchesi, G., Camurri Piloni, A., Nicolin, V., Turco, G., & Di Lenarda, R. (2021). Chairside CAD/CAM materials: Current trends of clinical uses. *Biology (Basel)*, 10(11), 1170. DOI: 10.3390/biology10111170 PMID: 34827163

Martin, N., & Mulligan, S. (2022). Environmental sustainability through good-quality oral healthcare. *international dental journal, 72*(1), 26.

Martin, N., & Mulligan, S. (2022). Environmental sustainability through good-quality oral healthcare. *international dental journal, 72*(1), 26.

Martin, N., Mulligan, S., Fuzesi, P., Webb, T. L., Baird, H., Spain, S., & Hatton, P. V. (2020, August). Waste plastics in clinical environments: a multi-disciplinary challenge. In *Creative Circular Economy Approaches to Eliminate Plastics Waste* (pp. 86–91). UK Research and Innovation and UK Circular Plastics Network.

Mulligan, S., Smith, L., & Martin, N. (2021). Sustainable oral healthcare and the environment: Challenges. *Dental Update*, 48(6), 493–501. DOI: 10.12968/denu.2021.48.6.493

Nakre, P. D., & Harikiran, A. G. (2013). Effectiveness of oral health education programs: A systematic review. *Journal of International Society of Preventive & Community Dentistry*, 3(2), 103–115. DOI: 10.4103/2231-0762.127810 PMID: 24778989

Nghayo, H. A., Palanyandi, C. E., Ramphoma, K. J., & Maart, R. (2024). Oral health community engagement programs for rural communities: A scoping review. *PLoS One*, 19(2), e0297546. DOI: 10.1371/journal.pone.0297546 PMID: 38319914

O'Meara Dental. The first carbon neutral dentist in Australia. *Coolplanet*. available at https://www.coolplanet.com.au/case-studies/omeara-dental-the-first-carbonneutral-dentist-in-australia/

Samra, A. P. B., Morais, E., Mazur, R. F., Vieira, S. R., & Rached, R. N. (2016). CAD/CAM in dentistry–a critical review. *Revista Odonto Ciência*, 31(3), 140–144. DOI: 10.15448/1980-6523.2016.3.21002

Saxena, V., Datla, A., Deheriya, M., Tiwari, N., Shoukath, S., & Bhargava, A. (2024). EcoSmile: A Comprehensive Analysis of Sustainable Dental Practices Using Mixed Methodology. *Advances in Human Biology*, 10-4103.

Sheiham, A., & James, W. P. T. (2015). Diet and dental caries: The pivotal role of free sugars reemphasized. *Journal of Dental Research*, 94(10), 1341–1347. DOI: 10.1177/0022034515590377 PMID: 26261186

Steinbach, I., Stancliffe, R., Berners-Lee, M., & Duane, B. (2018). Carbon modelling within dentistry. Towards a sustainable future. In . Public Health England.

âncu, A. M. C., Didilescu, A. C., Pantea, M., Sfeatcu, R., & Imre, M. (2023, May). Aspects Regarding Sustainability among Private Dental Practitioners from Bucharest, Romania: A Pilot Study. [). MDPI.]. *Health Care*, 11(9), 1326. PMID: 37174868

Tenelanda-López, D., Valdivia-Moral, P., & Castro-Sánchez, M. (2020). Eating habits and their relationship to oral health. *Nutrients*, 12(9), 2619. DOI: 10.3390/nu12092619 PMID: 32867393

Top Green Dental Initiatives in. 2024. (2024). Available at https://adit.com/top-green-dental-initiatives-2024

US Department of Health and Human Services Federal Panel on Community Water Fluoridation. (2015). US Public Health Service recommendation for fluoride concentration in drinking water for the prevention of dental caries. *Public Health Reports*, 130(4), 318–331. DOI: 10.1177/003335491513000408 PMID: 26346489

Walsh, T., Worthington, H. V., Glenny, A. M., Marinho, V. C., & Jeroncic, A. (2019). Fluoride toothpastes of different concentrations for preventing dental caries. *Cochrane database of systematic reviews*, (3). World Dental Federation FDI, The Carbon Neutral Dental Office. *FDI* available at https://www.fdiworlddental.org/sites/default/files/2024-04/FDI%20World%20Dental%20Federation%20-%20The%20Carbon%20Neutral%20Dental%20Office%20%20Case%20Study.pdf

World Dental Federation FDI. Sustainability in dentistry. available at https://www.fdiworlddental.org/sustainability-dentistry

World Health Organization. (2022). *Global oral health status report: towards universal health coverage for oral health by 2030*. World Health Organization.

Yeung, C. A., Chong, L. Y., & Glenny, A. M. (2015). *Fluoridated milk for preventing dental caries*. Cochrane. DOI: 10.1002/14651858.CD003876.pub3

Compilation of References

Fouad, M. (2021). Mastering the risky business of public-private partnerships in infrastructure.

Mahzouni, A. (2008). Participatory local governance for sustainable community-driven development.

Sachs, J. D., Woo, W. T., Yoshino, N., & Taghizadeh-Hesary, F. (2019). Why is green finance important?

Aayog, N. I. T. I. (2024), Release of SDG India Index 2023-24, India Accelerates Progress towards the SDGs Despite Global Headwinds, https://pib.gov.in/PressReleasePage.aspx?PRID=2032857

Abbas, S., Alnoor, A., Sin Yin, T., Mohammed Sadaa, A., Raad Muhsen, Y., Wah Khaw, K., & Ganesan, Y. (2023). Antecedents of trustworthiness of social commerce platforms: A case of rural communities using multi group SEM & MCDM methods. *Electronic Commerce Research and Applications*, 62, 101322. DOI: 10.1016/j.elerap.2023.101322

Abedin, M. A., Collins, A. E., Habiba, U., & Shaw, R. (2018). Climate Change, Water Scarcity, and Health Adaptation in Southwestern Coastal Bangladesh. *International Journal of Disaster Risk Science*, 10(1), 28–42. DOI: 10.1007/s13753-018-0211-8

Abedin, M. A., & Jahiruddin, M. (2015). Waste generation and management in Bangladesh: An overview. *Asian Journal of Medical and Biological Research*, 1(1), 114–120. https://www.ebupress.com/journal/ajmbr. DOI: 10.3329/ajmbr.v1i1.25507

Abraham, B. (2018). Video Game Visions of Climate Futures: ARMA 3 and Implications for Games and Persuasion. *Games and Culture*, 13(1), 71–91. DOI: 10.1177/1555412015603844

Acharya, B., Dey, S., & Zidan, M. (2024). *IoT-Based Smart Waste Management for Environmental Sustainability*. CRC Press.

Adhikari, R. C. (2022). Investigation on solid waste management in developing countries. *Journal of Research and Development (Srinagar)*, 5(1), 42–52. DOI: 10.3126/jrdn.v5i1.50095

Afifa, , Arshad, K., Hussain, N., Ashraf, M. H., & Saleem, M. Z. (2024). Air pollution and climate change as grand challenges to sustainability. *The Science of the Total Environment*, 928, 172370. DOI: 10.1016/j.scitotenv.2024.172370 PMID: 38604367

African Climate Summit. (2023, September). *Africa Climate Week promotes regional initiatives for sustainable growth ahead of COP28*. Retrieved from https://www.africaclimatesummit.org

Agarwal, A., de los Angeles, M. S., Bhatia, R., Chéret, I., Davila-Poblete, S., Falkenmark, M., . . . Wright, A. (2000). *Integrated water resources management* (pp. 1-67). Stockholm: Global water partnership.

Agarwal, Anil, Marian S. de los Angeles, Ramesh Bhatia, Ivan Chéret, Sonia Davila-Poblete, Malin Falkenmark, F. Gonzalez Villarreal et al. (2000). *Integrated water resources management*. Stockholm: Global water partnership.

Agrawal, A., & Gibson, C. C. (1999). Enchantment and Disenchantment: The Role of Community in Natural Resource Conservation. *World Development*, 27(4), 629–649. DOI: 10.1016/S0305-750X(98)00161-2

Ahlawat, A., & Singh, G. (2019). Challenges in Public Private Partnership Infrastructure Project-A Case Study. In *Conference Paper May*.

Ahmed, F., Hasan, S., Rana, M. S., & Sharmin, N. (2022). A conceptual framework for zero waste management in Bangladesh. *International Journal of Environmental Science and Technology*. Advance online publication. DOI: 10.1007/s13762-022-04127-6

Ahmed, M. F., Ahuja, S., Alauddin, M., Hug, S. J., Lloyd, J. R., Pfaff, A., Pichler, T., Saltikov, C., Stute, M., & van Geen, A. (2006). Ensuring safe drinking water in Bangladesh. *Science*, 314(5806), 1687–1688. DOI: 10.1126/science.1133146 PMID: 17170279

Ahmed, R. (2024). Innovative Waste Management Solutions: A Global Perspective Challenges and Opportunities and the Bangladesh Context. *Preprints*. https://doi.org/DOI: 10.20944/preprints202407.2617.v1

Ahmed, R. R., Akbar, W., Aijaz, M., Channar, Z. A., Ahmed, F., & Parmar, V. (2023). The role of green innovation on environmental and organizational performance: Moderation of human resource practices and management commitment. *Heliyon*, 9(1), e12679. DOI: 10.1016/j.heliyon.2022.e12679 PMID: 36660461

Ahmed, S. A., & Ali, S. M. (2006). People as partners: Facilitating people's participation in public–private partnerships for solid waste management. *Habitat International*, 30(4), 781–796. DOI: 10.1016/j.habitatint.2005.09.004

Ahsan, A., Alamgir, M., El-Sergany, M. M., Shams, S., Rowshon, M. K., & Nik Daud, N. N. (2014). Assessment of municipal solid waste management system in a developing country. *Chinese Journal of Engineering*, 2014, 561935. Advance online publication. DOI: 10.1155/2014/561935

Ajulor, O., & Ibikunle, B. (2016). Theories of Local Government and their Relevance to Nigeria Experience. *Abuja Journal of Administration and Management*, 9(2), 76–89.

Akanda, M. G., Farhana, M., & Hafiza Nazneen Labonno, A. S. (2022). M Mahmudul Hasan Rifat, Mst. Nazneen Sultana. An Assessment to Solid Waste Management System in the Rajshahi City Vodra Railway Slum Through Community Participation. International. *Journal of Environmental Protection and Policy.*, 10(2), 22–30. DOI: 10.11648/j.ijepp.20221002.12

Akkerman, T., Hajer, M., & Grin, J. (2004). The Interactive State: Democratisation from Above? *Political Studies*, 52(1), 82–95. DOI: 10.1111/j.1467-9248.2004.00465.x

Akoto, O., Gyamfi, O., Darko, G., & Barnes, V. R. (2014). Changes in water quality in the Owabi water treatment plant in Ghana. *Applied Water Science*, 7(1), 175–186. DOI: 10.1007/s13201-014-0232-4

Akter, S., & Wamba, S. F. (2016). Big data analytics in e-commerce: A systematic review and agenda for future research. *Electronic Markets*, 26(2), 173–194. DOI: 10.1007/s12525-016-0219-0

Al Amin, M. (August 07, 2022). Waste Management System of Rural and Urban Areas in Bangladesh. The Asian Age. Accessed March 08, 2023 from https://dailyasianage.com/news/291268/waste-management-system-of-rural-and-urban-areas-in-bangladesh

Al Masum, A. (2014). Ground Water Quality Assessment of Different Educational Institutions in Rajshahi City Corporation, Bangladesh. *American Journal of Environmental Protection*, 3(2), 64. DOI: 10.11648/j.ajep.20140302.14

Alam, S. (2021). SDGs in Bangladesh: Implementation Challenges and Way Forward. *Data Science and SDGs: Challenges, Opportunities and Realities*, 1-14.

Alamgir, M., & Ahsan, N. (2007). Municipal solid waste and recovery potential: Bangladesh perspective. *Iranian Journal of Environmental Health Science and Engineering (IJEHSE)*, 4(2), 67-76. SID. https://sid.ir/paper/539035/en

Alamgir, M., & Cheng, M.-C. (2023). Do Green Bonds Play a Role in Achieving Sustainability? *Sustainability (Basel)*, 15(13), 10177. DOI: 10.3390/su151310177

Alam, M. S., Kabir, E., & Chowdhury, M. A. K. (2004). Power sector reform in Bangladesh: Electricity distribution system. *Energy*, 29(11), 1773–1783. DOI: 10.1016/j.energy.2004.03.005

Alam, M. Z., Rahman, M. A., & Al Firoz, M. A. (2013). Water supply and sanitation facilities in urban slums: A case study of Rajshahi City corporation slums. *American Journal of Civil Engineering and Architecture*, 1(1), 1–6. DOI: 10.12691/ajcea-1-1-1

Alcivar-Falcones, J., & Miranda-Pichucho, F. (2024). Contributions of Participatory Governance in Public Management. An Opportunity for Sustainable Local Development. *Digital Publisher CEIT*, 9(1), 746–761. DOI: 10.33386/593dp.2024.1.2237

Al-Dulaimi, G. A., & Younes, M. K. (2017). Assessment of potable water quality in Baghdad City, Iraq. *Air, Soil and Water Research*, 10, 1178622117733441. DOI: 10.1177/1178622117733441

Aleluia, J., & Ferrão, P. (2016). Characterization of Urban Waste Management Practices in Developing Asian Countries: A New Analytical Framework Based on Waste Characteristics and Urban Dimension. *Waste Management (New York, N.Y.)*, 58, 415–429. DOI: 10.1016/j.wasman.2016.05.008 PMID: 27220609

Ali, T. (2019, March 21). *Waste of water, way to disaster*. The Daily Star. https://www.thedailystar.net/supplements/news/waste-water-way-disaster-1718767

Ali, M., & Harper, M. (2004). *Sustainable Composting. Water, Engineering and Development Centre (WEDC)*. Loughborough University.

Alotaibi, I., Abido, M. A., Khalid, M., & Savkin, A. V. (2020). A comprehensive review of recent advances in smart grids: A sustainable future with renewable energy resources. *Energies*, 13(23), 6269. DOI: 10.3390/en13236269

Alsop, R., Bertelsen, M., & Holland, J. (2006). *Empowerment in Practice: From Analysis to Implementation*. World Bank.

Alves, E. C., Steiner, A. Q., & Amaral, A. M. F. (2023). Environmental Governance and International Relations: A Systematic Review of Theories, Methods, and Issues in Latin American Publications. *Revista de Relaciones Internacionales*, 96(2), 87–121. DOI: 10.15359/ri.96-2.4

American Council for an Energy-Efficient Economy. (2020). State and Local Policy Database. Retrieved from https://database.aceee.org/state/energy-code

Ames, B., ... (2001), Macroeconomic Policy and Poverty Reduction, prepared by the International Monetary Fund and the World Bank, https://www.imf.org/external/pubs/ft/exrp/macropol/eng/

Anderson, W., Taylor, C., McDermid, S., Ilboudo-Nébié, E., Seager, R., Schlenker, W., Cottier, F., De Sherbinin, A., Mendeloff, D., & Markey, K. (2021). Violent conflict exacerbated drought-related food insecurity between 2009 and 2019 in sub-Saharan Africa. *Nature Food*, 2(8), 603–615. DOI: 10.1038/s43016-021-00327-4 PMID: 37118167

Ando, K., Sugiura, J., Ohnuma, S., Tam, K.-P., Hübner, G., & Adachi, N. (2019). Persuasion Game: Cross Cultural Comparison. *Simulation & Gaming*, 50(5), 532–555. DOI: 10.1177/1046878119880236

Andresen, S. (2001). Global Environmental Governance: UN Fragmentation and Co-ordination. In *Yearbook of International Cooperation on Environment and Development 2001/2002* (pp. 19–26). Earthscan Publications.

Andrew, C., & Goldsmith, M. (1998). From local government to local governance—And beyond? *International Political Science Review*, 19(2), 101–117. DOI: 10.1177/019251298019002002

Angeles, G., Lance, P., Barden-O'Fallon, J., Islam, N., Mahbub, A. Q. M., & Nazem, N. I. (2009). The 2005 census and mapping of slums in Bangladesh: Design, select results and application. *International Journal of Health Geographics*, 8(1), 1–19. DOI: 10.1186/1476-072X-8-32 PMID: 19505333

Anitha, K. (2024). Emerging Trends in Sustainability: A Conceptual Exploration. In Kulkarni, S., & Haghi, A. K. (Eds.), *Global Sustainability. World Sustainability Series*. Springer., DOI: 10.1007/978-3-031-57456-6_2

Anne, P. M. Velenturf, Phil Purnell (2021) Principles for a sustainable circular economy, Sustainable Production and Consumption, Volume 27, 2021, Pages 1437-1457, ISSN 2352-5509, https://doi.org/DOI: 10.1016/j.spc.2021.02.018

Anonymous, . (2024). Big data analytics and corporate sustainability strategy: A management-oriented perspective. *Strategic Direction*, 40(5), 31–33. DOI: 10.1108/SD-05-2024-0054

Ansell, C. (2011). *Pragmatist Governance: Re-Imagining Institutions and Democracy*. Oup Usa.

Apollo, A., & Mbah, M. F. (2021). Challenges and opportunities for climate change education (Cce) in East Africa: A critical review. *Climate (Basel)*, 9(6), 93. DOI: 10.3390/cli9060093

Ara, S., Khatun, R., & Uddin, M. S. (2021). Urbanization challenge: Solid waste management in Sylhet city, Bangladesh. *International Journal of Engineering Applied Sciences and Technology*, 5(10), 20–28. DOI: 10.33564/IJEAST.2021.v05i10.004

Ardoin, N. M., Bowers, A. W., & Gaillard, E. (2020). Environmental education outcomes for conservation: A systematic review. *Biological Conservation*, 24, 108224. DOI: 10.1016/j.biocon.2019.108224

Arshad, A., Shahzad, F., Rehman, I. U., & Sergi, B. S. (2023). A systematic literature review of blockchain technology and environmental sustainability: Status quo and future research. *International Review of Economics & Finance*, 88, 1602–1622. DOI: 10.1016/j.iref.2023.07.044

Arteaga, C., Silva, J., & Yarasca-Aybar, C. (2023). Solid waste management and urban environmental quality of public space in Chiclayo, Peru. *City and Environment Interactions*, 20, 100112. DOI: 10.1016/j.cacint.2023.100112

Asheim, B., Grillitsch, M., & Trippl, M. (2017). Introduction: Combinatorial knowledge bases, regional innovation, and development dynamics. *Economic Geography*, 93(5), 429–435. DOI: 10.1080/00130095.2017.1380775

Ashraf, S., Nazemi, A., & AghaKouchak, A. (2021). Anthropogenic drought dominates groundwater depletion in Iran. *Scientific Reports*, 11(1), 9135. DOI: 10.1038/s41598-021-88522-y PMID: 33911120

Ashton, K. (2009). That 'Internet of Things' thing. *RFID Journal*, 22(7), 97–114.

Asian Development Bank. (2018). *Public-Private Partnerships in Infrastructure Development: Lessons Learned from Case Studies in Asia and the Pacific*. Asian Development Bank.

Aswathy, Y. (2022), Environmental Governance in India: Problems and Prospects, *Rajasthali Journal*, Vol. 1, Issue. 2, 22-26, https://rajasthali.marudharacollege.ac.in/papers/Volume-1/Issue-2/02-04.pdf

Austin, D., & Macauley, M. K. (2001). Cutting through Environmental Issues: Technology as a Double-Edged Sword. *The Brookings Review*, 19(1), 24–27. DOI: 10.2307/20080956

Awino, F. B., & Apitz, S. E. (2023). Solid waste management in the context of the waste hierarchy and circular economy frameworks: An international critical review. *Integrated Environmental Assessment and Management*, 20(1), 9–35. DOI: 10.1002/ieam.4774 PMID: 37039089

Aydin, M., Söğüt, Y., & Erdem, A. (2024). The role of environmental technologies, institutional quality, and globalization on environmental sustainability in European Union countries: New evidence from advanced panel data estimations. *Environmental Science and Pollution Research International*, 31(7), 10460–10472. Advance online publication. DOI: 10.1007/s11356-024-31860-x PMID: 38200188

Ayvazyan, S. A., Afanase, M. Y., & Rudenko, V. A. (2012). Some questions of specification of three-factor models of the company's production potential, taking into account intellectual capital. *Applied Economics*, 3(27), 56–66.

Azevedo, B. D., Scavarda, L. F., Caiado, R. G. G., & Fuss, M. (2021). Improving urban household solid waste management in developing countries based on the German experience. *Waste Management (New York, N.Y.)*, 120, 772–783. DOI: 10.1016/j.wasman.2020.11.001 PMID: 33223248

Azhoni, A., Jude, S., & Holman, I. (2018). Adapting to climate change by water management organisations: Enablers and barriers. *Journal of Hydrology (Amsterdam)*, 559, 736–748. DOI: 10.1016/j.jhydrol.2018.02.047

Aziz, H. A., Omar, F. M., Halim, H. A., & Hung, Y. T. (2022). Health-care waste management. In L. K. Wang, M. H. S. Wang, & Y. T. Hung (Eds.), *Solid waste engineering and management*. Handbook of environmental engineering (Vol. 25). Springer. https://doi.org/DOI: 10.1007/978-3-030-96989-9_4

Aziz, M. A., Majumder, M. A. K., Kabir, M. S., Hossain, M. I., Rahman, N. M. F., Rahman, F., & Hosen, S. (2015). Groundwater depletion with expansion of irrigation in Barind tract: A case study of Rajshahi district of Bangladesh. *Int J Geol Agric Environ Sci*, 3, 32–38.

Babenko, V.; Perevozova, I.; Mandych, O.; Kvyatko, T.; Maliy, O. & Mykolenko, I. (2019). World informatization in conditions of international globalization: factors of influence. *Global. J. Environ. Sci. Manage.*, 5(SI): 172-179, 2019. DOI: DOI: 10.22034/gjesm.2019.SI.19

Baburaj, A. (2021). Artificial intelligence v. intuitive decision making how far can it transform corporate governance? *GNLU L. Rev.*, 8, 233.

Bäckstrand, K., Khan, J., Kronsell, A., & Lövbrand, E. (2021). The promise of participatory environmental governance: Participatory institutions and the challenge of democratic environmental governance. *Environmental Politics*, 30(1), 1–20. DOI: 10.1080/09644016.2020.1816557

Bai, X., (2020). Overcoming barriers to urban green infrastructure: Lessons from Singapore. *Journal of Environmental Management*, 266, 110605.

Bakker, K., & Ritts, M. (2022). Environmental Governance in a Wired World. *The Nature of Data: Infrastructures, Environments. Politics*, 61.

Bamunuarachchige, T. C., & de Zoysa, H. K. S. (Eds.). (2022). *Waste technology for emerging economies* (1st ed.). CRC Press., DOI: 10.1201/9781003132349

Bandyopadhyay, J., & Shiva, V. (1988). Political Economy of Ecological Movements. *Economic and Political Weekly*, 23(24), 1223–1232.

Bandyopadhyay, P., & Sen, J. (2021). Innovating IoT and digital marketing for sustainable environmental governance. *Journal of Sustainable Development*, 9(1), 45–56.

Bandyopadhyay, S., (2021). Access to finance for sustainable infrastructure development: Insights from India. *International Journal of Sustainable Development and World Ecology*, 28(2), 116–130.

Bandyopadhyay, S., & Sen, J. (2019). Internet of Things: Applications and challenges in technology and standardization. *Wireless Personal Communications*, 108(1), 1–14.

Bandyopadhyay, S., & Sen, J. (2021). Challenges in Internet of Things (IoT) integration with cloud computing: A systematic review. *Journal of Ambient Intelligence and Humanized Computing*, 12(2), 2267–2285.

Banerjee, M., & Sinha, S. (2021). Understanding the role of digital marketing in promoting sustainable consumer behavior. *Journal of Strategic Marketing*, 29(5), 420–439.

Banerjee, S., & Sinha, S. (2021). Digital marketing and sustainability: A review and agenda for future research. *Journal of Business Research*, 131, 323–336. DOI: 10.1016/j.jbusres.2021.07.037

Bansard, Jennifer. & Schröder, Mika. (2021). Deep Dive - The Sustainable Use of Natural Resources: The Governance Challenge, https://www.iisd.org/articles/deep-dive/sustainable-use-natural-resources-governance-challenge

Banuri, T., & Najam, A. (2002). *Civic Entrepreneurship: A Civil Society Perspective on Sustainable Development*. Gandhara Academic Press. Also sponsored by Stockholm Environment Institute, United Nations Environment Program, and the RING Regional and International Networking Group.

Barau, A. S., & Al Hosani, N. (2015). Prospects of environmental governance in addressing sustainability challenges of seawater desalination industry in the Arabian Gulf. *Environmental Science & Policy*, 50, 145–154. DOI: 10.1016/j.envsci.2015.02.008

Barca, F., McCann, P., & Rodríguez-Pose, A. (2012). The case for regional development intervention: Place-based versus place-neutral approaches. *Journal of Regional Science*, 52(1), 134–152. DOI: 10.1111/j.1467-9787.2011.00756.x

Barros, M. V., Salvador, R., do Prado, G. F., Carlos de Francisco, A., & Piekarski, C. M. (2021). Circular economy as a driver to sustainable businesses, Cleaner Environmental Systems, Volume 2, 2021, 100006, ISSN 2666-7894, https://doi.org/ DOI: 10.1016/j.cesys.2020.100006

Bartoli, A., Delnevo, G., Oggioni, A., & Verticale, G. (2021). Smart city services: Integrating IoT sensors and digital marketing for environmental monitoring. In 2021 IEEE Conference on Computer Communications Workshops (INFOCOM WKSHPS) (pp. 1-6). IEEE.

Bartoli, M., Martín-Martín, A., Álvarez-Álvarez, S., & Palazzi, C. E. (2021). Smart Citizen: A concept for citizen-centered governance of urban and environmental issues. *Sensors (Basel)*, 21(8), 2863. Advance online publication. DOI: 10.3390/s21082863 PMID: 33921782

Barua, P., & Rahman, S. H. (2021). Urban management in Bangladesh. In *The Palgrave Encyclopedia of Urban and Regional Futures*. Palgrave Macmillan., DOI: 10.1007/978-3-030-51812-7_147-1

Bashir, M. F., Ma, B., Bilal, F., Komal, B., & Bashir, M. A. (2021). Analysis of environmental taxes publications: A bibliometric and systematic literature review. *Environmental Science and Pollution Research International*, 28(16), 20700–20716. DOI: 10.1007/s11356-020-12123-x PMID: 33405155

Batsford, H., Shah, S., & Wilson, G. J. (2022). A changing climate and the dental profession. *British Dental Journal*, 232(9), 603–606. DOI: 10.1038/s41415-022-4202-1 PMID: 35562450

Bauer, N., Megyesi, B., Halbac-Cotoara-Zamfir, R., & Halbac-Cotoara-Zamfir, C. (2020). Attitudes and environmental citizenship. *Conceptualizing environmental citizenship for 21st century education*, 4, 97-111.

Bauman, Z. (2000). *Liquid Modernity*. Polity.

Bayne, A., Knudson, A., Garg, A., & Kassahun, M. (2013). Promising practices to improve access to oral health care in rural communities. *Rural Evaluation Brief*, 7, 1–6.

Beck, U., & Beck-Gernsheim, E. (2002). *Individualization: Institutionalized Individualism and Its Social and Political Consequences* (1st ed.). SAGE Publications Ltd.

Bedoya, Franco, Mani, Sebastian & Muthukumara. (2020-2011) World Bank Group, The Drivers of Firms' Compliance to Environmental Regulations: The Case of India, https://openknowledge.worldbank.org/entities/publication/3aa0e778-1cfe-5329-a208-59d6a1c2ee12

Behera, M. R., Pradhan, H. S., Behera, D., Jena, D., & Satpathy, S. K. (2021). Achievements and challenges of India's sanitation campaign under clean India mission: A commentary. *Journal of Education and Health Promotion*, 10(1), 350. DOI: 10.4103/jehp.jehp_1658_20 PMID: 34761036

Benito, B., Guillamón, M. D., Martínez-Córdoba, P. J., & Ríos, A. M. (2021). Influence of selected aspects of local governance on the efficiency of waste collection and street cleaning services. *Waste Management (New York, N.Y.)*, 126, 800–809. DOI: 10.1016/j.wasman.2021.04.019 PMID: 33895563

Benson, M. H., & Craig, R. K. (Eds.). (2020). *The End of Sustainability: Resilience and the Future of Environmental Governance in the Anthropocene*. University of Kansas Press.

Benson, R., Conerly, O. D., Sander, W., Batt, A. L., Boone, J. S., Furlong, E. T., Glassmeyer, S. T., Kolpin, D. W., Mash, H. E., Schenck, K. M., & Simmons, J. E. (2017). Human health screening and public health significance of contaminants of emerging concern detected in public water supplies. *The Science of the Total Environment*, 579, 1643–1648. DOI: 10.1016/j.scitotenv.2016.03.146 PMID: 28040195

Benzaghta, M. A., Elwalda, A., & Mous, M. M. (2021). SWOT analysis applications: An integrative literature review. *Journal of Global Business Insights*, 6(1), 1–21. https://digitalcommons.usf.edu/cgi/viewcontent.cgi?article=1148&context=globe. DOI: 10.5038/2640-6489.6.1.1148

Berkes, F., Colding, J., & Folke, C. (Eds.). (2000). *Navigating social-ecological systems: Building resilience for complexity and change*. Cambridge University Press.

Bermeo Tinoco, E. O., & Ortiz Cañares, S. N. (2024) *Estrategias ambientales y prácticas sostenibles en pymes: caso Aeroagripac S.A.* (trabajo de titulación). UTMACH, Facultad de Ciencias Empresariales, Machala, Ecuador. http://repositorio.utmachala.edu.ec/handle/48000/22631

Bernabe, E., Marcenes, W., Hernandez, C. R., Bailey, J., Abreu, L., & Arora, A. (2020). GBD 2017 oral disorders collaborators. global, regional, and national levels and trends in burden of oral conditions from 1990 to 2017: A systematic analysis for the global burden of disease 2017 study. *Journal of Dental Research*, 99(4), 362–373. DOI: 10.1177/0022034520908533 PMID: 32122215

Beşiroğlu, S., Tağtekin, D., & Beşiroğlu, Ş. (2021). Sustainability In Dentistry. *European Journal of Research in Dentistry*, 5(2), 91–98.

Beula, D., & Sureshkumar, M. (2021). A review on the toxic E-waste killing health and environment–Today's global scenario. *Materials Today: Proceedings*, 47, 2168–2174. DOI: 10.1016/j.matpr.2021.05.516

Bhavnani, D., Goldstick, J. E., Cevallos, W., Trueba, G., & Eisenberg, J. N. (2014). Impact of rainfall on diarrheal disease risk associated with unimproved water and sanitation. *The American Journal of Tropical Medicine and Hygiene*, 90(4), 705–711. DOI: 10.4269/ajtmh.13-0371 PMID: 24567318

Bhuiyan, S. H. (2010). A crisis in governance: Urban solid waste management in Bangladesh. *Habitat International*, 34(1), 125–133. DOI: 10.1016/j.habitatint.2009.08.002

Bhutta, U. S., Tariq, A., Farrukh, M., Raza, A., & Iqbal, M. K. (2022). Green bonds for sustainable development: Review of literature on development and impact of green bonds. *Technological Forecasting and Social Change*, 175, 121378. DOI: 10.1016/j.techfore.2021.121378

Biermann, F. (2005). The Rationale for a World Environment Organization. In Biermann, F., & Bauer, S. (Eds.), *A World Environment Organization: Solution or Threat for Effective Environmental Governance?* (pp. 117–144). Ashgate.

Biermann, F. (Eds.). (2009). *Global Environmental Governance Reconsidered*. MIT Press.

Biermann, F., & Pattberg, P. (2008). Global environmental governance: Taking stock, moving forward. *Annual Review of Environment and Resources*, 33(1), 277–294. DOI: 10.1146/annurev.environ.33.050707.085733

Biermann, F., & Pattberg, P. (2012). Global environmental governance revisited. *Environmental Politics*, 21(1), 115–134. DOI: 10.1080/09644016.2012.738749

Bieser, J. C., & Hilty, L. M. (2018). Assessing indirect environmental effects of information and communication technology (ICT): A systematic literature review. *Sustainability (Basel)*, 10(8), 2662. DOI: 10.3390/su10082662

Billions of people will lack access to safe water, sanitation and hygiene in 2030 unless progress quadruples – warn WHO, UNICEF. (n.d.). https://www.unicef.org/bangladesh/en/press-releases/billions-people-will-lack-access-safe-water-sanitation-and-hygiene-2030-unless

Blair, A. M., & Hitchcock, D. H. (2001). *Environment and business*. Routledge.

Blanchard, O., & Buchs, A. (2015). Clarifying Sustainable Development Concepts Through Role-Play. *Simulation & Gaming*, 46(6), 697–712. DOI: 10.1177/1046878114564508

Blatz, W. E. (1966). *Human Security: Some Reflections*. University of Toronto Press. DOI: 10.3138/9781442632134

Bodansky, D. (2010). The history of the global climate change regime. *Review of European, Comparative & International Environmental Law*, 19(1), 63–77.

Bodansky, D. (2016). *The Art and Craft of International Environmental Law*. Harvard University Press.

Bodansky, D. (2016). The Paris Climate Agreement: A New Hope? *The American Journal of International Law*, 110(2), 288–319. DOI: 10.5305/amerjintelaw.110.2.0288

Boiral, O., Heras-Saizarbitoria, I., & Brotherton, M.-C. (2020). Improving environmental management through Indigenous peoples' involvement. *Environmental Science & Policy*, 103, 10–20. DOI: 10.1016/j.envsci.2019.10.006

Bolan, S., Padhye, L. P., Jasemizad, T., Govarthanan, M., Karmegam, N., Wijesekara, H., Amarasiri, D., Hou, D., Zhou, P., Biswal, B. K., Balasubramanian, R., Wang, H., Siddique, K. H. M., Rinklebe, J., Kirkham, M. B., & Bolan, N. (2024). Impacts of climate change on the fate of contaminants through extreme weather events. *The Science of the Total Environment*, 909, 168388. DOI: 10.1016/j.scitotenv.2023.168388 PMID: 37956854

Bookchin, M. (1962). *Our Synthetic Environment*. Harper & Row.

Booth, A. (2022). Carbon footprint modelling of national health systems: Opportunities, challenges and recommendations. *The International Journal of Health Planning and Management*, 37(4), 1885–1893. DOI: 10.1002/hpm.3447 PMID: 35212060

Boulding, C., & Wampler, B. (2010). Voice, Votes, and Resources: Evaluating the Effect of Participatory Democracy on Well-being. *World Development*, 38(1), 125–135. DOI: 10.1016/j.worlddev.2009.05.002

Bradu, P., Biswas, A., Nair, C., Sreevalsakumar, S., Patil, M., Kannampuzha, S., Mukherjee, A. G., Wanjari, U. R., Renu, K., Vellingiri, B., & Gopalakrishnan, A. V. (2023). Recent Advances in Green Technology and Industrial Revolution 4.0 for a Sustainable Future. *Environmental Science and Pollution Research International*, 30(60), 124488–124519. DOI: 10.1007/s11356-022-20024-4 PMID: 35397034

Brauch, H. G. (2009). Human Security Concepts in Policy and Science. In *Facing Global Environmental Change: Environment, Human, Energy, Food, Health, and Water Security Concepts* (pp. 965-989). Springer. DOI: 10.1007/978-3-540-68488-6_74

Brauch, H. G. (2008). Introduction. In *Globalization and Environmental Challenges: Reconceptualizing security in the 21st Century* (pp. 27–44). Springer. DOI: 10.1007/978-3-540-75977-5_1

Brickson, L., Zhang, L., Vollrath, F., Douglas-Hamilton, I., & Alexander, J. (2023, November). Titus Elephants and algorithms: A review of the current and future role of AI in elephant monitoring. *Journal of the Royal Society, Interface*, 15(208), 20230367. Advance online publication. DOI: 10.1098/rsif.2023.0367 PMID: 37963556

Bright, M., & Matsuura, K. (2017). *1001 Natural wonders you must see before you die*. Chartwell Books.

Brockhaus, M., Di Gregorio, M., & Mardiah, S. (2021). Multi-level governance and its role in environmental sustainability. *Environmental Policy and Governance*, 31(1), 65–79.

Broughton, E. (2005). The Bhopal disaster and its aftermath: A review. *Environment & Health*, 4(1), 6. https://www.ncbi.nlm.nih.gov/pmc/articles/PMC1142333/. DOI: 10.1186/1476-069X-4-6 PMID: 15882472

Brown, C. (2008). Emergent Sustainability: The Concept of Sustainable Development in a Complex World. In *Globalization and Environmental Challenges: Reconceptualizing security in the 21st Century* (pp. 141-150). Springer.

Brown, A., & Jones, B. (2019). Regulatory frameworks for sustainable investment: A comparative analysis. *Journal of Sustainable Development*, 12(3), 45–58.

Brown, C., (2018). Capacity building strategies for private sector engagement in sustainability: Lessons from emerging economies. *Sustainable Development Journal*, 15(2), 78–91.

Buck, M., Sturzaker, J., & Mell, I. (2023). Playing games around climate change - new ways of working to develop climate change resilience. *Journal of Planning Literature*, 38(4), 622–622.

Bull, J., Jobstvogt, N., Böhnke-Henrichs, A., Mascarenhas, A., Sitas, N., Baulcomb, C., Lambini, C. K., Rawlins, M., Baral, H., Zähringer, J., Carter-Silk, E., Balzan, M. V., Kenter, J. O., Häyhä, T., Petz, K., & Koss, R. (2016). Strengths, Weaknesses, Opportunities and Threats: A SWOT analysis of the ecosystem services. *Ecosystem Services*, 17, 99–111. DOI: 10.1016/j.ecoser.2015.11.012

Bundesministerium für Wirtschaft und Energie (BMWi). (2017). G20—Shaping digitalization at global level. Organisation for Economic Co-Operation and Development: Paris, France. Available online: https://www.de.digital/DIGITAL/Redaktion/EN/Dossier/g20-shaping-digitalisation-at-global-level.html (accessed on 22 July, 2024).

Burck, J., & Hagen, U. HÃúhne, N., Nascimento, L., & Bals, C. (2019). Climate change performance index. *Bonn: Germanwatch, NewClimate Institute and Climate Action Network.*

Busch, J., Engelmann, J., Fisher, B., & Sanchez-Azofeifa, G. A. (2023). The rise of global environmental treaties in the Anthropocene: How to understand the new frameworks. *Environmental Science & Policy*, 145, 345–357.

Busumtwi-Sam, J. (2002). Development and Human Security: Whose Security, and from What? *International Journal (Toronto, Ont.)*, 57(2), 253–272. DOI: 10.1177/002070200205700207

Byravan, S., & Rajan, S. C. (2013), An Evaluation of India's National Action Plan on Climate Change, DOI:DOI: 10.2139/ssrn.2195819

Cabrera, M., Bedi, R., & Lomazzi, M. (2024). The Public Health Approach to Oral Health: A Literature Review. *Oral*, 4(2), 231–242. DOI: 10.3390/oral4020019

Caillois, R. (2001). *Man, play and games.* University of Illinois Press.

Callahan, K. (2007). *Elements of Effective Governance.* Taylor&Francis Group CRC Press.

Cañas, D. (2023). *Estrategias para la sostenibilidad de las Pequeñas y Medianas Empresas – MiPyMES – en Santander, basado en la experiencia de las empresas de Triple Impacto.* Universidades Tecnológicas de Santander. Pp. 1-69. http://repositorio.uts.edu.co:8080/xmlui/handle/123456789/12730

Carayannis, E. G., & Campbell, D. F. J. (2010). Triple Helix, Quadruple Helix and Quintuple Helix and how do knowledge, innovation and the environment relate to each other? A proposed framework for a trans-disciplinary analysis of sustainable development and social ecology. *International Journal of Social Ecology and Sustainable Development*, 1(1), 41–69. https://www.igi-global.com/bookstore/article.aspx?titleid=41959. DOI: 10.4018/jsesd.2010010105

Carayannis, E. G., & Campbell, D. F. J. (2011). Open Innovation Diplomacy and a 21st Century Fractal Research, Education, and Innovation (FREIE) Ecosystem: Building on the Quadruple and Quintuple Helix Innovation Concepts and the "Mode 3" Knowledge Production System. *Journal of the Knowledge Economy*, 2(3), 327–372. DOI: 10.1007/s13132-011-0058-3

Carbon modelling within dentistry: towards a sustainable future. (2018). *Public Health England & The Centre for Sustainable Healthcare.* Available at: https://assets.publishing.service.gov.uk/government/uploads/system/uploads/attachment_data/file/724777/Carbon_modelling_within_dentistry.pdf

Carley, M., Smith, H., & Jenkins, P. (2013). *Urban development and civil society: The role of communities in sustainable cities.* Routledge. DOI: 10.4324/9781315071725

Carrasco, L., Valdivieso, P., & Herrera, A. (2023). Estructura de capital y rentabilidad de las empresas ecuatorianas amigables con el medio ambiente en el período 2015 – 2021. *Iberian Journal of Information Systems and Technologies,* (60). 240-252. https://media.proquest.com/media/hms/PFT/1/6a1EV?_s=v3kFLwTSzfZnTOK49nTKNry4nzI%3D

Carson, R. (1962). *Silent Spring.* Fawcett.

Castells, M. (1997). The Information Age. Economy, Society and Culture.: Vol. II. *Power and Identity.* Blackwell.

Catley-Carlsonreflects, M. (2019). The non-stop waste of water. *Nature,* •••, 565.

Chaffey, D., & Ellis-Chadwick, F. (2019). *Digital marketing: Strategy, implementation and practice* (7th ed.). Pearson Education Limited.

Chaffin, B. C., Garmestani, A. S., Gunderson, L. H., Benson, M. H., Angeler, D. G., Arnold, C. A., Cosens, B., Craig, R. K., Ruhl, J. B., & Allen, C. R. (2016). Transformative environmental governance. *Annual Review of Environment and Resources,* 41(1), 399–423. DOI: 10.1146/annurev-environ-110615-085817 PMID: 32607083

Chai, S.-K. (1998). Endogenous Ideology Formation and Economic Policy in Former Colonies. *Economic Development and Cultural Change,* 46(2), 263–290. DOI: 10.1086/452338

Chandler, J. A. (2001). *Local government today.* Manchester University Press.

Chang, A. Y. (2019). *Playing Nature: Ecology in Video Games.* University of Minnesota Press. DOI: 10.5749/j.ctvthhd94

Chang, C. P., Wen, J., & Zheng, M. (2022). Environmental governance and innovation: An overview. *Environmental Science and Pollution Research International,* •••, 1–2. PMID: 34981403

Chang, Y.-L., & Ke, J. (2024). Socially Responsible Artificial Intelligence Empowered People Analytics: A Novel Framework Towards Sustainability. *Human Resource Development Review,* 23(1), 88–120. DOI: 10.1177/15344843231200930

Chasek, P. S.. (2017). *Global Environmental Politics* (7th ed.). Westview Press.

Chasek, P. S., Downie, D. L., & Brown, J. W. (2013). *Global environmental politics* (6th ed.). Routledge.

Chatterjee, S., & Dutta, S. (2019). Financing public-private partnerships for sustainable development: Evidence from India. *Journal of Infrastructure Development*, 11(2), 164–180.

Cheberyako, O. V., Varnalii, Z. S., Borysenko, O. A., & Miedviedkova, N. S. (2021). "Green" Finance as a Modern Tool for Social and Environmental Security. IOP Conference Series: Earth and Environmental Science. DOI DOI: 10.1088/1755-1315/915/1/012017

Chen, Y., Xu, G., & Jiang, T. (2017). Design and implementation of air quality monitoring system based on IoT technology. In 2017 IEEE 2nd Information Technology, Networking, Electronic and Automation Control Conference (ITNEC) (pp. 2367-2370). IEEE. https://doi.org/DOI: 10.1109/ITNEC.2017.8265640

Chen, C. F., & Chang, Y. Y. (2021). Exploring green innovation strategy through digital marketing. *Sustainability*, 13(12), 6573. Advance online publication. DOI: 10.3390/su13126573

Chen, C. H., Wang, C. C., & Chen, Y. Z. (2021). Intelligent brushing monitoring using a smart toothbrush with recurrent probabilistic neural network. *Sensors (Basel)*, 21(4), 1238. DOI: 10.3390/s21041238 PMID: 33578684

Chen, D., Shams, S., Carmona-Moreno, C., & Leone, A. (2010). Assessment of open source GIS software for water resources management in developing countries. *Journal of Hydro-environment Research*, 4(3), 253–264. DOI: 10.1016/j.jher.2010.04.017

Chen, H., & Chang, Y. (2021). Digital marketing strategies for promoting sustainable consumption patterns. *Journal of Business Research*, 78, 197–206.

Chen, K., & Zhou, J. L. (2014). Occurrence and behavior of antibiotics in water and sediments from the Huangpu River, Shanghai, China. *Chemosphere*, 95, 604–612. DOI: 10.1016/j.chemosphere.2013.09.119 PMID: 24182403

Chen, L., Li, J., Xu, M., & Xing, W. (2024). Navigating urban complexity: The role of GIS in spatial planning and urban development. *Applied and Computational Engineering.*, 65(1), 282–287. DOI: 10.54254/2755-2721/65/20240519

Chen, M., Wan, J., Gonzalez, S. G., Liao, X., & Lei, Y. (2017). A survey of fog computing: Concepts, applications and issues. *Internet of Things : Engineering Cyber Physical Human Systems*, 1, 27.

Chen, S., Li, H., & Liu, X. (2023). Blockchain and IoT integrated smart ESG reporting: A case study of a multinational corporation. *Journal of Cleaner Production*, 316, 128258. Advance online publication. DOI: 10.1016/j.jclepro.2021.128258

Chen, S., & Liu, N. (2022). Research on citizen participation in government ecological environment governance based on the research perspective of "dual carbon target". *Journal of Environmental and Public Health*, 2022(1), 2022. DOI: 10.1155/2022/5062620 PMID: 35769833

Chowdhury, M. T. A., Ahmed, K. J., Ahmed, M. N. Q., & Haq, S. M. A. (2021). How do teachers' perceptions of climate change vary in terms of importance, causes, impacts and mitigation? A comparative study in Bangladesh. *SN Social Sciences*, 1(7), 1–35. DOI: 10.1007/s43545-021-00194-7 PMID: 34693329

CHS (Commission on Human Security). (2003). *Human Security Now*. https://reliefweb.int/organization/commission-human-security

Clark, R., Reed, J., & Sunderland, T. (2018). Bridging funding gaps for climate and sustainable development: Pitfalls, progress and potential of private finance. *Land Use Policy*, 71, 335–346. DOI: 10.1016/j.landusepol.2017.12.013

Clark, W. C., & Dickson, N. M. (2003). Sustainability science: The emerging research program. *Proceedings of the National Academy of Sciences of the United States of America*, 100(14), 8059–8061. DOI: 10.1073/pnas.1231333100 PMID: 12794187

Clausen, A., Vu, H. H., & Pedrono, M. (2011). An evaluation of the environmental impact assessment system in Vietnam: The gap between theory and practice. *Environmental Impact Assessment Review*, 31(2), 136–143. DOI: 10.1016/j.eiar.2010.04.008

Coenen, L., Hansen, T., & Rekers, J. V. (2015). Innovation policy for grand challenges: An economic geography perspective. *Geography Compass*, 9(9), 483–496. DOI: 10.1111/gec3.12231

Coenen, L., & Morgan, K. (2019). Evolving geographies of innovation: Existing paradigms, critiques, and possible alternatives. *Norsk Geografisk Tidsskrift*.

Conca, K., & Dabelko, G. D. (Eds.). (2017). *Green Planet Blues: Environmental Politics from Stockholm to Johannesburg* (4th ed.). Westview Press.

Connor, R., Renata, A., Ortigara, C., Koncagül, E., Uhlenbrook, S., Lamizana-Diallo, B. M., ... & Brdjanovic, D. (2017). The United Nations world water development report 2017. Wastewater: the untapped resource. *The United Nations world water development report*.

Cook, E., Woolridge, A., Stapp, P., Edmondson, S., & Velis, C. A. (2022). Medical and healthcare waste generation, storage, treatment and disposal: A systematic scoping review of risks to occupational and public health. *Critical Reviews in Environmental Science and Technology*, 53(15), 1452–1477. DOI: 10.1080/10643389.2022.2150495

Cosbey, A. 2005. International Agreements and Sustainable Development: Achieving the Millenium Development Goals. Winnipeg, Manitoba, Canada: International Institute for Environment and Development (IISD). Accessed at: http://www.iisd.org/pdf/2005/ investment_iias.pdf in March 2006

Costanza, R., Cumberland, J. H., Daly, H. E., Goodland, R., & Norgaard, R. B. (1997). *An Introduction to Ecological Economics*. CRC Press. DOI: 10.1201/9781003040842

Cotrim, C. D. S., Semedo, A., & Lemos, G. (2022). Brazil wave climate from a high-resolution wave hindcast. *Climate (Basel)*, 10(4), 53. DOI: 10.3390/cli10040053

Crafts, N., & Fearson, P. (2010). Lessons from 1930s Great Depression. *Oxford Review of Economic Policy*, 26(3), 285–317. DOI: 10.1093/oxrep/grq030

Creighton, J. L. (2005). *The Public Participation Handbook: Making Better Decisions Through Citizen Involvement*. Jossey-Bass.

Cronen, V. E. (1995). Practical Theory and The Tasks Ahead for Social Approaches to Communication. In Leeds-Hurwitz, L. (Ed.), *Social Approaches to Communication* (pp. 217–242). Guilford Press.

Crowley, E. J., Silk, M. J., & Crowley, S. L. (2021). The educational value of virtual ecologies in Red Dead Redemption 2 [Article]. *People and Nature*, 3(6), 1229–1243. DOI: 10.1002/pan3.10242

Cullen-Knox, C., Eccleston, R., Haward, M., Lester, E., & Vince, J. (2017). Contemporary Challenges in Environmental Governance: Technology, governance and the social licence. *Environmental Policy and Governance*, 27(1), 3–13. DOI: 10.1002/eet.1743

D'avila Garcez, C. M. (2000). Sistemas Locais de Inovação na Economia do Aprendizado: Uma abordagem conceitual. [Dezembro]. *Revista do BNDES*, 14(7), 351–366.

Dabo, Shuaibu & Jibril, Mohammed & Suberu, Habibat & Murjanatu, & Ayawa, Garba. (2024). The Role of Geographic Information Systems (GIS) as an Effective Tool in Environmental Planning and Management.

Dagdeviren, H., & Robertson, S. A. (2009). *Access to Water in the Slums of the Developing World* (No. 57). Working paper.

Dalby, S. (2008). Security and environment linkages revisited. In *Globalization and environmental challenges: Reconceptualizing security in the 21st century* (pp. 165–172). Springer. DOI: 10.1007/978-3-540-75977-5_9

Daly, H. E., & Farley, J. (2004). *Ecological Economics: Principles and Applications*. Island Press.

Daly, H. E., & Farley, J. (2011). *Ecological economics: Principles and applications* (2nd ed.). Island Press.

Dandona, L. (2020). Health and economic impact of air pollution in the States of India: The Global Burden of Disease Study 2019. *The Lancet. Planetary Health*, 5(1), e25–e-38. DOI: 10.1016/S2542-5196(20)30298-9 PMID: 33357500

Das, A., (2020). Regulatory reforms for public-private partnerships in infrastructure: Lessons from India. *Utilities Policy*, 64, 101032.

Das, M. K., Islam, A. S., Karmakar, S., Khan, M. J. U., Mohammed, K., Islam, G. T., Bala, S. K., & Hopson, T. M. (2020). Synoptic flow patterns and large-scale characteristics of flash flood-producing rainstorms over northeast Bangladesh. *Meteorology and Atmospheric Physics*, 132(5), 613–629. DOI: 10.1007/s00703-019-00709-1

David, B. Olawade, Ojima Z. Wada, Aanuoluwapo Clement David-Olawade, Oluwaseun Fapohunda, Abimbola O. Ige, Jonathan Ling (2024). Artificial intelligence potential for net zero sustainability: Current evidence and prospects,Next Sustainability,Volume 4,2024,100041,ISSN 2949-8236, https://doi.org/DOI: 10.1016/j.nxsust.2024.100041

Davis, J. (2004). Corruption in public service delivery: Experience from South Asia's water and sanitation sector. *World Development*, 32(1), 53–71. DOI: 10.1016/j.worlddev.2003.07.003

Davoodi, S. M. M., & Mocherniak, S. (2021). Legal Issues and Challenges in Public-Private Partnerships: An International Perspective. *Journal of Legal Affairs and Dispute Resolution in Engineering and Construction*, 13(1), 04521001.

De Croo, A. (2015). Why digital is key to sustainable growth. World Economic Forum: Cologny, Switzerland. Available online: https://www.weforum.org/agenda/2015/03/why-digital-is-key-to-sustainable-growth/ (accessed on 22 July, 2024)

de Jong, E., & Vijge, M. J. (2021). From Millennium to Sustainable Development Goals: Evolving discourses and their reflection in policy coherence for development. *Earth System Governance*, 7, 100087. DOI: 10.1016/j.esg.2020.100087 de Leon, M. L. (2020). Barriers to environmentally sustainable initiatives in oral health care clinical settings. *Canadian Journal of Dental Hygiene : CJDH = Journal Canadien de l'Hygiene Dentaire : JCHD*, 54(3), 156. PMID: 33240375

de Melo, M. A. C. (1997). *Processo de Planejamento e as Inovações Tecnológicas e Sociais: uma perspectiva sócio-ecológica*. Anais do 5o. Seminário de Modernização Tecnológica.

de Melo, M.A.C. (1977b). Articulated incrementalism: a strategy for planning (with special reference to the design of an information system as an articulative task). Tese de Doutorado, University of Pennsylvania, Filadélfia.

de Melo, M.A.C. (2002a). *Enriquecendo a Atuação de Incubadora de Emrpesas. Tecnologia e Inovação: experiências de gestão nas micro e pequenas empresas*. PGT/USP. de Melo, M.A.C. (2002b). Inovação e Modernização Tecnológica e Organizacional MPMEs: o domínio interorganizacional. *Seminário Internacional: Políticas para Sistemas Produtivos Locais de MPME*: 2002

de Melo, A. (2001). *The Innovation Systems of Latin America and the Caribbean*. Inter-American Development Bank. Agosto.

de Oliveira, M. C. C., Machado, M. C., Jabbour, C. J. C., & de Sousa Jabbour, A. B. L. (2019). Paving the way for the circular economy and more sustainable supply chains: Shedding light on formal and informal governance instruments used to induce green networks. *Management of Environmental Quality*, 30(5), 1095–1113. DOI: 10.1108/MEQ-01-2019-0005

de Oliveira, U. R., Menezes, R. P., & Fernandes, V. A. (2024). A systematic literature review on corporate sustainability: Contributions, barriers, innovations and future possibilities. *Environment, Development and Sustainability*, 26(2), 3045–3079. DOI: 10.1007/s10668-023-02933-7 PMID: 36687736

Deb, U. K. (2016). Performance of the agriculture sector. *Routledge Handbook of Contemporary Bangladesh*, 08.

Delivering better oral health: an evidence-based toolkit for prevention. (2021). *UK Government*. Available at https://www.gov.uk/government/publications/delivering-better-oral-health-an-evidence-based-toolkit-for-prevention

Department of Environment. (2010). *National 3R Strategy for Waste Management*. Ministry of Environment and Forests, Government of the People's Republic of Bangladesh. https://doe.portal.gov.bd/sites/default/files/files/doe.portal.gov.bd/publications/7dc258b2_4501_400d_b066_5e56d84f439f/National_3R_Strategy.pdf

Department of Health and Public Health England. (2017). Delivering better oral health: an evidence-based toolkit for prevention. *PHE Gateway Number*: 2016224. https://assets.publishing.service.gov.uk/government/uploads/system/uploa

Deterding, S., Dixon, D., Khaled, R., & Nacke, L. (2011). From game design elements to gamefulness: defining "gamification" *Proceedings of the 15th International Academic MindTrek Conference: Envisioning Future Media Environments*, Tampere, Finland. DOI: 10.1145/2181037.2181040

Dey, A. B. (2021, September 24). *What a waste of water!* The Daily Star. https://www.thedailystar.net/news/bangladesh/news/what-waste-water-2182951

Dhanapal, G. (2023). Barriers and opportunities in achieving climate and sustainable development goals in India: A multilevel analysis. *Journal of Integrative Environmental Sciences*, 20(1), 1–16. Advance online publication. DOI: 10.1080/1943815X.2022.2163665

Dhirani, L. L., Mukhtiar, N., Chowdhry, B. S., & Newe, T. (2023). Ethical Dilemmas and Privacy Issues in Emerging Technologies: A Review. *Sensors (Basel)*, 23(3), 1151. DOI: 10.3390/s23031151 PMID: 36772190

Diamond, J. (2005). *Collapse: How societies choose to fail or succeed*. Penguin Books.

Dias, A. B., Rosentha, L. D., & de Melo, M. A. C. (2000) Enriquecendo a Atuação de Incubadoras de Empresas. *XXI Simpósio de Gestão da Inovação Tecnológica. Co-autores: Adriano Batista Dias e David Rosenthal*. São Paulo: 2000.

Diaz, D., & Moore, F. (2017). 2017/11/01). Quantifying the economic risks of climate change. *Nature Climate Change*, 7(11), 774–782. DOI: 10.1038/nclimate3411

Dilek, M. Ş., & Furuncu, Y. (2019). *Bitcoin mining and its environmental effects*. Atatürk Üniversitesi İktisadi ve İdari Bilimler Dergisi.

Dimitrova, A., ... (2013). Literature review on fundamental concepts and definitions, objectives, and policy goals as well as instruments relevant for socio-ecological transition, *WWWforEurope Working Paper*, No. 40, WWWforEurope, Vienna.

Dixon, J., Field, J., Gibson, E., & Martin, N. (2024). Curriculum content for environmental sustainability in dentistry. *Journal of Dentistry*, 147, 105021. DOI: 10.1016/j.jdent.2024.105021 PMID: 38679135

Domingo, S. N., & Manejar, A. J. A. (2021). *An analysis of regulatory policies on solid waste management in the Philippines: Ways forward* (No. 2021-02). PIDS Discussion Paper Series.

Dominko, M., Primc, K., Slabe-Erker, R., & Kalar, B. (2023). A bibliometric analysis of circular economy in the fields of business and economics: Towards more action-oriented research. *Environment, Development and Sustainability*, 25(7), 5797–5830. DOI: 10.1007/s10668-022-02347-x PMID: 35530441

Douglas, B. D., & Brauer, M. (2021). Gamification to prevent climate change: A review of games and apps for sustainability. *Current Opinion in Psychology*, 42, 89–94. DOI: 10.1016/j.copsyc.2021.04.008 PMID: 34052619

Draetta, L., Dabbicco, M., & Bisogno, M. (2021). Blockchain and carbon markets: A new horizon for environmental sustainability. *Technological Forecasting and Social Change*, 170, 120930.

Droj, G. (2012). GIS and remote sensing in environmental management. *Journal of Environmental Protection and Ecology*, •••, 2.

Druilhe, C., & Garnsey, E. (2000), 'The incubation of Academic Spin-off Companies', Proceedings of the High Technology Small Firm Conference 2000, 22-23 May, University of Twente, Enschede, The Netherlands (published as ' Academic spin-off ventures: a resource-opportunity approach' in During, W., Oakey, R. and Kauser, S. Eds. 2001, New Technology Based Firms in the New Millennium, Oxford: Pergamon).

Duane, B., & Fennell-Wells, A. (2023). *Clinical guidelines for environmental sustainability in dentistry*. Trinity College Dublin.

Duane, B., Lyne, A., Parle, R., & Ashley, P. (2022). The environmental impact of community caries prevention-part 3: Water fluoridation. *British Dental Journal*, 233(4), 303–307. DOI: 10.1038/s41415-022-4251-5 PMID: 36028695

Duane, B., Stancliffe, R., Miller, F. A., Sherman, J., & Pasdeki-Clewer, E. (2020). Sustainability in dentistry: A multifaceted approach needed. *Journal of Dental Research*, 99(9), 998–1003. DOI: 10.1177/0022034520919391 PMID: 32392435

Duane, B., Steinbach, I., Ramasubbu, D., Stancliffe, R., Croasdale, K., Harford, S., & Lomax, R. (2019). Environmental sustainability and travel within the dental practice. *British Dental Journal*, 226(7), 525–530. DOI: 10.1038/s41415-019-0115-z PMID: 30980009

Dubey, U. K. B., & Kothari, D. P. (2022). Research Methodology. In *Research Methodology*. DOI: 10.1201/9781315167138

Dubey, R., Gunasekaran, A., Childe, S. J., Papadopoulos, T., Luo, Z., Wamba, S. F., & Roubaud, D. (2019). Can big data and predictive analytics improve social and environmental sustainability? *Technological Forecasting and Social Change*, 144, 534–545. DOI: 10.1016/j.techfore.2017.06.020

Dyer-Witheford, N., & de Peuter, G. (2021, May). Postscript: Gaming While Empire Burns. *Games and Culture, 16*(3), 371-380. *Article*, 1555412020954998. Advance online publication. DOI: 10.1177/1555412020954998

Dyer-Witheford, N., & Peuter, G. d. (2009). *Games of empire: Global capitalism and video games*. University of Minnesota Press.

Earth Negotiations Bulletin, 2005., Volume 15 Number 129, 2 October. On final decision UNEP/FAO/RC/COP.2/CRP.5/Rev.1

Ehrlich, P. R. (1968). *The Population Bomb*. Ballantine Books.

Eisenack, K. (2013). A Climate Change Board Game for Interdisciplinary Communication and Education. *Simulation & Gaming*, 44(2-3), 328–348. DOI: 10.1177/1046878112452639

Elkington, J. (1999). *Cannibals with Forks: The Triple Bottom Line of 21st Century Busine*. Oxford University Press.

Ellen MacArthur Foundation. (2021). The circular economy in detail. Retrieved from https://www.ellenmacarthurfoundation.org/explore/the-circular-economy-in-detail

Elliot, L. (2005). Expanding the Mandate of the United Nations Security Council. In Chambers, W. B., & Green, J. F. (Eds.), *Reforming International Environmental Governance: From Institutional Limits to Innovative Reforms* (pp. 204–226). United Nations University Press.

Enayat, A. Moallemi, Shirin Malekpour, Michalis Hadjikakou, Rob Raven, Katrina Szetey, Dianty Ningrum, Ahmad Dhiaulhaq, Brett A. Bryan (2020). Achieving the sustainable Development Goals Requires Trans disciplinary Innovation at the Local Scale, One Earth, Volume 3, Issue 3, Pages 300-313, ISSN 2590-3322, https://doi.org/DOI: 10.1016/j.oneear.2020.08.006

Engel, K., Jokiel, A., & Kraljevic, M. Geiger & Smith, K. (2011). *Big Cities. Big Water. Big Challenges: Water in an Urbanizing World*. http://www.wwf.se/source.php/1390895/Big Cities_Big Water_Big Challenges_2011.pdf

Engl, (2013). *Partizipative Governance, 130ff .85. Committee of the Regions, EGTC Monitoring Report 2013*, 99.EEF379.indb174EEF 379.indb 17408/11/201601:0908/11/2016

Erickson, J. J., Smith, C. D., Goodridge, A., & Nelson, K. L. (2017). Water quality effects of intermittent water supply in Arraiján, Panama. *Water Research*, 114, 338–350. DOI: 10.1016/j.watres.2017.02.009 PMID: 28279879

Erismann, S., Pesantes, M. A., Beran, D., Leuenberger, A., Farnham, A., Berger Gonzalez de White, M., Labhardt, N. D., Tediosi, F., Akweongo, P., Kuwawenaruwa, A., Zinsstag, J., Brugger, F., Somerville, C., Wyss, K., & Prytherch, H. (2021). How to bring research evidence into policy? Synthesizing strategies of five research projects in low-and middle-income countries. *Health Research Policy and Systems*, 19(1), 1–13. DOI: 10.1186/s12961-020-00646-1 PMID: 33676518

Esguerra, J. (2003). *The Corporate Muddle of Manila's Water Concessions*. https://www.wateraid.org/~/media/Publications/privatesector-participation-manila.pdf

Estevez, E., Janowski, T., & Estevez, J. (2013). Electronic government and the information systems perspective: Third International Conference, EGOVIS 2012, Vienna, Austria, September 3-6, 2012. Proceedings. Springer.

Estevez, E., Janowski, T., & Dzhusupova, Z. (2013). Electronic governance for sustainable development: How EGOV solutions contribute to SD goals? In *Proceedings of the 14th Annual International Conference on Digital Government Research*, Quebec City, QC, Canada, 17–20 June 2013; ACM: New York, NY, USA; pp. 92–101. DOI: 10.1145/2479724.2479741

Etim, E. E., Odoh, R., Itodo, A. U., Umoh, S. D., & Lawal, U. (2013). Water quality index for the assessment of water quality from different sources in the Niger Delta Region of Nigeria. *Frontiers in Science*, 3(3), 89–95.

Etkind, E., & de Vries, B. J. (2023). Co-creating a sustainability transition through community empowerment and education: A case study from Israel. *Journal of Cleaner Production*, 382, 135194. DOI: 10.1016/j.jclepro.2022.135194

Etzion, D., & Aragon-Correa, J. A. (2016). Big Data, Management, and Sustainability: Strategic Opportunities Ahead. *Organization & Environment*, 29(2), 147–155. DOI: 10.1177/1086026616650437

European Commission. (2012). *Guide to Research and Innovation Strategies for Smart Specialisations (RIS 3)*. Publications Office of the European Union.

European Investment Bank. (2021). *Green Infrastructure Financing Strategies: Lessons Learned from Successful Projects*. European Investment Bank.

Evangelista, P., Ibarra-Yunez, A., & Prieto-Martinez, J. (2014). *Development beyond digital divides: Towards digital opportunities*. Springer.

Evangelista, R., Guerrieri, P., & Meliciani, V. (2014). The economic impact of digital technologies in Europe. [CrossRef]. *Economics of Innovation and New Technology*, 23(8), 802–824. DOI: 10.1080/10438599.2014.918438

Fagerberg, J. (2018). Mobilizing innovation for sustainability transitions: A comment on transformative innovation policy. *Research Policy*, 47(9), 1568–1576. DOI: 10.1016/j.respol.2018.08.012

Fairbanks, S. D., Pramanik, S. K., Thomas, J. A., Das, A., & Martin, N. (2021). The management of mercury from dental amalgam in wastewater effluent. *Environmental Technology Reviews*, 10(1), 213–223. DOI: 10.1080/21622515.2021.1960642

Fang, B., Yu, J., Chen, Z., & Chen, Z. (2023). Artificial intelligence for waste management in smart cities: A review. *Environmental Chemistry Letters*, 21(4), 1959–1989. DOI: 10.1007/s10311-023-01604-3 PMID: 37362015

Fannin, A. L., & Hickel, J. (2023). Compensation for Atmospheric Appropriation. *Nature Sustainability*, 6(9), 1077–1086. DOI: 10.1038/s41893-023-01130-8

Farooq, S., Sharif, T. T., & Alam, S. (2019). *Study on Drinking Water Quality Served in Restaurants and Tea Stalls in Gazipur Area* (Doctoral dissertation, Department of Civil and Environmental Engineering, Islamic University of Technology, Gazipur, Bangladesh).

Fattah, M., Rimi, R. A., & Morshed, S. R. (2022). Knowledge, behavior, and drivers of residents' willingness to pay for a sustainable solid waste collection and management system in Mymensingh City, Bangladesh. *Journal of Material Cycles and Waste Management*, 24(4), 1551–1564. DOI: 10.1007/s10163-022-01422-9

Fazey, I., Carmen, E., Ross, H., Rao-Williams, J., Hodgson, A., Searle, B. A., & Thankappan, S. (2021). *Social dynamics of community resilience building in the face of climate change: The case of three Scottish communities* (Vol. 16). Springer Japan.

Fazey, I., Hughes, C., Schäpke, N. A., Leicester, G., Eyre, L., Goldstein, B. E., Hodgson, A., Mason-Jones, A. J., Moser, S. C., Sharpe, B., & Reed, M. S. (2021). Renewing universities in our climate emergency: Stewarding system change and transformation. *Frontiers in Sustainability*, 2, 677904. DOI: 10.3389/frsus.2021.677904

Felzmann, H., Fosch-Villaronga, E., Lutz, C., & Tamò-Larrieux, A. (2020). Towards transparency by design for artificial intelligence. *Science and Engineering Ethics*, 26(6), 3333–3361. DOI: 10.1007/s11948-020-00276-4 PMID: 33196975

Fernandez Galeote, D., Rajanen, M., Rajanen, D., Legaki, N.-Z., Langley, D. J., & Hamari, J. (2021, June). Gamification for climate change engagement: Review of corpus and future agenda. *Environmental Research Letters*, 16(6), 063004. Advance online publication. DOI: 10.1088/1748-9326/abec05

Fernández-Navarro, P., Villanueva, C. M., García-Pérez, J., Boldo, E., Goñi-Irigoyen, F., Ulibarrena, E., Rantakokko, P., García-Esquinas, E., Pérez-Gómez, B., Pollán, M., & Aragonés, N. (2017). Chemical quality of tap water in Madrid: Multicase control cancer study in Spain (MCC-Spain). *Environmental Science and Pollution Research International*, 24(5), 4755–4764. DOI: 10.1007/s11356-016-8203-y PMID: 27981479

Fernando, R. L. S. (2019). Solid waste management of local governments in the Western Province of Sri Lanka: An implementation analysis. *Waste Management (New York, N.Y.)*, 84, 194–203. DOI: 10.1016/j.wasman.2018.11.030 PMID: 30691892

Ferreira, G. (2015) Interview by Peter Ulrich, Porto (Portugal), 26th February 2015.

Ferronato, N., & Torretta, V. (2019). Waste mismanagement in developing countries: A review of global issues. *International Journal of Environmental Research and Public Health*, 16(6), 1060. DOI: 10.3390/ijerph16061060 PMID: 30909625

Fischer, F. (2006). Participatory Governance as Deliberative Empowerment. The Cultural politics of Discursive Space. *American Review of Public Administration*, 36(1), 19–40. DOI: 10.1177/0275074005282582

Fischer, M., (2020). Stakeholder Perspectives on Private Sector Participation in Sustainable Development: Insights from Interviews. *Journal of Sustainable Development*, 15(3), 78–92.

Flood, S., Cradock-Henry, N. A., Blackett, P., & Edwards, P. (2018). 2018/06/01). Adaptive and interactive climate futures: Systematic review of 'serious games' for engagement and decision-making. *Environmental Research Letters*, 13(6), 063005. DOI: 10.1088/1748-9326/aac1c6

Folke, C., Hahn, T., Olsson, P., & Norberg, J. (2005). Adaptive governance of social-ecological systems. *Annual Review of Environment and Resources*, 30(1), 441–473. DOI: 10.1146/annurev.energy.30.050504.144511

Forss, K. 2004. Strengthening the Governance of UNEP: A Discussion of Reform Issues. Commissioned by the Ministry for Foreign Affairs, Sweden. http://www.sweden.gov.se/content/1/c6/03/24/92/ f60748c8.pdf accessed March 2006

Foster, J. B. (2000). *Marx's ecology: Materialism and Nature*. Monthly Review Press.

Fox, C. J., & Miller, H. T. (1996). *Postmodern public administration: Towards discourse*. Thousand Oaks (etc.). *Sage (Atlanta, Ga.)*.

Francis. (2015). *Laudato Sì*. Libreria Editrice Vaticana.

Froehlich, J., Findlater, L., & Landay, J. (2022). The importance of energy-use feedback in energy conservation: A cost-effective approach. [TOCHI]. *ACM Transactions on Computer-Human Interaction*, 29(2), 10.

Froehlich, J., Findlater, L., & Landay, J. A. (2022). The design and evaluation of eco-feedback technology. [TOCHI]. *ACM Transactions on Computer-Human Interaction*, 29(1), 1. Advance online publication. DOI: 10.1145/3492203

Frondel, M., Horbach, J., & Rennings, K. (2007). End-of-pipe or cleaner production? An empirical comparison of environmental innovation decisions across OECD countries. *Business Strategy and the Environment*, 16(8), 571–584. DOI: 10.1002/bse.496

Galaz, V., Centeno, M. A., Callahan, P. W., Causevic, A., Patterson, T., Brass, I., Baum, S., Farber, D., Fischer, J., Garcia, D., McPhearson, T., Jimenez, D., King, B., Larcey, P., & Levy, K. (2021). Artificial intelligence, systemic risks, and sustainability. *Technology in Society*, 67, 101741. DOI: 10.1016/j.techsoc.2021.101741

Ganguly, Sunayana. (2016), Deliberating Environmental Policy in India: Participation and the Role of Advocacy, DOI: 10.4324/9781315744476

Garcia, L. E. (2008). 'Integrated water resources management: A 'small' step for conceptualists, a giant step for practitioners'. *International Journal of Water Resources Development*, 24(1), 23–36. DOI: 10.1080/07900620701723141

Garzoni, A., (2020). *Digital transformation and human-centricity in times of crisis*. Springer.

Garzoni, A., De Turi, I., Secundo, G., & Del Vecchio, P. (2020). Fostering digital transformation of SMEs: A four levels approach. *Management Decision*, 58(8), 1543–1562. DOI: 10.1108/MD-07-2019-0939

Gasper, D., & Gómez, O. A. (2023). Solidarity and Human Insecurity: Interpreting and Extending the HDRO's 2022 Special Report on Human Security. *Journal of Human Development and Capabilities*, 24(2), 263–273. DOI: 10.1080/19452829.2022.2161491

Gaventa J and G Barrett, (2010) *So What Difference Does It Make? Mapping the Outcomes of Citizen Engagement*, Development Research Centre: Citizenship, Participation and Accountability, London, DFID.

Gaventa, J. (2006). Triumph, Deficit or Contestation? Deepening the" Deepening Democracy" Debate, *Working Paper 264*, Brighton, Institute of Development Studies.

Gay, E. F. (1932). The Great Depression. *Foreign Affairs*, 10(4), 529–540. DOI: 10.2307/20030459

Geels, F. (2002). Technological transitions as evolutionary reconfiguration processes: A multilevel perspective and a case-study. *Research Policy*, 31(8-9), 1257–1274. DOI: 10.1016/S0048-7333(02)00062-8

Geertz, C. (1973). *The Interpretation of Culture*. Basic Books.

Geissdoerfer, M., Pieroni, M. P. P., Pigosso, D. C. A., & Soufani, K. (2020).Circular business models: A review, Journal of Cleaner Production, Volume 277, 123741, ISSN 0959-6526, https://doi.org/DOI: 10.1016/j.jclepro.2020.123741

Geissdoerfer, M., Savaget, P., Bocken, N. M. P., & Hultink, E. J. (2017). The circular economy–A new sustainability paradigm? *Journal of Cleaner Production*, 143, 757–768. DOI: 10.1016/j.jclepro.2016.12.048

Gemmill-Herren, B., & Bamidele-Izu, A. (2002), The role of NGOs and Civil Society in Global Environmental Governance, https://www.researchgate.net/publication/228786506_The_role_of_NGOs_and_Civil_Society_in_Global_Environmental_Governance

George, G., (2020). *Sustainability and digitalization: Opportunities and challenges*. Springer.

George, G., Merrill, R. K., & Schillebeeckx, S. J. (2020). Digital sustainability and entrepreneurship: How digital innovations are helping tackle climate change and sustainable development. [CrossRef]. *Entrepreneurship Theory and Practice*.

Gerber, A., Ulrich, M., Wäger, F. X., Roca-Puigròs, M., Gonçalves, J. S. V., & Wäger, P. (2021). Games on Climate Change: Identifying Development Potentials through Advanced Classification and Game Characteristics Mapping. *Sustainability (Basel)*, 13(4), 1997. DOI: 10.3390/su13041997

Gerlak, A. K., Heikkila, T., & Newig, J. (2020). Learning in environmental governance: Opportunities for translating theory to practice. *Journal of Environmental Policy and Planning*, 22(5), 653–666. DOI: 10.1080/1523908X.2020.1776100

Gerrard, M. (2001). Public-private partnerships. Finance and Development, 38(3), 48-51. Monteiro, R., Ferreira, J. C., & Antunes, P. (2020). Green infrastructure planning principles: An integrated literature review. *Land (Basel)*, 9(12), 525.

Ghisellini, P., Cialani, C., & Ulgiati, S. (2016). A review on circular economy: The expected transition to a balanced interplay of environmental and economic systems. *Journal of Cleaner Production*, 114, 11–32. DOI: 10.1016/j.jclepro.2015.09.007

Ghosh, S., (2020). Capacity building for public-private partnerships in sustainable development: A case study of India. *Sustainable Development*, 28(4), 988–999.

Ghosh, S., & Nanda, P. (2017). *Green Infrastructure Financing: Institutional Investors, PPPs and Bankable Projects.* Palgrave Macmillan.

Gibson, C., Head, L., & Carr, C. (2015). From incremental change to radical disjuncture: Rethinking everyday household sustainability practices as survival skills. *Annals of the Association of American Geographers*, 105(2), 416–424. DOI: 10.1080/00045608.2014.973008

Giddens, A. (1993). Profiles and Critiques in Social Theory, in Cassel P. (ed), 1993. *The Giddens Reader*, London, Macmillan, pp 218-225.

Gil, G. N. (2015). Environmental Justice in India: The National Green Tribunal and Expert Members, Translational. *Environmental Law (Northwestern School of Law)*, 5(1), •••. https://www.cambridge.org/core/journals/transnational-environmental-law/article/environmental-justice-in-india-the-national-green-tribunal-and-expert-members/2E26B50742FFB8BB743557132DC7DD66

Glass, L.-M., & Newig, J. (2019). Governance for achieving the Sustainable Development Goals: How important are participation, policy coherence, reflexivity, adaptation, and democratic institutions? *Earth System Governance*, 2, 100031. DOI: 10.1016/j.esg.2019.100031

Gleick, P. H., Wolff, G. H., & Cushing, K. K. (2003). Waste not, want not: The potential for urban water conservation in California.

Glenn, J. C. and Gordon, T., 1999. 2050 Global Normative Scenarios. American Council for the United Nations University in cooperation with The Foundation for the Future

Global Impact Investing Network. (2021). Impact Investment Case Studies. Retrieved from https://thegiin.org/research/publication/impact-investment-case-studies

Goetz, A. M., & Gaventa, J. (2001). FromConsultation to Influence: Bringing citizenvoice and client focus into service delivery. *IDS Working Paper*, •••, 138.

Goguichvili S., Linenberger A., Gillette A., The Global Legal Landscape of Space: Who Writes the Rules on the Final Frontier? Wilson Centre, 1 October 2021

González Álvarez, M. D. (1997). *Processos de Planejamento nos Pólos Tecnológicos: um enfoque adaptativo. Tese de Doutorado, Departamento de Engenharia Industrial.* PUC-Rio.

Gosling, S. N., & Arnell, N. W. (2013). A global assessment of the impact of climate change on water scarcity. *Climatic Change*, 134(3), 371–385. DOI: 10.1007/s10584-013-0853-x

Gouverneur, J., & Netzer, N. (2014). Take the Wheel and Steer! Trade Unions and the Just Transition. In *State of the World 2014. State of the World*. Island Press., DOI: 10.5822/978-1-61091-542-7_21

Gráda, C. (2007). Making Famine History. *Journal of Economic Literature*, 45(1), 5–38. DOI: 10.1257/jel.45.1.5

Green dentistry: a way forward for oral-health. *European Federation of Periodontology,* accessed at https://www.efp.org/publications/perio-insight/green-dentistry-a-way-forward-for-oral-health-professionals/

Grindle, M. (2007). *Going Local: Decentralization, Participation, and the Promise of Good Governance*. Princeton University Press.

Gubbi, J., Buyya, R., Marusic, S., & Palaniswami, M. (2013). Internet of Things (IoT): A vision, architectural elements, and future directions. *Future Generation Computer Systems*, 29(7), 1645–1660. DOI: 10.1016/j.future.2013.01.010

Guerrero, L. A., Maas, G., & Hogland, W. (2013). Solid Waste Management Challenges for Cities in Developing Countries. *Waste Management (New York, N.Y.)*, 33(1), 220–232. DOI: 10.1016/j.wasman.2012.09.008 PMID: 23098815

Guha, S., Jana, R. K., & Sanyal, M. K. (2022). Artificial neural network approaches for disaster management: A literature review (2010–2021). *International Journal of Disaster Risk Reduction*, 81, 103276. DOI: 10.1016/j.ijdrr.2022.103276

Gulati, M., & Rao, M. Y. (2019). Financing India's renewable energy vision: The case of solar parks. *Energy Policy*, 128, 158–164.

Gulati, R., (2021). Enhancing stakeholder engagement in public-private partnerships for sustainable development: Insights from India. *Sustainable Cities and Society*, 70, 102885.

Gulia, S., Shukla, N., Padhi, L., Bosu, P., Goyal, S. K., & Kumar, R. (2022). Evolution of air pollution management policies and related research in India. *Environmental Challenges*, 6, 100431. Advance online publication. DOI: 10.1016/j.envc.2021.100431

Gunningham, N. (2011, July). Enforcing Environmental Regulation. *Journal of Environmental Law*, 23(2), 169–201. DOI: 10.1093/jel/eqr006

Guo, W., Li, K., Fang, Z., Feng, T., & Shi, T. (2022). A sustainable recycling process for end-of-life vehicle plastics: A case study on waste bumpers, Waste Management, Volume 154, Pages 187-198, ISSN 0956-053X, https://doi.org/DOI: 10.1016/j.wasman.2022.10.006

Gupta, J., Arts, K. Achieving the 1.5 °C objective: just implementation through a right to (sustainable) development approach. *International Environmental Agreements: Politics, Law and Economics*, 18(1), 11-28. DOI: 10.1007/s10784-017-9376-7

Gupta, J., & Ringius, L. (2001). The EU's Climate Leadership: Reconciling Ambition and Reality. [Springer Science Business Media B.V]. *International Environmental Agreement: Politics, Law and Economics*, 1(2), 281–299. DOI: 10.1023/A:1010185407521

Gupta, R., & Sharma, S. (2020). Transparency and accountability in public-private partnerships for sustainable development: Evidence from India. *Journal of Public Affairs*, 20(3), e2162.

Gupta, R., & Sharma, S. (2022). Public-private partnerships in sustainable infrastructure development: A systematic review. *Journal of Environmental Policy and Planning*, 25(1), 112–128.

Gustafson, P., & Hertting, N. (2017). Understanding Participatory Governance: An Analysis of Participants' Motives for Participation. *American Review of Public Administration*, 47(5), 538–549. DOI: 10.1177/0275074015626298

GWP. (2002) Integrated Water Resources Management, TAC Background Paper #4, Global Water Partnership Technical Advisory Committee, Stockholm: 2000.

GWP. (2007) *IWRM Explained*. Available at: http://www.gwptoolbox.org/learn/iwrm-explained#:~:text=of%20GWP's%20strategy-What%20is%20IWRM%3F,vital%20ecosystems%20and%20the%20environment.

Gwynne, R. N. (2009). Modernization Theory. In Kitchin, R., & Thrift, N. (Eds.), *International Encyclopedia of Human Geography* (pp. 164–168). Elsevier. DOI: 10.1016/B978-008044910-4.00108-5

Haberl, H., Winiwarter, V., Andersson, K., Ayres, R. U., Boone, C., Castillo, A., Cunfer, G., Fischer-Kowalski, M., Freudenburg, W. R., Furman, E., Kaufmann, R., Krausmann, F., Langthaler, E., Lotze-Campen, H., Mirtl, M., Redman, C. L., Reenberg, A., Wardell, A., Warr, B., & Zechmeister, H. (2006). From LTER to LTSER: Conceptualizing the socio-economic dimension of long-term socio-ecological research. *Ecology and Society*, 11(2), 13. Retrieved September 3, 2010, from www.ecologyandsociety.org/vol11/iss2/art13/. DOI: 10.5751/ES-01786-110213

Habiba, U., Abedin, M. A., & Shaw, R. (2013). Disaster education in Bangladesh: opportunities and challenges. *Disaster Risk Reduction Approaches in Bangladesh*, 307–330.

Hackley, D. M. (2021). Climate change and oral health. *International Dental Journal*, 71(3), 173–177. DOI: 10.1111/idj.12628 PMID: 34024327

Haider, M. Z., & Riaz, M. R. (2021). Municipal solid waste management in Bangladesh: A study of X municipality. *Khulna University Studies*, 18(1), 37–43. DOI: 10.53808/KUS.2021.18.01.2101-S

Hajer, M., & Underhill, G. (2003) Rethinking Politics: Transnational Society, Network Interaction, and Democratic Governance. Working document, University of Amsterdam.

Hajjaji, Y., Boulila, W., Farah, I. R., Romdhani, I., & Hussain, A. (2021). Big data and IoT-based applications in smart environments: A systematic review, Computer Science Review, Volume 39, 2021, 100318, ISSN 1574-0137, https://doi.org/DOI: 10.1016/j.cosrev.2020.100318

Hao, L.-N., Umar, M., Khan, Z., & Ali, W. (2021). Green growth and low carbon emission in G7 countries: How critical the network of environmental taxes, renewable energy and human capital is? *The Science of the Total Environment*, 752, 141853. DOI: 10.1016/j.scitotenv.2020.141853 PMID: 32889278

Haque, M. (2019). Urban water governance: pricing of water for the slum dwellers of Dhaka metropolis. *Urban drought: Emerging water challenges in Asia*, 385-397.

Harford, S., & Duane, B. (2018). Sustainable dentistry: how-to guide for dental practice s sustainable dentistry how to guide for dental practices sustainable dentistry: howto guide for dental practices. *Cent. Sustain. Healthc*.

Hariram, N. P., Mekha, K. B., Suganthan, V., & Sudhakar, K. (2023). Sustainalism: An Integrated Socio-Economic-Environmental Model to Address Sustainable Development and Sustainability. *Sustainability (Basel)*, 15(13), 10681. DOI: 10.3390/su151310682

Harker-Schuch, I. E. P., Mills, F. P., Lade, S. J., & Colvin, R. M. (2020). CO2peration-Structuring a 3D interactive digital game to improve climate literacy in the 12-13-year-old age group. *Computers & Education*, 144, 103705. DOI: 10.1016/j.compedu.2019.103705

Harper, R. (2023). *Toward Climate-Smart Education Systems: A 7-Dimension Framework for Action*. Global Partnership for Education.

Harrington, W., & Fisher, A. C. (1982). Endangered Species — A Global Threat. *Resources*. https://www.resources.org/archives/endangered-species-a-global-threat/

Harvard Business Review (2023). How to make generative AI greener; 2023. https://hbr.org/2023/07/how-to-make-generative-ai-greener

Hasan, M. K., Shahriar, A., & Jim, K. U. (2019). Water pollution in Bangladesh and its impact on public health. *Heliyon*, 5(8), e02145. DOI: 10.1016/j.heliyon.2019.e02145 PMID: 31406938

Hasan, M. M., Pal, T. K., Paul, S., & Alam, M. A. (2021). Quality Measurement of Drinking Water in Rajshahi City Corporation and Effectiveness of Different Water Purifiers. *IOSR Journal of Applied Chemistry*, 14, 38–45. DOI: 10.9790/5736-1408013845

Hassan, S. S., Williams, G. A., & Jaiswal, A. K. (2019). Moving towards the second generation of lignocellulosic biorefineries in the EU: Drivers, challenges, and opportunities. *Renewable & Sustainable Energy Reviews*, 101, 590–599. DOI: 10.1016/j.rser.2018.11.041

Hattingh, J., Claassen, M., Leaner, J., Maree, G., Strydom, W., Turton, A. R., Van Wyk, E., & Moolman, M. (2005a) Towards innovative ways to ensure successful implementation of government tools. Report ENV-P-I 2004-083. Pretoria, CSIR.

Heets, V. M., & Seminozhenko, V. P. (2006). Innovative Prospects of Ukraine, Kh.: Constant.

Heikkila, T., & Weible, C. M. (2018). A semiautomated approach to analyzing polycentricity. *Environmental Policy and Governance*, 28(4), 308–318. DOI: 10.1002/eet.1817

Helmer, R., & Hespanhol, I. (1997). *Water pollution control: a guide to the use of water quality management principles*. CRC Press. DOI: 10.4324/9780203477540

Helsper, E. (2021). The digital disconnect: The social causes and consequences of digital inequalities. *The Digital Disconnect*, 1-232.

Henderson, R. (1994). Managing Innovation in the Information Age. *Harvard Business Review*, 72(01), 100–107.

Hernández Guzmán, D., & Hernández García de Velazco, J. (2023). Global Citizenship: Towards a Concept for Participatory Environmental Protection. *Global Society*, 38(2), 269–296. DOI: 10.1080/13600826.2023.2284150

Hertting, N., & Klijn, E. H. (2018). Institutionalization of Local Participatory Governance in France, the Netherlands, and Sweden: Three Arguments Reconsidered. N. Hertting and C. Kugelberg (Eds.) *Local Participatory Governance and Representative Democracy.* New York and London, Routledge.

Hodgson, C. (2015). Can the digital revolution be environmentally sustainable? The Guardian: London, UK. Available online: https://www.theguardian.com/global/blog/2015/nov/13/digital-revolution-environmental-sustainable (accessed on 5 July, 2024).

Hodgson, P. (2015). Environmental impacts of digitalization. In Steffen, P., Jahn, A., & Froese, J. (Eds.), *Advances and new trends in environmental and energy informatics* (pp. 189–199). Springer.

Hognogi, G. G., Pop, A. M., Marian-Potra, A. C., & Some fălean, T. (2021). The role of UAS–GIS in digital Era governance. A systematic literature review. *Sustainability (Basel)*, 13(19), 11097. DOI: 10.3390/su131911097

Holden, E., Linnerud, K., & Banister, D. (2016). The imperatives of sustainable development. *Sustainable Development (Bradford)*, 25(3), 213–226. DOI: 10.1002/sd.1647

Hollands, R. G. (2008). Will the real smart city please stand up? Intelligent, progressive, or entrepreneurial? *City (London, England)*, 12(3), 303–320. DOI: 10.1080/13604810802479126

Homer-Dixon, T. F. (1994). Environmental Scarcities and Violent Conflict: Evidence from Cases. *International Security*, 19(1), 5–40. DOI: 10.2307/2539147

Hondo, D., Arthur, L., & Gamaralalage, P. J. D. (2020). Solid Waste Management in Developing Asia: Prioritizing Waste Separation.

Hoogendoorn, B., van der Zwan, P., & Thurik, R. (2017). Sustainable entrepreneurship: The role of perceived barriers and risk. [CrossRef]. *Journal of Business Ethics*, 175, 1133–1154.

Hoogendoorn, M., (2017). Business models for sustainable innovation: State-of-the-art and steps towards a research agenda. *Journal of Cleaner Production*, 140(Part 1), 155–166. DOI: 10.1016/j.jclepro.2016.04.105

Hoornweg, D., & Bhada-Tata, P. (2012). What a Waste: A Global Review of Solid Waste

Hoosain, M. S., Paul, B. S., Kass, S., & Ramakrishna, S. (2023). Tools towards the sustainability and circularity of data centers. *Circular Economy and Sustainability*, 3(1), 173–197. DOI: 10.1007/s43615-022-00191-9 PMID: 35791435

Hosagrahar, J. (2011). Landscapes of water in Delhi: negotiating global norms and local cultures. *Megacities: Urban form, governance, and sustainability*, 111-132.

Hossain, I., Haque, A. K. M., & Ullah, S. M. (2023b). Role of City Governance Institutions in Ensuring Sustainable Water: A Study on Gazipur City of Bangladesh. SSRN *Electronic Journal*. DOI: 10.2139/ssrn.4480760

Hossain, I., Mahmudul Haque, A. K. M., & Akram Ullah, S. M. (2024). Socio-economic Dimensions of Climate Change in Urban Bangladesh: A Focus on the Initiatives of Local Governing Agencies. In *Climate Crisis, Social Responses and Sustainability: Socio-Ecological Study on Global Perspectives* (pp. 293-316). Springer. DOI: 10.1007/978-3-031-58261-5_13

Hossain, I., & Haque, A. M. (2023). Role of Responsible Institution in Urban Water Governance and Sanitation Management: A Focus on Metropolitan Areas of Bangladesh. *Research Square*. DOI: 10.21203/rs.3.rs-3064713/v1

Hossain, I., Haque, A. M., Kılıç, Z., Ullah, S. A., Azrour, M., & Mabrouki, J. (2024b). Exploring Household Waste Management Practices and IoT Adoption in Barisal City Corporation. In *Smart Internet of Things for Environment and Healthcare* (pp. 27–45). Springer Nature Switzerland. DOI: 10.1007/978-3-031-70102-3_2

Hossain, I., Haque, A. M., & Ullah, S. A. (2023a). Assessment of Domestic Water Usage and Wastage in Urban Bangladesh: A Study of Rajshahi City Corporation. *The Journal of Indonesia Sustainable Development Planning*, 4(2), 109–121. DOI: 10.46456/jisdep.v4i2.462

Hossain, I., Haque, A. M., & Ullah, S. A. (2023d). Assessing Sustainable Waste Management Practices in Rajshahi City Corporation: An Analysis for Local Government Enhancement using IoT and AI Technology. *Research Square*. https://doi.org/DOI: 10.21203/rs.3.rs-3397290/v1

Hossain, I., Haque, A. M., & Ullah, S. A. (2024a). Assessing sustainable waste management practices in Rajshahi City Corporation: An analysis for local government enhancement using IoT, AI, and Android technology. *Environmental Science and Pollution Research International*, •••, 1–19. DOI: 10.1007/s11356-024-33171-7 PMID: 38581631

Hossain, I., Haque, A. M., & Ullah, S. A. (2024e). Policy evaluation and performance assessment for sustainable urbanization: A study of selected city corporations in Bangladesh. *Frontiers in Sustainable Cities*, 6, 1377310. DOI: 10.3389/frsc.2024.1377310

Hossain, I., Haque, A. M., Ullah, S. A., Azrour, M., Mabrouki, J., & Kılıç, Z. (2024b). Sustainable Water Management at City Corporation Level in Bangladesh: A Comparative Analysis between SCC and BCC. In *Sustainable and Green Technologies for Water and Environmental Management* (pp. 215–237). Springer Nature Switzerland. DOI: 10.1007/978-3-031-52419-6_16

Hossain, I., Mahmudul Haque, A. K. M., & Akram Ullah, S. M. (2023). Role of Government Institutions in Promoting Sustainable Development in Bangladesh: An Environmental Governance Perspective. *Journal of Current Social and Political Issues*, 1(2), 42–53. DOI: 10.15575/jcspi.v1i2.485

Hossain, I., Nekmahmud, M., & Fekete-Farkas, M. (2022). How do Environmental Knowledge, Eco-label Knowledge, and Green Trust Impact Consumers' Pro-Environmental Behaviour for Energy Efficient Household Appliances? *Sustainability (Basel)*, 14(11), 6513. DOI: 10.3390/su14116513

Hossain, I., Ullah, S. A., & Haque, A. M. (2023c). Water and Sanitation Services at the Local Government Level in Bangladesh: An Analysis of SDG 6 Implementation Status and Way Forward. *Asia Social Issues*, 17(3), e265358. DOI: 10.48048/asi.2024.265358

Hossen, M. S., Haque, A. M., Hossain, I., Haque, M. N., & Hossain, M. K. (2024). Towards comprehensive urban sustainability: Navigating predominant urban challenges and assessing their severity differential in Bangladeshi city corporations. *Urbanization. Sustainability Science*, 1(1), 1–17.

Howson, P., Oakes, S., & Baynham-Hughes, J. (2021). Blockchain's role in sustainability: Transformative or problematic? *Environment, Development and Sustainability*, 23(1), 613–631.

Hsu, C. L., Chang, K. C., & Chen, M. C. (2020). Applying green marketing strategy to enhance green purchase intention: The moderation effects of collaborative consumption and switching costs. *Sustainability*, 12(11), 4402. Advance online publication. DOI: 10.3390/su12114402

Hsu, C., (2020). Leveraging digital platforms for promoting sustainability: The case of eco-friendly products. *Journal of Marketing Management*, 36(5-6), 463–482.

Hügel, S., & Davies, A. R. (2024). Expanding adaptive capacity: Innovations in education for place-based climate change adaptation planning. *Geoforum*, 150, 103978. DOI: 10.1016/j.geoforum.2024.103978

Huizinga, J. (1949). *Homo ludens: A study of the play-element in culture*. Routledge.

Hurley, S., & White, S. (2018). *Carbon Modelling within Dentistry: Towards a Sustainable Future*. Public Health England.

Hurlimann, A., Cobbinah, P. B., Bush, J., & March, A. (2021). Is climate change in the curriculum? An analysis of Australian urban planning degrees. *Environmental Education Research*, 27(7), 970–991. DOI: 10.1080/13504622.2020.1836132

Hussain, M., Balsara, K. P., & Nagral, S. (2012). Reuse of single-use devices: Looking back, looking forward. *The National Medical Journal of India*, 25(3), 151. PMID: 22963293

Hussain, S., & Reza, M. (2023). Environmental Damage and Global Health: Understanding the Impacts and Proposing Mitigation Strategies. *Journal of Big-Data Analytics and Cloud Computing*, 8(2), 1–21.

Hutchinson, G. E. (1957). *Concluding remarks*. On WWW at http://www.gypsymoth.ento.vt.edu/sharov/PopEcol/lec11/niche.html. Accessed 11.11.2005

Hwang, J., (2021). Impact of digital marketing on sustainable behaviors: Insights from environmental campaigns. *Journal of Environmental Management*, 277, 111367.

Hwang, J., Choi, J., & Lee, S. (2021). Environmental sustainability campaign effectiveness: The roles of interactivity and environmental message features in mobile digital advertising. *Journal of Advertising*, 50(4), 398–415. DOI: 10.1080/00913367.2021.1967326

Hwang, J., Han, S., Koo, C., & Lee, C. (2021). A study on the digital marketing strategies of global fashion brands focused on sustainability. *Journal of Fashion Marketing and Management*, 25(2), 191–207.

Hwang, J., Kim, Y., & Kim, M. (2021). Integrating Internet of Things (IoT) technology and digital marketing for sustainable environmental governance. *Journal of Environmental Management*, 280, 111789. DOI: 10.1016/j.jenvman.2020.111789

Hyderabad Metro Rail Limited. (2021). Hyderabad Metro Rail Project: Transforming Urban Mobility. Retrieved from https://hmrl.co.in/project-overview/

ICLEI - Local Governments for Sustainability. (2021). Sustainability Planning. Retrieved from https://icleiusa.org/programs/planning/sustainability-planning/

Iglesias, A., Garrote, L., Flores, F., & Moneo, M. (2007). Challenges to manage the risk of water scarcity and climate change in the Mediterranean. *Water Resources Management*, 21(5), 775–788. DOI: 10.1007/s11269-006-9111-6

Iheozor-Ejiofor, Z., Worthington, H. V., Walsh, T., O'Malley, L., Clarkson, J. E., Macey, R., Alam, R., Tugwell, P., Welch, V., & Glenny, A. M. (2015). Water fluoridation for the prevention of dental caries. *Cochrane Database of Systematic Reviews*, 2015(9), 6. DOI: 10.1002/14651858.CD010856.pub2 PMID: 26092033

Iñiguez-Gallardo, V., & López-Rodríguez, F. (2024). Gobernanza participativa para manglares en Ecuador. *Madera y Bosques*, 30(4), e3042612. DOI: 10.21829/myb.2024.3042612

Intergovernmental Panel on Climate Change. IPCC (2021). Climate change 2021: The physical science basis. Contribution of Working Group I to the Sixth Assessment Report of the Intergovernmental Panel on Climate Change. Cambridge University Press. https://www.ipcc.ch/report/ar6/wg1/

International Finance Corporation. (2019). Unlocking Private Investment for Sustainable Development: Lessons from Investors. Retrieved from https://www.ifc.org/wps/wcm/connect/topics_ext_content/ifc_external_corporate_site/sustainability-at-ifc/publications/publications_handbook_unlockingprivateinvestment

International Renewable Energy Agency. (2020). *Renewable Energy Policies in a Time of Transition: Focus on Africa*. International Renewable Energy Agency.

Intriligator, M. D. (1994). Global Security after the End of Cold War. *Conflict Management and Peace Science*, 13(2), 101–111. DOI: 10.1177/073889429401300201

IPCC. (2014). [*Synthesis Report. Contribution of Working Groups I, II and III to the Fifth Assessment Report of the Intergovernmental Panel on Climate Change.* Geneva, Switzerland: IPCC.]. *Climatic Change*, •••, 2014.

Irfan, M. (2017). *Cybersecurity in the digital age: Tools, techniques, and implications for government*. Springer.

Islam, M. R. (1999). *Solid waste culture and its impact on health of the residents of Dhaka City*. Observer Magazine.

Islam, M. R. (2020). Environment and disaster education in the secondary school curriculum in Bangladesh. *SN Social Sciences*, 1(1), 23. DOI: 10.1007/s43545-020-00025-1

Ivanova, M. (2021). *The Untold Story of the World's Leading Environmental Institution: UNEP at Fifty*. MIT Press. DOI: 10.7551/mitpress/12373.001.0001

J. Abraham, B. (2022). *Digital games after climate change*. Palgrave Macmillan.

Jager, N. W., Newig, J., Challies, E., & Kochskämper, E. (2020). Pathways to implementation: Evidence on how participation in environmental governance impacts on environmental outcomes. *Journal of Public Administration: Research and Theory*, 30(3), 383–399. DOI: 10.1093/jopart/muz034

Jänicke, M., & Jörgens, H. (2020). New approaches to environmental governance. In *The ecological modernisation reader* (pp. 156–189). Routledge. DOI: 10.4324/9781003061069-13

Jasanoff, S. (2004a) post-sovereign science and global nature. Harvard University, Environmental Politics/ Colloquium Papers.

Jasanoff, S. (1997). Civilization and Madness: The Great BSE Scare of 1996. *Public Understanding of Science (Bristol, England)*, 6(3), 221–232. DOI: 10.1088/0963-6625/6/3/002

Jasanoff, S. (2004b). The idiom of co-production. In Jasanoff, S. (Ed.), *States of Knowledge. The coproduction of science and social order* (pp. 1–13). Routledge. DOI: 10.4324/9780203413845-6

Jasmine, B. (2022), Environmental Governance in India: Issues, Concerns and Opportunities, https://www.teriin.org/article/environmental-governance-india-issues-concerns-and-opportunities

Javaid, M., Haleem, A., Singh, R. P., Suman, R., & Gonzalez, E. S. (2022). Understanding the adoption of Industry 4.0 technologies in improving environmental sustainability. *Sustainable Operations and Computers*, 3, 203–217. DOI: 10.1016/j.susoc.2022.01.008

Javed, S., & Malik, F. (2022). Urban solid waste management. *American Journal of Environment Studies*, 5(2), 11–25. DOI: 10.47672/ajes.1268

Javier, M. S., Sara, M. Z., & Hugh, T. (2017). Water pollution from agriculture: a global review. *The Food and Agriculture Organization of the United Nations Rome and the International Water Management Institute on behalf of The Water Land and Ecosystems research program Colombo.*

Jeevanasai, S. A., Saole, P., Rath, A. G., Singh, S., Rai, S., & Kumar, M. (2023). Shades & shines of gender equality with respect to sustainable development goals (SDGs): The environmental performance perspectives. *Total Environment Research Themes*, 8, 100082. DOI: 10.1016/j.totert.2023.100082

Jena, L., & Nayak, U. (2020). Theories of Career Development: An analysis. *Indian Journal of Natural Sciences*, 10(60).

Jenkins, H. (2006). *Fans, bloggers, and games: Exploring participatory culture*. New York University Press.

Jensen, M. B., Johnson, B., Lorenz, E., & Lundvall, B. Å. (2007). Forms of knowledge and modes of innovation. *Research Policy*, 36(5), 680–693. DOI: 10.1016/j.respol.2007.01.006

Jeon, B., Oh, J., & Son, S. (2021, March). Effects of tooth brushing training, based on augmented reality using a smart toothbrush, on oral hygiene care among people with intellectual disability in Korea. [). MDPI.]. *Health Care*, 9(3), 348. PMID: 33803836

Jerin, D. T., Sara, H. H., Radia, M. A., Hema, P. S., Hasan, S., Urme, S. A., Audia, C., Hasan, M. T., & Quayyum, Z. (2022). An overview of progress towards implementation of solid waste management policies in Dhaka, Bangladesh. *Heliyon*, 8(2), e08918. Advance online publication. DOI: 10.1016/j.heliyon.2022.e08918 PMID: 35243053

Jha, V., & Singh, A. (2016). Green Infrastructure in Indian Cities: Policies and Practices. *Public Affairs and Administration: Concepts, Methodologies, Tools, and Applications*, 3, 1315–1329.

Jiménez, A., Saikia, P., Giné, R., Avello, P., Leten, J., Liss Lymer, B., Schneider, K., & Ward, R. (2020). Unpacking water governance: A framework for practitioners. *Water (Basel)*, 12(3), 827. DOI: 10.3390/w12030827

Jin, H., Zhang, E., & Espinosa, H. D. (2023, November). Recent Advances and Applications of Machine Learning in Experimental Solid Mechanics: A Review. ASME. *Applied Mechanics Reviews*, 75(6), 061001. DOI: 10.1115/1.4062966

Johnson, H., South, N., & Walters, R. (2016). The commodification and exploitation of fresh water: Property, human rights and green criminology. *International Journal of Law, Crime and Justice*, 44, 146–162. DOI: 10.1016/j.ijlcj.2015.07.003

Johnson, T., & Wang, L. (2020). The role of financial incentives in promoting private sector investment in renewable energy projects. *Renewable Energy*, 45(4), 201–215.

Jones, J. F. (2004). Human Security and Social Development. *Denver Journal of International Law and Policy*, 33(1), 92–103.

Jørgensen, U. (2001). Greening of Technology and Ecotechnology. *International Encyclopaedia of the Social and Behavioural Sciences*, 6393-6396.

Juujärvi, S., & Pesso, K. (2013). Actor Roles in an Urban Living Lab: What Can We Learn from Suurpelto, Finland? *Technology Innovation Management Review*, 11(3), 22–27. http://timreview.ca/article/742. DOI: 10.22215/timreview/742

Juul, J. (2003). *The Game, the Player, the World: Looking for a Heart of Gameness.* https://www.jesperjuul.net/text/gameplayerworld/

Juul, J. (2005). *Half real: Video games between real rules and fictional worlds.* The MIT Press.

Juul, J. (2010). *A casual revolution: Reinventing video games and their players.* The MIT Press.

Kabil, N. S., & Eltawil, S. (2017). Prioritizing the risk factors of severe early childhood caries. *Dentistry Journal*, 5(1), 4. DOI: 10.3390/dj5010004 PMID: 29563410

Kaijim, W., & Sharma, G. (2023).Closing the sustainability data and insights gap. https://www.tcs.com/what-we-do/services/data-and-analytics/white-paper/enterprise-sustainability-data-insights

Kamble, S., Gunasekaran, A., & Arha, H. (2019). Understanding the blockchain technology adoption in supply chains-Indian context. *International Journal of Production Research*, 50(7), 2009–2033. DOI: 10.1080/00207543.2018.1518610

Kaplan, A. M., & Haenlein, M. (2010). Users of the world, unite! The challenges and opportunities of Social Media. *Business Horizons*, 53(1), 59–68. DOI: 10.1016/j.bushor.2009.09.003

Kapp, R. W., Jr. (2014) Safe Drinking Water Act, Encyclopedia of Toxicology, https://www.sciencedirect.com/topics/agricultural-and-biological-sciences/clean-water-act

Karvonen, A., Cugurullo, F., & Caprotti, F. (Eds.). (2019). *Inside Smart Cities: Place, Politics and Urban Innovation.* Routledge.

Kashbour, W., Gupta, P., Worthington, H. V., & Boyers, D. (2020). Pit and fissure sealants versus fluoride varnishes for preventing dental decay in the permanent teeth of children and adolescents. *Cochrane Database of Systematic Reviews*, 11. PMID: 33142363

Kassim, S. M., & Ali, M. (2006). Solid waste collection by the private sector: Households' perspectives findings from a study in Dar-es-Salaam city, Tanzania. *Habitat International*, 30(4), 769–780. DOI: 10.1016/j.habitatint.2005.09.003

Kates, R. W., Clark, W. C., Corell, R., Hall, J. M., Jaeger, C. C., Lowe, I., McCarthy, J. J., Schellnhuber, H. J., Bolin, B., Dickson, N. M., Faucheux, S., Gallopin, G. C., Grübler, A., Huntley, B., Jäger, J., Jodha, N. S., Kasperson, R. E., Mabogunje, A., Matson, P., & Svedin, U. (2001). Sustainability science. *Science*, 292(5517), 641–642. DOI: 10.1126/science.1059386 PMID: 11330321

Kauneckis, D. L., & Auer, M. R. (2013). A Simulation of International Climate Regime Formation. *Simulation & Gaming*, 44(2-3), 302–327. DOI: 10.1177/1046878112470542

Kaur, D., & Diwakar, S. K. (2022). A Review on Environmental Impact due to Technological Advancement in Agriculture. *International Journal of Innovative Research in Computer Science & Technology*, 10(2), 136–139.

Keeso, A. (2014), "Big Data and Environmental Sustainability: a Conversation Starter", Smith School of Enterprise and the Environment, Working Paper Series, pp. 14-04.

Kemp, R., Loorbach, D., & Rotmans, J. (2007). Transition management as a model for managing processes of co-evolution towards sustainable development. *International Journal of Sustainable Development and World Ecology*, 14(1), 78–91. DOI: 10.1080/13504500709469709

Kennedy, D. M. (2009). What the New Deal Did. *Political Science Quarterly*, 124(2), 251–268. DOI: 10.1002/j.1538-165X.2009.tb00648.x

Kenney, M. (Ed.). (2000). *Understanding Silicon Valley: the anatomy of an entrepreneurial region*. Stanford University Press. DOI: 10.1515/9781503618381

Khalid, R. M., & Maidin, A. J. (2022). *Introduction: Governance issues towards achieving SDGs in Southeast Asia. Good Governance and the Sustainable Development Goals in Southeast Asia*. Routledge. DOI: 10.4324/9781003230724-1

Khan, A. U., Hassan, M., & Atkins, P. (2014). International curriculum transfer in geography in higher education: An example. *Journal of Geography in Higher Education*, 38(3), 348–360. DOI: 10.1080/03098265.2014.912617

Khan, M. M., Siddique, M., Yasir, M., Qureshi, M. I., Khan, N., & Safdar, M. Z. (2022). The Significance of Digital Marketing in Shaping Ecotourism Behavior through Destination Image. *Sustainability (Basel)*, 14(12), 7395. DOI: 10.3390/su14127395

Khan, R., Chaudhry, I. S., & Farooq, F. (2019). Impact of human capital on employment and economic growth in developing countries. *Review of Economics and Development Studies*, 5(3), 487–496. DOI: 10.26710/reads.v5i3.701

Khan, S. A. R., Yu, Z., & Umar, M. (2021). How environmental awareness and corporate social responsibility practices benefit the enterprise? An empirical study in the context of emerging economy. *Management of Environmental Quality*, 32(5), 863–885. DOI: 10.1108/MEQ-08-2020-0178

Khatun, S., Shaon, S. M., & Sadekin, N. (2021). Impact of poverty and inequality on economic growth of Bangladesh. *Journal of Economics and Sustainable Development*, 12(10), 107–120.

Khong, Y. F. (2001). Human Security: A Shotgun Approach to Alleviating Human Misery? *Global Governance*, 7(3), 231–236. DOI: 10.1163/19426720-00703003

Kim, H.-E. (2011). Defining Green Technology. In *The Role of the Patent System in Stimulating Innovation and Technology Transfer for Climate Change: Including Aspects of Licencing and Competition Law* (pp. 15-20). Nomos Verlagsgesellschaft mbH.

Kimani-Murage, E. W., & Ngindu, A. M. (2007). Quality of water the slum dwellers use: The case of a Kenyan slum. *Journal of Urban Health*, 84(6), 829–838. DOI: 10.1007/s11524-007-9199-x PMID: 17551841

Kim, R. E. (2015). Transnational networks and global environmental governance: The cities for climate protection program. *Global Environmental Politics*, 15(1), 89–107.

King, G., & Murray, C. (2001-02). Rethinking Human Security. *Political Science Quarterly*, 116(4), 585–610. DOI: 10.2307/798222

Kingsbury, D., Remenyi, J., McKay, J., & Hunt, J. (2004). *Key Issues in Development*. Palgrave Macmillan.

Kirchherr, J., Yang, N.-H. N., Schulze-Spüntrup, F., Heerink, M. J., & Hartley, K. (2023). Conceptualizing the Circular Economy (Revisited): An Analysis of 221 Definitions, Resources, Conservation and Recycling, Volume 194, 107001, ISSN 0921-3449, https://doi.org/DOI: 10.1016/j.resconrec.2023.107001

Kirk, A. (2001). Appropriating Technology: The Whole Earth Catalog and Counterculture Environmental Policies. *Environmental History*, 6(3), 374–394. DOI: 10.2307/3985660

Kisinger, C., & Matsui, K. (2021). Responding to climate-induced displacement in bangladesh: A governance perspective. *Sustainability (Basel)*, 13(14), 7788. DOI: 10.3390/su13147788

Kitchin, R. (2014). The real-time city? Big data and smart urbanism. *GeoJournal*, 79(1), 1–14. DOI: 10.1007/s10708-013-9516-8

Kivimaa, P., Hildén, M., Huitema, D., Jordan, A., & Newig, J. (2017). Experiments in climate governance–A systematic review of research on energy and built environment transitions. *Journal of Cleaner Production*, 169, 17–29. DOI: 10.1016/j.jclepro.2017.01.027

Kivimaa, P., & Martiskainen, M. (2019). Innovation and adoption in the water sector: a systematic literature review of barriers and facilitators. Environmental Innovation and Societal Transitions, 33, 165-183. Bryson, J. M., et al. (2020). Coproduction and Public-Private Partnership: A Comparative Analysis of Governance Challenges. *Public Administration Review*, 80(2), 319-328.

Klabbers, J. H. G., Swart, R. J., Janssen, R., Van Vellinga, P., & van Ulden, A. P. (1996). Climate science and climate policy: Improving the science/policy interface. *Mitigation and Adaptation Strategies for Global Change*, 1(1), 73–93. DOI: 10.1007/BF00625616

Klebanova, T.S. & Kizima, N.A. (2010). Models of estimation, analysis and forecasting of social and economic systems, *Monogr. Kh.:* FOP Liborkina L.M., VD "INZHEK".

Kleiner, G. B. (2013). System Economics and System-Oriented Modeling. *Economics and Mathematical Methods*, 3, 71–93.

Klohe, K., Koudou, B. G., Fenwick, A., Fleming, F., Garba, A., Gouvras, A., Harding-Esch, E. M., Knopp, S., Molyneux, D., D'Souza, S., Utzinger, J., Vounatsou, P., Waltz, J., Zhang, Y., & Rollinson, D. (2021). A systematic literature review of schistosomiasis in urban and peri-urban settings. *PLoS Neglected Tropical Diseases*, 15(2), e0008995. DOI: 10.1371/journal.pntd.0008995 PMID: 33630833

Kloppenburg, S., Gupta, A., Kruk, S. R., Makris, S., Bergsvik, R., Korenhof, P., Solman, H., & Toonen, H. M. (2022). Scrutinizing environmental governance in a digital age: New ways of seeing, participating, and intervening. *One Earth*, 5(3), 232–241. DOI: 10.1016/j.oneear.2022.02.004

Kneese, A. V. (1979). Development and Environment. *Third World Quarterly*, 11(1), 84–90. DOI: 10.1080/01436597908419408

Knippschild, R. (2011). Cross-Border Spatial Planning: Understanding. Designing and Managing Cooperation processes in the German-Polish-Czech Borderland. *European Planning Studies*, 19(4), 629–645. DOI: 10.1080/09654313.2011.548464

Kofoworola, O. F. (2007). Recovery and recycling practices in municipal solid waste management in Lagos, Nigeria. *Waste Management (New York, N.Y.)*, 27(9), 1139–1143. DOI: 10.1016/j.wasman.2006.05.006 PMID: 16904308

Korten, D. (1989). *Getting to the 21st Century: Voluntary Action and the Global Agenda*. Kumarian Press.

Kostka, G., Zhang, X., & Shin, K. (2020). Information, technology, and digitalization in China's environmental governance. *Journal of Environmental Planning and Management*, 63(1), 1–13. DOI: 10.1080/09640568.2019.1681386

Kotler, P., Kartajaya, H., & Setiawan, I. (2010). *Marketing 3.0: From products to customers to the human spirit*. John Wiley & Sons. DOI: 10.1002/9781118257883

Kowarsch, M., & Jabbour, J. (2017). Solution-oriented global environmental assessments: Opportunities and challenges. *Environmental Science & Policy*, 77, 187–192. DOI: 10.1016/j.envsci.2017.08.013

Krausmann, F., Fischer-Kowalski, M., Grunbuhel, C., & Krausmann, F.. (2008). Socio-ecological regime transitions in Austria and the United Kingdom. *Ecological Economics*, 65(1), 187–201. DOI: 10.1016/j.ecolecon.2007.06.009

Kraus, S., Jones, P., Kailer, N., Weinmann, A., Chaparro-Banegas, N., & Roig-Tierno, N. (2021). Digital transformation: An overview of the current state of the art of research. *SAGE Open*, 11(3), 21582440211047576. DOI: 10.1177/21582440211047576

Kshetri, N. (2021). Blockchain Technology for Improving Transparency and Citizen's Trust. In Arai, K. (Ed.), *Advances in Information and Communication. FICC 2021. Advances in Intelligent Systems and Computing* (Vol. 1363). Springer., DOI: 10.1007/978-3-030-73100-7_52

Kücklich, J. R. (2005). Precarious Playbour: Modders and the Digital Games Industry. *The Fibreculture Journal*(5).

Kuhlman, T., & Farrington, J. (2010). What is sustainability? *Sustainability (Basel)*, 2(11), 3436–3448. DOI: 10.3390/su2113436

Kumar Panday, P. (2006). Central-local relations, inter-organisational coordination and policy implementation in urban Bangladesh. *Asia Pacific Journal of Public Administration*, 28(1), 41–58. DOI: 10.1080/23276665.2006.10779314

Kumar, K. D., Rane, D. D., Muralidhar, A., Goundar, S., & Reddy, P. V. (2024). E-Waste Management: A Significant Solution for Green Computing. In *Sustainable Solutions for E-Waste and Development* (pp. 56-73). IGI Global. DOI: 10.4018/979-8-3693-1018-2.ch005

Kumar, A., & Sarangi, S. (2023). Artificial intelligence in sustainable development of municipal solid waste management. [IJRASET]. *International Journal for Research in Applied Science and Engineering Technology*, 11(V), 6744–6751. DOI: 10.22214/ijraset.2023.53247

Kumari, T., & Raghubanshi, A. S. (2023). Waste management practices in the developing nations: Challenges and opportunities. In *Waste management and resource recycling in the developing world* (pp. 773–797). Elsevier., DOI: 10.1016/B978-0-323-90463-6.00017-8

Kumar, P., Sahani, J., Rawat, N., Debele, S., Tiwari, A., Emygdio, A. P. M., Abhijith, K. V., Kukadia, V., Holmes, K., & Pfautsch, S. (2023). Using empirical science education in schools to improve climate change literacy. *Renewable & Sustainable Energy Reviews*, 178, 113232. DOI: 10.1016/j.rser.2023.113232

Kumar, R., & Bansal, P. (2018). Public Private Partnership (PPP) in Infrastructure: Case Study of India. *Journal of Infrastructure Development*, 10(1), 77–88.

Kumar, S., & Patel, R. (2019). Capacity building for sustainable entrepreneurship: Insights from developing countries. *Journal of Entrepreneurship Development*, 8(2), 87–102.

Kundu, Amitabh & Varghese, K. (2010), Regional Inequality and 'Inclusive Growth' in India under Globalization: Identification of Lagging States for Strategic Intervention, Oxfam India, Oxfam India working papers series, OIWPS – VI.

Kundzewicz, Z. W., Mata, L. J., Arnell, N. W., Doll, P., Kabat, P., Jimenez, B., ... & Shiklomanov, I. (2007). Freshwater resources and their management.

Kurniawan, T. A., Lo, W., Singh, D., Othman, M. H. D., Avtar, R., Hwang, G. H., Albadarin, A. B., Kern, A. O., & Shirazian, S. (2021). A societal transition of MSW management in Xiamen (China) toward a circular economy through integrated waste recycling and technological digitization. *Environmental Pollution*, 277, 116741. DOI: 10.1016/j.envpol.2021.116741 PMID: 33652179

Kwok, R. (2019). Can climate change games boost public understanding? *Proceedings of the National Academy of Sciences of the United States of America*, 116(16), 7602–7604. DOI: 10.1073/pnas.1903508116 PMID: 30992396

Kyriakopoulos, G. L. (2022). Energy Communities Overview: Managerial Policies, Economic Aspects, Technologies, and Models. *Journal of Risk and Financial Management*, 15(11), 521. DOI: 10.3390/jrfm15110521

Lachman, D. (2013). A survey and review of approaches to study transitions. *Energy Policy*, 58, 269–276. DOI: 10.1016/j.enpol.2013.03.013

Laclau, E. (1990). *New reflections on the revolution of our time*. Verso.

Lago, X. (2015) Interview by Peter Ulrich, Vigo (Spain), 26th february 2015

Lalkaka, R., & Abetti, P. (1999). Business Incubation and Entreprise Support Systems in Restructuring Countries. *Creativity and Innovation Management*, 8(3), 197–209. DOI: 10.1111/1467-8691.00137

Lamia, F., Kafy, A., & Poly, S. (2018). Assessment of Water supply system and water quality of Rajshahi WASA in Rajshahi City Corporation (RCC) area, Bangladesh. 1st National Conference on Water Resources Engineering (NCWRE 2018), CUET.

Lanjouw, J. O., & Mody, A. (1996). Innovation and the international diffusion of environmentally responsive technology. *Research Policy*, 25(4), 549–571. DOI: 10.1016/0048-7333(95)00853-5

Lanoie, P., Patry, M., & Lajeunesse, R. (2008). Environmental regulation and productivity: Testing the porter hypothesis. *Journal of Productivity Analysis*, 30(2), 121–128. DOI: 10.1007/s11123-008-0108-4

Laszlo, A., Blachfellner, S., Bosch, O., Nguyen, N., Bulc, V., Edson, M., Wilby, J., & Pór, G. (2012). *Proceedings of the International Federation for Systems Research Conversations,* 14-19 April 2012, Linz, Austria / G. Chroust, G. Metcalf (eds.); pp.41-65https://hdl.handle.net/2440/77156

Latour, B. (1987). *Science in Action*. Harvard University Press.

Latour, B. (1988). *The Pasteurization of France*. Harvard University Press.

Latour, B., & Woolgar, S. (1979). *Laboratory Life: The Social Construction of Scientific Facts*. Sage.

Leal Filho, W., Sima, M., Sharifi, A., Luetz, J. M., Salvia, A. L., Mifsud, M., Olooto, F. M., Djekic, I., Anholon, R., Rampasso, I., Kwabena Donkor, F., Dinis, M. A. P., Klavins, M., Finnveden, G., Chari, M. M., Molthan-Hill, P., Mifsud, A., Sen, S. K., & Lokupitiya, E. (2021). Handling climate change education at universities: An overview. *Environmental Sciences Europe*, 33(1), 1–19. DOI: 10.1186/s12302-021-00552-5 PMID: 34603904

Lebel, L., Anderies, J. M., Campbell, B., Folke, C., Hatfield-Dodds, S., Hughes, T. P., & Wilson, J. (2006). Governance and the capacity to manage resilience in regional social-ecological systems. *Ecology and Society*, 11(1), art19. DOI: 10.5751/ES-01606-110119

Lee, H., & Smith, J. (2017). Public-private partnerships for sustainable development: A review of models and best practices. *Journal of Sustainable Infrastructure*, 10(4), 189–204.

Lee, J. I., & Wolf, S. A. (2018). Critical assessment of implementation of the Forest Rights Act of India. *Land Use Policy*, 79, 834–844. https://www.sciencedirect.com/science/article/abs/pii/S0264837717311705. DOI: 10.1016/j.landusepol.2018.08.024

Lee, J. J., Ceyhan, P., Jordan-Cooley, W., & Sung, W. (2013). GREENIFY: A Real-World Action Game for Climate Change Education. *Simulation & Gaming*, 44(2-3), 349–365. DOI: 10.1177/1046878112470539

Leese, L. AI for Earth: How NASA's Artificial Intelligence and Open Science Efforts Combat Climate Change, The National Aeronautics and Space Administration, updated on 23 May 2024.

Lee, T.-C., Anser, M. K., Nassani, A. A., Haffar, M., Zaman, K., & Muhammad, M. Q. A. (2021). Managing Natural Resources through Sustainable Environmental Actions: A Cross-Sectional Study of 138 Countries. *Sustainability (Basel)*, 13(22), 12475. DOI: 10.3390/su132212475

Leeuwis, C., & van den Ban, A. (2004). Communication for Innovation in Agricultural and Rural Resource Management. *Building on the Tradition of Agricultural Extension.* Oxford: Blackwell Science (in cooperation with CTA).

Lehdonvirta, V., & Castronova, E. (2014). *Virtual economies: Design and analysis*. The MIT Press. DOI: 10.7551/mitpress/9525.001.0001

Lens M. GDPR & Convention 108: Adequate Protection in a Big Data Era? Tilburg University, 8 June 2018

Lenschow, A., Newig, J., & Challies, E. (2016). Globalization's limits to the environmental state? Integrating telecoupling into global environmental governance. *Environmental Politics*, 25(1), 136–159. DOI: 10.1080/09644016.2015.1074384

Lester, R. K. (2005). Universities, Innovation, and the Competitiveness of Local Economies A Summary Report from the Local Innovation Systems Project – Phase I Industrial Performance Center Massachusetts Institute of Technology 13 December 2005 MIT Industrial Performance Center Working Paper 05-010

Le, T., Ha, P., Tram, V., Thuy, N., Phuong, D., Tran, T. N., & Nguyen, L. (2023). GIS Application in Environmental Management: A Review. *VNU Journal of Science: Earth and Environmental Sciences.*, 39(2). Advance online publication. DOI: 10.25073/2588-1094/vnuees.4957

Leuenberger, D., Engesser, M., Herrmann, T., Nagel, P., Wyttenbach, M., & Reimann, S. (2018). *A wireless sensor network for monitoring environmental parameters in a chocolate factory. In 2018 Global Internet of Things Summit (GIoTS)*. IEEE., DOI: 10.1109/GIOTS.2018.8534556

Lidskog, R., Mol, A. P., & Oosterveer, P. (2015). Towards a global environmental sociology? Legacies, trends and future directions. *Current Sociology*, 63(3), 339–368. DOI: 10.1177/0011392114543537 PMID: 25937642

Li, E., Endter-Wada, J., & Li, S. (2015). Characterizing and contextualizing the water challenges of megacities. *Journal of the American Water Resources Association*, 51(3), 589–613. DOI: 10.1111/1752-1688.12310

Li, M., (2021). IoT applications in water management: A review. *Water Research*, 200, 117249.

Linkov, I., Trump, B. D., Poinsatte-Jones, K., & Florin, M.-V. (2018). Governance strategies for a sustainable digital world. *Sustainability (Basel)*, 10(2), 440. Advance online publication. DOI: 10.3390/su10020440

Linnenluecke, M. K. (2022). Environmental, social and governance (ESG) performance in the context of multinational business research. *Multinational Business Review*, 30(1), 1–16. DOI: 10.1108/MBR-11-2021-0148

Li, S., Xu, L. D., & Zhao, S. (2021). The internet of things: A survey. *Information Systems Frontiers*, 25(2), 417–455.

Liu, H. Y., Grossberndt, S., & Kobernus, M. (2017). Citizen science and citizens' observatories: Trends, roles, challenges and development needs for science and environmental governance. *Mapping and the Citizen Sensor*, 351-376.

Liu, J., Ren, H., Ye, X., Wang, W., Liu, Y., Lou, L., Cheng, D., He, X., Zhou, X., Qiu, S., Fu, L., & Hu, B. (2017). Bacterial community radial-spatial distribution in biofilms along pipe wall in chlorinated drinking water distribution system of East China. *Applied Microbiology and Biotechnology*, 101(2), 749–759. DOI: 10.1007/s00253-016-7887-8 PMID: 27761636

Liu, K., Tan, Q., Yu, J., & Wang, M. (2023). A global perspective on e-waste recycling. *Circular Economy*, 2(1), 100028. DOI: 10.1016/j.cec.2023.100028

Liu, Q., (2020). A green Internet of Things for sustainable development. *Environmental Research Letters*, 15(3), 035004. Advance online publication. DOI: 10.1088/1748-9326/ab6e5d

Liu, Y. (2023). Ecological Thinking about the Metaverse from a Posthumanist Perspective. *Critical Arts*, •••, 1–18. DOI: 10.1080/02560046.2023.2276423

Liu, Y., Yuan, S., Zhu, Y., Ren, L., Chen, R., Zhu, X., & Xia, R. (2023). The patterns, magnitude, and drivers of unprecedented 2022 mega-drought in the Yangtze River Basin, China. *Environmental Research Letters*, 18(11), 114006. DOI: 10.1088/1748-9326/acfe21

Li, X., Wang, H., Xia, H., & Sun, L. (2021). Development of a smart water meter based on IoT technology. *Water Resources Management*, 35, 3143–3154. DOI: 10.1007/s11269-021-02990-4

Lockwood, M., Davidson, J., & Curtis, A. (2021). Governance principles for natural resource management. *Environmental Management*, 57(2), 317–332.

Loewenberg, S. (2007). Scientists tackle water contamination in Bangladesh. *Lancet*, 370(9586), 471–472. DOI: 10.1016/S0140-6736(07)61214-8 PMID: 17695063

Löfsten, H., & Lindelof, P. (2001). Science parks in Sweden — Industrial renewal and development? *R & D Management*, 5(3), 309–322. DOI: 10.1111/1467-9310.00219

Lohri, C. R., Camenzind, E. J., & Zurbrügg, C. (2014). Financial sustainability in municipal solid waste management – Costs and revenues in Bahir Dar, Ethiopia. *Waste Management (New York, N.Y.)*, 34(2), 542–552. DOI: 10.1016/j.wasman.2013.10.014 PMID: 24246579

López, C. (2020). *Relación entre la innovación tecnológica y el proceso de internacionalización de las pymes exportadoras de café verde de la selva central* [Tesis de Licenciatura]. Universidad Continental, Huancayo, Perú. https://hdl.handle.net/20.500.12394/8468

Lotz-Sisitka, H., Mandikonza, C., Misser, S., & Thomas, K. (2021). Making sense of climate change in a national curriculum. *Teaching and Learning for Change*, 92.

Lueckenhoff, D., & Brown, S. (2015). *Public–Private Partnerships Beneficial for Implementing Green Infrastructure*. Water Law & Policy Monitor.

Luoma, P., Rauter, R., Penttinen, E., & Toppinen, A. (2023, November). The value of data for environmental sustainability as perceived by the customers of a tissue-paper supplier. *Corporate Social Responsibility and Environmental Management*, 30(6), 3110–3123. DOI: 10.1002/csr.2541

Luo, X., & Bhattacharya, C. B. (2006). Corporate social responsibility, customer satisfaction, and market value. *Journal of Marketing*, 70(4), 1–18. DOI: 10.1509/jmkg.70.4.001

Luo, Y., Salman, M., & Lu, Z. (2021). Heterogeneous impacts of environmental regulations and foreign direct investment on green innovation across different regions in China. *The Science of the Total Environment*, 759, 143744. DOI: 10.1016/j.scitotenv.2020.143744 PMID: 33341514

Lupova-Henry, E., & Dotti, N. F. (2019). Governance of sustainable innovation: Moving beyond the hierarchy-market-network trichotomy? A systematic literature review using the 'who-how-what' framework. *Journal of Cleaner Production*, 210, 738–748. DOI: 10.1016/j.jclepro.2018.11.068

Lynch, M. (1991). Laboratory space and the technological complex: An investigation of topical contextures. *Science in Context*, 4(1), 51–78. DOI: 10.1017/S0269889700000156

Macnamara, J. (2018). *Public relations, activism, and social change: Speaking up*. Routledge., DOI: 10.4324/9781351213206

Maertens, L. (2022). The untold story of the world's leading environmental institution: UNEP at fifty by Maria Ivanova. *Global Environmental Politics*, 22(3), 200–202. DOI: 10.1162/glep_r_00669

Mahajan, Y., Patil, R., Pattanaik, S., Firake, T. S., Kodulkar, R., Damre, S. S., & Uplaonkar, D. (2024). A Review of Techniques and Applications for Machine Learning and Deep Learning. *International Journal of Intelligent Systems and Applications in Engineering*, 12(16s), 182–187. https://ijisae.org/index.php/IJISAE/article/view/4804

Maldonado-Guzmán, G., Molina-Morejón, V., & Juárez-del Toro, R. (2024). Efectos de las regulaciones medioambientales en la eco-innovación y el rendimiento sustentable en la industria automotriz mexicana. La granja. *Revista de Ciências da Vida*, 39(1), 78–91. DOI: 10.17163/lgr.n39.2024.05

Malik, A., Padget, M., Carter, S., Wakiyama, T., Maitland-Scott, I., Vyas, A., Boylan, S., Mulcahy, G., Li, M., Lenzen, M., Charlesworth, K., & Geschke, A. (2021). Environmental impacts of Australia's largest health system. *Resources, Conservation and Recycling*, 169, 105556. DOI: 10.1016/j.resconrec.2021.105556

Mallik, S. (2023). Colonial Biopolitics and Great Bengal Famine of 1943. *GeoJournal*, 88(3), 3205–3221. DOI: 10.1007/s10708-022-10803-4 PMID: 36531534

Mallk, K. (2023). Global Development by Public Participation: An Approach to Achieve SDGs. *Journal of Public Administration and Governance*, 13(1). Advance online publication. https://www.researchgate.net/publication/369110515_Global_Development_by_Public_Participation_An_Approach_to_Achieve_SDGs. DOI: 10.5296/jpag.v13i1.20590

Maltais, A., & Nykvist, B. (2020). Understanding the role of green bonds in advancing sustainability. *Journal of Sustainable Finance & Investment*, •••, 1–20. DOI: 10.1080/20430795.2020.1724864

Manjakkal, L., Mitra, S., Petillot, Y. R., Shutler, J., Scott, E. M., Willander, M., & Dahiya, R. (2021). Connected sensors, innovative sensor deployment, and intelligent data analysis for online water quality monitoring. *IEEE Internet of Things Journal*, 8(18), 13805–13824. DOI: 10.1109/JIOT.2021.3081772

Marchesi, G., Camurri Piloni, A., Nicolin, V., Turco, G., & Di Lenarda, R. (2021). Chairside CAD/CAM materials: Current trends of clinical uses. *Biology (Basel)*, 10(11), 1170. DOI: 10.3390/biology10111170 PMID: 34827163

Marshall, R. E., & Farahbakhsh, K. (2013). Systems approaches to integrated solid waste management in developing countries. *Waste Management (New York, N.Y.)*, 33(4), 988–1003. DOI: 10.1016/j.wasman.2012.12.023 PMID: 23360772

Martin, N., & Mulligan, S. (2022). Environmental sustainability through good-quality oral healthcare. *international dental journal, 72*(1), 26.

Martin, P. (2018). The intellectual structure of game research. *Game Studies, 18*(1). https://gamestudies.org/1801/articles/paul_martin

Martin, A., Armijos, M. T., Coolsaet, B., Dawson, N., & Edwards, AS, G., Few, R., ... & White, C. S. (. (2020). Environmental justice and transformations to sustainability. *Environment*, 62(6), 19–30. DOI: 10.1080/00139157.2020.1820294

Martinich, J., & Crimmins, A. (2019). 2019/05/01). Climate damages and adaptation potential across diverse sectors of the United States. *Nature Climate Change*, 9(5), 397–404. DOI: 10.1038/s41558-019-0444-6 PMID: 31031825

Martin, N., Mulligan, S., Fuzesi, P., Webb, T. L., Baird, H., Spain, S., & Hatton, P. V. (2020, August). Waste plastics in clinical environments: a multi-disciplinary challenge. In *Creative Circular Economy Approaches to Eliminate Plastics Waste* (pp. 86–91). UK Research and Innovation and UK Circular Plastics Network.

Marxt, C., & Brunner, C. (2012). *Analyzing and improving the national innovation system of highly developed countries: the case of Switzerland Technol*. Forecast. Soc. Chang., DOI: 10.1016/j.techfore.2012.07.008

Mathur, V. N., Thakur, P., & Rajadhyaksha, N. (2014). Public-Private Partnerships for Sustainable Infrastructure: Exploring the Challenges faced by Local Governments in India. *Environment, Development and Sustainability*, 16(2), 335–351.

Matter, A., Dietschi, M., & Zurbrügg, C. (2013). Improving the informal recycling sector through segregation of waste in the household – The case of Dhaka Bangladesh. *Habitat International*, 38, 150–156. DOI: 10.1016/j.habitatint.2012.06.001

Matzner, N., & Herrenbrück, R. (2017). Simulating a Climate Engineering Crisis:Climate Politics Simulated by Students in Model United Nations. *Simulation & Gaming*, 48(2), 268–290. DOI: 10.1177/1046878116680513

Mauter, M. S., Zucker, I., Perreault, F., Werber, J. R., Kim, J. H., & Elimelech, M. (2018). The role of nanotechnology in tackling global water challenges. *Nature Sustainability*, 1(4), 166–175. DOI: 10.1038/s41893-018-0046-8

Mayer, B., & Randazzo, M. (2023). Environmental justice in the global south: Policy frameworks and indigenous contributions. *Journal of Environmental Law*, 35(3), 467–495.

McCann, P., & Ortega-Argilés, R. (2013). Modern regional innovation policy. *Cambridge Journal of Regions, Economy and Society*, 6(2), 187–216. DOI: 10.1093/cjres/rst007

McQuail, D., & Windahl, S. (2015). *Communication models for the study of mass communications* (2nd ed.). Longman. DOI: 10.4324/9781315846378

Meadowcroft, J. (2021). Climate change governance: The challenge of steering social-ecological systems towards sustainability. *Environmental Politics*, 30(1), 1–20.

Medeiros, E. (2011). (Re)defining the Euroregion Concept. *European Planning Studies*, 19(1), 141–158. DOI: 10.1080/09654313.2011.531920

Mee, L. D. (2005). The Role of UNEP and UNDP in Multilateral Environmental Agreements. *International Environmental Agreement: Politics, Law and Economics*, 5(1), 227–263. DOI: 10.1007/s10784-005-3805-8

Mejean, A., Pottier, A., Fleurbaey, M., & Zuber, S. (2020, November). Catastrophic climate change, population ethics and intergenerational equity [Article]. *Climatic Change*, 163(2), 873–890. DOI: 10.1007/s10584-020-02899-9

Mekonnen, M. M., & Hoekstra, A. Y. (2016). Four billion people facing severe water scarcity. *Science Advances*, 2(2), e1500323. DOI: 10.1126/sciadv.1500323 PMID: 26933676

Mendelow, A. L. (1991) 'Environmental Scanning: The Impact of the Stakeholder Concept'. Proceedings From the Second International Conference on Information Systems 407-418. Cambridge, MA.

Mensah, J., & Ricart Casadevall, S. (2019). Sustainable development: Meaning, history, principles, pillars, and implications for human action: Literature review. *Cogent Social Sciences*, 5(1), 1653531. Advance online publication. DOI: 10.1080/23311886.2019.1653531

Mérida, J., & Telleria, I. (2021). ¿Una nueva forma de hacer política? Modos de gobernanza participativa y «Ayuntamientos del Cambio» en España (2015-2019). *Gestión y Análisis de Políticas Públicas*, (26), 92–110. DOI: 10.24965/gapp.i26.10841

Mesjasz-Lech, A. (2014). Municipal waste management in context of sustainable urban development. *Procedia: Social and Behavioral Sciences*, 151, 244–256. DOI: 10.1016/j.sbspro.2014.10.023

Meya, J. N., & Eisenack, K. (2018). Effectiveness of gaming for communicating and teaching climate change. *Climatic Change*, 149(3-4), 319–333. DOI: 10.1007/s10584-018-2254-7

Mikati, I., Benson, A. F., Luben, T. J., Sacks, J. D., & Richmond-Bryant, J. (2018). Disparities in Distribution of Particulate Matter Emission Sources by Race and Poverty Status. *American Journal of Public Health*, 108(4), 480–485. DOI: 10.2105/AJPH.2017.304297 PMID: 29470121

Minghua, Z., Xiumin, F., Rovetta, A., Qichang, H., Vicentini, F., Bingkai, L., Giusti, A., & Yi, L. (2009). Municipal solid waste management in Pudong New Area, China. *Journal of Waste Management*, 29(3), 1227–1233. DOI: 10.1016/j.wasman.2008.07.016 PMID: 18951780

Ministry of Environment Forest and Climate Change. M. (2022). *Climate Change Initiatives of Bangladesh Achieving Climate Resilience*. www.moef.gov.bd, Molthan-Hill, P., Worsfold, N., Nagy, G. J., Leal Filho, W., & Mifsud, M. (2019). Climate change education for universities: A conceptual framework from an international study. *Journal of Cleaner Production, 226*, 1092–1101.

Miorandi, D., Sicari, S., De Pellegrini, F., & Chlamtac, I. (2022). Internet of Things: Vision, applications and research challenges. *Ad Hoc Networks*, 10(7), 1497–1516. DOI: 10.1016/j.adhoc.2012.02.016

Mishra, P. S., Kumar, A., & Das, N. (2020). Corporate sustainability practices in polluting industries: Evidence from India China and USA. *Problemy Ekorozwoju*, 15(1), 161–168. DOI: 10.35784/pe.2020.1.17

Mishra, V., Tiwari, A. D., Aadhar, S., Shah, R., Xiao, M., Pai, D. S., & Lettenmaier, D. (2019). Drought and Famine in India, 1870-2016. *Geophysical Research Letters*, 46(4), 2075–2083. DOI: 10.1029/2018GL081477

Misleh, D., Dziumla, J., De La Garza, M., & Guenther, E. (2024). Sustainability against the logics of the state: Political and institutional barriers in the Chilean infrastructure sector. *Environmental Innovation and Societal Transitions*, 51, 100842. DOI: 10.1016/j.eist.2024.100842

Mitchell, R. B. (2003). International environmental agreements: A survey of their features, formation, and effects. *Annual Review of Environment and Resources*, 28(1), 429–461. DOI: 10.1146/annurev.energy.28.050302.105603

Mitchell, R. B.. (2015). *Global Environmental Politics: Dilemmas in World Politics* (7th ed.). CQ Press.

Mittling, D. (2004). Reshaping local democracy through participatory governance. Environment & Urbanization.16:1 Participatory Governance, April 2004, IIED: London, UK. Also available online at:www.iied.org/urban/pubs/eu_briefs.htmlBox

Mobin, M. N., Islam, M. S., Mia, M. Y., & Bakali, B. (2014). Analysis of physico-chemical properties of the Turag River water, Tongi, Gazipur in Bangladesh. *Journal of Environmental Science & Natural Resources*, 7(1), 27–33.

Moe, W. W., & Moon, Y. (2018). Developing and testing a survey instrument for measuring user-perceived website service quality. *JMR, Journal of Marketing Research*, 40(2), 196–209. DOI: 10.1509/jmkr.40.2.196.19227

Mohammed, M., Shafiq, N., Abdallah, N. A. W., Ayoub, M., & Haruna, A. (2020, April). A review on achieving sustainable construction waste management through application of 3R (reduction, reuse, recycling): A lifecycle approach. []. IOP Publishing.]. *IOP Conference Series. Earth and Environmental Science*, 476(1), 012010. DOI: 10.1088/1755-1315/476/1/012010

Moh, Y. (2017). Solid waste management transformation and future challenges of source separation and recycling practice in Malaysia. *Resources, Conservation and Recycling*, 116, 1–14. DOI: 10.1016/j.resconrec.2016.09.012

Mol, A. P. J. (2006). From environmental sociologies to environmental sociology? A comparison of U.S. and European environmental sociology. *Organization & Environment*, 19(1), 5–27. DOI: 10.1177/1086026605285643

Mol, A. P. J., & Buttel, F. H. (Eds.). (2002). *The Environmental State under Pressure*. Elsevier. DOI: 10.1016/S0196-1152(2002)10

Monck, C. S. P., Porter, R. B., Quintas, P. R., & Storey, D. J. (1988). *Science Parks and the Growth of High Technology Firms*. Routledge.

Moon, K., Blackman, D., Brewer, T. D., & Sarre, S. D. (2017). Environmental governance for urgent and uncertain problems. *Biological Invasions*, 19(3), 785–797. DOI: 10.1007/s10530-016-1351-7

Morgan, B. T., (2019). Fragmented infrastructure governance: Challenges for the strategic integration of urban infrastructure. *Land Use Policy*, 86, 401–413.

Morgera, E., & Kulovesi, K. (Eds.). (2016). *Research Handbook on International Law and Climate Change*. Edward Elgar Publishing.

Mori, S. (2004). Institutionalization of NGO Involvement in Policy Functions for Global Environmental Governance. In Kanie, N., & Haas, P. M. (Eds.), *Emerging Forces in Environmental Governance* (pp. 157–175). United Nations University Press.

Mosse, D. (2000). Irrigation and Statecraft in Zamindari India. In Fuller, C. J., & Benei, V. (Eds.), *The Everyday State and Society in Modern India* (pp. 163–193). Social Science Press.

Mostafavi, Nariman. et al., (2021). Resilience of environmental policy amidst the rise of conservative populism. *J Environ Stud Sci.* 2022; 12(2): 311–326. DOI: 10.1007/s13412-021-00721-1

Mostakim, K., Arefin, M. A., Islam, M. T., Shifullah, K. M., & Islam, M. A. (2021). Harnessing energy from the waste produced in Bangladesh: Evaluating potential technologies. *Heliyon*, 7(10), e08221. DOI: 10.1016/j.heliyon.2021.e08221 PMID: 34729441

Mott, G., Razo, C., & Hamwey, R. (2021). Carbon emissions anywhere threaten development everywhere. https://unctad.org/news/carbon-emissions-anywhere-threaten-development-everywhere#:~:text=Developed%20countries%20must%20accelerate%20the,This%20is%20in%20everyone's%20interest

Muguruza, C. C. (2017). Human Security as a Policy Framework: Critics and Challenges. *Deusto Journal of Human Rights*, 4(4), 15–35. DOI: 10.18543/aahdh-4-2007pp15-35

Mukhtarov, F., & Karakaya, E. (2020). Regulatory challenges to public-private partnerships for sustainable development: Evidence from a systematic literature review. *Sustainability Science*, 15(2), 535–552.

Mulligan, S., Smith, L., & Martin, N. (2021). Sustainable oral healthcare and the environment: Challenges. *Dental Update*, 48(6), 493–501. DOI: 10.12968/denu.2021.48.6.493

Munévar-Quintero, C., Valencia-Hernández, J. G., Hernández-Gómez, N., Aguirre-Fajardo, A. M., & Ramírez-Ríos, M. (2023). Gobernanza participativa en la conformación del Sistema Nacional de Áreas Protegidas de la ecorregión Eje Cafetero, Colombia. *Jurídicas*, 20(1), 139–157. DOI: 10.17151/jurid.2023.20.1.7

Munoz Cabre, M., Thrasher, R., & Najam, A. (2009). Measuring the negotiation burden of multilateral environmental agreements. *Global Environmental Politics*, 9(4), 1–13. DOI: 10.1162/glep.2009.9.4.1

Mutsuddi, I. (2024). *Ethical Considerations in the Intersection of AI and Environmental Science. In the edited book "Maintaining a Sustainable World in the Nexus of Environmental Science and AI" by Singh, B., Kaunert, C., Vig, K., Dutta, S*. IGI Global.

Nahar, K., Ottom, M. A., Alshibli, F., & Shquier, M. A. (2020). Air quality index using machine learning—A jordan case study. COMPUSOFT. *International Journal of Advancements in Computing Technology*, 9(9), 3831–3840.

Nakre, P. D., & Harikiran, A. G. (2013). Effectiveness of oral health education programs: A systematic review. *Journal of International Society of Preventive & Community Dentistry*, 3(2), 103–115. DOI: 10.4103/2231-0762.127810 PMID: 24778989

Narasimhan, H., & Biswas, P. (2019). The Barefoot Approach to Empowerment: Solar Mamas' Contributions to Sustainable Development. *International Journal of Gender and Entrepreneurship*, 11(4), 377–391. DOI: 10.1108/IJGE-06-2019-0089

NASA. (2022). *GISS Surface Temperature Analysis (v4)*https://data.giss.nasa.gov/gistemp/maps/

Nations, U. (2007). *From Stockholm to Kyoto: A Brief History of Climate Change*. https://www.un.org/en/chronicle/article/stockholm-kyoto-brief-history-climate-change

Navarro, Z. (1998). Participation, democratis-ing practices and the formation of a modernpolity – the case of 'participatory budgeting in Porto Alegre, Brazil (1989-1998)'. *Development*, 41(3), 68–71.

Nazrul Islam, A. K. M., Deb, U. K., Al Amin, M., Jahan, N., Ahmed, I., Tabassum, S., Ahamad, M. G., Nabi, A., Singh, N. P., Byjesh, K., & others. (2013). *Vulnerability to Climate Change: Adaptation Strategies and Layers of Resilience–Quantifying Vulnerability to Climate Change in Bangladesh.*

Nelles, J., & Durand, F. (2014). Political rescaling and metropolitan governance in cross-border regions: Comparing the cross-border metropolitan areas of Lille and Luxembourg. *European Urban and Regional Studies*, 21(1), 104–122. DOI: 10.1177/0969776411431103

Neset, T. S., Andersson, L., Uhrqvist, O., & Navarra, C. (2020). Serious Gaming for Climate Adaptation-Assessing the Potential and Challenges of a Digital Serious Game for Urban Climate Adaptation. *Sustainability (Basel)*, 12(5), 1789. DOI: 10.3390/su12051789

Newman, D. (2006). Borders and bordering: Towards an interdisciplinary dialogue. *European Journal of Social Theory*, 9(2), 171–186. DOI: 10.1177/1368431006063331

Newman, E. (2010). Critical Human Security Studies. *Review of International Studies*, 36(1), 77–94. DOI: 10.1017/S0260210509990519

Nghayo, H. A., Palanyandi, C. E., Ramphoma, K. J., & Maart, R. (2024). Oral health community engagement programs for rural communities: A scoping review. *PLoS One*, 19(2), e0297546. DOI: 10.1371/journal.pone.0297546 PMID: 38319914

Nguyen, T. M. (2021). A systematic review of the use of IoT in environmental governance. *Journal of Cleaner Production*, 302, 126628. Advance online publication. DOI: 10.1016/j.jclepro.2021.126628

Nisar, Q. A., Nasir, N., Jamshed, S., Naz, S., Ali, M., & Ali, S. (2021). Big data management and environmental performance: Role of big data decision-making capabilities and decision-making quality. *Journal of Enterprise Information Management*, 34(4), 1061–1096. DOI: 10.1108/JEIM-04-2020-0137

Nowotny, H., Scott, P., & Gibbons, M. (2001). *Re-thinking science: knowledge and the public in an age of uncertainty*. Polity Press.

O'Meara Dental. The first carbon neutral dentist in Australia. *Coolplanet*. available at https://www.coolplanet.com.au/case-studies/omeara-dental-the-first-carbonneutral-dentist-in-australia/

O'Neill, K., Weinthal, E., Marion Suiseeya, K. R., Bernstein, S., Cohn, A., Stone, M. W., & Cashore, B. (2013). Methods and global environmental governance. *Annual Review of Environment and Resources*, 38(1), 441–471. DOI: 10.1146/annurev-environ-072811-114530

Obaideen, K., Albasha, L., Iqbal, U., & Mir, H. (2024). Wireless power transfer: Applications, challenges, barriers, and the role of AI in achieving sustainable development goals-A bibliometric analysis. *Energy Strategy Reviews*, 53, 101376. DOI: 10.1016/j.esr.2024.101376

Oberthür, S., & Gehring, T. (2006). Reforming international environmental governance: An institutional perspective. *International Environmental Agreement: Politics, Law and Economics*, 6(4), 359–376. DOI: 10.1007/s10784-004-3095-6

Oberthür, S., & Groen, L. (2017). Explaining goal achievement in international negotiations: The EU and the Paris Agreement. *Journal of European Public Policy*, 24(10), 1481–1500.

Obuljen Koržinek, N., Žuvela, A., Jelinčić, D. A., & Polić, M. (2014) *Strategija razvoja kulture Grada Dubrovnika 2015. – 2025*. Dubrovnik: Grad Dubrovnik. Available online at: https://dura.hr/user_files/admin/strateski%20dokumenti/Kulturna%20strategija%20grada%20Dubrovnika.pdf (20/9/2017).

Ochoa, B., López, M., & Hernández, M. (2023). Gobernanza participativa en la implementación de modelos educativos flexibles (MEF*). Estudio de caso: Institución Educativa San José del Trigal del municipio de San José de Cúcuta*. Universidad Simón Bolivar. https://hdl.handle.net/20.500.12442/14384

OECD. (1999a). *Managing National Innovation Systems*. OECD.

OECD. (2001a). *Innovative Clusters: drivers of national innovation systems*. Paris: OECD, 2001a

OECD. (2001b). *Drivers of Growth: information technology, innovation, and entrepreneurship. Science and Industry Outlook*. OECD.

OECD. (2001c). *Innovative Networks: co-operation in national innovation systems*. OECD.

OECD. (2008). The Challenge of Capacity Development. *OECD Journal on Development*, 8(3), 233–276. DOI: 10.1787/journal_dev-v8-art40-en

OECD. (2011). *Regions and innovation policy, OECD Reviews of regional innovation*. Author.

OECD. (2020). *Public-Private Partnerships for Green Infrastructure: Policy Considerations and Good Practices*. OECD Publishing.

Ojha, H. R., ... (2019), Governance: Key for Environmental Sustainability in the Hindu Kush Himalaya, *The Hindu Kush Himalaya Assessment,* https://link.springer.com/chapter/10.1007/978-3-319-92288-1_16

Ojha, H., Nightingale, A. J., Gonda, N., Muok, B. O., Eriksen, S., Khatri, D., & Paudel, D. (2022). Transforming environmental governance: Critical action intellectuals and their praxis in the field. *Sustainability Science*, 17(2), 621–635. DOI: 10.1007/s11625-022-01108-z PMID: 35222728

Okusimba, George. (2019). Role of GIS as a Tool for Environmental Planning and Management. 2454-9444. .DOI: 10.20431/2454-9444.0501002

Oladapo, B. I., Ismail, S. O., Afolalu, T. D., Olawade, D. B., & Zahedi, M. (2021). Review on 3D printing: Fight against COVID-19. *Materials Chemistry and Physics*, 2021(258), 123943. DOI: 10.1016/j.matchemphys.2020.123943 PMID: 33106717

Olukanni, D. O., Pius-Imue, F. B., & Joseph, S. O. (2020). Public perception of solid waste management practices in Nigeria: Ogun State experience. *Recycling*, 5(2), 8. DOI: 10.3390/recycling5020008

Ong, P. (2012). Environmental Justice and Green-Technology Adoption. *Journal of Policy Analysis and Management*, 31(3), 578–597. DOI: 10.1002/pam.21631

Ordieres-Meré, J., Prieto Remón, T., & Rubio, J. (2020). Digitalization: An opportunity for contributing to sustainability from knowledge creation. [CrossRef]. *Sustainability (Basel)*, 12(4), 1460. DOI: 10.3390/su12041460

Organization for Economic Co-Operation and Development (OECD). (2017a). Key issues for digital transformation in the G20. Report prepared for a joint G20 German Presidency/OECD Conference. Available online: https://www.oecd.org/g20/key-issues-for-digital-transformation-in-the-g20.pdf (accessed on 10 July, 2024).

Organization for Economic Co-Operation and Development (OECD). (2017b). *Secretary-General's Report to Ministers*. OECD Publishing. [CrossRef]

Osman, O., & Yassin, E. E. H. (2024). Fostering environmental and resources management in Sudan through geo-information systems: A prospective approach for sustainability. *Journal of Degraded and Mining Lands Management*, 11(3), 5647–5657. DOI: 10.15243/jdmlm.2024.113.5647

Ostro Energy. (2021). Ostro Energy: Powering Sustainable Growth. Retrieved from https://www.ostroenergy.in/sustainability/

Ostrom, E. (2006). *Upravljanje zajedničkim dobrima. Evolucija institucija za kolektivno djelovanje*. Naklada Jesenski i Turk.

Ostrom, E. (2009). A general framework for analyzing sustainability of social-ecological systems. *Science*, 325(5939), 419–422. DOI: 10.1126/science.1172133 PMID: 19628857

Oteng-Ababio, M., Arguello, J. E. M., & Gabbay, O. (2013). Solid waste management in African cities: Sorting the facts from the fads in Accra, Ghana. *Habitat International*, 39, 96–104. DOI: 10.1016/j.habitatint.2012.10.010

Ouariachi, T., Gutiérrez-Pérez, J., & Olvera-Lobo, M.-D. (2018). 2018/09/02). Can serious games help to mitigate climate change? Exploring their influence on Spanish and American teenagers' attitudes / ¿Pueden los serious games ayudar a mitigar el cambio climático? Una exploración de su influencia sobre las actitudes de los adolescentes españoles y estadounidenses. *PsyEcology*, 9(3), 365–395. DOI: 10.1080/21711976.2018.1493774

Ouariachi, T., Olvera-Lobo, M. D., Gutiérrez-Pérez, J., & Maibach, E. (2019). 2019/05/04). A framework for climate change engagement through video games. *Environmental Education Research*, 25(5), 701–716. DOI: 10.1080/13504622.2018.1545156

Pabón, I. (2024). *Parameters for the Formulation of a Public Policy for Water Sanitation of the Bogotá River with a Participatory Governance Approach: Involving Key Actors in Decision Making* [Master's Thesis]. http://hdl.handle.net/11634/56126

Palacios, J. P., Toledo-Córdova, M. F., Miranda-Aburto, E., & Flores Farro, A. (2021). Políticas públicas y gobernanza participativa local. *Revista Venezolana de Gerencia*, 26(95), 564–577. DOI: 10.52080/rvgluz.27.95.8

Palmer, P. M., Wilson, L. R., O'Keefe, P., Sheridan, R., King, T., & Chen, C. Y. (2008). Sources of pharmaceutical pollution in the New York City Watershed. *The Science of the Total Environment*, 394(1), 90–102. DOI: 10.1016/j.scitotenv.2008.01.011 PMID: 18280543

Panday, P. K. (2007). Policy implementation in urban Bangladesh: role of intra-organizational coordination. *Public Organization Review: A Global Journal*, 7(3), 237–259.

Pandey, K. (2018), Urbanisation: India loses natural resources to economic growth: report, *Down To Earth*, Retrieved from: https://www.downtoearth.org.in/urbanisation/india-loses-natural-resources-to-economic-growth-report-61836

Pandey, S. K., & Tiwari, S. (2009). Physico-chemical analysis of ground water of selected area of Ghazipur city-A case study. *Nature and Science*, 7(1), 17–20.

Pandit, A. B., & Kumar, J. K. (2015). Clean Water for Developing Countries. *Annual Review of Chemical and Biomolecular Engineering*, 6(1), 217–246. DOI: 10.1146/annurev-chembioeng-061114-123432 PMID: 26247291

Pandve, H. T. (2009). India's National Action Plan on Climate Change. *Indian Journal of Occupational and Environmental Medicine*, 13(1), 17–19. https://www.ncbi.nlm.nih.gov/pmc/articles/PMC2822162/. DOI: 10.4103/0019-5278.50718 PMID: 20165607

Papachristos, G., Sofianos, A., & Adamides, E. (2013). System Interactions in Socio-technical Transitions: Extending the Multi-level Perspective. *Environmental Innovation and Societal Transitions*, 7, 53–69. DOI: 10.1016/j.eist.2013.03.002

Papadopoulou, P., (2021). Exploring the impact of Internet of Things (IoT) technology on sustainable marketing. *Journal of Cleaner Production*, 299, 126743.

Paramati, S. R., Mo, D., & Huang, R. (2021). The role of financial deepening and green technology on carbon emissions: Evidence from major OECD economies. *Finance Research Letters*, 41, 101794. DOI: 10.1016/j.frl.2020.101794

Pardikar, R. (2020). Global North is Responsible for 92% of Excess Emissions. *Eos*, https://eos.org/articles/global-north-is-responsible-for-92-of-excess-emissions

Parhamfar, M., Sadeghkhani, I., & Adeli, A. M. (2024). Towards the net zero carbon future: A review of blockchain-enabled peer-to-peer carbon trading. *Energy Science & Engineering*, 12(3), 1242–1264. DOI: 10.1002/ese3.1697

Pariatamby, A., & Fauziah, S. H. (2014). Sustainable 3R practice in the Asia and Pacific Regions: the challenges and issues. *Municipal solid waste management in Asia and the Pacific Islands: Challenges and strategic solutions*, 15-40.

Paris, R. (2004). Still an Inscrutable Concept. *Security Dialogue*, 35(3), 370–372. DOI: 10.1177/096701060403500327

Park, A., & Li, H. (2021). The Effect of Blockchain Technology on Supply Chain Sustainability Performances. *Sustainability (Basel)*, 13(4), 1726. DOI: 10.3390/su13041726

Parry, M. L., Rosenzweig, C., Iglesias, A., Livermore, M., & Fischer, G. (2004, April). Effects of climate change on global food production under SRES emissions and socio-economic scenarios. *Global Environmental Change*, 14(1), 53–67. DOI: 10.1016/j.gloenvcha.2003.10.008

Partelow, S., Schlüter, A., Armitage, D., Bavinck, M., Carlisle, K., Gruby, R. L., ... & Van Assche, K. (2020). Environmental governance theories: a review and application to coastal systems.

Parveen, M. (2024). Role of Environmental NGOs in Raising Awareness and Promoting Environmental Campaigns in Bangladesh. In *Multi-Stakeholder Contribution in Asian Environmental Communication* (pp. 90–102). Routledge. DOI: 10.4324/9781032670508-10

Patel, R., Taneja, S., Singh, J., & Sharma, S. (2024). Modelling of surface run-off using SWMM and GIS for efficient stormwater management. *Current Science*, 126, 463–469. DOI: 10.18520/cs/v126/i4/463-469

Pateman, C. (2012). Participatory Democracy Revised. *Perspectives on Politics*, 10(1), 7–19. DOI: 10.1017/S1537592711004877

Pathak, N. (2020). Chipko Movement: Lessons Learned for Community-based Conservation. *Environmental Development*, 34, 100516. DOI: 10.1016/j.envdev.2020.100516

Pathan, M. S., Richardson, E., Galvan, E., & Mooney, P. (2023). The Role of Artificial Intelligence within Circular Economy Activities—A View from Ireland. *Sustainability (Basel)*, 15(12), 9451. DOI: 10.3390/su15129451

Patil RM, Dinde HT, Powar SK (2020) A literature review on prediction of air quality index and forecasting ambient air pollutants using machine learning algorithms 5(8):1148–1152.

Patil, N. A., Tharun, D., & Laishram, B. (2016). Infrastructure development through PPPs in India: Criteria for sustainability assessment. *Journal of Environmental Planning and Management*, 59(4), 708–729. DOI: 10.1080/09640568.2015.1038337

Pattberg, P., & Widerberg, O. (2015). Theorizing global environmental governance: Key findings and future questions. *Millennium*, 40(2), 421–448. DOI: 10.1177/0305829814561773

Patwa, N., Sivarajah, U., Seetharaman, A., Sarkar, S., Maiti, K., & Hingorani, K. (2021). Towards a circular economy: An emerging economies context,Journal of Business Research,Volume 122, Pages 725-735, ISSN 0148-2963, https://doi.org/ DOI: 10.1016/j.jbusres.2020.05.015

Pedrycz, W., Deveci, M., & Chen, Z.-S. (2024). BIM-based building performance assessment of green buildings -A case study from China. *Applied Energy*, 373, 123977. DOI: 10.1016/j.apenergy.2024.123977

Pei, T., Xu, J., Liu, Y., Huang, X., Zhang, L., Dong, W., Qin, C., Song, C., Gong, J., & Zhou, C. (2021). GI Science and remote sensing in natural resource and environmental research: Status quo and future. *Geography and Sustainability*, 2(3), 207–215. DOI: 10.1016/j.geosus.2021.08.004

Pelto, P. J. (1973) The Snowmobile revolution: technology and social change in the arctic. Cummings Publishing Company, Menlo Park, CA (revised Waveland Press, Prospects Heights, IL 1987 Percival RV, Miller AS, Schroeder CH, Leape JP (1992) *Environ-mental regulation: law, science, and policy*. Little, Brown, and Company, Boston, USA.

Peng, N. (2023). Application of Machine learning techniques in environmental governance: A review. *Advances in Engineering Technology Research.*, 7(1), 528. DOI: 10.56028/aetr.7.1.528.2023

Perinchery, A. (2024), Climate Action, and Also Contradictions: NITI Aayog's SDG India Index 2024 Report is a Mixed Bag, *WIRE*, Retrieved from: https://thewire.in/environment/climate-action-and-also-contradictions-niti-aayogs-sdg-india-index-2024-report-is-a-mixed-bag

Perkmann, M. (2007). Policy entrepreneurship and multi-level governance: A comparative study of European cross-border regions. *Environment and Planning. C, Government & Policy*, 25(6), 861–879. DOI: 10.1068/c60m

Perrings, Charles. & Halkos, George. (2015). *Environmental Research. Letters.* Vol. 10, No.9,1-10, https://iopscience.iop.org/article/10.1088/1748-9326/10/9/095015

Peters, B. Guy & Pierre, J. (eds.) (2001) *Politicians, bureaucrats, and administrative reform.* London: Routledge.

Phipps, M., & Burbach, M. E. (2010). Strategic sustainability and employee attitudes: The role of organizational culture. *Business Strategy and the Environment*, 19(1), 1–17. DOI: 10.1002/bse.670

Pierre, J. (2005). Comparative urban governance. Uncovering complex causalities. *Urban Affairs Review*, 40(4), 446–462. DOI: 10.1177/1078087404273442

Pilbeam, C., Alvarez, G., & Wilson, H. (2012). The governance of supply networks: A systematic literature review. *Supply Chain Management*, 17(4), 358–376. DOI: 10.1108/13598541211246512

Pimenta-Bueno, J.A. (1999). Parque de Inovação Tecnológica e Cultural da Gávea: A Visão da PUC-Rio. Documento Interno, 23 págs, 1999.

Pizzol, M., & Andersen, M. S. (2022). Green tech for green growth? Insights from Nordic environmental innovation. In *Circular Economy Oriented Business Model Innovations: A European Perspective.* Springer. DOI: 10.1007/978-3-031-08313-6_8

Plúas-Arias, N., & Mejía-Caguana, D. (2024). Relación entre innovación tecnológica y sostenibilidad ambiental: Hacia un futuro resiliente y verde. *CIENCIAMATRIA*, 10(2), 482–500. DOI: 10.35381/cm.v10i2.1377

Porter, M. (1980). *Competitive Strategy: techniques for analyzing industries and competitors*. Free Press.

Porter, M. (2001). *Clusters of Innovation: regional foundations of U.S. competitiveness. Council on Competitiveness*. Monitor Group.

Porter, M. E., & Heppelmann, J. E. (2014). How smart, connected products are transforming competition. *Harvard Business Review*, 92(11), 64–88.

Porter, M. E., & Heppelmann, J. E. (2015). How smart, connected products are transforming companies. *Harvard Business Review*, 93(10), 96–114.

Postel, S. (2000). ENTERING AN ERA OF WATER SCARCITY: THE CHALLENGES AHEAD. *Ecological Applications*, 10(4), 941–948. DOI: 10.1890/1051-0761(2000)010[0941:EAEOWS]2.0.CO;2

Pouladi, P., Badiezadeh, S., Pouladi, M., Yousefi, P., Farahmand, H., Kalantari, Z., Yu, D. J., & Sivapalan, M. (2021). Interconnected governance and social barriers impeding the restoration process of Lake Urmia. *Journal of Hydrology (Amsterdam)*, 598, 126489. DOI: 10.1016/j.jhydrol.2021.126489

Pramudawardhani, R., (2021). Policy framework analysis for sustainable green infrastructure development in Indonesia. *Sustainable Cities and Society*, 75, 103206.

Price, R. M. (1996). Technology and Strategic Advantage. *California Management Review*, 38(3), 38–56. DOI: 10.2307/41165842

Principles Relating to Remote Sensing of the Earth from Outer Space, United Nations Office for Outer Space Affairs, www.unoosa.org/oosa/en/ourwork/spacelaw/principles/remote-sensing-principles.html

Puppim de Oliveira, J. A., Doll, C. N., Siri, J., Dreyfus, M., Farzaneh, H., & Capon, A. (2018). Urban Governance and the Systems Approaches to Health-Environment Co-Benefits in Cities. *Cadernos de Saude Publica*, 34, e00037518.

Putnam, R. D. (1993). *Making Democracy Work: Civic Traditions in Modern Italy*. Princeton University Press.

Qadir, S. A., Al-Motairi, H., Tahir, F., & Al-Fagih, L. (2021). Incentives and strategies for financing the renewable energy transition: A review. *Energy Reports*, 7, 3590–3606. DOI: 10.1016/j.egyr.2021.06.041

Quayson, F. O. (2018). *The Feasibility of Establishing A Private International Virtual High School In Ghanathe-feasibility-of-establishing-a-private-international-virtual-high-school-in-ghana. 1*(1).

Rabin, B. M., Laney, E. B., & Philipsborn, R. P. (2020). The unique role of medical students in catalyzing climate change education. *Journal of Medical Education and Curricular Development*, 7, 2382120520957653. DOI: 10.1177/2382120520957653 PMID: 33134547

Rachinger, M., Rauter, R., Müller, C., Vorraber, W., & Schirgi, E. (2019). Digitalization and its influence on business model innovation. [CrossRef]. *Journal of Manufacturing Technology Management*, 30(8), 1143–1160. DOI: 10.1108/JMTM-01-2018-0020

Rafiq, M., Cong Li, Y., Cheng, Y., Rahman, G., Zhao, Y., & Khan, H. U. (2023). Estimation of regional meteorological aridity and drought characteristics in Baluchistan province, Pakistan. *PLoS One*, 18(11), e0293073. DOI: 10.1371/journal.pone.0293073 PMID: 38033048

Rahman, T., & Khan, N. A. (2008). Rethinking corruption. New Age, a Vernacular English daily. www.newagebd.com

Rahman, A. S., Kamruzzaman, M., Jahan, C. S., Mazumder, Q. H., & Hossain, A. (2016). Evaluation of spatio-temporal dynamics of water table in NW Bangladesh: An integrated approach of GIS and Statistics. *Sustainable Water Resources Management*, 2(3), 297–312. DOI: 10.1007/s40899-016-0057-4

Raja, G. B. (2021). Impact of internet of things, artificial intelligence, and blockchain technology in Industry 4.0. Internet of Things, Artificial Intelligence and Blockchain Technology, 157-178.

Rajamani, L. (2006). The changing nature of international environmental law. *International Environmental Agreement: Politics, Law and Economics*, 6(2), 115–139.

Rajmohan, K. V. S., Ramya, C., Viswanathan, M. R., & Varjani, S. (2019). Plastic pollutants: Effective waste management for pollution control and abatement. *Current Opinion in Environmental Science & Health*, 12, 72–84. DOI: 10.1016/j.coesh.2019.08.006

Rakhecha, P. R. (2020). Water environment pollution with its impact on human diseases in India. *Int J Hydro*, 4(4), 152–158. DOI: 10.15406/ijh.2020.04.00240

Raman, R., Pattnaik, D., Lathabai, H. H., Kumar, C., Govindan, K., & Nedungadi, P. (2024). Green and sustainable AI research: An integrated thematic and topic modeling analysis. *Journal of Big Data*, 11(1), 55. DOI: 10.1186/s40537-024-00920-x

Ramazanov, S., Antoshkina, L., Babenko, V., & Akhmedov, R. (2019). Integrated model of stochastic dynamics for control of a socio-ecological-oriented innovation economy. *Periodicals of Engineering and Natural Sciences*, 7 (2), August 2019, pp.763-773 Available online at: http://pen.ius.edu.ba

Ramazanov, S. K. (2018). *Prediction and management of innovative economics based on the integral static model in phase space.*

Ramos, S., & Pires, L. (2021). Sustainable smart cities and the Internet of Things (IoT): What role for digital marketing? In 2021 IEEE International Smart Cities Conference (ISC2) (pp. 1-6). IEEE. https://doi.org/DOI: 10.1109/ISC252465.2021.9642998

Randolph, B., & Troy, P. (2008). Attitudes to conservation and water consumption. *Environmental Science & Policy*, 11(5), 441–455. DOI: 10.1016/j.envsci.2008.03.003

Rashid, A., Baloch, N., Rasheed, R. and Ngah,A.H. (2024), "Big data analytics-artificial intelligence and sustainable performance through green supply chain practices in manufacturing firms of a developing country, *Journal of Science and Technology Policy Management*, Vol. ahead-of-print No. ahead-of-print. https://doi.org/DOI: 10.1108/JSTPM-04-2023-0050

Raskin, P., Gleick, P., Kirshen, P., Pontius, G., & Strzepek, K. (1997). *Water futures: assessment of long-range patterns and problems. Comprehensive assessment of the freshwater resources of the world.* SEI.

Rasmussen, W. D., & Porter, J. M. (1981). Strategies for Dealing with World Hunger: Post-World War II Policies. *American Journal of Agricultural Economics*, 63(5), 810–818. DOI: 10.2307/1241249

Rastegar, H., Eweje, G., & Sajjad, A. (2024). The impact of environmental policy on renewable energy innovation: A systematic literature review and research directions. *Sustainable Development (Bradford)*, 32(4), 3859–3876. DOI: 10.1002/sd.2884

Rasul, M. T., & Jahan, M. S. (2010). Quality of ground and surface water of Rajshahi city area for sustainable drinking water source. *Journal of Scientific Research*, 2(3), 577–577. DOI: 10.3329/jsr.v2i3.4093

Reckien, D., & Eisenack, K. (2013). Climate Change Gaming on Board and Screen: A Review. *Simulation & Gaming*, 44(2-3), 253–271. DOI: 10.1177/1046878113480867

Reis, J., & Ballinger, R. C. (2020). Creating a climate for learning-experiences of educating existing and future decision-makers about climate change. *Marine Policy*, 111, 111. DOI: 10.1016/j.marpol.2018.07.007

Rentschler, L., & Leonova, N. (2023). Global Air Pollution Exposure and Poverty. *Nature Communications*, 14(4432), 1–11. PMID: 37481598

Reportlinker (2022). IoT Market with COVID-19 analysis by Component, Organization Size, Focus Area and Region - Global Forecast to 2026, https://www.reportlinker.com/p04944762/?utm_source=GNW

Rijsberman, F. R., Molden, D. J., & Makin, I. W. (2002). World Water Supplies: are they adequate. In *3rd rosenberg international forum on water policy, Canberra* (pp. 7-11).

Rijsberman, F. R. (2006). Water scarcity: Fact or fiction? *Agricultural Water Management*, 80(1–3), 5–22. DOI: 10.1016/j.agwat.2005.07.001

Riyad, A. S. M., Hassan, M., Rahman, M. A., Alam, M., & Akid, A. S. M. (2014). Characteristics and Management of Commercial Solid Waste in Khulna City of Bangladesh. *International Journal of Renewable Energy and Environmental Engineering*, 2.

Rizvi, A., & Sclar, E. (2014). Implementing bus rapid transit: A tale of two Indian cities. *Research in Transportation Economics*, 48, 194–204. DOI: 10.1016/j.retrec.2014.09.043

Roberts, M., (2021). Assessing the impact of environmental regulations on private sector investment in clean technologies. *Environmental Science & Policy*, 18(3), 112–127.

Robinne, F. N., Bladon, K. D., Miller, C., Parisien, M. A., Mathieu, J., & Flannigan, M. D. (2018). A spatial evaluation of global wildfire-water risks to human and natural systems. *The Science of the Total Environment*, 610, 1193–1206. DOI: 10.1016/j.scitotenv.2017.08.112 PMID: 28851140

Robinson, C., Von Broemssen, M., Bhattacharya, P., Häller, S., Bivén, A., Hossain, M., & Thunvik, R. (2011). Dynamics of arsenic adsorption in the targeted arsenic-safe aquifers in Matlab, south-eastern Bangladesh: Insight from experimental studies. *Applied Geochemistry*, 26(4), 624–635. DOI: 10.1016/j.apgeochem.2011.01.019

Rockström, J., Schellnhuber, H. J., Hoskins, B., Ramanathan, V., Schlosser, P., Brasseur, G. P., Gaffney, O., Nobre, C., Meinshausen, M., Rogelj, J., & Lucht, W. (2016). The world's biggest gamble. *Earth's Future*, 10(10), 465–470. DOI: 10.1002/2016EF000392

Rockström, J., Steffen, W., Noone, K., Persson, Å., Chapin, F. S.III, Lambin, E. F., Lenton, T. M., Scheffer, M., Folke, C., Schellnhuber, H. J., Nykvist, B., de Wit, C. A., Hughes, T., van der Leeuw, S., Rodhe, H., Sörlin, S., Snyder, P. K., Costanza, R., Svedin, U., & Foley, J. A. (2009). A safe operating space for humanity. *Nature*, 461(7263), 472–475. DOI: 10.1038/461472a PMID: 19779433

Rodrigues, R. Legal and human rights issues of AI: Gaps, challenges and vulnerabilities, Journal of Responsible Technology, Volume 4, 2020, 100005, ISSN 2666-6596, https://doi.org/DOI: 10.1016/j.jrt.2020.100005

Rofman A., Toscani M. d. l. P. y Ferrari Mango C. (2024). Transformaciones en la gobernanza participativa local de la política social en municipios del Gran Buenos Aires en tiempos pandémicos. Geopolítica(s). *Revista de estudios sobre espacio y poder, 15*(1), 145-166. DOI: 10.5209/geop.93330

Rogers, E., & Weber, E. P. (2010). Thinking Harder About Outcomes for Collaborative Governance Arrangements. *American Review of Public Administration*, 40(5), 546–567. DOI: 10.1177/0275074009359024

Rogoff, I., & Schneider, F. (2008). Productive Anticipation. In Held, D., & Moore, H. L. (Eds.), *Cultural Politics in a Global Age. Uncertainty, Solidarity, and Innovation* (pp. 346–358). Oneworld Publications.

Rohr, J. A. (2002). *Civil servants and their constitutions*. Studies in Government & Public.

Rolnick, D., Donti, P. L., & Kaack, L. H. (2021). Tackling climate change with machine learning. *Nature Communications*, 12(1), 1–14. PMID: 33397941

Rolnick, D., Donti, P. L., Kaack, L. H., Kochanski, K., Lacoste, A., Sankaran, K., Ross, A. S., Milojevic-Dupont, N., Jaques, N., Waldman-Brown, A., Luccioni, A. S., Maharaj, T., Sherwin, E. D., Mukkavilli, S. K., Kording, K. P., Gomes, C. P., Ng, A. Y., Hassabis, D., Platt, J. C., & Bengio, Y. (2022). Tackling climate change with machine learning. *ACM Computing Surveys*, 55(2), 1–96. DOI: 10.1145/3485128

Romanelli, L., (2020). The impact of digital marketing on sustainability: A systematic review. *Journal of Cleaner Production*, 242, 118208.

Rosenberg, J. (2020). Stakeholder perspectives on barriers to implementing green infrastructure projects through public-private partnerships. *Environmental Policy and Governance*, 30(6), 423–437.

Rosenboom, J. G., Langer, R., & Traverso, G. (2022). Bioplastics for a circular economy. *Nature Reviews. Materials*, 7(2), 117–137. DOI: 10.1038/s41578-021-00407-8 PMID: 35075395

Round, C., & Visseren-Hamakers, I. (2022). Blocked chains of governance: Using blockchain technology for carbon offset markets? *Frontiers in Blockchain.*, 5, 957316. DOI: 10.3389/fbloc.2022.957316

Roy, D. K. (2006). Governance, competitiveness and growth: The challenges for Bangladesh. ADB Institute Discussion Paper No. 53. Manila: Asian Development Bank.

Roy, H., Alam, S. R., Bin-Masud, R., Prantika, T. R., Pervez, M. N., Islam, M. S., & Naddeo, V. (2022). A review on characteristics, techniques, and waste-to-energy aspects of municipal solid waste management: Bangladesh perspective. *Sustainability (Basel)*, 14(16), 10265. DOI: 10.3390/su141610265

Roy, S. (1996). Development, Environment and Poverty: Some Issues for Discussion. *Economic and Political Weekly*, 31(4), 29–41.

Roznai, Y. (2014). The Insecurity of Human Security. *Wisconsin International Law Journal*, 32(1), 95–141.

Ruiz-Mallén, I., Satorras, M., March, H., & Baró, F. (2022). Community climate resilience and environmental education: Opportunities and challenges for transformative learning. *Environmental Education Research*, 28(7), 1088–1107. DOI: 10.1080/13504622.2022.2070602

Rukuni, R. (2021). The Legacy of the Green Belt Movement in Kenya: Lessons for Sustainable Development. *Journal of Environmental Policy and Planning*, 23(5), 652–670. DOI: 10.1080/1523908X.2021.1927870

Rumore, D., Schenk, T., & Susskind, L. (2016). Role-play simulations for climate change adaptation education and engagement. *Nature Climate Change*, 6(8), 745–750. DOI: 10.1038/nclimate3084

Rutherford, J. (2011). Rethinking the relational socio-technical materialities of cities and ICTs. *Journal of Urban Technology*, 18(1), 21–33. DOI: 10.1080/10630732.2011.578407

Saha, H. N., Mandal, A., & Sinha, A. (2017). Recent trends in the Internet of Things. In 2017 8th Annual Industrial Automation and Electromechanical Conference (IEMECON) (pp. 252-256). IEEE. DOI: 10.1109/CCWC.2017.7868439

Sahdev, I., Kumar, S., & Sahu, T. (2024), E-waste management in India, Sustainability, *Agri, Food and Environmental Research*, Vol. 12., http://dx.doi.org/

Saigal, A., & Bawa, D. (2024), Role of Local Leadership in attaining Sustainable Development Goals, *Terra Green (TERI)*, Vol. 16, https://shaktifoundation.in/role-of-local-leadership-in-attaining-sustainable-development-goals/

Saini, K., Mummoorthy, A., Chandrika, R., & Gowri Ganesh, N. S. (Eds.). (2023). *AI, IoT, and Blockchain Breakthroughs in E-governance*. IGI Global. DOI: 10.4018/978-1-6684-7697-0

Saito, K. (2017). *Karl Marx's ecosocialism: Captialism, nature, and the unfinished critique of political economy*. Monthly Review Press. DOI: 10.2307/j.ctt1gk099m

Salinas, I., Guerrero, G., Satlov, M., & Hidalgo, P. (2022). Climate change in chile's school science curriculum. *Sustainability (Basel)*, 14(22), 15212. DOI: 10.3390/su142215212

Samad, G., & Manzoor, R. (2011). Green Growth: An Environmental Technology Approach. *Pakistan Development Review*, 50(4), 471–490.

Samra, A. P. B., Morais, E., Mazur, R. F., Vieira, S. R., & Rached, R. N. (2016). CAD/CAM in dentistry–a critical review. *Revista Odonto Ciência*, 31(3), 140–144. DOI: 10.15448/1980-6523.2016.3.21002

Sands, P. (2017). *Principles of International Environmental Law* (4th ed.). Cambridge University Press.

Sanger, C. (1972/73). Environment and Development. *International Journal (Toronto, Ont.)*, 28(1), 103–120. DOI: 10.2307/40201094

Sapiains, R., Ibarra, C., Jiménez, G., O'Ryan, R., Blanco, G., Moraga, P., & Rojas, M. (2021). Exploring the contours of climate governance: An interdisciplinary systematic literature review from a southern perspective. *Environmental Policy and Governance*, 31(1), 46–59. DOI: 10.1002/eet.1912

Sarangi, S. (2020). *Sustainable communication: Cultural and environmental perspectives*. Routledge.

Sarkar, Debnarayan. (2011), The implementation of the forest rights act in India: Critical issues Economic Affairs 31(2):25-29, 1111/j.1468-0270.2011.02097.x

Sarkar, N. (2024), Delhi gets the attention - but Kolkata's air pollution is just as dangerous, https://news.mongabay.com/2024/04/delhi-gets-the-attention-but-kolkatas-air-pollution-is-just-as-dangerous/

Sarker, I. H. (2021). Context-aware learning and big data analytics for smart IoT-based environment: A survey. *Knowledge and Information Systems*, 63(2), 357–385. DOI: 10.1007/s10115-020-01496-2

Sarker, M. N. I., Peng, Y., Khatun, M. N., Alam, G. M. M., Shouse, R. C., & Amin, M. R. (2022). Climate finance governance in hazard prone riverine islands in Bangladesh: Pathway for promoting climate resilience. *Natural Hazards*, 110(2), 1115–1132. DOI: 10.1007/s11069-021-04983-4

Sarker, M. R. A., & Lu, W. (2020). Public-private partnership (PPP) for sustainable green infrastructure development: A review. *Journal of Cleaner Production*, 249, 119315.

Sarker, M. R. A., & Lu, W. (2021). Institutional capacity building for public-private partnerships in infrastructure: A case study of India. *International Journal of Public Administration*, 44(5), 370–383.

Satterthwaite, D., Archer, D., Colenbrander, S., Dodman, D., Hardoy, J., Mitlin, D., & Patel, S. (2020). Building Resilience to Climate Change in Informal Settlements. *One Earth*, 2(2), 143–156. DOI: 10.1016/j.oneear.2020.02.002

Sauvé, S., Bernard, S., & Sloan, P. (2016). Environmental Sciences, Sustainable Development and Circular Economy: Alternative Concepts for Trans-disciplinary Research. *Environmental Development*, 17, 48–56. DOI: 10.1016/j.envdev.2015.09.002

Saxena, V., Datla, A., Deheriya, M., Tiwari, N., Shoukath, S., & Bhargava, A. (2024). EcoSmile: A Comprehensive Analysis of Sustainable Dental Practices Using Mixed Methodology. *Advances in Human Biology*, 10-4103.

Saxena, A., Singh, R., Gehlot, A., Akram, S. V., Twala, B., Singh, A., Montero, E. C., & Priyadarshi, N. (2023). Technologies Empowered Environmental, Social, and Governance (ESG): An Industry 4.0 Landscape. *Sustainability (Basel)*, 15(1), 309. DOI: 10.3390/su15010309

Saxenian, A. (1994). *Regional Advantage: culture and competition in Silicon Valley and Route 128*. Harvard Business Press.

Scarlat, N., Motola, V., Dallemand, J. F., Monforti, F., & Mofor, L. (2015). Evaluation of Energy Potential of Municipal Solid Waste from African Urban Areas. *Renewable & Sustainable Energy Reviews*, 50, 1269–1286. DOI: 10.1016/j.rser.2015.05.067

Schlapfer, A., & Marinova, D. (2001). Local Innovation Systems: nature, importance and role. *Conference Proceedings: International Summer Academy on Technological Studies:* user involvement in technological innovation. Deutschlandsberg. Austria.

Schoormann, T., Strobel, G., Möller, F., Petrik, D., & Zschech, P. (2023). Artificial Intelligence for Sustainability—A Systematic Review of Information Systems Literature. Communications of the Association for Information Systems, 52, pp-pp. Retrieved from https://aisel.aisnet.org/cais/vol52/iss1/8

Schot, J., & Rip, A. (1997). The past and future of constructive technology assessment. *Technological Forecasting and Social Change*, 54(2–3), 251–268. DOI: 10.1016/S0040-1625(96)00180-1

Schot, J., & Steinmueller, W. E. (2018). Three frames for innovation policy: R&D, systems of innovation and transformative change. *Research Policy*, 47(9), 1554–1567. DOI: 10.1016/j.respol.2018.08.011

Sciolla, L., & Morgera, E. (2021). Balancing ecological integrity, social equity, and economic viability in sustainable environmental governance. *Ecological Economics*, 179, 106857.

Seadon, J. K. (2010). Sustainable waste management systems. *Journal of Cleaner Production*, 18(16-17), 1639–1651. DOI: 10.1016/j.jclepro.2010.07.009

Seckler, D., Barker, R., & Amarasinghe, U. (1999). Water Scarcity in the Twenty-first Century. *International Journal of Water Resources Development*, 15(1–2), 29–42. DOI: 10.1080/07900629948916

Secretary General, U. N. (2005). *Larger Freedom: Towards Development, Security, and Human Rights for all. Report of the UN Secretary General to the General Assembly, A/59/2005*.

Seebauer, S., & Ieee. (2013). Measuring climate change knowledge in a social media game with a purpose. Paper presented at the *5th International Conference on Games and Virtual Worlds for Serious Applications (VS-GAMES)*, Bournemouth Univ, Bournemouth, ENGLAND. DOI: 10.1109/VS-GAMES.2013.6624236

Seebauer, S. (2014). Validation of a social media quiz game as a measurement instrument for climate change knowledge. *Entertainment Computing*, 5(4), 425–437. https://doi.org/https://doi.org/10.1016/j.entcom.2014.10.007. DOI: 10.1016/j.entcom.2014.10.007

Sellers, J. M., & Litström, A. (2007). Decentralization, Local Government, and the Welfare State. *Governance: An International Journal of Policy, Administration and Institutions*, 20(4), 609–632. DOI: 10.1111/j.1468-0491.2007.00374.x

Sen, A. (1981). *Poverty and Famines: An Essay on Entitlement and Deprivation*. Clarendon Press.

Sen, A. (1999). *Development as Freedom*. Oxford University Press.

Seritan, A. L., Coverdale, J., & Brenner, A. M. (2022). Climate change and mental health curricula: Addressing barriers to teaching. *Academic Psychiatry*, 46(5), 551–555. DOI: 10.1007/s40596-022-01625-0 PMID: 35314961

Shabalala, Nonkanyiso Pamella. (2023), Environmental Education as a Catalyst to Teach Students About Their Economy and Politics, *Journal Pendidikan Indonesia Gemilang* 3(2): 306-322, . v3i2.229DOI: 10.53889/jpig

Shafique, M. N., Rashid, A., Bajwa, I. S., Kazmi, R., Khurshid, M. M., & Tahir, W. A. (2018). Effect of IoT capabilities and energy consumption behavior on green supply chain integration. *Applied Sciences (Basel, Switzerland)*, 8(12), 2481. DOI: 10.3390/app8122481

Shandas, V., & Messer, W. B. (2008). Fostering green communities through civic engagement: Community-based environmental stewardship in the Portland area. *Journal of the American Planning Association*, 74(4), 408–418. DOI: 10.1080/01944360802291265

Sharif, N. (1992). Technological dimensions of international cooperation and sustainable development. *Technological Forecasting and Social Change*, 42(4), 367–383. DOI: 10.1016/0040-1625(92)90080-D

Sharma, A., & Singh, S. (2021). Promoting innovation in public-private partnerships for sustainable development: The role of government policies. *Journal of Public Procurement*, 21(2), 127–143.

Sharma, M., & Dikshit, O. (2021). Air Pollution and Health Management in Delhi: Current Challenges and Prospects. In *Air Pollution Sources, Statistics and Health Effects*. Nova Science Publishers.

Sharma, N. (2021). Sustainable digital marketing: Integrating environmental concerns in marketing strategies. *Journal of Marketing Management*, 37(9-10), 811–831. DOI: 10.1080/0267257X.2021.1927960

Sharma, R. (2023). Civil society organizations' institutional climate capacity for community-based conservation projects: Characteristics, factors, and issues. *Current Research in Environmental Sustainability*, 5, 100218. DOI: 10.1016/j.crsust.2023.100218

Sharma, R. (2023). Plastic Waste Management: Challenges and Potential Solution With the application of AI. *International Journal of Sustainable Development Through AI. ML and IoT*, 2(1), 1–27.

Sharma, S. K. (2019). Role of Remote Sensing and GIS in Integrated Water Resources Management (IWRM). In Ray, S. (Ed.), *Ground Water Development - Issues and Sustainable Solutions*. Springer., DOI: 10.1007/978-981-13-1771-2_13

Shearmur, R., Carrincazeaux, C., & Doloreux, D. (Eds.). (2016). *Handbook on the Geographies of Innovation*. Edward Elgar. DOI: 10.4337/9781784710774

Sheheli, S. (2014). Waste Disposal and Management System in Rural Areas of Mymensingh. *Progressive Agriculture*, 18(2), 241–246. DOI: 10.3329/pa.v18i2.18278

Sheiham, A., & James, W. P. T. (2015). Diet and dental caries: The pivotal role of free sugars reemphasized. *Journal of Dental Research*, 94(10), 1341–1347. DOI: 10.1177/0022034515590377 PMID: 26261186

Sheng, J., Zhou, W., & Zhu, B. (2020). La coordinación de los intereses de las partes interesadas en la regulación ambiental: Lessons from China's environmental regulation policies from the perspective of the evolutionary game theory. *Journal of Cleaner Production*, 249, 119385. DOI: 10.1016/j.jclepro.2019.119385

Shixuan, C., Yuchen, G., Yizhou, H., Lilai, S., Qiming, Z., Yaru, P., & Shulin, Z. (2023). Advances and applications of machine learning and deep learning in environmental ecology and health, Environmental Pollution, Volume 335, 2023, 122358, ISSN 0269-7491, https://doi.org/DOI: 10.1016/j.envpol.2023.122358

Shohel, M. M. C., Ashrafuzzaman, M., Islam, M. T., Shams, S., & Mahmud, A. (2021). 'Blended teaching and learning in higher education: Challenges and opportunities', *Handbook of research on developing a post-pandemic paradigm for virtual technologies in higher education*, pp.27-50. DOI: 10.4018/978-1-7998-6963-4.ch002

Shrivastava, A., & Kumar, R. (2015). Green Infrastructure Financing and Institutional Investors. *Public Policy and Administration Research*, 5(8), 84–93.

Shunglu, R., Köpke, S., Kanoi, L., Nissanka, T. S., Withanachchi, C. R., Gamage, D. U., Dissanayake, H. R., Kibaroglu, A., Ünver, O., & Withanachchi, S. S. (2022). Barriers in Participative Water Governance: A Critical Analysis of Community Development Approaches. *Water (Basel)*, 14(5), 762. DOI: 10.3390/w14050762

Sicangco, M. C. T. (2020). *Climate change, coming soon to a court near you: International climate change legal frameworks*. Asian Development Bank., DOI: 10.22617/TCS200365-2

Silva, G. M. E., Campos, D. F., Brasil, J. A. T., Tremblay, M., Mendiondo, E. M., & Ghiglieno, F. (2022). Advances in technological research for online and in situ water quality monitoring—A review. *Sustainability (Basel)*, 14(9), 5059. DOI: 10.3390/su14095059

Silva, W. D. O., & Morais, D. C. (2021). Transitioning to a circular economy in developing countries: A collaborative approach for sharing responsibilities in solid waste management of a Brazilian craft brewery. *Journal of Cleaner Production*, 319, 128703. DOI: 10.1016/j.jclepro.2021.128703

Silvestre, B. S., & Ţîrcă, D. M. (2019). Innovations for sustainable development: Moving toward a sustainable future. *Journal of Cleaner Production*, 208, 325–332. https://www.sciencedirect.com/science/article/abs/pii/S0959652618329834. DOI: 10.1016/j.jclepro.2018.09.244

Singh, R. (2024), Environmental Policies in India, Environmental Problems, Protection and Policies, 299-317, https://www.researchgate.net/publication/378590048 _ENVIRONMENTAL_POLICIES_IN_INDIA

Singh, A. P., & Misra, D. C.Ajit Pratap SinghDinesh Chandra Misra. (2022). Waste management issue and solutions using IOT. *International Journal of Science and Research Archive*, 7(1), 260–282. DOI: 10.30574/ijsra.2022.7.1.0209

Singh, A., & Prasad, S. M. (2015). Remediation of Heavy Metal Contaminated Ecosystem: An Overview on Technology Advancement. *International Journal of Environmental Science and Technology*, 12(1), 353–366. DOI: 10.1007/s13762-014-0542-y

Singh, P. K., & Maheswaran, R. (2024). Analysis of social barriers to sustainable innovation and digitisation in supply chain. *Environment, Development and Sustainability*, 26(2), 5223–5248. DOI: 10.1007/s10668-023-02931-9 PMID: 36687733

Singh, V., Singh, S., & Biswal, A. (2021). Exceedances and trends of particulate matter (PM2.5) in five Indian megacities. *The Science of the Total Environment*, 750, 141461. DOI: 10.1016/j.scitotenv.2020.141461 PMID: 32882489

Sira, M. (2024). Potential of Advanced Technologies for Environmental Management Systems. *Management Systems in Production Engineering.*, 32(1), 33–44. DOI: 10.2478/mspe-2024-0004

Skains, R. L., Rudd, J. A., Horry, R., & Ross, H. (2022, January 27). Playing for Change: Teens' Attitudes Towards Climate Change Action as Expressed Through Interactive Digital Narrative Play. *Frontiers in Communication*, 6, 789824. Advance online publication. DOI: 10.3389/fcomm.2021.789824

Smelser, N. J., & Baltes, P. B. (Eds.). (2001). *International encyclopedia of the social & behavioral sciences* (Vol. 11). Elsevier.

Smil, V. (1999). China's Great Famine: 40 Years Later. *BMJ (Clinical Research Ed.)*, 319(7225), 1619–1621. DOI: 10.1136/bmj.319.7225.1619 PMID: 10600969

Smith, A. H., Lingas, E. O., & Rahman, M. (2000). Contamination of drinking-water by arsenic in Bangladesh: A public health emergency. *Bulletin of the World Health Organization*, 78(9), 1093–1103. PMID: 11019458

Smith, A., & Raven, R. (2012). What is a protective space? Reconsidering niches in transitions to sustainability. *Research Policy*, 41(6), 1025–1036. DOI: 10.1016/j.respol.2011.12.012

Smith, K. (2018). The role of financial incentives in promoting private sector engagement in sustainable development projects. *Journal of Sustainable Finance & Investment*, 22(1), 45–58.

Sofoulis, Z. (2011). Smart metering and water end use data: Conservation for the digital age. [TOCHI]. *ACM Transactions on Computer-Human Interaction*, 18(1), 1.

Solou, R. (2002). Theory of growth. Panorama of the economic thought of the end of the twentieth century, Ed. D. Greenway, M. Blini, I. Stewart, *SPb.: Econ. Shk.*, vol. 1. pp. 479-506, 2002.

Soma, K., Termeer, C. J., & Opdam, P. (2016). Informational governance–A systematic literature review of governance for sustainability in the Information Age. *Environmental Science & Policy*, 56, 89–99. DOI: 10.1016/j.envsci.2015.11.006

Somanathan, T. V. The Administrative and Regulatory State, Retrieved from: https://icpp.ashoka.edu.in/wp-content/uploads/2024/04/Somanathan-The-Administrative-and-Regulatory-State-1.pdf

Song, M., Wang, S., & Zhang, H. (2020). Could environmental regulation and R&D tax incentives affect green product innovation? *Journal of Cleaner Production*, 258, 120849. DOI: 10.1016/j.jclepro.2020.120849

SPACE LAW & POLICY. CRITICAL STUDY OF INTERNATIONAL LEGAL REGULATION, https://legalvidhiya.com/space-law-policy-critical-study-of-international-legal-regulation/

Spiezia, V., & Tscheke, J. International agreements on cross-border data flows and international trade: A statistical analysis, OECD Science, Technology, and Industry Working Papers, 2020

Staszewski, G. (2008). Reason-giving and accountability. *Minnesota Law Review*, 93, 1253.

Steblianskaia, E., Vasiev, M., Denisov, A., Bocharnikov, V., Steblyanskaya, A., & Wang, Q. (2023). Environmental-social-governance concept bibliometric analysis and systematic literature review: Do investors becoming more environmentally conscious? *Environmental and Sustainability Indicators*, 17, 100218. DOI: 10.1016/j.indic.2022.100218

Steffen, W.. (2015). *Global Change and the Earth System: A Planet Under Pressure*. Springer.

Steinbach, I., Stancliffe, R., Berners-Lee, M., & Duane, B. (2018). Carbon modelling within dentistry. Towards a sustainable future. In . Public Health England.

Stieglitz, S., Lattemann, C., Robra-Bissantz, S., Zarnekow, R., & Brockmann, T. (Eds). (2017). *Gamification: Using Game Elements in Serious Contexts*. Springer.

Stiglitz, J., & Wallsten, S. (2021). Lessons learned from public-private partnerships in green infrastructure: Insights from project stakeholders. *Journal of Infrastructure Development*, 13(2), 135–151.

Strat, A. L. (2010). Paris: Local Authorities Regain Control of Water Management. *The Transnational Institute (TNI)*. http://www. tni.org/sites/www.tni.org/files/Paris Chapter by Anne Le Strat En_ final.pdf

Suhaili, W. H., Patchmuthu, R. K., & Shams, S. (2023). October. The Internet of Things in the Rearing of Giant Freshwater Prawn: A Pilot Study. In *2023 6th International Conference on Applied Computational Intelligence in Information Systems (ACIIS)* (pp. 1-6). IEEE.

Suits, B. H. (1978). *The grasshopper: Games, life and utopia*. University of Toronto Press. DOI: 10.3138/9781487574338

Sujauddin, M., Huda, S. M. S., & Hoque, A. T. M. R. (2008). Household solid waste characteristics and management in Chittagong, Bangladesh. *Waste Management (New York, N.Y.)*, 28(9), 1688–1695. DOI: 10.1016/j.wasman.2007.06.013 PMID: 17845843

Sultanovich, M. D. (2023). The Main Directions of Poverty Reduction in our Country. *Galaxy International Interdisciplinary Research Journal*, 11(2), 164–171.

Sustainable Development Goals. United Nations, www.un.org/sustainabledevelopmentTong, W.W.; Zhang, J.M. Can Environmental Regulation Promote Technological Innovation. Financ. Sci. 2012, 11, 66–74

Sutton-Smith, B. (1997). *The ambiguity of play*. Harvard University Press.

Su, Y. S. (2021). Challenges and opportunities of financing green public-private partnership projects: A critical review. *Sustainability*, 13(5), 2709.

Syeed, M. M. M., Hossain, M. S., Karim, M. R., Uddin, M. F., Hasan, M., & Khan, R. H. (2023). Surface water quality profiling using the water quality index, pollution index and statistical methods. *Critical Review*, 18(June), 100247. DOI: 10.1016/j.indic.2023.100247

Tadjbakhsh, S., & Chenoy, A. M. (2007). *Human Security: Concept and Implications*. Routledge. DOI: 10.4324/9780203965955

Takagi, H. (2019). Statistics on typhoon landfalls in Vietnam: Can recent increases in economic damage be attributed to storm trends? *Urban Climate*, 30, 100506. DOI: 10.1016/j.uclim.2019.100506

Takinana, A., & Baars, R. C. (2023). *Climate change education in the South Pacific: Resilience for whom? Asia Pacific Viewpoint*. Business Source Index.

Talaviya, T., Shah, D., Patel, N., Yagnik, H., & Shah, M. (2020). Implementation of artificial intelligence in agriculture for optimisation of irrigation and application of pesticides and herbicides, Artificial Intelligence in Agriculture, Volume 4, 2020, Pages 58-73, ISSN 2589-7217, https://doi.org/DOI: 10.1016/j.aiia.2020.04.002

Taleb, M., & Kadhum, H. (2024). *The Role of Artificial Intelligence in Promoting the Environmental, Social and Governance (ESG)*. Practices., DOI: 10.1007/978-3-031-63717-9_17

Tamayo-Vera, D., Wang, X., & Mesbah, M. (2024). A Review of Machine Learning Techniques in Agroclimatic Studies. *Agriculture*, 14(3), 481. DOI: 10.3390/agriculture14030481

âncu, A. M. C., Didilescu, A. C., Pantea, M., Sfeatcu, R., & Imre, M. (2023, May). Aspects Regarding Sustainability among Private Dental Practitioners from Bucharest, Romania: A Pilot Study. []. MDPI.]. *Health Care*, 11(9), 1326. PMID: 37174868

Tan, S. Y., & Taeihagh, A. (2020). Smart city governance in developing countries: A systematic literature review. *Sustainability (Basel)*, 12(3), 899. DOI: 10.3390/su12030899

Tasnim, T., Annum, F., Curtis, A. E., Braun, M., Shuvo, S. R. S., Anjum, B., Amjath-Babu, T. S., & Khanam, F. (2023). *Bridging the climate information gap for adaptation: Mainstreaming climate services into higher education in Bangladesh*.

Tenelanda-López, D., Valdivia-Moral, P., & Castro-Sánchez, M. (2020). Eating habits and their relationship to oral health. *Nutrients*, 12(9), 2619. DOI: 10.3390/nu12092619 PMID: 32867393

Testa, F., Iraldo, F., & Frey, M. (2022). The effect of environmental regulation on firms' competitive performance: The case of the building & construction sector in some EU regions. *Journal of Environmental Management*, 92(9), 2136–2144. DOI: 10.1016/j.jenvman.2011.03.039 PMID: 21524840

Thackway, R., & Olsson, K. (1999). Public/private partnerships and protected areas: Selected Australian case studies. *Landscape and Urban Planning*, 44(2–3), 87–97. DOI: 10.1016/S0169-2046(99)00003-1

Thakur, Ramesh (2005). From National Security to Human Security. *The Japan Times*, 13 October 2005.

Tharsanee, R. M., Soundariya, S., & Vishnupriya, B. (2020). Machine Learning and Data Analytics for Environmental Science: A Review, Prospects and Challenges. *IOP Conference Series. Materials Science and Engineering*, 955(1), 012107. DOI: 10.1088/1757-899X/955/1/012107

The European Sustainable Development Strategy (2001). *The potential for sustainable development of 2001*. A sustainable Europe for a better World // enterprise of chemical industry. M.: Publishing house Europe – Sustainable Development. SDS 2005-2010.

The Rio Conventions | UNFCCC; United Nations Climate Change, www.unfccc.int

Thompson, S. C., & Barton, M. A. (1994). Ecocentric and Anthropocentric Attitudes towards the Environment. *Journal of Environmental Psychology*, 14(2), 149–157. DOI: 10.1016/S0272-4944(05)80168-9

Tidd, J.. (1997). *Managing Innovation: integrating technological, market and organizational change*. John Wiley & Sons.

Tironi, M., & Lisboa, D. (2023). Artificial intelligence in the new forms of environmental governance in the Chilean State: Towards an eco-algorithmic governance. *Technology in Society*, 74, 102264. DOI: 10.1016/j.techsoc.2023.102264

Tödtling, F., & Trippl, M. (2005). One size fits all? Towards a differentiated regional innovation policy. *Research Policy*, 34(8), 203–1219. DOI: 10.1016/j.respol.2005.01.018

Tol, R. S. J. (2009). Spr). The Economic Effects of Climate Change. *The Journal of Economic Perspectives*, 23(2), 29–51. DOI: 10.1257/jep.23.2.29

Tomor, Z., Meijer, A., Michels, A., & Geertman, S. (2019). Smart governance for sustainable cities: Findings from a systematic literature review. *Journal of Urban Technology*, 26(4), 3–27. DOI: 10.1080/10630732.2019.1651178

Top Green Dental Initiatives in. 2024. (2024). Available at https://adit.com/top-green-dental-initiatives-2024

Tortajada, C., & Biswas, A. K. (2018). Achieving universal access to clean water and sanitation in an era of water scarcity: Strengthening contributions from academia. *Current Opinion in Environmental Sustainability*, 34, 21–25. DOI: 10.1016/j.cosust.2018.08.001

Trist, E. L. (1976). *A Concept of Organizational Ecology: an invited address to the three Melbourne universities.*

Tulloch, R., & Wilkins, M. (2020). Smart water management: Leveraging IoT for water conservation. *Journal of Water Resources Planning and Management*, 146(6), 04020045. Advance online publication. DOI: 10.1061/(ASCE)WR.1943-5452.0001211

Turton, A. R., & Meissner, R. (2000) Analysis of the influence from the international hydropolitical environment, driving the process of change in the water sector (Appendix F). In: Report of the institutional support task team – shared rivers initiative of the Incomati river. Scottsville and Harare, Institute of Natural Resources (INR) and Swedish International Development Agency (SIDA). Available online at https://www.up.ac.za/academic/libarts/polsci/awiru

Turton, A. R., Hattingh, J., Maree, G. A., Roux, D. J., Claassen, M., & Strydom, W. F. (2007). *'Governance as a Trialogue: Government-Society-Science in Transition', Water Resources Development and Management Series, ISBN-10 3-540-46265-1.* Springer-Verlag Berlin Heidelberg. DOI: 10.1007/978-3-540-46266-8

Tu, Y., & Wu, W. (2020). ¿Cómo mejora la innovación ecológica la ventaja competitiva de las empresas? The Role of Organizational Learning. *Sustainable Production and Consumption.* Advance online publication. DOI: 10.1016/j.spc.2020.12.031

U.S. DEPArtment of Energy. (2020). Federal Incentives for Renewable Energy. Retrieved from https://www.energy.gov/eere/federal-incentives-renewable-energy

U.S. Environmental Protection Agency. (2021). Renewable Portfolio Standards. Retrieved from https://www.EPA.gov/statelocalenergy/renewable-portfolio-standards

Uddin, N. (2018). Assessing urban sustainability of slum settlements in Bangladesh: Evidence from Chittagong city. *Journal of urban management*, 7(1), 32-42. https://doi.org/DOI: 10.1016/j.jum.2018.03.002

Uitto, J. I., & Biswas, A. K. (2000). Water for urban areas: Challenges and perspectives. *Journal - American Water Works Association*, 92(12), 136.

ul Haq, M. (1995a). *Reflections on Human Development.* Oxford University Press.

ul Haq, M. (1995b). New Imperatives of Human Security. *World Affairs* 4(1), 68-73.

UN General Assembly. (2005). 2005 World Summit Outcome, A/60/150. https://www.un.org/en/development/desa/population/migration/generalassembly/docs/globalcompact/A_RES_60_1.pdf

UN. (2004). A More Secure World – Our Shared Responsibility – Report of the High-Level Panel on Threats, Challenges and Change. https://www.un.org/peacebuilding/sites/www.un.org.peacebuilding/files/documents/hlp_more_secure_world.pdf

UNDP (2009) Capacity Development: A UNDP Primer, Available at https://iwrmactionhub.org/resource/capacity-development-undp-primer.

UNDP. (1990). *Human Development Report.* https://hdr.undp.org/content/human-development-report-1990

UNDP. (1994). *Human Development Report.* https://hdr.undp.org/content/human-development-report-1994

UNDP. (1995). *Human Development Report.* https://hdr.undp.org/content/human-development-report-1995

UNDP. (1997). *Human Development Report.* https://hdr.undp.org/content/human-development-report-1997

UNDP. (1998). *Human Development Report.* https://hdr.undp.org/content/human-development-report-1998

UNDP. (1999). *Human Development Report.* https://hdr.undp.org/content/human-development-report-1999

UNDP. (2011). *Human Development Report.* https://hdr.undp.org/content/human-development-report-2011

UNDP. (2020). *Human Development Report.* https://hdr.undp.org/content/human-development-report-2020

UNDP. (2022). *Human Development Report.* https://hdr.undp.org/content/human-development-report-2021-22

UNEP (2018). Municipal solid waste: Is it garbage or gold?" *UNEP Global Environmental Alert Service (GEAS)*, October

UNEP (United Nations Environment Program). (2011). *Towards a Green Economy: Pathways to Sustainable Development and Poverty Eradication.* UNEP.

UNEP, 'Policy and Strategy for Gender Equality and the Environment 2014-2017', February 2015, para. 21.

Ungar, M. (2018). Systemic resilience: Principles and processes for a science of change in contexts of adversity. *Ecology and Society*, 23(4), art34. https://www.jstor.org/stable/26796886. DOI: 10.5751/ES-10385-230434

UNICEF & WHO. (2019). *Progress on Household Drinking Water, Sanitation and Hygiene 2000–2017*. Special Focus on Inequalities. UNICEF & WHO.

UNICEF & WHO. (2020). *Progress on Drinking Water, Sanitation and Hygiene in Schools: Special Focus on COVID-19*. UNICEF & WHO.

United Nations (UN). (2017). Sustainable development goals: 17 goals to transform our world. United Nations: New York, NY, USA. Available online: https://www.un.org/sustainabledevelopment/sustainable-development-goals/ (accessed on 22 June 2024).

United Nations Environment Programme (UNEP). (2015). Global Waste Management Outlook. UNEP, International Solid Waste Association (ISWA). Nairobi, Kenya: United Nations Environment Programme. Available at: https://www.unep.org/resources/report/global-waste-management-outlook

United Nations Environment Programme. (2018). Green Public Procurement: A Global Review. Retrieved from https://www.unep.org/resources/report/green-public-procurement-global-review

United Nations Environment Programme. (2023, June). *Global pact on plastic pollution: steps toward an inclusive, legally binding agreement. World Environment Day*. Retrieved from https://www.unep.org/plastics-instrument

United Nations. (2015) *Water and Sanitation - United Nations Sustainable Development*. Available at: https://www.un.org/sustainabledevelopment/water-and-sanitation/

United Nations. (2018) *Water Action Decade - United Nations Sustainable Development, United Nations*. Available at: https://www.un.org/sustainabledevelopment/water-action-decade/

United Nations. 'The Millennium Development Goals Report 2015', 2015, p.5. Available at: https://www.undp.org/content/dam/undp/library/MDG/english/UNDP_MDG_Report_2015.pdf p. 5

Urho, N., Ivanova, M., Dubrova, A., & Escobar-Pemberthy, N. (2019). *International environmental governance: Accomplishments and way forward*. Nordic Council of Ministers.

Uriarte-Gallastegi, N., Arana-Landín, G., Landeta-Manzano, B., & Laskurain-Iturbe, I. (2024). The Role of AI in Improving Environmental Sustainability: A Focus on Energy Management. *Energies*, 17(3), 649. DOI: 10.3390/en17030649

US Department of Health and Human Services Federal Panel on Community Water Fluoridation. (2015). US Public Health Service recommendation for fluoride concentration in drinking water for the prevention of dental caries. *Public Health Reports*, 130(4), 318–331. DOI: 10.1177/003335491513000408 PMID: 26346489

van Beek, L., Milkoreit, M., Prokopy, L., Reed, J. B., Vervoort, J., Wardekker, A., & Weiner, R. (2022). 2022/02/16). The effects of serious gaming on risk perceptions of climate tipping points. *Climatic Change*, 170(3), 31. DOI: 10.1007/s10584-022-03318-x

van Strien, J., Gelderman, C. J., & Semeijn, J. (2019). Performance-based contracting in military supply chains and the willingness to bear risks. *Journal of Defense Analytics and Logistics*, 3(1), 83–107. DOI: 10.1108/JDAL-10-2017-0021

Vargas, E., Montes, J., & Zartha, J. (2017). *Prospectiva sobre innovación ambiental. Un estudio para mipymes hoteleras de Colombia y México*. Pp. 167-187. Instituto de Ciencias Económico Administrativas. https://repository.uaeh.edu.mx/books/163/ti.pdf#page=169

Varis, O., & Vakkilainen, P. (2001). China's 8 challenges to water resources management in the first quarter of the 21st Century. *Geomorphology*, 41(2-3), 93–104. DOI: 10.1016/S0169-555X(01)00107-6

Vaz, N. (2021). *Digital business transformation: How established companies sustain competitive advantage from now to next*. John Wiley & Sons.

Velmurugan, S., & Shanmugavel, S. (2021). Sustainable environmental governance through digital marketing and IoT: A review. *Journal of Environmental Management*, 278, 111549.

Verhoeven, I. (2004). Veranderend politiek burgerschap en democratie. In *E.R. Engelen & M. Sie Dhian Ho (red.) De staat van de democratie. Democratie voorbij de staat. WRR verkenningen*. Amsterdam University Press.

Verma, A., & Saurabh, M. (2020), An Analysis of Poverty Alleviation Programmes in India with Special Reference to Sustainable Development Goals, https://papers.ssrn.com/sol3/papers.cfm?abstract_id=3637927

Vishwanath, S. (2013, February 15). *How much water does an urban citizen need?* The Hindu. https://www.thehindu.com/features/homes-and-gardens/how-much-water-does-an-urban-citizen-need/article4393634.ece

Visvanathan, C., Adhikari, R., & Ananth, A. P. (2007). 3R practices for municipal solid waste management in Asia. *Linnaeus Eco-Tech*, 11-22.

Viswanathan, K. (2021). Integrating IoT and digital marketing for sustainable environmental practices. *Journal of Environmental Management*, 277, 111354.

Voigt, C. (2016). The Paris Agreement: What is the standard of conduct for parties? *Questions of International Law*, 24, 17–28.

von Schomberg, R. (2012). Prospects for technology assessment in a framework of responsible research and innovation. Dusseldorp, M. & Beecroft, R. (eds.) *Technikfolgen abschätzen lehren*, 39–61. Wiesbaden: VS Verlag für Sozialwissenschaften. DOI: 10.1007/978-3-531-93468-6_2

Vörösmarty, C. J., Green, P., Salisbury, J., & Lammers, R. B. (2000). Global Water Resources: Vulnerability from Climate Change and Population Growth. *Science*, 289(5477), 284–288. DOI: 10.1126/science.289.5477.284 PMID: 10894773

Waddington, D. I., & Fennewald, T. (2018). Grim FATE: Learning About Systems Thinking in an In-Depth Climate Change Simulation. *Simulation & Gaming*, 49(2), 168–194. DOI: 10.1177/1046878117753498

Walker, W. S., Baccini, A., Schwartzman, S., Ríos, S., Oliveira-Miranda, M. A., Augusto, C., & Smith, P. (2020). The role of forest conversion, degradation, and disturbance in the carbon dynamics of Amazon indigenous territories and protected areas. *Proceedings of the National Academy of Sciences of the United States of America*, 117(6), 3015–3025. DOI: 10.1073/pnas.1913321117 PMID: 31988116

Wals, A. E. J. (2015). Social Learning-Oriented Capacity-Building for Critical Transitions Towards Sustainability. In *Schooling for Sustainable Development in Europe* (pp. 87–107). Springer International Publishing., DOI: 10.1007/978-3-319-09549-3_6

Walsh, T., Worthington, H. V., Glenny, A. M., Marinho, V. C., & Jeroncic, A. (2019). Fluoride toothpastes of different concentrations for preventing dental caries. *Cochrane database of systematic reviews*, (3). World Dental Federation FDI, The Carbon Neutral Dental Office. *FDI* available at https://www.fdiworlddental.org/sites/default/files/2024-04/FDI%20World%20Dental%20Federation%20-%20The%20Carbon%20Neutral%20Dental%20Office%20%20Case%20Study.pdf

Walumbwa, F. O., Hartnell, C. A., & Misati, E. (2017). Does ethical leadership enhance group learning behavior? Examining the mediating influence of group ethical conduct, justice climate, and peer justice. *Journal of Business Research*, 72, 14–23. DOI: 10.1016/j.jbusres.2016.11.013

Wampler, B. (2012). Entering the State: Civil Society Activisim and Participatory Governance in Brazil. *Political Studies*, 60(2), 341–362. DOI: 10.1111/j.1467-9248.2011.00912.x

Wang, H., & Guo, J. (2024). New way out of efficiency-equity dilemma: Digital technology empowerment for local government environmental governance. *Technological Forecasting and Social Change*, 200, 123184. DOI: 10.1016/j.techfore.2023.123184

Wang, H., & Ran, B. (2021). Network governance and collaborative governance: A thematic analysis on their similarities, differences, and entanglements. *Public Management Review*, 25(6), 1187–1211. DOI: 10.1080/14719037.2021.2011389

Wang, L., & Murdock, S. (2018). Community engagement in green infrastructure projects: Lessons from case studies. *Journal of Environmental Planning and Management*, 61(5-6), 865–881.

Wang, Q., Su, M., & Li, R. (2020). The impact of the digital economy on the efficiency of environmental governance in China. *Journal of Cleaner Production*, 276, 123716. DOI: 10.1016/j.jclepro.2020.123716

Wang, S., Li, J., & Razzaq, A. (2023). Do environmental governance, technology innovation and institutions lead to lower resource footprints: An imperative trajectory for sustainability. *Resources Policy*, 80, 103142. DOI: 10.1016/j.resourpol.2022.103142

Wang, W., Li, Y., Lu, N., Wang, D., Jiang, H., & Zhang, C. (2020). Does increasing carbon emissions lead to accelerated eco-innovation? Empirical evidence from China. *Journal of Cleaner Production*, 251, 119690. DOI: 10.1016/j.jclepro.2019.119690

Wang, W., Sun, X., & Zhang, Y. (2021). Leveraging IoT and digital marketing for promoting sustainable environmental practices. *Journal of Cleaner Production*, 295, 126409.

Wang, Z., Yu, Y., Roy, K., Gao, C., & Huang, L. (2023). The Application of Machine Learning: Controlling the Preparation of Environmental Materials and Carbon Neutrality. *International Journal of Environmental Research and Public Health*, 20(3), 1871. DOI: 10.3390/ijerph20031871 PMID: 36767237

Warner, J. F. (2006). More sustainable participation? Multi-stakeholder platforms for integrated catchment management. *International Journal of Water Resources Development*, 22(1), 15–35. DOI: 10.1080/07900620500404992

Water, U. N. (2018). Policy Brief on Water Governance. https://shorturl.at/acozK

WCED (World Commission on Environment and Development). (1987). *Report of the World Commission on Environment and Development: Our Common Future*. United Nations.

Wehn de Montalvo, U., & Alaerts, G. (2013). Leadership in knowledge and capacity development in the water sector: A status review. *Water Policy*, 15(S2), 1–14. DOI: 10.2166/wp.2013.109

Weill, P., & Woerner, S. L. (2018). Is your company ready for a digital future? *MIT Sloan Management Review*, 59(2), 21–25.

Weiss, T. G., & Wilkinson, R. (Eds.). (2023). *International organization and global governance* (3rd ed.). Routledge., DOI: 10.4324/9781003266365

Whig, P., Sharma, P., Aneja, N., Elngar, A. A., & Silva, N. (Eds.). (2024). *Artificial Intelligence and Machine Learning for Sustainable Development: Innovations, Challenges, and Applications* (1st ed.). CRC Press., DOI: 10.1201/9781003497189

WHO & UNICEF. (2019). WASH in Health Care Facilities: Global Baseline Report 2019. WHO & UNICEF, Geneva.

WHO. (2018). Burden of disease from ambient air pollution for 2016. https://cdn.who.int/media/docs/default-source/air-pollution-documents/air-quality-and-health/aap_bod_results_may2018_final.pdf

WHO. (2024). Health Consequences of air population on populations. https://www.who.int/news/item/25-06-2024-what-are-health-consequences-of-air-pollution-on-populations

WHO/UNICEF Joint Water Supply & Sanitation Monitoring Programme. (2015). *Progress on sanitation and drinking water: 2015 update and MDG assessment*. World Health Organization.

Wichai-Utcha, N., & Chavalparit, O. (2019). 3Rs Policy and plastic waste management in Thailand. *Journal of Material Cycles and Waste Management*, 21(1), 10–22. DOI: 10.1007/s10163-018-0781-y

Wiedenhofer, D., Guan, D., Liu, Z., Meng, J., Zhang, N., & Wei, Y.-M. (2017). 2017/01/01). Unequal household carbon footprints in China. *Nature Climate Change*, 7(1), 75–80. DOI: 10.1038/nclimate3165 PMID: 29375673

Wiig, H., & Wood, M. (1995). *What Comprises a Regional Innovation System? An empirical study*. Studies in Technology, Innovation and Economic Policy Group.

Wijman, T. (2022). *The Games Market Will Show Strong Resilience in 2022, Growing by 2.1% to Reach $196.8 Billion* Newzoo. Retrieved Aug 6th from https://newzoo.com/insights/articles/the-games-market-will-show-strong-resilence-in-2022

Willem, A., & Lucidarme, S. (2014). Pitfalls and challenges for trust and effectiveness Designed to Succeed 21 The Korean Journal of Policy Studies on collaborative networks. *Public Management Review*, 16(5), 733–760. DOI: 10.1080/14719037.2012.744426

Williamson, O. E. (1985). *The Economic institutions of capitalism*. Oxford University Press.

Williamson, O. E. (1991). Comparative economic organization: The analysis of discretestructural alternatives. *Administrative Science Quarterly*, 36(2), 269–296. DOI: 10.2307/2393356

Williamson, O. E. (1996). *The mechanisms of governance*. Oxford UniversityPress. DOI: 10.1093/oso/9780195078244.001.0001

Williams, T., Boucher, J., & Bowers, C. (2021). Integrating IoT technology for environmental sustainability: Insights from a case study. *Journal of Environmental Management*, 280, 111855. Advance online publication. DOI: 10.1016/j.jenvman.2021.111855

Wilson, D. C., Velis, C. A., & Rodic, L. (2013). Integrated Sustainable Waste Management in Developing Countries. *Proceedings of the Institution of Civil Engineers: Waste Management Resource*, 166, 52–68.

WIR (The World Inequality Report). (2022). https://wir2022.wid.world/.

Wolfert, S., Ge, L., Verdouw, C., & Bogaardt, M. J. (2017). Big data in smart farming – A review. *Agricultural Systems*, 153, 69–80. DOI: 10.1016/j.agsy.2017.01.023

Woo, Junghoon & Asutosh, Ashish & Li, Jiaxuan & Ryor, Wolfgang & Charles, & Kibert, Charles & Shojaei, Alireza. (2020). Blockchain: A Theoretical Framework for Better Application of Carbon Credit Acquisition to the Building Sector. .DOI: 10.1061/9780784482858.095

Woodhill, J., Kishore, A., Njuki, J., Jones, K., & Hasnain, S. (2022). Food systems and rural wellbeing: Challenges and opportunities. *Food Security*, 14(5), 1099–1121. DOI: 10.1007/s12571-021-01217-0 PMID: 35154517

World Bank Group. (2019). *Public-Private Partnerships in Infrastructure: Lessons Learned from Recent Experience*. World Bank Group.

World Bank. (2006). *Alandur Sewerage Project: The Tamil Nadu Experience. Water and Sanitation Program Field Note*. Washington, DC: World Bank. JICA. (2016). *Berhampur Solid Waste Management Project: A PPP Model for Sustainable Urban Development*. Japan International Cooperation Agency.

World Bank. (2018). *Enhancing opportunities for clean and resilient growth in urban Bangladesh: Country environmental analysis 2018*. The World Bank Group.

World Bank. (2018). *Water scarce cities: Thriving in a finite world*. World Bank.

World Bank. (2018). *What a Waste 2.0: A Global Snapshot of Solid Waste Management to 2050*. Urban Development Series. World Bank., DOI: 10.1596/978-1-4648-1329-0

World Bank. (2019). *Regulatory Governance for Infrastructure Sector Investment: Synthesis Report*. World Bank.

World Bank. (2021). State and Trends of Carbon Pricing 2021. Retrieved from https://openknowledge.worldbank.org/handle/10986/36316

World Dental Federation FDI. Sustainability in dentistry. available at https://www.fdiworlddental.org/sustainability-dentistry

World Economic Forum. (2019). Public-Private Collaboration for Sustainable Infrastructure: Principles and Toolkit. Retrieved from http://www3.weforum.org/docs/WEF_Public_Private_Collaboration_Sustainable_Infrastructure_Report_2019.pdf

World Economic Forum. (2020) Global gender gap report 2020, Retrieved from https://www.weforum.org/reports/gender-gap-2020-report-100-years-pay-equality

World Health Organization. (2022). *Global oral health status report: towards universal health coverage for oral health by 2030*. World Health Organization.

World Water Forum. (2000) Final report: second world water forum and ministerial conference. The Hague, Ministry of Foreign Affairs.

Wright, C., & Nyberg, D. (2021). Climate change, capitalism, and corporations: Processes of creative self-destruction. *Organization Studies*, 42(5), 747–766.

Wright, N. S., & Reames, T. G. (2020). Unraveling the links between organizational factors and perceptions of community sustainability performance: An empirical investigation of community-based nongovernmental organizations. *Sustainability (Basel)*, 12(12), 4986. DOI: 10.3390/su12124986

Wu, J. S., & Lee, J. J. (2015). 2015/05/01). Climate change games as tools for education and engagement. *Nature Climate Change*, 5(5), 413–418. DOI: 10.1038/nclimate2566

Wu, J., Wang, S., Yang, X., & Li, S. (2022). Blockchain and IoT-based smart ESG reporting platform architecture: A conceptual framework. *Sustainability*, 14(5), 2453. Advance online publication. DOI: 10.3390/su14052453

Wu, P., & Tan, M. (2012). Challenges for sustainable urbanization: A case study of water shortage and water environment changes in Shandong, China. *Procedia Environmental Sciences*, 13, 919–927. DOI: 10.1016/j.proenv.2012.01.085

Wu, Y., & Tham, J. (2023). The impact of environmental regulation, Environment, Social and Government Performance, and technological innovation on enterprise resilience under a green recovery. *Heliyon*, 9(10), e20278. DOI: 10.1016/j.heliyon.2023.e20278 PMID: 37767495

Xiaoman, Z., Shanbing, L., & Shengchao, Y. (2023). How does the digitization of government environmental governance affect environmental pollution? spatial and threshold effects, Journal of Cleaner Production, Volume 415,2023,137670,ISSN 0959-6526,https://doi.org/DOI: 10.1016/j.jclepro.2023.137670

Xing, G., Zhang, Y., & Guo, J. (2023, March 10). Environmental Regulation in Evolution and Governance Strategies. *International Journal of Environmental Research and Public Health*, 20(6), 4906. DOI: 10.3390/ijerph20064906 PMID: 36981813

Xinzhong, Y. (2017). Thinking Environmentally: Introduction to the Special Issue on Environmental Ethics. *Frontiers of Philosophy in China*, 12(2), 191–194.

Xu, C., Chen, X., & Dai, W. (2022). Effects of Digital Transformation on Environmental Governance of Mining Enterprises: Evidence from China. *International Journal of Environmental Research and Public Health*, 19(24), 16474. DOI: 10.3390/ijerph192416474 PMID: 36554353

Xu, J. Y. (2021). Impact of urbanization on ecological efficiency in China: An empirical analysis based on provincial panel data. *Ecological Indicators*, 129, 107827. DOI: 10.1016/j.ecolind.2021.107827

Xu, J., She, S., & Liu, W. (2022). Role of digitalization in environment, social and governance, and sustainability: Review-based study for implications. *Frontiers in Psychology*, 13, 961057. DOI: 10.3389/fpsyg.2022.961057 PMID: 36533022

Xu, L. D., He, W., & Li, S. (2014). Internet of Things in industries: A survey. *IEEE Transactions on Industrial Informatics*, 10(4), 2233–2243. DOI: 10.1109/TII.2014.2300753

Yang, H., Ma, M., Thompson, J. R., & Flower, R. J. (2018). Waste management, informal recycling, environmental pollution and public health. *Journal of Epidemiology and Community Health*, 72(3), 237–243. DOI: 10.1136/jech-2016-208597 PMID: 29222091

Yan, Z., Yu, Y., Du, K., & Zhang, N. (2024). How does environmental regulation promote green technology innovation? Evidence from China's total emission control policy. *Ecological Economics*, 219, 108137. DOI: 10.1016/j.ecolecon.2024.108137

Yeo, M. A., (2021). Public-private partnerships for sustainable urban green infrastructure: A case study of Seoul, South Korea. *Sustainability*, 13(1), 381.

Yeow, K. E., & Ng, S.-H. (2021). The impact of green bonds on corporate environmental and financial performance. *Managerial Finance*, 47(10), 1486–1510. DOI: 10.1108/MF-09-2020-0481

Yeung, C. A., Chong, L. Y., & Glenny, A. M. (2015). *Fluoridated milk for preventing dental caries*. Cochrane. DOI: 10.1002/14651858.CD003876.pub3

Young, O. R.. (2014). *The Institutional Dimensions of Global Environmental Change: Science, Economics, Politics, and Ethics*. MIT Press.

Yousaf, Z., Radulescu, M., Sinisi, C. I., Serbanescu, L., & Păunescu, L. M. (2021). Towards Sustainable Digital Innovation of SMEs from the Developing Countries in the Context of the Digital Economy and Frugal Environment. *Sustainability (Basel)*, 13(10), 5715. DOI: 10.3390/su13105715

Yu, C.-Y., & Chiang, Y.-C. (2017). Designing a climate-resilient environmental curriculum—A transdisciplinary challenge. *Sustainability (Basel)*, 10(1), 77. DOI: 10.3390/su10010077

Yue, P., Wei, W., Shangsong, Z., & Yunqiang, L. (2024). Does digitalization help green consumption? Empirical test based on the perspective of supply and demand of green products, Journal of Retailing and Consumer Services, Volume 79,2024,103843,ISSN 0969-6989,https://doi.org/DOI: 10.1016/j.jretconser.2024.103843

Yuen, K. W., Switzer, A. D., Teng, P. P., & Lee, J. S. H. (2024). Statistics on Typhoon Intensity and Rice Damage in Vietnam and the Philippines. *GeoHazards*, 5(1), 22–37. DOI: 10.3390/geohazards5010002

Zameer H, Yasmeen H (2022) *Green innovation and environmental awareness driven green purchase intentions*. Mark Intell Plan (ahead-of-print). Rodrigo Ramos Hospodar Felippe Valverde https://doi.org/DOI: 10.11606/issn.2179-0892.geousp.2023.194172.es

Zanella, A., Bui, N., Castellani, A., Vangelista, L., & Zorzi, M. (2014). Internet of Things for smart cities. *IEEE Internet of Things Journal*, 1(1), 22–32. DOI: 10.1109/JIOT.2014.2306328

Zarsky, Lyuba, & Tay, Simon SC. (2000), Civil Society and the Future of Environmental Governance in Asia, https://nautilus.org/eassnet/civil-society-and-the-future-of-environmental-governance-in-asia/

Zeng, Y., Zhang, Y., & Yang, J. (2021). The role of artificial intelligence in sustainable agriculture: A review. *Journal of Cleaner Production*, 307, 127164.

Zhang, J., Qin, Q., Li, G., & Tseng, C. H. (2021). Sustainable municipal waste management strategies through life cycle assessment method: A review. *Journal of Environmental Management*, 287, 112238. DOI: 10.1016/j.jenvman.2021.112238 PMID: 33714044

Zhang, L., & White, M. (2021). Green finance: A pathway to sustainable development? *Journal of Sustainable Finance & Investment*, 11(3), 205–217. DOI: 10.1080/20430795.2020.1868752

Zhang, N., Deng, J., Ahmad, F., Draz, M. U., & Abid, N. (2023). The dynamic association between public environmental demands, government environmental governance, and green technology innovation in China: Evidence from panel VAR model. *Environment, Development and Sustainability*, 25(9), 9851–9875. DOI: 10.1007/s10668-022-02463-8

Zhang, X., Aranguiz, M., Xu, D., Zhang, X., & Xu, X. (2018). Utilizing blockchain for better enforcement of green finance law and regulations. In *Transforming climate finance and green investment with blockchains* (pp. 289–301). Academic Press. DOI: 10.1016/B978-0-12-814447-3.00021-5

Zhang, Y., (2019). Barriers to the development of green infrastructure: Evidence from China. *Sustainability*, 11(2), 481.

Zhao, H. (2008). Empirical Analysis on the Impact of Environmental Regulation on Technological Innovation of Chinese Enterprises. *Manag. Mod.*, 3, 3–5.

Zhou, L., & Luo, Y. (2021). Digital marketing strategies for sustainable consumption: A review of existing research and future directions. *Journal of Business Research*, 127, 131–142. DOI: 10.1016/j.jbusres.2021.01.012

Zhou, X., & Cui, Y. (2019). Green bonds, corporate performance, and corporate social responsibility. *Sustainability (Basel)*, 11(23), 6881. DOI: 10.3390/su11236881

Zou, L., Xu, R., Wang, H., Wang, Z., Sun, Y., & Li, M. (2023). Chemical recycling of polyolefins: a closed-loop cycle of waste to olefins. China Science Publishing & Media. 10. DOI: 10.1093/nsr/nwad207

Zúniga-González, C. A. (2021). 2021, The impact of economic and political reforms on environmental performance in developing countries. *PLoS One*, 16(10), e0257631. DOI: 10.1371/journal.pone.0257631 PMID: 34610016

About the Contributors

Imran Hossain is a dedicated academician and researcher with a profound commitment to sustainable development and environmental governance. Currently, Imran serves as a lecturer in Political Science at Varendra University, Bangladesh. Beyond his academic role, Imran has delved into extensive research on the SDGs, collaborating with both national and renowned international organizations. Imran's research portfolio is diverse and impactful, encompassing projects on various aspects of sustainable development, including environmental governance, waste management, climate change, water management, and sustainable cities. Notably, he is at the forefront of exploring the intersection between sustainable development and Industry 4.0 tools, aiming to utilize technology to achieve sustainable solutions. His research interests span sustainable governance, smart technology, sustainable waste management, and IoT-based solutions for environmental sustainability. Imran actively engages in scholarly discourse by participating in national and international conferences. As a testament to his expertise, he serves as a reviewer for various scientific journals indexed in Scopus. Imran has published a number of research articles and book chapters indexed in Scopus and other renowned databases. Imran's dedication to advancing knowledge extends to editorial roles, where he currently serves as the editor of several scientific books published by IGI Global International and CRC Press, Taylor and Francis Group. Beyond academia, Imran is a prolific writer and columnist for national dailies, where he shares insights and perspectives on sustainable development, environmental governance, and smart technology.

A.K.M. Mahmudul Haque is a Professor at the Department of Political Science, University of Rajshahi. He did his PhD at the Institute of Bangladesh Studies, University of Rajshahi where he examined the policies, practices, and challenges of environmental governance of Bangladesh. Haque is a Visiting Scholar at the University of Victoria, Canada. He also attended a PhD Internship Program at the Centre for Asia Pacific Initiatives (CAPI) of the University of Victoria, Canada. His areas of specialization are environmental governance, climate change, sustainable development, and local governance. His key contributions focus on evaluating local government's environmental governance. Haque's dedication to advancing knowledge extends to editorial roles, where he currently serves as the editor of several scientific books published by IGI Global International and CRC Press, Taylor, and Francis Group. Dr. Haque has already carried out a number of research projects funded by different national and international organizations. Dr. Haque published more than 60 scientific papers in various national and international journals. He frequently contributes columns to national dailies in both Bengali and English. He supervised/is supervising the research of thirteen students, including three PhD fellows (already awarded their PhD). He has delivered numerous lectures mainly on environmental governance and sustainable development in about 20 conferences and seminars held at various universities domestically and overseas.

S.M. Akram Ullah is a prominent researcher and professor at the University of Rajshahi, specializing in political science with a strong focus on governance, sustainable development, and public policy in Bangladesh. His research portfolio is diverse and extensive, addressing pressing issues such as sustainable water management, environmental governance, human rights, and local government efficacy. His work often provides a comparative lens, analyzing policy and governance structures in Bangladesh within broader regional and global contexts. Dr. Ullah's publications appear in reputable international journals and edited volumes, many of which are indexed in databases like SCOPUS. His recent work explores the role of city governance institutions in water and sanitation management, linking these services to the broader goals of sustainable development. In addition to journal articles, Dr. Ullah has contributed book chapters in international series by publishers such as Springer and Palgrave Macmillan, focusing on sustainable waste management, climate change, and socio-economic impacts on urban environments. His research collaborations have led to critical insights into the practical challenges of implementing Sustainable Development Goals (SDGs) at the local government level, particularly in water and sanitation services. Dr. Ullah's work not only advances academic understanding of governance issues but also provides practical recommendations for policymakers in Bangladesh and other developing nations. His research emphasizes the importance of integrating technology and participatory governance to improve service delivery and enhance the accountability of local government bodies. Through his publications and ongoing research, Dr. Akram Ullah contributes significantly to both the academic field of political science and the practical landscape of governance in Bangladesh.

***

Divya Bansal, Associate Professor cum Deputy Director at Amity University Online, Noida, has over 19 years of experience in teaching Business Management at UG and PG levels. She is a dedicated educator passionate about student success. Her teaching areas include General Management, Entrepreneurship, Strategic Management, Marketing Management, and Organisational Behaviour, Dr. Bansal has presented research at national and international conferences and published articles in leading journals. Her research focuses on Big Data Analytics, Supply Chain Management, Blockchain Technology, and Cultural Marketing. She explores AI, disruptive innovation, and digital transformation in Business Management and is interested in new statistical tools. She has organized guest lectures, conferences, and workshops.

Naboshree Bhattacharya is an accomplished economist with a specialization in healthcare economics and a broader research focus on development economics. With a rich academic background, she holds alumni status from prestigious institutions including St. Xavier's College, Ranchi University, and Jadavpur University. Over the course of her career spanning 12 years, Naboshree has garnered extensive experience in both academic and industrial settings. Currently serving as an Assistant Professor at Amity University Jharkhand, she brings a wealth of knowledge and expertise to her role. Naboshree's contributions to the field extend beyond the classroom. She is a prolific author, having penned six books on various topics within economics, collaborating with both national and international authors. Additionally, her research output includes numerous papers and book chapters published in esteemed journals and publications, both domestically and internationally.

Md. Ikhtiar Uddin Bhuiyan is an Associate Professor in the Department of Government and Politics at Jahangirnagar University, Dhaka, Bangladesh. Right now, he is pursuing his PhD degree in Governance and Development under Graduate School of Public Administration, National Institute of Development Administration (NIDA), Bangkok, Thailand. Apart from this academic rule, he served as an advisor for Transparency International of Bangladesh (TIB), Yes Group, and Jahangirnagar University. Mr. Ikhtiar joined the department in 2016 as a lecturer. Since joining as a lecturer, he has been actively involved in research, teaching, and administrative activities with sincerity and dedication. Prior to joining the department, he worked in the Bangladesh Civil Service (BCS) since 2014, and before joining BCS, he also served in the Customs Department as a Customs Officer. In addition, he also worked in the Ministry of Information, Communication, and Technology (ICT) as a protocol officer. Moreover, he worked for the Investment Corporation of Bangladesh as a senior officer. In 2018, Mr. Ikhtiar went to Japan for his advanced-level second Master's degree on public policy and returned in September 2019 after successfully completing his Master's degree from the National Graduate Institute for Policy Studies (GRIPS), Tokyo, Japan. His main areas of research are Public Policy, South Asian Dynamics of Politics, and governance issues. He has published papers in journals such as the Springer Nature, USBED Journal of Turkey, The Jahangirnagar Review, the CENRAPS Journal of Social Science, and the Asian Studies Journal. He also presented seven research papers at different international conferences, including the International Political Science Association (IPSA) Congress in Portugal. Currently, he is a member of the International Political Science Association (IPSA) and the China-Bangladesh Mutual Friendship Trade Center Limited. Moreover, he is also a member of the Japanese Universities Alumni Association of Bangladesh (JUAAB).

A.N. Bushra is a dedicated young researcher who has excelled in her academic journey in Public Administration. With both Bachelor's and Master's degrees achieved with distinction, Bushra demonstrates outstanding analytical prowess and a strong dedication to addressing social disparities, particularly in gender equality, public policy, e-governance in developing nations, public healthcare, and social justice. Her current research focuses on pivotal areas shaping the future of public administration, including data-driven decision-making, the impact of Artificial Intelligence and automation on public services, smart governance initiatives, and government innovation. Published in prestigious journals, her work has ignited significant discourse and holds the potential to influence policy-making. As Bushra embarks on her career, her intellect, perseverance, and commitment position her as a promising scholar ready to contribute substantially to the social sciences.

Ruchi Gupta is an Energy Advisor in GIZ India. She is holding a Ph.D. in Environmental Science. Her focus area of work is renewable energy primarily solar and biomass energy. Her expertise in energy, biomass, and environmental management is widely recognized. Dr. Gupta's dedication to advancing knowledge and practices in these areas underscores her commitment to sustainable development and environmental preservation. Her significant academic and professional achievements position her as a prominent figure in her field, reflecting a strong commitment to sustainable solutions and environmental stewardship.

Shafiul Islam After completing undergraduate with honours and post-graduation in Public Administration from the University of Dhaka, Dr. Shafiul Islam joined at the Department of Public Administration, University of Rajshahi in 2006 and presently working as a Professor in the same department. He obtained his Doctor of Philosophy (Ph. D.) degree in 2015 for focusing on good governance in Bangladesh. He has participated in many national and international seminars, workshops and conferences at home and abroad and presented scholarly papers. About 30 book chapters and articles have been published in national and international peer-reviewed journals and books. In different capacities, He has already conducted as many as 30 research projects funded by the World Bank (WB), UNESCO, Ministry of Public Administration, Ministry of Information and Broadcasting, Ministry of Education, NAEM, UGC, Local Government Institute (LGI) under the Ministry of LGRD & Cooperative and University of Rajshahi Bangladesh. He has presented about 40 scholarly research papers at international conferences at home and abroad. Before joining the university, he was engaged in practical and professional journalism for more than 10 years and worked in different capacities such as Dhaka University Correspondent, and Staff Correspondent at different national Daily English and Bengali newspapers published from Dhaka. His research interest areas are public administration, public policy, governance, development, waste management, higher education, SDG, 4iR, media, youth development, social development, rural development, gender development, disaster management and governance, local government, public-private partnership, etc.

Manas Kumar Jha, holding an M.Sc. in Environment Management, is currently the Delivery Head at Hexagon Geosystem in India. A qualified University Grant Commission scholar, he has made significant contributions to the field through numerous research papers and books. His expertise in environmental management is well-recognized, reflecting a strong commitment to advancing knowledge and practices in this crucial area. Jha's academic and professional achievements underscore his dedication to sustainable development and environmental preservation, positioning him as a prominent figure in his field.

Jipson Joseph has a PhD in Ethics from DVK Bangalore. Currently, he is a Research Scholar in the School of Law at Christ University, Bangalore. His research interests are in the fields of ethics, gender studies, constitutional law, human rights, human security, environmental justice, developmental policies and global poverty.

Anitha.K presently serves as the Head of the Department of Management Studies at Meenakshi Academy of Higher Education and Research (MAHER), located in India. In 2014, she earned her UGC NET Management & HRM qualification. Her research pursuits encompass AI applications in HRM, marketing, and sustainability. Dr. Anitha.K has authored numerous chapters in books published by Springer, Taylor & Francis CRC Press, Nova Science Publishers, USA and IGI Global, USA. She has 6 years of teaching experience and 4 years of industry experience and has actively participated in various conferences, consistently securing Best Paper awards for her contributions. She is the author of 1 Book and serves as an editor in one of the upcoming book "Fintech Innovation – Practise and Progress in AI and ML Techniques for Finance" by Bentham Science Publishers, Singapore. She is a Life Member and executive member of the Education Research and Development Association (ERDA). Additionally, she has served as a resource person for faculty development programs and guest lectures. Within MAHER, she holds dual roles as Placement Officer and Assistant Director of Online Education. Notably, she has been honored with the Research Excellence Award consecutively since 2020 at MAHER.

Yigang Liu, who holds a PhD in Design Theory from Shanghai University, is a lecturer at Art and Design College, Nanjing Forestry University. Dr Yigang is currently a board member of CDiGRA (Chinese Digital Game Research Association). His research centres on the academic discipline of creative industry studies, encompassing areas such as digital media art and design, game culture and industry, and social media studies.

Dilip Kumar Markandey, Associate Director at the Central Pollution Control Board, is a distinguished scientist with a Ph.D. in Environmental Microbiology. His expertise in environmental management is highly regarded, demonstrating a profound commitment to advancing sustainable practices and knowledge. Dr. Markandey's academic and professional achievements highlight his dedication to environmental preservation and sustainable development, positioning him as a prominent figure in his field. His contributions reflect a strong commitment to protecting the environment and promoting sustainable practices, making significant impacts on environmental policies and management strategies.

Pranavi Mishra is a final-year MSc Biotechnology student whose recent academic work includes a project focused on the green synthesis, characterization, and dermatological applications of silver nanoparticles. . In addition to her scientific endeavors, Pranavi is an emerging fictional author, seamlessly integrating her analytical skills and creative vision to craft narratives that resonate with both scientific and literary audiences.

Moriom Akter is an advocate at Dhaka Judge Court and a lecturer in the Department of Law at Daffodil International University, Dhaka. She holds an LL.B and LL.M from Daffodil University and is currently pursuing a PhD in the Department of Government and Politics at Jahangirnagar University. With a research focus on Public, Community, and Environmental Art, Moriom actively publishes articles on environmental issues, child sexual abuse, cyberbullying, and human rights in high-impact journals. Known for her dedication to enhancing student employability, Moriom aims to transform Bangladesh's academic landscape to better equip students for the modern job market.

Indranil Mutsuddi is PhD in Management Studies from Amity University Uttar Pradesh. He is presently associated with the Department of Management Studies, JIS University, Kolkata as HOD and Associate Professor in HR. Dr. Indranil has more than 20 years of teaching experience as a faculty member in Human Resources Management, Organizational Behaviour and Behavioural Sciences. He had published 61 research papers, authored 3 text books in Human Resource Management and contributed 10 chapters in edited books.

Madhusudan Narayan is a seasoned marketing scholar with a Ph.D., an MBA in Marketing, and a PGD in International Business. His expertise spans Digital Marketing, Marketing Analytics, and emerging technologies such as Blockchain, AI & ML, IoT, and AR & VR. Dr. Narayan's extensive research, published in prestigious journals indexed in Scopus, Web of Science, and UGC-Care, underscores his commitment to academic excellence. He also plays a vital role in the academic community through his work on Editorial Boards and as a reviewer for national and international journals. With years of teaching experience, Dr. Narayan's dedication to advancing knowledge and driving innovation in Marketing and Emerging Technologies is evident in both his research and his contributions to the field.

Ananya Pandey is a Research Scholar in School of Law at Christ University, Bangalore. She has her research interests in the field of Constitutional Law, Human Rights, Intellectual Property Law and Jurisprudence.

Shashwata Sahu is a Guest Faculty at the University Law College, Utkal University, Bhubaneswar, India. She is pursuing her Ph.D. in Constitutional Law from the School of Law, Kalinga Institute of Industrial Technology (KIIT) University, Bhubaneswar, where she also awarded her LL.M Degree with a Constitutional and Administrative Law specialization. She has an extensive publication record, including research papers on climate change, green crime, and healthcare during the COVID-19 pandemic. Additionally, she has authored several law-related books and presented papers at numerous national and international conferences. Her research interests include Constitutional Law, Health Law, Environmental Law, and Digital Governance.

Sawmeem Sajia is a master's student in the Department of Government and Politics at Jahangirnagar University, Dhaka. Her main research interests include Education Policy, South Asian Politics, Foreign Policy, and Public Service Delivery. She is passionate about understanding how these areas affect people's lives and hopes to contribute to improving governance and policy in the region. Through her work, she aims to make a positive impact on public service and education systems in South Asia.

Ramesh Chandra Sethi, an Assistant Professor of Political Science and Public Policy at the Kalinga Institute of Industrial Technology (KIIT) University, Bhubaneswar, India, is known for his research that has practical applications. Before joining KIIT University, he worked as an Assistant Professor at the Central University of Jharkhand, Ranchi. He teaches Political Science to Undergraduate students and Systems of Governance and Public Policy to Postgraduate and Ph.D. students at the KIIT University. From 2013 to 2016, he worked at the Centre for Innovations in Public Systems (CIPS), an Autonomous Centre of the Administrative Staff College of India (ASCI), Hyderabad, established and funded by the Government of India. While working at CIPS, Dr. Sethi worked directly with the Department of Health and Family Welfare of various State Governments in India. His research focuses on the theoretical-empirical study of the policy process, systems of governance, public policy, health policy, and rural development. Dr. Sethi has published research papers in reputed national and international journals. Some of his research publications span Law and Public Policy, Public Health, Indigenous People and the Environment, Climate Change, and Sustainable Development Goals.

Abrar Shafi is an MBBS student at the Sir Salimullah Medical College, Dhaka, Bangladesh. He has already participated in many research seminars, workshops and conferences. He is keen on medical practice and research.

Shahriar Shams is currently working as an Assistant Professor in Civil Engineering Programme Area, Faculty of Engineering, Universiti Teknologi Brunei (UTB) since December, 2012. Before joining UTB, he worked as an Assistant Professor in the Department of Civil and Environmental Engineering at Islamic University of Technology (IUT), a subsidiary organ of OIC. He obtained his PhD from The University of Manchester (UK) in 2008 and his M.Sc. in Environmental Engineering and Sustainable Infrastructure from Royal Institute of Technology (KTH) Sweden sponsored by Swedish International Development Agency (SIDA). He has published over 50 refereed journals and conference papers, including Book Chapters. His research interest includes SWM, IWRM and Climate Change.

Index

Symbols

3R Strategy 122, 144, 465, 466, 467, 468, 479, 481

A

Artificial Intelligence 34, 47, 52, 145, 238, 240, 241, 256, 257, 260, 263, 266, 267, 270, 271, 272, 275, 278, 282, 284, 285, 289, 294, 295, 299, 304, 305, 317, 319, 321, 381, 382, 384, 385, 389, 394, 403, 406, 407, 409, 410, 411, 441

Awareness 10, 12, 16, 19, 21, 33, 48, 53, 107, 121, 122, 127, 128, 129, 131, 132, 133, 136, 138, 139, 141, 142, 146, 155, 160, 194, 195, 212, 215, 217, 218, 221, 222, 224, 226, 252, 253, 283, 291, 311, 349, 362, 365, 391, 399, 400, 402, 403, 414, 417, 418, 420, 424, 425, 427, 464, 468, 470, 472, 473, 477, 479, 482, 483, 491, 495, 500

B

Bangladesh 31, 95, 107, 114, 119, 120, 121, 122, 123, 125, 126, 128, 129, 135, 136, 137, 138, 139, 142, 143, 144, 145, 146, 181, 182, 183, 184, 185, 188, 189, 190, 191, 193, 194, 195, 196, 197, 198, 199, 200, 201, 202, 203, 204, 205, 206, 207, 208, 209, 323, 330, 355, 356, 357, 381, 407, 414, 430, 465, 466, 467, 473, 477, 478, 480, 485, 486, 487, 488, 489

Big Data 228, 234, 235, 240, 266, 267, 268, 270, 271, 272, 277, 278, 279, 282, 283, 284, 287, 294, 298, 300, 303, 304, 308, 310, 312, 319, 382, 385, 387, 400, 459

C

Capacity Development 413, 418, 419, 421, 423, 424, 425, 427, 428, 429, 431, 432, 433

Case Studies 11, 29, 34, 48, 49, 82, 149, 151, 152, 153, 154, 161, 162, 167, 168, 174, 175, 178, 211, 398, 424, 452

Challenges 1, 2, 4, 8, 9, 10, 12, 13, 14, 15, 17, 18, 22, 24, 26, 29, 34, 35, 36, 37, 41, 42, 49, 50, 51, 52, 53, 54, 55, 71, 73, 74, 76, 81, 89, 112, 113, 115, 117, 119, 121, 122, 123, 125, 128, 129, 131, 132, 133, 134, 135, 136, 137, 138, 139, 140, 141, 142, 143, 145, 150, 151, 152, 153, 154, 155, 156, 161, 163, 164, 165, 167, 168, 169, 172, 174, 176, 178, 181, 182, 183, 184, 185, 186, 188, 189, 190, 192, 194, 195, 197, 198, 199, 201, 202, 203, 204, 205, 206, 207, 208, 209, 211, 212, 213, 214, 215, 216, 218, 219, 220, 221, 223, 224, 226, 228, 230, 231, 233, 237, 238, 239, 240, 241, 242, 243, 245, 248, 249, 250, 251, 252, 253, 255, 256, 259, 260, 261, 262, 265, 266, 267, 270, 272, 274, 279, 281, 285, 289, 292, 294, 295, 299, 300, 302, 305, 306, 309, 310, 311, 312, 315, 323, 324, 325, 326, 327, 329, 331, 334, 336, 338, 341, 342, 344, 345, 349, 352, 354, 355, 357, 362, 365, 373, 378, 381, 382, 383, 384, 385, 386, 387, 388, 389, 390, 395, 396, 399, 400, 401, 402, 404, 408, 409, 410, 413, 414, 416, 418, 422, 425, 426, 427, 428, 429, 431, 436, 438, 453, 457, 465, 466, 467, 470, 471, 477, 478, 479, 480, 481, 482, 486, 487, 488, 489, 498, 507, 509

Circular Economy 34, 52, 54, 55, 57, 61, 97, 116, 144, 214, 260, 262, 267, 318, 381, 382, 383, 385, 386, 388, 393, 394, 395, 396, 397, 400, 404, 405, 406, 407, 408, 409, 410, 462, 468, 486, 487, 489, 509

City Corporation 124, 145, 183, 184, 186, 189, 191, 192, 193, 196, 201, 203, 204, 205, 206, 407, 465, 467, 473, 474, 475, 482, 486, 487

Climate Change 1, 2, 7, 8, 10, 11, 19, 22, 24, 25, 27, 28, 31, 32, 35, 36, 37, 38, 39, 40, 41, 42, 43, 44, 45, 47, 52, 53, 54, 55, 56, 57, 58, 59, 60, 92, 98, 110, 114, 120, 150, 186, 197, 198, 199, 200, 203, 205, 209, 212, 217, 230, 231, 238, 245, 248, 250, 260, 263, 267, 274, 275, 278, 279, 288, 289, 292, 296, 297, 298, 300, 302, 307, 313, 315, 319, 321, 323, 324, 325, 326, 329, 334, 344, 346, 348, 349, 350, 352, 354, 355, 356, 357, 359, 360, 361, 362, 363, 364, 365, 366, 367, 368, 369, 371, 372, 374, 375, 376, 377, 378, 379, 382, 383, 384, 385, 386, 387, 389, 392, 396, 399, 402, 404, 409, 410, 411, 413, 414, 425, 430, 467, 491, 492, 493, 494, 498, 505, 508

Climate Change Mitigation 2, 10, 38, 41, 52, 53, 307, 359, 361

Climate Resilience 38, 42, 323, 324, 325, 326, 327, 328, 329, 330, 331, 332, 333, 334, 335, 336, 337, 338, 339, 340, 341, 342, 343, 344, 345, 346, 347, 348, 349, 350, 351, 352, 353, 356, 357

Collective Responsibility 84

Consultation 63, 64, 65, 66, 76, 77, 83, 166, 168, 170, 497

Curriculum 19, 323, 324, 325, 327, 328, 330, 332, 333, 334, 335, 336, 337, 338, 339, 340, 341, 342, 343, 344, 346, 350, 351, 352, 353, 354, 355, 356, 357, 491, 501, 508

D

Data Analytics 18, 212, 228, 234, 240, 241, 242, 265, 266, 267, 268, 271, 273, 277, 278, 279, 280, 282, 284, 285, 287, 288, 290, 294, 297, 298, 300, 303, 304, 308, 310, 315, 382, 385, 400, 414, 415, 487

Digital Marketing 211, 212, 213, 215, 216, 217, 218, 219, 220, 221, 222, 223, 224, 225, 226, 227, 228, 229, 231, 232, 233, 234, 235

Digital Technologies 214, 230, 240, 242, 249, 270, 314, 382, 389, 400, 403, 404, 419

E

Ecological Movements 91, 112

Emission 29, 115, 251, 255, 292, 319, 387, 390, 391, 392, 494, 504

Environment 2, 3, 4, 5, 6, 7, 8, 10, 11, 13, 14, 15, 16, 18, 19, 24, 25, 27, 28, 29, 32, 33, 35, 36, 37, 41, 44, 45, 46, 48, 49, 55, 56, 57, 58, 59, 60, 67, 75, 87, 88, 91, 92, 94, 95, 96, 97, 100, 101, 102, 103, 104, 105, 106, 108, 109, 110, 111, 112, 113, 115, 116, 117, 118, 120, 121, 144, 145, 146, 150, 152, 153, 155, 157, 158, 160, 161, 163, 164, 165, 172, 176, 177, 178, 184, 185, 192, 194, 201, 204, 207, 208, 209, 213, 233, 234, 235, 238, 250, 259, 260, 261, 262, 264, 268, 271, 272, 276, 277, 278, 279, 280, 282, 283, 288, 289, 290, 291, 295, 296, 297, 298, 299, 301, 302, 307, 309, 315, 316, 317, 318, 319, 327, 329, 346, 348, 355, 356, 362, 366, 372, 383, 390, 391, 392, 395, 397, 398, 399, 400, 401, 406, 407, 408, 410, 416, 417, 418, 419, 420, 421, 424, 429, 430, 431, 432, 435, 437, 440, 443, 446, 447, 449, 452, 453, 454, 457, 461, 466, 467, 468, 470, 472, 474, 475, 477, 478, 479, 481, 482, 483, 484, 486, 491, 492, 493, 494, 495, 497, 502, 503, 505, 509

Environmental Governance 1, 2, 3, 8, 9, 10, 11, 12, 13, 14, 15, 16, 17, 18, 19, 20, 21, 22, 24, 26, 29, 32, 33, 34, 35, 36, 37, 41, 42, 44, 45, 46, 47, 49, 50, 54, 55, 56, 57, 58, 59, 60, 83, 114, 119, 120, 122, 211, 212, 213, 214, 215, 216, 219, 220, 221, 222, 223,

224, 225, 226, 228, 231, 233, 234, 235, 237, 238, 239, 240, 242, 243, 245, 248, 249, 252, 253, 254, 255, 256, 257, 258, 259, 260, 261, 262, 264, 265, 268, 269, 270, 271, 273, 274, 275, 276, 280, 284, 285, 287, 288, 289, 290, 291, 292, 294, 295, 296, 297, 298, 300, 301, 302, 306, 307, 309, 310, 311, 312, 313, 314, 315, 316, 317, 318, 319, 321, 381, 382, 383, 384, 385, 386, 387, 389, 398, 400, 402, 403, 404, 405, 411, 413, 417, 418, 419, 425, 429, 491, 494
Environmental Impact 20, 82, 98, 114, 128, 142, 160, 161, 214, 223, 224, 225, 238, 248, 249, 254, 256, 257, 275, 301, 306, 315, 386, 393, 394, 400, 405, 466, 468, 480, 494, 495, 496, 504, 505, 508
Environmental Law 26, 40, 52, 58, 59
ESG 25, 26, 52, 217, 229, 234, 235, 268, 269, 270, 271, 272, 273, 283, 321, 391, 392

F

Future Trends 52, 381, 387

G

Gamification 359, 364, 365, 366, 367, 368, 369, 371, 372, 375, 376, 379
Global Governance 114, 146, 252, 253
Governance 1, 2, 3, 8, 9, 10, 11, 12, 13, 14, 15, 16, 17, 18, 19, 20, 21, 22, 24, 25, 26, 27, 29, 32, 33, 34, 35, 36, 37, 39, 41, 42, 43, 44, 45, 46, 47, 49, 50, 51, 52, 53, 54, 55, 56, 57, 58, 59, 60, 61, 63, 64, 65, 66, 67, 68, 69, 70, 71, 72, 73, 74, 75, 76, 77, 78, 79, 80, 81, 82, 83, 84, 85, 86, 87, 88, 89, 110, 114, 119, 120, 122, 129, 133, 134, 135, 137, 139, 141, 142, 144, 146, 154, 155, 162, 164, 165, 169, 176, 177, 178, 184, 185, 189, 199, 203, 204, 205, 206, 208, 209, 211, 212, 213, 214, 215, 216, 217, 218, 219, 220, 221, 222, 223, 224, 225, 226, 227, 228, 229, 231, 232, 233, 234, 235, 237, 238, 239, 240, 242, 243, 245, 248, 249, 252, 253, 254, 255, 256, 257, 258, 259, 260, 261, 262, 263, 264, 265, 268, 269, 270, 271, 272, 273, 274, 275, 276, 280, 283, 284, 285, 287, 288, 289, 290, 291, 292, 294, 295, 296, 297, 298, 300, 301, 302, 305, 306, 307, 308, 309, 310, 311, 312, 313, 314, 315, 316, 317, 318, 319, 320, 321, 356, 357, 381, 382, 383, 384, 385, 386, 387, 388, 389, 390, 391, 397, 398, 400, 401, 402, 403, 404, 405, 407, 408, 409, 411, 413, 414, 415, 417, 418, 419, 421, 423, 424, 425, 426, 429, 431, 432, 449, 467, 469, 470, 471, 472, 475, 483, 485, 486, 491, 494
Government Policies 149, 150, 151, 152, 158, 168, 170, 172, 173, 177, 424, 447, 454
Green Finance 34, 52, 381, 382, 386, 388, 390, 391, 392, 400, 403, 404, 410, 411, 442, 443, 455
Green Socio-Technological Innovation 435, 436, 438, 440, 452, 453

H

Human Rights 40, 66, 104, 106, 109, 110, 111, 115, 116, 205, 304, 410
Human Security 91, 92, 94, 95, 99, 100, 101, 103, 104, 105, 106, 107, 108, 109, 110, 111, 112, 113, 114, 115, 116, 117

I

Inadequate Infrastructure 121, 122, 133, 181, 185, 186, 188, 199, 467
Institutional Capacity 14, 164, 166, 170, 172, 177, 188, 199, 475
Integrated Water Resources Management 200, 413, 414, 415, 417, 425, 428, 430, 431
International Agreements 33, 36, 38, 39, 41, 52, 120, 293, 302, 303, 317, 321,

383, 384, 400, 402
IoT 52, 145, 146, 204, 211, 212, 213, 214, 215, 216, 217, 218, 219, 220, 221, 222, 223, 224, 225, 226, 227, 228, 229, 230, 231, 232, 233, 234, 235, 240, 263, 265, 266, 268, 270, 274, 275, 276, 277, 278, 280, 282, 284, 295, 298, 299, 300, 301, 303, 304, 306, 307, 308, 310, 312, 315, 329, 347, 348, 400, 407, 415, 486, 487, 489

L

Local Governance 74, 75, 78, 139, 142, 308, 408, 467, 469, 470, 471, 472, 485
Low-Carbon Green Infrastructure 150, 151, 152, 153, 154, 155, 156, 173

M

Machine Learning 263, 265, 266, 267, 270, 271, 274, 275, 276, 279, 280, 282, 283, 284, 285, 289, 294, 295, 299, 300, 304, 308, 385, 389, 394, 400, 410, 415
Magic Circle 359, 362, 363, 364, 368, 369, 372
Metabolism Framework 368, 370, 371, 372
Municipality 82, 125, 130, 131, 134, 145, 146, 479

N

National Legislation 41
Non-State Actors 84, 105

P

Play Theory 362
Policy Design 22, 91, 93, 101, 254
Policy Framework 95, 99, 102, 103, 105, 106, 109, 111, 115, 150, 151, 177, 392
Policy Implementation 15, 19, 20, 34, 44, 50, 51, 152, 155, 206, 207, 289, 471
Policymaking 19, 45, 56, 248, 265, 295, 296, 382, 435, 436, 438, 444, 445, 448, 453

Pollution 1, 2, 4, 5, 6, 7, 8, 12, 13, 15, 19, 20, 21, 22, 24, 25, 26, 28, 29, 31, 32, 35, 37, 41, 42, 43, 44, 45, 52, 53, 54, 57, 60, 94, 95, 96, 97, 100, 101, 105, 106, 107, 110, 112, 116, 118, 120, 121, 131, 133, 140, 145, 182, 192, 197, 203, 204, 205, 207, 213, 215, 216, 217, 221, 222, 226, 259, 260, 263, 264, 270, 276, 277, 279, 280, 284, 285, 287, 288, 290, 291, 292, 293, 299, 300, 307, 317, 382, 384, 389, 390, 391, 392, 393, 396, 404, 407, 410, 414, 429, 477, 480, 487, 491, 492, 494, 495, 497, 500, 501, 504
Private Sector Investment 135, 153, 155, 157, 158, 159, 160, 161, 162, 163, 166, 171, 172, 176, 177
Public Participation 9, 10, 16, 20, 22, 27, 33, 38, 63, 64, 65, 66, 67, 68, 69, 70, 71, 72, 73, 74, 75, 76, 77, 78, 79, 80, 84, 85, 155, 156, 166, 242, 287, 427, 469
Public Participation Spaces 63, 64, 70

R

Regulations 2, 3, 10, 11, 16, 19, 22, 24, 32, 33, 35, 40, 41, 43, 48, 52, 55, 67, 77, 79, 111, 125, 133, 139, 149, 152, 154, 163, 165, 166, 167, 177, 194, 195, 213, 221, 251, 290, 293, 297, 298, 300, 303, 310, 313, 315, 384, 387, 390, 392, 397, 403, 411, 417, 424, 425, 440, 442, 443, 444, 451, 452, 454, 455, 461, 469, 478, 483, 502

S

Safe Water Unavailability 196
Semi-urban 119, 121, 125, 128, 129, 130, 135, 136, 140, 141, 142
Socio-Ecology 440
Stakeholder Collaboration 47, 388
Stakeholder Engagement 16, 20, 44, 50, 151, 154, 155, 156, 163, 164, 166, 168, 170, 172, 175, 215, 220, 226, 245, 249, 388
Strategy 34, 38, 48, 50, 56, 122, 144, 156,

229, 231, 233, 271, 272, 278, 282, 300, 304, 318, 325, 329, 330, 334, 369, 383, 391, 409, 416, 421, 432, 438, 440, 441, 442, 443, 445, 457, 462, 463, 465, 466, 467, 468, 479, 481, 502

Structural Factors 63, 64, 74, 83

Sustainability 2, 3, 10, 11, 14, 15, 17, 19, 20, 21, 22, 24, 26, 27, 28, 32, 34, 35, 36, 37, 38, 42, 43, 46, 48, 49, 50, 52, 53, 54, 55, 56, 57, 61, 75, 101, 102, 103, 105, 108, 109, 110, 112, 113, 114, 122, 125, 128, 129, 142, 145, 146, 153, 157, 158, 159, 160, 161, 162, 163, 168, 174, 175, 176, 177, 178, 179, 186, 191, 199, 203, 204, 205, 206, 208, 209, 212, 215, 216, 217, 218, 220, 221, 223, 224, 226, 227, 228, 229, 230, 231, 232, 233, 234, 235, 240, 241, 242, 245, 248, 249, 252, 253, 254, 255, 256, 258, 259, 260, 262, 263, 264, 265, 266, 267, 268, 269, 271, 272, 273, 277, 278, 280, 282, 283, 285, 290, 300, 315, 316, 317, 320, 321, 324, 325, 333, 344, 356, 357, 373, 375, 376, 378, 382, 383, 385, 386, 387, 388, 390, 391, 392, 393, 400, 401, 402, 403, 405, 406, 407, 408, 409, 410, 411, 412, 415, 417, 421, 424, 432, 433, 436, 437, 440, 441, 442, 444, 446, 448, 451, 452, 453, 454, 458, 466, 474, 481, 483, 486, 487, 489, 491, 493, 498, 501, 502, 505, 507, 508, 509, 510, 511

Sustainable Development 1, 2, 3, 4, 6, 7, 8, 9, 10, 13, 15, 17, 18, 20, 21, 23, 25, 26, 28, 29, 31, 33, 36, 37, 39, 40, 41, 42, 45, 47, 48, 49, 51, 53, 54, 55, 60, 82, 91, 92, 98, 99, 101, 102, 103, 108, 111, 112, 114, 116, 117, 121, 145, 150, 151, 154, 163, 166, 170, 171, 172, 174, 175, 176, 177, 178, 183, 185, 186, 199, 203, 204, 205, 211, 214, 215, 228, 229, 230, 232, 234, 263, 266, 267, 270, 276, 278, 280, 285, 288, 290, 291, 295, 297, 302, 304, 309, 317, 319, 321, 324, 325, 351, 374, 375, 381, 382, 383, 384, 385, 386, 387, 388, 390, 391, 392, 397, 398, 399, 400, 401, 402, 403, 404, 405, 406, 409, 410, 411, 416, 417, 423, 429, 431, 432, 433, 437, 440, 441, 445, 446, 452, 454, 457, 459, 463, 468, 486, 489

Sustainable Development Goals 3, 8, 17, 25, 26, 28, 29, 49, 121, 151, 199, 211, 214, 234, 266, 267, 276, 280, 302, 319, 321, 324, 351, 386, 388, 392, 401, 406, 409, 416, 423, 431, 445, 446, 468

Sustainable Waste Management 123, 126, 132, 134, 135, 136, 138, 139, 141, 142, 145, 204, 407, 465, 467, 468, 469, 470, 477, 483, 484, 487, 488, 489

Systematic Literature Review 176, 237, 239, 243, 256, 259, 260, 262, 263, 264, 406, 487

T

Technology 20, 29, 31, 44, 47, 49, 52, 53, 54, 55, 93, 96, 97, 112, 114, 115, 116, 138, 143, 144, 145, 158, 172, 202, 204, 212, 214, 220, 225, 228, 229, 230, 231, 232, 233, 235, 237, 238, 239, 240, 241, 242, 243, 245, 248, 250, 253, 255, 256, 257, 259, 260, 261, 262, 263, 264, 266, 268, 270, 271, 275, 277, 278, 280, 282, 283, 284, 287, 288, 289, 290, 294, 295, 297, 298, 301, 302, 305, 306, 309, 310, 311, 312, 313, 314, 315, 316, 319, 320, 321, 347, 352, 381, 383, 389, 393, 398, 400, 403, 404, 407, 408, 410, 414, 415, 429, 436, 437, 438, 439, 441, 442, 443, 444, 445, 446, 447, 448, 449, 450, 451, 454, 458, 459, 460, 461, 462, 463, 464, 465, 483, 487, 499, 504, 505, 508

U

UNDP 59, 92, 94, 95, 96, 98, 99, 100, 102, 103, 104, 105, 106, 107, 108, 109, 110, 111, 117, 121, 401, 419, 432

Universality 92, 107

W

Waste 4, 5, 6, 22, 28, 34, 46, 47, 52, 57, 61, 92, 97, 119, 120, 121, 122, 123, 124, 125, 126, 127, 128, 129, 130, 131, 132, 133, 134, 135, 136, 137, 138, 139, 140, 141, 142, 143, 144, 145, 146, 147, 149, 154, 155, 178, 181, 184, 185, 187, 188, 189, 191, 192, 193, 194, 195, 199, 200, 201, 202, 203, 204, 205, 206, 217, 222, 238, 250, 251, 257, 259, 261, 262, 263, 264, 268, 271, 275, 282, 298, 299, 300, 308, 313, 341, 382, 383, 392, 393, 394, 395, 396, 397, 400, 401, 407, 412, 415, 443, 452, 465, 466, 467, 468, 469, 470, 471, 472, 473, 474, 475, 476, 477, 478, 479, 480, 481, 482, 483, 484, 485, 486, 487, 488, 489, 492, 493, 494, 495, 498, 499, 500, 501, 502, 503, 504, 505, 509

Waste Management 4, 5, 6, 22, 28, 46, 119, 120, 121, 122, 123, 124, 125, 126, 127, 128, 129, 130, 131, 132, 133, 134, 135, 136, 137, 138, 139, 140, 141, 142, 143, 144, 145, 146, 147, 149, 154, 155, 178, 181, 184, 185, 187, 189, 191, 192, 193, 199, 200, 204, 206, 217, 251, 261, 263, 264, 268, 282, 300, 308, 313, 392, 396, 397, 400, 401, 407, 465, 466, 467, 468, 469, 470, 471, 472, 473, 474, 475, 476, 477, 478, 479, 480, 481, 482, 483, 484, 485, 486, 487, 488, 489, 492, 501, 502

Waste Reduction 47, 124, 136, 275, 382, 383, 396, 469, 471

Water 1, 2, 4, 5, 6, 7, 8, 12, 13, 14, 15, 22, 26, 28, 29, 32, 33, 35, 37, 41, 42, 88, 89, 92, 95, 96, 112, 120, 121, 131, 134, 143, 160, 176, 178, 181, 182, 183, 184, 185, 186, 187, 188, 189, 190, 191, 192, 194, 195, 196, 197, 198, 199, 200, 201, 202, 203, 204, 205, 206, 207, 208, 209, 212, 213, 216, 222, 232, 234, 248, 266, 272, 276, 277, 280, 290, 291, 293, 296, 297, 299, 301, 306, 308, 385, 390, 391, 392, 396, 400, 401, 413, 414, 415, 416, 417, 418, 419, 420, 421, 422, 423, 424, 425, 426, 427, 428, 429, 430, 431, 432, 433, 466, 467, 474, 486, 487, 491, 492, 494, 495, 496, 497, 498, 501, 502, 508, 509, 510

Water Contamination 197, 206